Introduction to Probability

This classroom-tested textbook is an introduction to probability theory, with the right balance between mathematical precision, probabilistic intuition, and concrete applications. *Introduction to Probability* covers the material precisely, while avoiding excessive technical details. After introducing the basic vocabulary of randomness, including events, probabilities, and random variables, the text offers the reader a first glimpse of the major theorems of the subject: the law of large numbers and the central limit theorem. The important probability distributions are introduced organically as they arise from applications. The discrete and continuous sides of probability are treated together to emphasize their similarities. Intended for students with a calculus background, the text teaches not only the nuts and bolts of probability theory and how to solve specific problems, but also why the methods of solution work.

David F. Anderson is a Professor of Mathematics at the University of Wisconsin-Madison. His research focuses on probability theory and stochastic processes, with applications in the biosciences. He is the author of over thirty research articles and a graduate textbook on the stochastic models utilized in cellular biology. He was awarded the inaugural Institute for Mathematics and its Applications (IMA) Prize in Mathematics in 2014, and was named a Vilas Associate by the University of Wisconsin-Madison in 2016.

Timo Seppäläinen is the John and Abigail Van Vleck Chair of Mathematics at the University of Wisconsin-Madison. He is the author of over seventy research papers in probability theory and a graduate textbook on large deviation theory. He is an elected Fellow of the Institute of Mathematical Statistics. He was an IMS Medallion Lecturer in 2014, an invited speaker at the 2014 International Congress of Mathematicians, and a 2015–16 Simons Fellow.

Benedek Valkó is a Professor of Mathematics at the University of Wisconsin-Madison. His research focuses on probability theory, in particular in the study of random matrices and interacting stochastic systems. He has published over thirty research papers. He has won a National Science Foundation (NSF) CAREER award and he was a 2017–18 Simons Fellow.

CAMBRIDGE MATHEMATICAL TEXTBOOKS

Cambridge Mathematical Textbooks is a program of undergraduate and beginning graduate level textbooks for core courses, new courses, and interdisciplinary courses in pure and applied mathematics. These texts provide motivation with plenty of exercises of varying difficulty, interesting examples, modern applications, and unique approaches to the material.

A complete list of books in the series can be found at
www.cambridge.org/mathematics
Recent titles include the following:

Chance, Strategy, and Choice: An Introduction to the Mathematics of Games and Elections, S. B. Smith
Set Theory: A First Course, D. W. Cunningham
Chaotic Dynamics: Fractals, Tilings, and Substitutions, G. R. Goodson
Introduction to Experimental Mathematics, S. Eilers & R. Johansen
A Second Course in Linear Algebra, S. R. Garcia & R. A. Horn
Exploring Mathematics: An Engaging Introduction to Proof, J. Meier & D. Smith
A First Course in Analysis, J. B. Conway
Introduction to Probability, D. F. Anderson, T. Seppäläinen & B. Valkó

Introduction to Probability

DAVID F. ANDERSON
University of Wisconsin-Madison

TIMO SEPPÄLÄINEN
University of Wisconsin-Madison

BENEDEK VALKÓ
University of Wisconsin-Madison

Shaftesbury Road, Cambridge CB2 8EA, United Kingdom

One Liberty Plaza, 20th Floor, New York, NY 10006, USA

477 Williamstown Road, Port Melbourne, VIC 3207, Australia

314–321, 3rd Floor, Plot 3, Splendor Forum, Jasola District Centre, New Delhi – 110025, India

103 Penang Road, #05–06/07, Visioncrest Commercial, Singapore 238467

Cambridge University Press is part of Cambridge University Press & Assessment, a department of the University of Cambridge.

We share the University's mission to contribute to society through the pursuit of education, learning and research at the highest international levels of excellence.

www.cambridge.org
Information on this title: www.cambridge.org/9781108415859

DOI: 10.1017/9781108235310

First published 2018

A catalogue record for this publication is available from the British Library

Library of Congress Cataloging-in-Publication data
Names: Anderson, David F., 1978– | Seppäläinen, Timo O., 1961– |
Valkó, Benedek, 1976–.
Title: Introduction to probability / David F. Anderson, University of Wisconsin, Madison, Timo Seppäläinen, University of Wisconsin, Madison, Benedek Valkó, University of Wisconsin, Madison.
Description: Cambridge: Cambridge University Press, [2018] | Series: Cambridge mathematical textbooks | Includes bibliographical references and index.
Identifiers: LCCN 2017018747 | ISBN 9781108415859
Subjects: LCSH: Probabilities–Textbooks.
Classification: LCC QA273 .A5534 2018 | DDC 519.2–dc23
LC record available at https://lccn.loc.gov/2017018747

ISBN 978-1-108-41585-9 Hardback

To our families

Contents

Preface

This text is an introduction to the theory of probability with a calculus background. It is intended for classroom use as well as for independent learners and readers. We think of the level of our book as "intermediate" in the following sense. The mathematics is covered as precisely and faithfully as is reasonable and valuable, while avoiding excessive technical details. Two examples of this are as follows.

- The probability model is anchored securely in a sample space and a probability (measure) on it, but recedes to the background after the foundations have been established.
- Random variables are defined precisely as functions on the sample space. This is important to avoid the feeling that a random variable is a vague notion. Once absorbed, this point is not needed for doing calculations.

Short, illuminating proofs are given for many statements but are not emphasized. The main focus of the book is on applying the mathematics to model simple settings with random outcomes and on calculating probabilities and expectations. Introductory probability is a blend of mathematical abstraction and hands-on computation where the mathematical concepts and examples have concrete real-world meaning.

The principles that have guided us in the organization of the book include the following.

(i) We found that the traditional initial segment of a probability course devoted to counting techniques is not the most auspicious beginning. Hence we start with the probability model itself, and counting comes in conjunction with sampling. A systematic treatment of counting techniques is given in an appendix. The instructor can present this in class or assign it to the students.
(ii) Most events are naturally expressed in terms of random variables. Hence we bring the language of random variables into the discussion as quickly as possible.
(iii) One of our goals was an early introduction of the major results of the subject, namely the central limit theorem and the law of large numbers. These are

covered for independent Bernoulli random variables in Chapter 4. Preparation for this influenced the selection of topics of the earlier chapters.

(iv) As a unifying feature, we derive the most basic probability distributions from independent trials, either directly or via a limit. This covers the binomial, geometric, normal, Poisson, and exponential distributions.

Many students reading this text will have already been introduced to parts of the material. They might be tempted to solve some of the problems using computational tricks picked up elsewhere. We warn against doing so. The purpose of this text is not just to teach the nuts and bolts of probability theory and how to solve specific problems, but also to teach you *why* the methods of solution work. Only armed with the knowledge of the "why" can you use the theory provided here as a tool that will be amenable to a myriad of applications and situations.

The sections marked with a diamond ♦ are optional topics that can be included in an introductory probability course as time permits and depending on the interests of the instructor and the audience. They can be omitted without loss of continuity.

At the end of most chapters is a section titled *Finer points* on mathematical issues that are usually beyond the scope of an introductory probability book. In the main text the symbol ♣ marks statements that are elaborated in the *Finer points* section of the chapter. In particular, we do not mention measure-theoretic issues in the main text, but explain some of these in the *Finer points* sections. Other topics in the *Finer points* sections include the lack of uniqueness of a density function, the Berry–Esséen error bounds for normal approximation, the weak versus the strong law of large numbers, and the use of matrices in multivariate normal densities. These sections are intended for the interested reader as starting points for further exploration. They can also be helpful to the instructor who does not possess an advanced probability background.

The symbol ▲ is used to mark the end of numbered examples, the end of remarks, and the end of proofs.

There is an exercise section at the end of each chapter. The exercises begin with a small number of warm-up exercises explicitly organized by sections of the chapter. Their purpose is to offer the reader immediate and basic practice after a section has been covered. The subsequent exercises under the heading *Further exercises* contain problems of varying levels of difficulty, including routine ones, but some of these exercises use material from more than one section. Under the heading *Challenging problems* towards the end of the exercise section we have collected problems that may require some creativity or lengthier calculations. But these exercises are still fully accessible with the tools at the student's disposal.

The concrete mathematical prerequisites for reading this book consist of basic set theory and some calculus, namely, a solid foundation in single variable calculus, including sequences and series, and multivariable integration. Appendix A gives a short list of the particular calculus topics used in the text. Appendix B reviews set theory, and Appendix D reviews some infinite series.

Sets are used from the get-go to set up probability models. Both finite and infinite geometric series are used extensively beginning already in Chapter 1. Single variable integration and differentiation are used from Chapter 3 onwards to work with continuous random variables. Computations with the Poisson distribution from Section 4.4 onwards require facility with the Taylor series of e^x. Multiple integrals arrive in Section 6.2 as we begin to compute probabilities and expectations under jointly continuous distributions.

The authors welcome feedback and will maintain a publicly available list of corrections.

We thank numerous anonymous reviewers whose comments made a real difference to the book, students who went through successive versions of the text, and colleagues who used the text and gave us invaluable feedback. Illustrations were produced with Wolfram Mathematica 11.

The authors gratefully acknowledge support from the National Science Foundation, the Simons Foundation, the Army Research Office, and the Wisconsin Alumni Research Foundation.

Madison, Wisconsin July, 2017

David F. Anderson
Timo Seppäläinen
Benedek Valkó

To the instructor

There is more material in the book than can be comfortably covered in one semester at a pace that is accessible to students with varying backgrounds. Hence there is room for choice by the instructor.

The list below includes all sections not marked with a ♦ or a ♣. It outlines one possible 15-week schedule with 150 minutes of class time per week.

Week 1. Axioms of probability, sampling, review of counting, infinitely many outcomes, review of the geometric series (Sections 1.1–1.3).

Week 2. Rules of probability, random variables, conditional probability (Sections 1.4–1.5, 2.1).

Week 3. Bayes' formula, independence, independent trials (Sections 2.2–2.4).

Week 4. Independent trials, birthday problem, conditional independence, probability distribution of a random variable (Sections 2.4–2.5, 3.1).

Week 5. Cumulative distribution function, expectation and variance (Sections 3.2–3.4).

Week 6. Gaussian distribution, normal approximation and law of large numbers for the binomial distribution (Sections 3.5 and 4.1–4.2).

Week 7. Applications of normal approximation, Poisson approximation, exponential distribution (Sections 4.3–4.5).

Week 8. Moment generating function, distribution of a function of a random variable (Sections 5.1–5.2).

Week 9. Joint distributions (Sections 6.1–6.2).

Week 10. Joint distributions and independence, sums of independent random variables, exchangeability (Sections 6.3 and 7.1–7.2).

Week 11. Expectations of sums and products, variance of sums (Sections 8.1–8.2).

Week 12. Sums and moment generating functions, covariance and correlation (Sections 8.3–8.4).

Week 13. Markov's and Chebyshev's inequalities, law of large numbers, central limit theorem (Sections 9.1–9.3).

Week 14. Conditional distributions (Sections 10.1–10.3).

Week 15. Conditional distributions, review (Sections 10.1–10.3).

The authors invest time in the computations with multivariate distributions in the last four chapters. The reason is twofold: this is where the material becomes more interesting and this is preparation for subsequent courses in probability and stochastic processes. The more challenging examples of Chapter 10 in particular require the students to marshal material from almost the entire course. The exercises under *Challenging problems* have been used for bonus problems and honors credit.

Often the Poisson process is not covered in an introductory probability course, and it is left to a subsequent course on stochastic processes. Hence the Poisson process (Sections 4.6 and 7.3) does not appear in the schedule above. One could make the opposite choice of treating the Poisson process thoroughly, with correspondingly less emphasis, for example, on exchangeability (Section 7.2) or on computing expectations with indicator random variables (Section 8.1). Note that the gamma distribution is introduced in Section 4.6 where it elegantly arises from the Poisson process. If Section 4.6 is skipped then Section 7.1 is a natural place to introduce the gamma distribution.

Other optional items include the transformation of a multivariate density function (Section 6.4), the bivariate normal distribution (Section 8.5), and the Monte Carlo method (Section 9.4).

This book can also accommodate instructors who wish to present the material at either a lighter or a more demanding level than what is outlined in the sample schedule above.

For a lighter course the multivariate topics can be de-emphasized with more attention paid to sets, counting, calculus details, and simple probability models.

For a more demanding course, for example for an audience of mathematics majors, the entire book can be covered with emphasis on proofs and the more challenging multistage examples from the second half of the book. These are the kinds of examples where probabilistic reasoning is beautifully on display. Some topics from the *Finer points* sections could also be included.

From gambling to an essential ingredient of modern science and society

Among the different parts of mathematics, probability is something of a newcomer. Its development into an independent branch of pure mathematics began in earnest in the twentieth century. The axioms on which modern probability theory rests were established by Russian mathematician Andrey Kolmogorov in 1933.

Before the twentieth century probability consisted mainly of solutions to a variety of applied problems. Gambling had been a particularly fruitful source of these problems already for a few centuries. The famous 1654 correspondence between two leading French mathematicians Pierre de Fermat and Blaise Pascal, prompted by a gambling question from a nobleman, is considered the starting point of systematic mathematical treatment of problems of chance. In subsequent centuries many mathematicians contributed to the emerging discipline. The first laws of large numbers and central limit theorems appeared in the 1700s, as did famous problems such as the birthday problem and gambler's ruin that are staples of modern textbooks.

Once the fruitful axiomatic framework was in place, probability could develop into the rich subject it is today. The influence of probability throughout mathematics and applications is growing rapidly but is still only in its beginnings. The physics of the smallest particles, insurance and finance, genetics and chemical reactions in the cell, complex telecommunications networks, randomized computer algorithms, and all the statistics produced about every aspect of life, are but a small sample of old and new application domains of probability theory. Uncertainty is a fundamental feature of human activity.

1

Experiments with random outcomes

The purpose of probability theory is to build mathematical models of experiments with random outcomes and then analyze these models. A random outcome is anything we cannot predict with certainty, such as the flip of a coin, the roll of a die, the gender of a baby, or the future value of an investment.

1.1. Sample spaces and probabilities

The mathematical model of a random phenomenon has standard ingredients. We describe these ingredients abstractly and then illustrate them with examples.

Definition 1.1. These are the ingredients of a probability model.

- The **sample space** Ω is the set of all the possible outcomes of the experiment. Elements of Ω are called **sample points** and typically denoted by ω.
- Subsets of Ω are called **events**. The collection of events in Ω is denoted by $\mathcal{F}$. ♣
- The **probability measure** (also called **probability distribution** or simply **probability**) P is a function from $\mathcal{F}$ into the real numbers. Each event A has a probability $P(A)$, and P satisfies the following axioms.
  (i) $0 \leq P(A) \leq 1$ for each event A.
  (ii) $P(\Omega) = 1$ and $P(\varnothing) = 0$.
  (iii) If $A_1, A_2, A_3, \dots$ is a sequence of pairwise disjoint events then

$$P\left(\bigcup_{i=1}^{\infty} A_i\right) = \sum_{i=1}^{\infty} P(A_i). \tag{1.1}$$

The triple $(\Omega, \mathcal{F}, P)$ is called a **probability space**. Every mathematically precise model of a random experiment or collection of experiments must be of this kind.

The three axioms related to the probability measure P in Definition 1.1 are known as *Kolmogorov's axioms* after the Russian mathematician Andrey Kolmogorov who first formulated them in the early 1930s.

A few words about the symbols and conventions. Ω is an upper case omega, and ω is a lower case omega. $\varnothing$ is the empty set, that is, the subset of Ω that contains no sample points. The only sensible value for its probability is zero. *Pairwise disjoint* means that $A_i \cap A_j = \varnothing$ for each pair of indices $i \neq j$. Another way to say this is that the events A_i are *mutually exclusive*. Axiom (iii) says that the probability of the union of mutually exclusive events is equal to the sum of their probabilities. Note that rule (iii) applies also to finitely many events.

Fact 1.2. If $A_1, A_2, \dots, A_n$ are pairwise disjoint events then

$$P(A_1 \cup \cdots \cup A_n) = P(A_1) + \cdots + P(A_n). \tag{1.2}$$

Fact 1.2 is a consequence of (1.1) obtained by setting $A_{n+1} = A_{n+2} = A_{n+3} = \cdots = \varnothing$. If you need a refresher on set theory, see Appendix B.

Now for some examples.

Example 1.3. We flip a fair coin. The sample space is $\Omega = \{\mathtt{H}, \mathtt{T}\}$ ($\mathtt{H}$ for heads and $\mathtt{T}$ for tails). We take $\mathcal{F} = \{\varnothing, \{\mathtt{H}\}, \{\mathtt{T}\}, \{\mathtt{H}, \mathtt{T}\}\}$, the collection of all subsets of Ω. The term "fair coin" means that the two outcomes are equally likely. So the probabilities of the singletons $\{\mathtt{H}\}$ and $\{\mathtt{T}\}$ are

$$P\{\mathtt{H}\} = P\{\mathtt{T}\} = \tfrac{1}{2}.$$

By axiom (ii) in Definition 1.1 we have $P(\varnothing) = 0$ and $P\{\mathtt{H}, \mathtt{T}\} = 1$. Note that the "fairness" of the coin is an assumption we make about the experiment. ▲

Example 1.4. We roll a standard six-sided die. Then the sample space is $\Omega = \{1, 2, 3, 4, 5, 6\}$. Each sample point ω is an integer between 1 and 6. If the die is fair then each outcome is equally likely, in other words

$$P\{1\} = P\{2\} = P\{3\} = P\{4\} = P\{5\} = P\{6\} = \tfrac{1}{6}.$$

A possible event in this sample space is

$$A = \{\text{the outcome is even}\} = \{2, 4, 6\}. \tag{1.3}$$

Then

$$P(A) = P\{2, 4, 6\} = P\{2\} + P\{4\} + P\{6\} = \tfrac{1}{2}$$

where we applied Fact 1.2 in the second equality. ▲

Some comments about the notation. In mathematics, sets are typically denoted by upper case letters A, B, etc., and so we use upper case letters to denote events. Like A in (1.3), events can often be expressed both in words and in mathematical symbols. The description of a set (or event) in terms of words or mathematical symbols is enclosed in braces $\{\ \}$. Notational consistency would seem to require that the probability of the event {2} be written as $P(\{2\})$. But it seems unnecessary to add the parentheses around the braces, so we simplify the expression to $P\{2\}$ or $P(2)$.

Example 1.5. (Continuation of Examples 1.3 and 1.4) The probability measure P contains our assumptions and beliefs about the phenomenon that we are modeling.

If we wish to model a flip of a biased coin we alter the probabilities. For example, suppose we know that heads is three times as likely as tails. Then we define our probability measure P_1 by $P_1\{\mathtt{H}\} = \frac{3}{4}$ and $P_1\{\mathtt{T}\} = \frac{1}{4}$. The sample space is again $\Omega = \{\mathtt{H}, \mathtt{T}\}$ as in Example 1.3, but the probability measure has changed to conform with our assumptions about the experiment.

If we believe that we have a loaded die and a six is twice as likely as any other number, we use the probability measure $\widetilde{P}$ defined by

$$\widetilde{P}\{1\} = \widetilde{P}\{2\} = \widetilde{P}\{3\} = \widetilde{P}\{4\} = \widetilde{P}\{5\} = \tfrac{1}{7} \quad \text{and} \quad \widetilde{P}\{6\} = \tfrac{2}{7}.$$

Alternatively, if we scratch away the five from the original fair die and turn it into a second two, the appropriate probability measure is

$$Q\{1\} = \tfrac{1}{6},\ Q\{2\} = \tfrac{2}{6},\ Q\{3\} = \tfrac{1}{6},\ Q\{4\} = \tfrac{1}{6},\ Q\{5\} = 0,\ Q\{6\} = \tfrac{1}{6}.$$ ▲

These examples show that to model different phenomena it is perfectly sensible to consider different probability measures on the same sample space. Clarity might demand that we distinguish different probability measures notationally from each other. This can be done by adding ornaments to the P, as in P_1 or $\widetilde{P}$ (pronounced "P tilde") above, or by using another letter such as Q. Another important point is that it is perfectly valid to assign a probability of zero to a nonempty event, as with Q above.

Example 1.6. Let the experiment consist of a roll of a pair of dice (as in the games of Monopoly or craps). We assume that the dice can be distinguished from each other, for example that one of them is blue and the other one is red. The sample space is the set of pairs of integers from 1 through 6, where the first number of the pair denotes the number on the blue die and the second denotes the number on the red die:

$$\Omega = \{(i,j) : i,j \in \{1,2,3,4,5,6\}\}.$$

Here (a, b) is a so-called *ordered pair* which means that outcome $(3, 5)$ is distinct from outcome $(5, 3)$. (Note that the term "ordered pair" means that order matters, not that the pair is in increasing order.) The assumption of fair dice would dictate equal probabilities: $P\{(i,j)\} = \frac{1}{36}$ for each pair $(i,j) \in \Omega$. An example of an event of interest would be

$$D = \{\text{the sum of the two dice is } 8\} = \{(2,6),(3,5),(4,4),(5,3),(6,2)\}$$

and then by the additivity of probabilities

$$P(D) = P\{(2,6)\} + P\{(3,5)\} + P\{(4,4)\} + P\{(5,3)\} + P\{(6,2)\}$$
$$= \sum_{(i,j):i+j=8} P\{(i,j)\} = 5 \cdot \tfrac{1}{36} = \tfrac{5}{36}.$$ ▲

Example 1.7. We flip a fair coin three times. Let us encode the outcomes of the flips as 0 for heads and 1 for tails. Then each outcome of the experiment is a sequence of length three where each entry is 0 or 1:

$$\Omega = \{(0,0,0),(0,0,1),(0,1,0),\ldots,(1,1,0),(1,1,1)\}. \tag{1.4}$$

This Ω is the set of ordered triples (or 3-tuples) of zeros and ones. Ω has $2^3 = 8$ elements. (We review simple counting techniques in Appendix C.) With a fair coin all outcomes are equally likely, so $P\{\omega\} = 2^{-3}$ for each $\omega \in \Omega$. An example of an event is

$$B = \{\text{the first and third flips are heads}\} = \{(0,0,0),(0,1,0)\}$$

with

$$P(B) = P\{(0,0,0)\} + P\{(0,1,0)\} = \tfrac{1}{8} + \tfrac{1}{8} = \tfrac{1}{4}.$$ ▲

Much of probability deals with repetitions of a simple experiment, such as the roll of a die or the flip of a coin in the previous two examples. In such cases *Cartesian product spaces* arise naturally as sample spaces. If $A_1, A_2, \ldots, A_n$ are sets then the Cartesian product

$$A_1 \times A_2 \times \cdots \times A_n$$

is defined as the set of ordered n-tuples with the ith element from A_i. In symbols

$$A_1 \times A_2 \times \cdots \times A_n = \{(x_1,\ldots,x_n) : x_i \in A_i \text{ for } i = 1,\ldots,n\}.$$

In terms of product notation, the sample space of Example 1.6 for a pair of dice can be written as

$$\Omega = \{1,2,3,4,5,6\} \times \{1,2,3,4,5,6\}$$

while the space for three coin flips in Example 1.7 can be expressed as

$$\Omega = \{0,1\} \times \{0,1\} \times \{0,1\} = \{0,1\}^3.$$

1.2. Random sampling

Sampling is choosing objects randomly from a given set. It can involve repeated choices or a choice of more than one object at a time. Dealing cards from a deck is an example of sampling. There are different ways of setting up such experiments which lead to different probability models. In this section we discuss three sampling mechanisms that lead to equally likely outcomes. This allows us to compute probabilities by counting. The required counting methods are developed systematically in Appendix C.

Before proceeding to sampling, let us record a basic fact about experiments with equally likely outcomes. Suppose the sample space Ω is a finite set and let $\#\Omega$ denote the total number of possible outcomes. If each outcome ω has the same probability then $P\{\omega\} = \frac{1}{\#\Omega}$ because probabilities must add up to 1. In this case probabilities of events can be found by counting. If A is an event that consists of elements $a_1, a_2, \dots, a_r$, then additivity and $P\{a_i\} = \frac{1}{\#\Omega}$ imply

$$P(A) = P\{a_1\} + P\{a_2\} + \cdots + P\{a_r\} = \frac{\#A}{\#\Omega}$$

where we wrote $\#A$ for the number of elements in the set A.

Fact 1.8. If the sample space Ω has finitely many elements and each outcome is equally likely then for any event $A \subset \Omega$ we have

$$P(A) = \frac{\#A}{\#\Omega}. \tag{1.5}$$

Look back at the examples of the previous section to check which ones were of the kind where $P\{\omega\} = \frac{1}{\#\Omega}$.

Remark 1.9. (Terminology) It should be clear by now that random outcomes do not have to be equally likely. (Look at Example 1.5 in the previous section.) However, it is common to use the phrase "an element is chosen at random" to mean that all choices are equally likely. The technically more accurate phrase would be "chosen *uniformly* at random." Formula (1.5) can be expressed by saying "when outcomes are equally likely, the probability of an event equals the number of favorable outcomes over the total number of outcomes." ▲

We turn to discuss sampling mechanisms. An *ordered sample* is built by choosing objects one at a time and by keeping track of the order in which these objects were chosen. After each choice we either replace (put back) or discard the just chosen object before choosing the next one. This distinction leads to *sampling with replacement* and *sampling without replacement.* An *unordered sample* is one where only the identity of the objects matters and not the order in which they came.

We discuss the sampling mechanisms in terms of an urn with numbered balls. An urn is a traditional device in probability (see Figure 1.1). You cannot see the contents of the urn. You reach in and retrieve one ball at a time without looking. We assume that the choice is uniformly random among the balls in the urn.

Figure 1.1. Three traditional mechanisms for creating experiments with random outcomes: an urn with balls, a six-sided die, and a coin.

Sampling with replacement, order matters

Suppose the urn contains n balls numbered $1, 2, \ldots, n$. We retrieve a ball from the urn, record its number, and put the ball back into the urn. (Putting the ball back into the urn is the *replacement* step.) We carry out this procedure k times. The outcome is the ordered k-tuple of numbers that we read off the sampled balls. Represent the outcome as $\omega = (s_1, s_2, \ldots, s_k)$ where s_1 is the number on the first ball, s_2 is the number on the second ball, and so on. The sample space Ω is a Cartesian product space: if we let $S = \{1, 2, \ldots, n\}$ then

$$\Omega = \underbrace{S \times S \times \cdots \times S}_{k \text{ times}} = S^k = \{(s_1, s_2, \ldots, s_k) : s_i \in S \text{ for } i = 1, \ldots, k\}. \tag{1.6}$$

How many outcomes are there? Each s_i can be chosen in n different ways. By Fact C.5 from Appendix C we have

$$\#\Omega = \underbrace{n \cdot n \cdots n}_{k \text{ times}} = n^k.$$

We assume that this procedure leads to equally likely outcomes, hence the probability of each k-tuple is $P\{\omega\} = n^{-k}$.

Let us illustrate this with a numerical example.

Example 1.10. Suppose our urn contains 5 balls labeled 1, 2, 3, 4, 5. Sample 3 balls with replacement and produce an ordered list of the numbers drawn. At each step we have the same 5 choices. The sample space is

$$\Omega = \{1, 2, 3, 4, 5\}^3 = \{(s_1, s_2, s_3) : \text{each } s_i \in \{1, 2, 3, 4, 5\}\}$$

and $\#\Omega = 5^3$. Since all outcomes are equally likely, we have for example

$$P\{\text{the sample is (2,1,5)}\} = P\{\text{the sample is (2,2,3)}\} = 5^{-3} = \frac{1}{125}. \qquad \blacktriangle$$

Repeated flips of a coin or rolls of a die are also examples of sampling with replacement. In these cases we are sampling from the set $\{\mathtt{H}, \mathtt{T}\}$ or $\{1, 2, 3, 4, 5, 6\}$.

(Check that Examples 1.6 and 1.7 are consistent with the language of sampling that we just introduced.)

Sampling without replacement, order matters

Consider again the urn with n balls numbered $1, 2, \dots, n$. We retrieve a ball from the urn, record its number, and put the ball aside, in other words not back into the urn. (This is the *without replacement* feature.) We repeat this procedure k times. Again we produce an ordered k-tuple of numbers $\omega = (s_1, s_2, \dots, s_k)$ where each $s_i \in S = \{1, 2, \dots, n\}$. However, the numbers $s_1, s_2, \dots, s_k$ in the outcome are distinct because now the same ball cannot be drawn twice. Because of this, we clearly cannot have k larger than n.

Our sample space is

$$\Omega = \{(s_1, s_2, \dots, s_k) : \text{each } s_i \in S \text{ and } s_i \neq s_j \text{ if } i \neq j\}. \tag{1.7}$$

To find $\#\Omega$, note that s_1 can be chosen in n ways, after that s_2 can be chosen in $n - 1$ ways, and so on, until there are $n - k + 1$ choices remaining for the last entry s_k. Thus

$$\#\Omega = n \cdot (n-1) \cdot (n-2) \cdots (n-k+1) = (n)_k. \tag{1.8}$$

Again we assume that this mechanism gives us equally likely outcomes, and so $P\{\omega\} = \frac{1}{(n)_k}$ for each k-tuple ω of distinct numbers. The last symbol $(n)_k$ of equation (1.8) is called the descending factorial.

Example 1.11. Consider again the urn with 5 balls labeled 1, 2, 3, 4, 5. Sample 3 balls without replacement and produce an ordered list of the numbers drawn. Now the sample space is

$$\Omega = \{(s_1, s_2, s_3) : \text{each } s_i \in \{1, 2, 3, 4, 5\} \text{ and } s_1, s_2, s_3 \text{ are all distinct}\}.$$

The first ball can be chosen in 5 ways, the second ball in 4 ways, and the third ball in 3 ways. So

$$P\{\text{the sample is (2,1,5)}\} = \frac{1}{5 \cdot 4 \cdot 3} = \frac{1}{60}.$$

The outcome $(2, 2, 3)$ is not possible because repetition is not allowed. ▲

Another instance of sampling without replacement would be a random choice of students from a class to fill specific roles in a school play, with at most one role per student.

If $k = n$ then our sample is a random ordering of all n objects. Equation (1.8) becomes $\#\Omega = n!$. This is a restatement of the familiar fact that a set of n elements can be ordered in $n!$ different ways.

Sampling without replacement, order irrelevant

In the previous sampling situations the order of the outcome was relevant. That is, outcomes $(1, 2, 5)$ and $(2, 1, 5)$ were regarded as distinct. Next we suppose that we do not care about order, but only about the set $\{1, 2, 5\}$ of elements sampled. This kind of sampling without replacement can happen when cards are dealt from a deck or when winning numbers are drawn in a state lottery. Since order does not matter, we can also imagine choosing the entire set of k objects at once instead of one element at a time.

Notation is important here. The ordered triple $(1, 2, 5)$ and the set $\{1, 2, 5\}$ must not be confused with each other. Consequently *in this context* we must not mix up the notations $(\,)$ and $\{\,\}$.

As above, imagine the urn with n balls numbered $1, 2, \dots, n$. Let $1 \le k \le n$. Sample k balls without replacement, but record only which balls appeared and not the order. Since the sample contains no repetitions, the outcome is a subset of size k from the set $S = \{1, 2, \dots, n\}$. Thus

$$\Omega = \{\omega \subset S : \#\omega = k\}.$$

(Do not be confused by the fact that an outcome ω is itself now a set of numbers.) The number of elements of Ω is given by the binomial coefficient (see Fact C.12 in Appendix C):

$$\#\Omega = \frac{n!}{(n-k)!k!} = \binom{n}{k}.$$

Assuming that the mechanism leads to equally likely outcomes, $P\{\omega\} = \binom{n}{k}^{-1}$ for each subset ω of size k.

Another way to produce an unordered sample of k balls without repetitions would be to execute the following three steps: (i) randomly order all n balls, (ii) take the first k balls, and (iii) ignore their order. Let us verify that the probability of obtaining a particular selection $\{s_1, \dots, s_k\}$ is $\binom{n}{k}^{-1}$, as above.

The number of possible orderings in step (i) is $n!$. The number of favorable orderings is $k!(n-k)!$, because the first k numbers must be an ordering of $\{s_1, \dots, s_k\}$ and after that comes an ordering of the remaining $n-k$ numbers. Then from the ratio of favorable to all outcomes

$$P\{\text{the selection is } \{s_1, \dots, s_k\}\} = \frac{k!(n-k)!}{n!} = \frac{1}{\binom{n}{k}},$$

as we expected.

The description above contains a couple of lessons.

(i) There can be more than one way to build a probability model to solve a given problem. But a warning is in order: once an approach has been chosen, it must be followed consistently. Mixing up different representations will surely lead to an incorrect answer.

(ii) It may pay to introduce additional structure into the problem. The second approach introduced order into the calculation even though in the end we wanted an outcome without order.

Example 1.12. Suppose our urn contains 5 balls labeled 1, 2, 3, 4, 5. Sample 3 balls without replacement and produce an unordered set of 3 numbers as the outcome. The sample space is

$$\Omega = \{\omega : \omega \text{ is a 3-element subset of } \{1,2,3,4,5\}\}.$$

For example

$$P(\text{the sample is } \{1,2,5\}) = \frac{1}{\binom{5}{3}} = \frac{2!3!}{5!} = \frac{1}{10}.$$

The outcome $\{2,2,3\}$ does not make sense as a set of three numbers because of the repetition. ▲

The fourth alternative, sampling with replacement to produce an unordered sample, does not lead to equally likely outcomes. This scenario will appear naturally in Example 6.7 in Chapter 6.

Further examples

The next example contrasts all three sampling mechanisms.

Example 1.13. Suppose we have a class of 24 children. We consider three different scenarios that each involve choosing three children.

(a) Every day a random student is chosen to lead the class to lunch, without regard to previous choices. What is the probability that Cassidy was chosen on Monday and Wednesday, and Aaron on Tuesday?

This is sampling with replacement to produce an ordered sample. Over a period of three days the total number of different choices is 24^3. Thus

$$P\{(\text{Cassidy, Aaron, Cassidy})\} = 24^{-3} = \frac{1}{13{,}824}.$$

(b) Three students are chosen randomly to be class president, vice president, and treasurer. No student can hold more than one office. What is the probability that Mary is president, Cory is vice president, and Matt treasurer?

Imagine that we first choose the president, then the vice president, and then the treasurer. This is sampling without replacement to produce an ordered sample. Thus

$$P\{\text{Mary is president, Cory is vice president, and Matt treasurer}\}$$
$$= \frac{1}{24 \cdot 23 \cdot 22} = \frac{1}{12{,}144}.$$

Suppose we asked instead for the probability that Ben is either president or vice president. We apply formula (1.5). The number of outcomes in which Ben ends up as president is $1 \cdot 23 \cdot 22$ (1 choice for president, then 23 choices for vice president, and finally 22 choices for treasurer). Similarly the number of ways in which Ben ends up as vice president is $23 \cdot 1 \cdot 22$. So

$$P\{\text{Ben is president or vice president}\} = \frac{1 \cdot 23 \cdot 22 + 23 \cdot 1 \cdot 22}{24 \cdot 23 \cdot 22} = \frac{1}{12}.$$

(c) A team of three children is chosen at random. What is the probability that the team consists of Shane, Heather and Laura?

A team means here simply a set of three students. Thus we are sampling without replacement to produce a sample without order.

$$P(\text{the team is } \{\text{Shane, Heather, Laura}\}) = \frac{1}{\binom{24}{3}} = \frac{1}{2024}.$$

What is the probability that Mary is on the team? There are $\binom{23}{2}$ teams that include Mary since there are that many ways to choose the other two team members from the remaining 23 students. Thus by the ratio of favorable outcomes to all outcomes,

$$P\{\text{the team includes Mary}\} = \frac{\binom{23}{2}}{\binom{24}{3}} = \frac{3}{24} = \frac{1}{8}.$$ ▲

Problems of unordered sampling without replacement can be solved either with or without order. The next two examples illustrate this idea.

Example 1.14. Our urn contains 10 marbles numbered 1 to 10. We sample 2 marbles without replacement. What is the probability that our sample contains the marble labeled 1? Let A be the event that this happens. However we choose to count, the final answer $P(A)$ will come from formula (1.5).

Solution with order. Sample the 2 marbles in order. As in (1.8), $\#\Omega = 10 \cdot 9 = 90$. The favorable outcomes are all the ordered pairs that contain 1:

$$A = \{(1, 2), (1, 3), \dots, (1, 10), (2, 1), (3, 1), \dots, (10, 1)\}$$

and we count $\#A = 18$. Thus $P(A) = \frac{18}{90} = \frac{1}{5}$.

Solution without order. Now the outcomes are subsets of size 2 from the set $\{1, 2, \dots, 10\}$ and so $\#\Omega = \binom{10}{2} = \frac{9 \cdot 10}{2} = 45$. The favorable outcomes are all the 2-element subsets that contain 1:

$$A = \{\{1, 2\}, \{1, 3\}, \dots, \{1, 10\}\}.$$

Now $\#A = 9$ so $P(A) = \frac{9}{45} = \frac{1}{5}$.

Both approaches are correct and of course they give the same answer. ▲

Example 1.15. Rodney packs 3 shirts for a trip. It is early morning so he just grabs 3 shirts randomly from his closet. The closet contains 10 shirts: 5 striped, 3 plaid, and 2 solid colored ones. What is the probability that he chose 2 striped and 1 plaid shirt?

To use the counting methods introduced above, the shirts need to be distinguished from each other. This way the outcomes are equally likely. So let us assume that the shirts are labeled, with the striped shirts labeled 1, 2, 3, 4, 5, the plaid ones 6, 7, 8, and the solid colored ones 9, 10. Since we are only interested in the set of chosen shirts, we can solve this problem with or without order.

If we solve the problem without considering order, then

$$\Omega = \{\{x_1, x_2, x_3\} : x_i \in \{1, \dots, 10\}, x_i \neq x_j\},$$

the collection of 3-element subsets of $\{1, \dots, 10\}$. $\#\Omega = \binom{10}{3} = \frac{8\cdot 9\cdot 10}{2\cdot 3} = 120$, the number of ways of choosing a set of 3 objects from a set of 10 objects. The set of favorable outcomes is

$$A = \{\{x_1, x_2, x_3\} : x_1, x_2 \in \{1, \dots, 5\},\ x_1 \neq x_2,\ x_3 \in \{6, 7, 8\}\}.$$

The number of favorable outcomes is $\#A = \binom{5}{2} \cdot \binom{3}{1} = 30$. This comes from the number of ways of choosing 2 out of 5 striped shirts (shirts labeled 1, 2, 3, 4, 5) times the number of ways of choosing 1 out of 3 plaid shirts (numbered 6, 7, 8). Consequently

$$P\{2 \text{ striped and } 1 \text{ plaid}\} = \frac{\#A}{\#\Omega} = \frac{\binom{5}{2}\binom{3}{1}}{\binom{10}{3}} = \frac{30}{120} = \frac{1}{4}.$$

We now change perspective and solve the problem with an ordered sample. To avoid confusion, we denote our sample space by $\widetilde{\Omega}$:

$$\widetilde{\Omega} = \{(x_1, x_2, x_3) : x_i \in \{1, \dots 10\}, x_1, x_2, x_3 \text{ distinct}\}.$$

We have $\#\widetilde{\Omega} = 10 \cdot 9 \cdot 8 = 720$. The event of interest, $\widetilde{A}$, consists of those triples that have two striped shirts and one plaid shirt.

The elements of $\widetilde{A}$ can be found using the following procedure: (i) choose the plaid shirt (3 choices), (ii) choose the position of the plaid shirt in the ordering (3 choices), (iii) choose the first striped shirt and place it in the first available position (5 choices), (iv) choose the second striped shirt (4 choices). Thus, $\#A = 3\cdot 3\cdot 5\cdot 4 = 180$, and $P(\widetilde{A}) = \frac{180}{720} = \frac{1}{4}$, which agrees with the answer above. ▲

1.3. Infinitely many outcomes

The next step in our development is to look at examples with infinitely many possible outcomes.

Example 1.16. Flip a fair coin until the first tails comes up. Record the number of flips required as the outcome of the experiment. What is the space Ω of possible outcomes? The number of flips needed can be any positive integer, hence Ω must contain all positive integers. We can also imagine the scenario where tails never comes up. This outcome is represented by ∞ (infinity). Thus

$$\Omega = \{\infty, 1, 2, 3, \dots\}.$$

The outcome is k if and only if the first $k-1$ flips are heads and the kth flip is tails. As in Example 1.7, this is one of the 2^k equally likely outcomes when we flip a coin k times, so the probability of this event is 2^{-k}. Thus

$$P\{k\} = 2^{-k} \quad \text{for each positive integer } k. \tag{1.9}$$

It remains to figure out the probability $P\{\infty\}$. We can derive it from the axioms of probability:

$$1 = P(\Omega) = P\{\infty, 1, 2, 3, \dots\} = P\{\infty\} + \sum_{k=1}^{\infty} P\{k\}.$$

From (1.9)

$$\sum_{k=1}^{\infty} P\{k\} = \sum_{k=1}^{\infty} 2^{-k} = 1, \tag{1.10}$$

which implies that $P\{\infty\} = 0$. ▲

Equation (1.9) defines the *geometric probability distribution* with success parameter $1/2$ on the positive integers. On line (1.10) we summed up a geometric series. If you forgot how to do that, turn to Appendix D.

Notice that the example showed us something that agrees with our intuition, but is still quite nontrivial: the probability that we never see tails in repeated flips of a fair coin is zero. This phenomenon gets a different treatment in Example 1.22 below.

Example 1.17. We pick a real number uniformly at random from the closed unit interval $[0, 1]$. Let X denote the number chosen. "Uniformly at random" means that X is equally likely to lie anywhere in $[0, 1]$. Obviously $\Omega = [0, 1]$. What is the probability that X lies in a smaller interval $[a, b] \subseteq [0, 1]$? Since all locations for X are equally likely, it appears reasonable to stipulate that the probability that X is in $[a, b]$ should equal the proportion of $[0, 1]$ covered by $[a, b]$:

$$P\{X \text{ lies in the interval } [a, b]\} = b - a \quad \text{for } 0 \le a \le b \le 1. \tag{1.11}$$

Equation (1.11) defines the *uniform probability distribution* on the interval $[0, 1]$. We meet it again in Section 3.1 as part of a systematic treatment of probability distributions. ♣ ▲

Example 1.18. Consider a dartboard in the shape of a disk with a radius of 9 inches. The bullseye is a disk of diameter $\frac{1}{2}$ inch in the middle of the board. What is the probability that a dart randomly thrown on the board hits the bullseye? Let us assume that the dart hits the board at a uniformly chosen random location, that is, the dart is equally likely to hit anywhere on the board.

The sample space is a disk of radius 9. For simplicity take the center as the origin of our coordinate system, so

$$\Omega = \{(x, y) : x^2 + y^2 \leq 9^2\}.$$

Let A be the event that represents hitting the bullseye. This is the disk of radius $\frac{1}{4}$: $A = \{(x, y) : x^2 + y^2 \leq (1/4)^2\}$. The probability should be uniform on the disk Ω, so by analogy with the previous example,

$$P(A) = \frac{\text{area of } A}{\text{area of } \Omega} = \frac{\pi \cdot (1/4)^2}{\pi \cdot 9^2} = \frac{1}{36^2} \approx 0.00077. \qquad \blacktriangle$$

There is a significant difference between the sample space of Example 1.16 on the one hand, and the sample spaces of Examples 1.17 and 1.18 on the other. The set $\Omega = \{\infty, 1, 2, 3, \dots\}$ is *countably infinite* which means that its elements can be arranged in a sequence, or equivalently, labeled by positive integers. A countably infinite sample space works just like a finite sample space. To specify a probability measure P, it is enough to specify the probabilities of the outcomes and then derive the probability of each event by additivity:

$$P(A) = \sum_{\omega : \omega \in A} P\{\omega\} \quad \text{for any event } A \subset \Omega.$$

Finite and countably infinite sample spaces are both called *discrete sample spaces*.

By contrast, the unit interval, and any nontrivial subinterval of the real line, is *uncountable*. No integer labeling can cover all its elements. This is not trivial to prove and we shall not pursue it here. But we can see that it is impossible to define the probability measure of Example 1.17 by assigning probabilities to individual points. To argue this by contradiction, suppose some real number $x \in [0, 1]$ has positive probability $c = P\{x\} > 0$. Since all outcomes are equally likely in this example, it must be that $P\{x\} = c$ for all $x \in [0, 1]$. But this leads immediately to absurdity. If A is a set with k elements then $P(A) = kc$ which is greater than 1 if we take k large enough. The rules of probability have been violated. We must conclude that the probability of each individual point is zero:

$$P\{x\} = 0 \quad \text{for each} \quad x \in [0, 1]. \tag{1.12}$$

The consequence of the previous argument is that the definition of probabilities of events on an uncountable space must be based on something other than individual points. Examples 1.17 and 1.18 illustrate how to use *length* and *area* to model a uniformly distributed random point. Later we develop tools for building models where the random point is not uniform.

The issue raised here is also intimately tied with the additivity axiom (iii) of Definition 1.1. Note that the axiom requires additivity only for a *sequence* of pairwise disjoint events, and not for an uncountable collection of events. Example 1.17 illustrates this point. The interval $[0, 1]$ is the union of all the singletons $\{x\}$ over $x \in [0, 1]$. But $P([0, 1]) = 1$ while $P\{x\} = 0$ for each x. So there is no conceivable way in which the probability of $[0, 1]$ comes by adding together probabilities of points. Let us emphasize once more that this does not violate axiom (iii) because $[0, 1] = \bigcup_{x\in[0,1]}\{x\}$ is an uncountable union.

1.4. Consequences of the rules of probability

We record some consequences of the axioms in Definition 1.1 that are worth keeping in mind because they are helpful for calculations. The discussion relies on basic set operations reviewed in Appendix B.

Decomposing an event

The most obviously useful property of probabilities is the additivity property: if $A_1, A_2, A_3, \dots$ are pairwise disjoint events and A is their union, then $P(A) = P(A_1) + P(A_2) + P(A_3) + \cdots$. Calculation of the probability of a complicated event A almost always involves decomposing A into smaller disjoint pieces whose probabilities are easier to find. The next two examples illustrate both finite and infinite decompositions.

Example 1.19. An urn contains 30 red, 20 green and 10 yellow balls. Draw two without replacement. What is the probability that the sample contains exactly one red or exactly one yellow? To clarify the question, it means the probability that the sample contains exactly one red, or exactly one yellow, *or both* (inclusive or). This interpretation of *or* is consistent with unions of events.

We approach the problem as we did in Example 1.15. We distinguish between the 60 balls for example by numbering them, though the actual labels on the balls are not important. This way we can consider an experiment with equally likely outcomes.

Having exactly one red or exactly one yellow ball in our sample of two means that we have one of the following color combinations: red-green, yellow-green or red-yellow. These are disjoint events, and their union is the event we are interested in. So

$$\begin{aligned} &P(\text{exactly one red or exactly one yellow}) \\ &\qquad = P(\text{red and green}) + P(\text{yellow and green}) + P(\text{red and yellow}). \end{aligned}$$

Counting favorable arrangements for each of the simpler events:

$$P(\text{red and green}) = \frac{30 \cdot 20}{\binom{60}{2}} = \frac{20}{59}, \qquad P(\text{yellow and green}) = \frac{10 \cdot 20}{\binom{60}{2}} = \frac{20}{177},$$

$$P(\text{red and yellow}) = \frac{30 \cdot 10}{\binom{60}{2}} = \frac{10}{59}.$$

This leads to

$$P(\text{exactly one red or exactly one yellow}) = \frac{20}{59} + \frac{20}{177} + \frac{10}{59} = \frac{110}{177}.$$

We used unordered samples, but we can get the answer also by using ordered samples. Example 1.24 below solves this same problem with inclusion-exclusion. ▲

Example 1.20. Peter and Mary take turns rolling a fair die. If Peter rolls 1 or 2 he wins and the game stops. If Mary rolls 3, 4, 5, or 6, she wins and the game stops. They keep rolling in turn until one of them wins. Suppose Peter rolls first.

(a) What is the probability that Peter wins and rolls at most 4 times?
To say that Peter wins and rolls at most 4 times is the same as saying that either he wins on his first roll, or he wins on his second roll, or he wins on his third roll, or he wins on his fourth roll. These alternatives are mutually exclusive. This is a fairly obvious way to decompose the event. So define events

$$A = \{\text{Peter wins and rolls at most 4 times}\}$$

and $A_k = \{\text{Peter wins on his } k\text{th roll}\}$. Then $A = \cup_{k=1}^4 A_k$ and since the events A_k are mutually exclusive, $P(A) = \sum_{k=1}^4 P(A_k)$.

To find the probabilities $P(A_k)$ we need to think about the game and the fact that Peter rolls first. Peter wins on his kth roll if first both Peter and Mary fail $k-1$ times and then Peter succeeds. Each roll has 6 possible outcomes. Peter's roll fails in 4 different ways and Mary's roll fails in 2 different ways. Peter's kth roll succeeds in 2 different ways. Thus the ratio of the number of favorable alternatives over the total number of alternatives gives

$$P(A_k) = \frac{(4 \cdot 2)^{k-1} \cdot 2}{(6 \cdot 6)^{k-1} \cdot 6} = \left(\frac{8}{36}\right)^{k-1} \frac{2}{6} = \left(\frac{2}{9}\right)^{k-1} \frac{1}{3}.$$

The probability asked is now obtained from a finite geometric sum:

$$\begin{aligned} P(A) &= \sum_{k=1}^{4} P(A_k) = \sum_{k=1}^{4} \left(\frac{2}{9}\right)^{k-1} \frac{1}{3} = \frac{1}{3} \sum_{j=0}^{3} \left(\frac{2}{9}\right)^{j} = \tfrac{1}{3} \cdot \frac{1-(\frac{2}{9})^4}{1-\frac{2}{9}} \\ &= \tfrac{3}{7}\left(1-(\tfrac{2}{9})^4\right). \end{aligned}$$

Above we changed the summation index to $j = k - 1$ to make the sum look exactly like the one in equation (D.2) in Appendix D and then applied formula (D.2).

(b) What is the probability that Mary wins?
If Mary wins, then either she wins on her first roll, or she wins on her second roll, or she wins on her third roll, etc., and these alternatives are mutually

exclusive. There is no a priori bound on how long the game can last. Hence we have to consider all the infinitely many possibilities.

Define the events $B = \{\text{Mary wins}\}$ and $B_k = \{\text{Mary wins on her } k\text{th roll}\}$. Then $B = \cup_{k=1}^{\infty} B_k$ is a union of pairwise disjoint events and the additivity of probability implies $P(B) = \sum_{k=1}^{\infty} P(B_k)$.

In order for Mary to win on her kth roll, first Peter and Mary both fail $k-1$ times, then Peter fails once more, and then Mary succeeds. Thus the ratio of the number of favorable alternatives over the total number of ways k rolls for both people can turn out gives

$$P(B_k) = \frac{(4 \cdot 2)^{k-1} \cdot 4 \cdot 4}{(6 \cdot 6)^k} = \left(\frac{8}{36}\right)^{k-1} \frac{16}{36} = \left(\frac{2}{9}\right)^{k-1} \frac{4}{9}.$$

The answer comes from a geometric series:

$$P\{\text{Mary wins}\} = P(B) = \sum_{k=1}^{\infty} P(B_k) = \sum_{k=1}^{\infty} \left(\frac{2}{9}\right)^{k-1} \frac{4}{9} = \frac{\frac{4}{9}}{1 - \frac{2}{9}} = \frac{4}{7}.$$

Note that we calculated the winning probabilities without defining the sample space. This will be typical going forward. Once we have understood the general principles of building probability models, it is usually not necessary to define explicitly the sample space in order to do calculations. ▲

Events and complements

Events A and A^c are disjoint and together make up Ω, no matter what the event A happens to be. Consequently

$$P(A) + P(A^c) = 1. \tag{1.13}$$

This identity tells us that to find $P(A)$, we can compute either $P(A)$ or $P(A^c)$. Sometimes one is much easier to compute than the other.

Example 1.21. Roll a fair die 4 times. What is the probability that some number appears more than once? A moment's thought reveals that this event contains a complicated collection of possible arrangements. Any one of 1 through 6 can appear two, three or four times. But also two different numbers can appear twice, and we would have to be careful not to overcount. However, switching to the complement provides an easy way out. If we set

$$A = \{\text{some number appears more than once}\}$$

then

$$A^c = \{\text{all rolls are different}\}.$$

By counting the possibilities $P(A^c) = \frac{6 \cdot 5 \cdot 4 \cdot 3}{6^4} = \frac{5}{18}$ and consequently $P(A) = \frac{13}{18}$. ▲

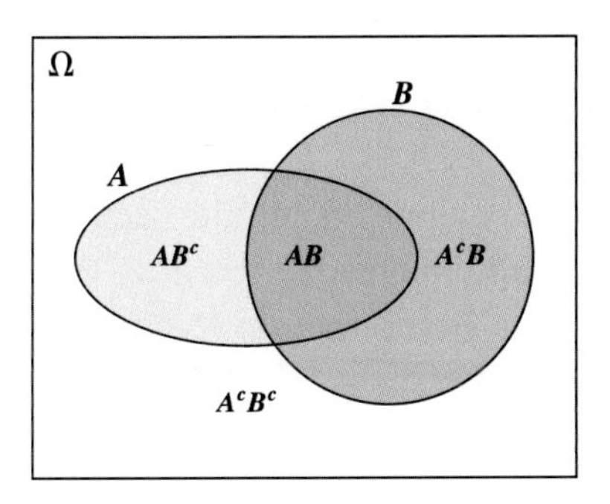

Figure 1.2. Venn diagram representation of two events A and B.

Equation (1.13) generalizes as follows. Intersecting with A and A^c splits any event B into two disjoint pieces $A \cap B$ and $A^c \cap B$, and so

$$P(B) = P(A \cap B) + P(A^c \cap B). \tag{1.14}$$

This identity is ubiquitous. Even in this section it appears several times.

There is an alternative way to write an intersection of sets: instead of $A \cap B$ we can write simply AB. Both will be used in the sequel. With this notation (1.14) is written as $P(B) = P(AB) + P(A^cB)$. The Venn diagram in Figure 1.2 shows a graphical representation of this identity.

Monotonicity of probability

Another intuitive and very useful fact is that a larger event must have larger probability:

$$\text{if } A \subseteq B \text{ then } P(A) \le P(B). \tag{1.15}$$

If $A \subseteq B$ then $B = A \cup A^cB$ where the two events on the right are disjoint. (Figure 1.3 shows a graphical representation of this identity.) Now inequality (1.15) follows from the additivity and nonnegativity of probabilities:

$$P(B) = P(A) + P(B \cap A^c) \ge P(A).$$

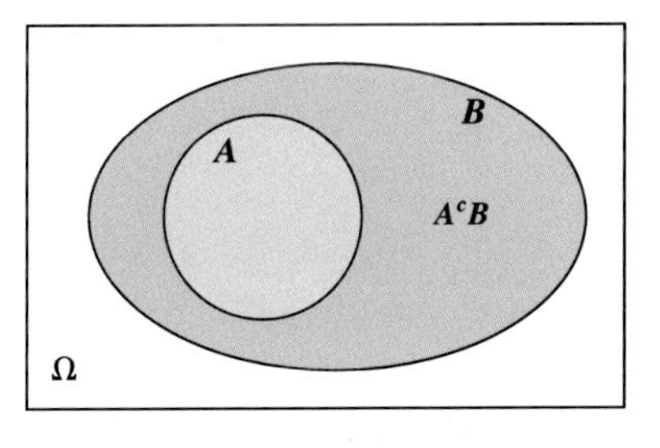

Figure 1.3. Venn diagram representation of the events $A \subseteq B$.

Example 1.22. Here is another proof of the fact, first seen in Example 1.16, that with probability 1 repeated flips of a fair coin eventually yield tails. Let A be the event that we never see tails and A_n the event that the first n coin flips are all heads. Never seeing tails implies that the first n flips must be heads, so $A \subseteq A_n$

and thus $P(A) \leq P(A_n)$. Now $P(A_n) = 2^{-n}$ so we conclude that $P(A) \leq 2^{-n}$. This is true for every positive n. The only nonnegative number $P(A)$ that can satisfy all these inequalities is zero. Consequently $P(A) = 0$.

The logic in this example is important. We can imagine an infinite sequence of flips all coming out heads. So this is not a logically impossible scenario. What we can say with mathematical certainty is that the heads-forever scenario has probability zero.

▲

Inclusion-exclusion

We move to inclusion-exclusion rules that tell us how to compute the probability of a union when the events are not mutually exclusive.

Fact 1.23. (Inclusion-exclusion formulas for two and three events)

$$P(A \cup B) = P(A) + P(B) - P(A \cap B). \tag{1.16}$$

$$\begin{aligned} P(A \cup B \cup C) = P(A) + P(B) + P(C) - P(A \cap B) - P(A \cap C) \\ - P(B \cap C) + P(A \cap B \cap C). \end{aligned} \tag{1.17}$$

We prove identity (1.16). Look at the Venn diagram in Figure 1.2 to understand the first and third step below:

$$\begin{aligned} P(A \cup B) &= P(AB^c) + P(AB) + P(A^cB) \\ &= \big(P(AB^c) + P(AB)\big) + \big(P(AB) + P(A^cB)\big) - P(AB) \\ &= P(A) + P(B) - P(AB). \end{aligned}$$

Identity (1.17) can be proved similarly with a Venn diagram for three events.

Note that (1.16) can be rearranged to express the probability of the intersection from the individual probabilities and the probability of the union:

$$P(A \cap B) = P(A) + P(B) - P(A \cup B). \tag{1.18}$$

Example 1.24. (Example 1.19 revisited) An urn contains 30 red, 20 green and 10 yellow balls. Draw two without replacement. What is the probability that the sample contains exactly one red or exactly one yellow?

We solved this problem in Example 1.19 by breaking up the event into smaller parts. Now apply first inclusion-exclusion (1.16) and then count favorable arrangements using unordered samples:

$$\begin{aligned} &P(\text{exactly one red or exactly one yellow}) \\ &\quad = P(\{\text{exactly one red}\} \cup \{\text{exactly one yellow}\}) \end{aligned}$$

$$= P(\text{exactly one red}) + P(\text{exactly one yellow})$$
$$- P(\text{exactly one red and exactly one yellow})$$
$$= \frac{30 \cdot 30}{\binom{60}{2}} + \frac{10 \cdot 50}{\binom{60}{2}} - \frac{30 \cdot 10}{\binom{60}{2}} = \frac{110}{177},$$

which gives the same probability we obtained in Example 1.19. ▲

Example 1.25. In a town 15% of the population is blond, 25% of the population has blue eyes and 2% of the population is blond with blue eyes. What is the probability that a randomly chosen individual from the town is not blond and does not have blue eyes? (We assume that each individual has the same probability to be chosen.)

In order to translate the information into the language of probability, we identify the sample space and relevant events. The sample space Ω is the entire population of the town. The important events or subsets of Ω are

$$A = \{\text{blond members of the population}\}, \qquad \text{and}$$
$$B = \{\text{blue-eyed members of the population}\}.$$

The problem gives us the following information:

$$P(A) = 0.15, \qquad P(B) = 0.25, \qquad \text{and} \qquad P(AB) = 0.02. \tag{1.19}$$

Our goal is to compute the probability of A^cB^c. At this point we could forget the whole back story and work with the following problem: suppose that (1.19) holds, find $P(A^cB^c)$.

By de Morgan's law (equation (B.1) on page 382) $A^cB^c = (A \cup B)^c$. Thus from the inclusion-exclusion formula we get

$$P(A^cB^c) = 1 - P(A \cup B) = 1 - (P(A) + P(B) - P(AB))$$
$$= 1 - (0.15 + 0.25 - 0.02) = 0.62.$$

Another way to get the same result is by applying (1.18) for A^c and B^c to express $P(A^cB^c)$ the following way

$$P(A^cB^c) = P(A^c) + P(B^c) - P(A^c \cup B^c).$$

We can compute $P(A^c)$ and $P(B^c)$ from $P(A)$ and $P(B)$. By de Morgan's law $A^c \cup B^c = (AB)^c$, so $P(A^c \cup B^c) = 1 - P(AB)$. Now we have all the ingredients and

$$P(A^cB^c) = (1 - 0.15) + (1 - 0.25) - (1 - 0.02) = 0.62.$$ ▲

The general formula is for any collection of n events $A_1, A_2, \dots, A_n$ on a sample space.

Fact 1.26. (General inclusion-exclusion formula)

$$\begin{aligned}P(A_1\cup\cdots\cup A_n) &= \sum_{i=1}^n P(A_i) - \sum_{1\le i_1<i_2\le n} P(A_{i_1}\cap A_{i_2}) \\ &+ \sum_{1\le i_1<i_2<i_3\le n} P(A_{i_1}\cap A_{i_2}\cap A_{i_3}) \\ &- \sum_{1\le i_1<i_2<i_3<i_4\le n} P(A_{i_1}\cap A_{i_2}\cap A_{i_3}\cap A_{i_4}) \\ &+\cdots + (-1)^{n+1}P(A_1\cap\cdots\cap A_n) \\ &= \sum_{k=1}^n (-1)^{k+1} \sum_{1\le i_1<\cdots<i_k\le n} P(A_{i_1}\cap\cdots\cap A_{i_k}). \end{aligned} \tag{1.20}$$

Note that summing over $1 \le i_1 < i_2 < i_3 \le n$ is exactly the same as summing over all triples of distinct indices from $\{1,\dots,n\}$. The number of such triples is $\binom{n}{3}$. The last line of equation (1.20) is a succinct way of expressing the entire formula: the first index k specifies the number of events put in each intersection, and then indices $i_1 < \cdots < i_k$ pick the k events out of $A_1,\dots,A_n$ that are intersected in each term of the last sum.

Our last example is a probability classic.

Example 1.27. Suppose n people arrive for a show and leave their hats in the cloakroom. Unfortunately, the cloakroom attendant mixes up the hats completely so that each person leaves with a random hat. Let us assume that all $n!$ assignments of hats are equally likely. What is the probability that no one gets his/her own hat? How does this probability behave as $n \to \infty$?

Define the events

$$A_i = \{\text{person } i \text{ gets his/her own hat}\}, \quad 1 \le i \le n.$$

The probability we want is

$$P\left(\bigcap_{i=1}^n A_i^c\right) = 1 - P\left(\bigcup_{i=1}^n A_i\right), \tag{1.21}$$

where we used de Morgan's law. We compute $P(A_1\cup\cdots\cup A_n)$ with the inclusion-exclusion formula (1.20). Let $i_1 < i_2 < \cdots < i_k$ and evaluate

$$\begin{aligned}P(A_{i_1}\cap A_{i_2}\cap\cdots\cap A_{i_k}) &= P\{\text{individuals } i_1, i_2,\dots,i_k \text{ get their own hats}\} \\ &= \frac{(n-k)!}{n!},\end{aligned} \tag{1.22}$$

where the formula comes from counting favorable arrangements: if we assign k individuals their correct hats, there are $(n-k)!$ ways to distribute the remaining hats. (Note that the event $A_{i_1} \cap A_{i_2} \cap \cdots \cap A_{i_k}$ does not say that these k are the *only* people who receive correct hats.) Thus

$$\sum_{i_1<i_2<\cdots<i_k} P(A_{i_1} \cap A_{i_2} \cap \cdots \cap A_{i_k}) = \binom{n}{k}\frac{(n-k)!}{n!} = \frac{1}{k!},$$

since there are $\binom{n}{k}$ terms in the sum. From (1.20)

$$P\left(\bigcup_{i=1}^{n} A_i\right) = 1 - \frac{1}{2!} + \frac{1}{3!} - \frac{1}{4!} + \cdots + (-1)^{n+1}\frac{1}{n!}$$

and then by (1.21)

$$P\left(\bigcap_{i=1}^{n} A_i^c\right) = 1 - 1 + \frac{1}{2!} - \frac{1}{3!} + \cdots + (-1)^n\frac{1}{n!} = \sum_{k=0}^{n}\frac{(-1)^k}{k!}. \qquad (1.23)$$

This is the beginning of the familiar series representation of the function e^x at $x = -1$. (See (D.3) in Appendix D for a reminder.) Thus the limit as $n \to \infty$ is immediate:

$$\lim_{n\to\infty} P(\text{no person among } n \text{ people gets the correct hat}) = \sum_{k=0}^{\infty}\frac{(-1)^k}{k!} = e^{-1}.$$

The reader with some background in algebra or combinatorics may notice that this example is really about the number of fixed points of a *random permutation*. A permutation of a set B is a bijective function $f : B \to B$. The fixed points of a permutation f are those elements x that satisfy $f(x) = x$. If we imagine that both the persons and the hats are numbered from 1 to n (with hat i belonging to person i) then we get a permutation that maps each person (or rather, her number) to the hat (or rather, its number) she receives. The result we derived says that as $n \to \infty$, the probability that a random permutation of n elements has no fixed points converges to e^{-1}. ▲

1.5. Random variables: a first look

In addition to the basic outcomes themselves, we are often interested in various numerical values derived from the outcomes. For example, in the game of Monopoly we roll a pair of dice, and the interesting outcome is the *sum* of the values of the dice. Or in a finite sequence of coin flips we might be interested in the total number of tails instead of the actual sequence of coin flips. This idea of attaching a number to each outcome is captured by the notion of a random variable.

Definition 1.28. Let Ω be a sample space. A **random variable** is a function from Ω into the real numbers. ♣

There are some conventions to get used to here. First the terminology: a random variable is not a variable but a function. Another novelty is the notation. In calculus we typically denote functions with lower case letters such as f, g and h. By contrast, random variables are usually denoted by capital letters such as X, Y and Z. The value of a random variable X at sample point ω is $X(\omega)$.

The study of random variables occupies much of this book. At this point we want to get comfortable with describing events in terms of random variables.

Example 1.29. We consider again the roll of a pair of dice (Example 1.6). Let us introduce three random variables: X_1 is the outcome of the first die, X_2 is the outcome of the second die, and S is the sum of the two dice. The precise definitions are these. For each sample point $(i,j) \in \Omega$,

$$X_1(i,j) = i, \quad X_2(i,j) = j, \quad \text{and} \quad S(i,j) = X_1(i,j) + X_2(i,j) = i + j.$$

To take a particular sample point, suppose the first die is a five and the second die is a one. Then the realization of the experiment is (5, 1) and the random variables take on the values

$$X_1(5,1) = 5, \quad X_2(5,1) = 1, \quad \text{and} \quad S(5,1) = X_1(5,1) + X_2(5,1) = 6.$$

In Example 1.6 we considered the event $D = \{\text{the sum of the two dice is 8}\}$ and computed its probability. In the language of random variables, $D = \{S = 8\}$ and

$$P\{S = 8\} = \sum_{(i,j):i+j=8} P\{X_1 = i, X_2 = j\} = \tfrac{5}{36}. \tag{1.24}$$

▲

A few more notational comments are in order. Recall that events are subsets of Ω. We write $\{S = 8\}$ for the set of sample points (i,j) such that $S(i,j) = 8$. The conventional full-fledged set notation for this is

$$\{(i,j) \in \Omega : S(i,j) = 8\}. \tag{1.25}$$

What we do in probability is drop all but the essential features of this notation. In most circumstances the expression $\{S = 8\}$ tells us all we need.

Inside events a comma is short for "and" which represents intersection. So the event on line (1.24) means the following:

$$\{X_1 = i, X_2 = j\} = \{X_1 = i \text{ and } X_2 = j\} = \{X_1 = i\} \cap \{X_2 = j\}.$$

If we wanted a union we would write "or":

$$\{X_1 = i \text{ or } X_2 = j\} = \{X_1 = i\} \cup \{X_2 = j\}.$$

Example 1.30. A die is rolled. If the outcome of the roll is 1, 2, or 3, the player loses \$1. If the outcome is 4, the player gains \$1, and if the outcome is 5 or 6, the player gains \$3. Let W denote the change in wealth of the player in one round of this game.

The sample space for the roll of the die is $\Omega = \{1, 2, 3, 4, 5, 6\}$. The random variable W is the real-valued function on Ω defined by

$$W(1) = W(2) = W(3) = -1, \quad W(4) = 1, \quad W(5) = W(6) = 3.$$

The probabilities of the values of W are

$$\begin{aligned} P\{W = -1\} = P\{1, 2, 3\} = \tfrac{1}{2}, \quad P\{W = 1\} = P\{4\} = \tfrac{1}{6}, \\ \text{and} \quad P\{W = 3\} = P\{5, 6\} = \tfrac{1}{3}. \end{aligned} \tag{1.26}$$

▲

Example 1.31. As in Example 1.17, select a point uniformly at random from the interval $[0, 1]$. Let Y be equal to twice the chosen point. The sample space is $\Omega = [0, 1]$, and the random variable is $Y(\omega) = 2\omega$. Let us compute $P\{Y \le a\}$ for $a \in [0, 2]$. By our convention discussed around (1.25), $\{Y \le a\} = \{\omega : Y(\omega) \le a\}$. Therefore

$$\{Y \le a\} = \{\omega : Y(\omega) \le a\} = \{\omega : 2\omega \le a\} = \{\omega : \omega \le a/2\} = [0, a/2],$$

where the second equality follows from the definition of Y, the third follows by algebra and the last is true because in this example the sample points ω are points in $[0, 1]$. Thus

$$P\{Y \le a\} = P([0, a/2]) = a/2 \qquad \text{for } a \in [0, 2]$$

by (1.11). If $a < 0$ then the event $\{Y \le a\}$ is empty, and consequently $P\{Y \le a\} = P(\varnothing) = 0$ for $a < 0$. If $a > 2$ then $P\{Y \le a\} = P(\Omega) = 1$. ▲

Example 1.32. A random variable X is *degenerate* if there is some real value b such that $P(X = b) = 1$. ▲

A degenerate random variable is in a sense not random at all because with probability 1 it has only one possible value. But it is an important special case to keep in mind. A real-valued function X on Ω is a *constant function* if there is some real value b such that $X(\omega) = b$ for all $\omega \in \Omega$. A constant function is a degenerate random variable. But a degenerate random variable does not have to be a constant function on all of Ω. Exercise 1.53 asks you to create an example.

As seen in all the examples above, events involving a random variable are of the form "the random variable takes certain values." The completely general form of such an event is

$$\{X \in B\} = \{\omega \in \Omega : X(\omega) \in B\}$$

where B is some subset of the real numbers. This reads "X lies in B."

Definition 1.33. Let X be a random variable. The **probability distribution** of the random variable X is the collection of probabilities $P\{X \in B\}$ for sets B of real numbers. ♣

The probability distribution of a random variable is an assignment of probabilities to subsets of $\mathbb{R}$ that satisfies again the axioms of probability (Exercise 1.54).

The next definition identifies a major class of random variables for which many exact computations are possible.

Definition 1.34. A random variable X is a **discrete random variable** if there exists a finite or countably infinite set $\{k_1, k_2, k_3, \dots\}$ of real numbers such that

$$\sum_i P(X = k_i) = 1 \tag{1.27}$$

where the sum ranges over the entire set of points $\{k_1, k_2, k_3, \dots\}$.

In particular, if the range of the random variable X is finite or countably infinite, then X is a discrete random variable. We say that those k for which $P(X = k) > 0$ are the *possible values* of the discrete random variable X.

In Example 1.29 the random variable S is the sum of two dice. The range of S is $\{2, 3, \dots, 12\}$ and so S is discrete.

In Example 1.31 the random variable Y is defined as twice a uniform random number from $[0, 1]$. For each real value a, $P(Y = a) = 0$ (see (1.12)), and so Y is *not* a discrete random variable.

The probability distribution of a discrete random variable is described completely in terms of its probability mass function.

Definition 1.35. The **probability mass function** (p.m.f.) of a discrete random variable X is the function p (or p_X) defined by

$$p(k) = P(X = k)$$

for possible values k of X.

The function p_X gives the probability of each possible value of X. Probabilities of other events of X then come by additivity: for any subset $B \subseteq \mathbb{R}$,

$$P(X \in B) = \sum_{k \in B} P(X = k) = \sum_{k \in B} p_X(k), \tag{1.28}$$

where the sum is over the possible values k of X that lie in B. A restatement of equation (1.27) gives $\sum_k p_X(k) = 1$ where the sum extends over all possible values k of X.

Example 1.36. (Continuation of Example 1.29) Here are the probability mass functions of the first die and the sum of the dice.

k	1	2	3	4	5	6
$p_{X_1}(k) = P(X_1 = k)$	$\frac{1}{6}$	$\frac{1}{6}$	$\frac{1}{6}$	$\frac{1}{6}$	$\frac{1}{6}$	$\frac{1}{6}$

k	2	3	4	5	6	7	8	9	10	11	12
$p_S(k) = P(S = k)$	$\frac{1}{36}$	$\frac{2}{36}$	$\frac{3}{36}$	$\frac{4}{36}$	$\frac{5}{36}$	$\frac{6}{36}$	$\frac{5}{36}$	$\frac{4}{36}$	$\frac{3}{36}$	$\frac{2}{36}$	$\frac{1}{36}$

Probabilities of events are obtained by summing values of the probability mass function. For example,

$$P(2 \le S \le 5) = p_S(2) + p_S(3) + p_S(4) + p_S(5) = \frac{1}{36} + \frac{2}{36} + \frac{3}{36} + \frac{4}{36} = \frac{10}{36}. \qquad \blacktriangle$$

Example 1.37. (Continuation of Example 1.30) Here are the values of the probability mass function of the random variable W defined in Example 1.30: $p_W(-1) = \frac{1}{2}$, $p_W(1) = \frac{1}{6}$, and $p_W(3) = \frac{1}{3}$. ▲

We finish this section with an example where the probability mass function of a discrete random variable is calculated with the help of a random variable whose range is an interval.

Example 1.38. We have a dartboard of radius 9 inches. The board is divided into four parts by three concentric circles of radii 1, 3, and 6 inches. If our dart hits the smallest disk, we get 10 points, if it hits the next region then we get 5 points, and we get 2 and 1 points for the other two regions (see Figure 1.4). Let X denote

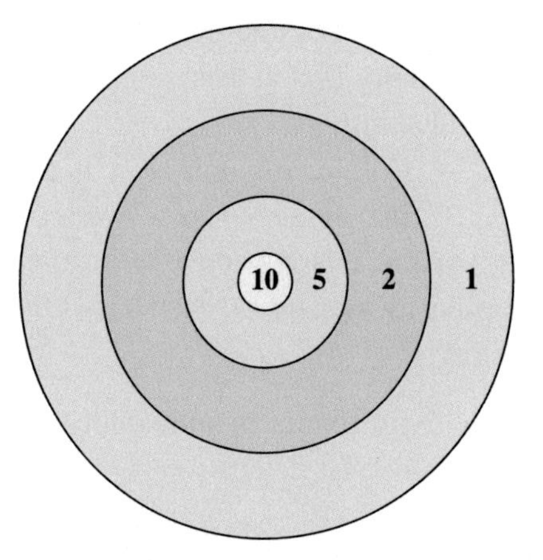

Figure 1.4. The dartboard for Example 1.38. The radii of the four circles in the picture are 1, 3, 6 and 9 inches.

the number of points we get when we throw a dart randomly (uniformly) at the board. How can we determine the distribution of X?

The possible values of X are $1, 2, 5$ and 10. We have to find the probabilities of these outcomes, that is, the probability mass function of X. It helps if we introduce another random variable: let $R \in [0, 9]$ denote the distance of the dart to the center. Then by Example 1.18 for any $a \in [0, 9]$ we have

$$P\{R \leq a\} = \frac{\pi \cdot a^2}{\pi \cdot 9^2} = \frac{a^2}{81}. \tag{1.29}$$

From this we can compute the probability mass function of X:

$$P\{X = 10\} = P\{R \leq 1\} = \frac{1}{81},$$

$$P\{X = 5\} = P\{1 < R \leq 3\} = P\{R \leq 3\} - P\{R \leq 1\} = \frac{8}{81},$$

$$P\{X = 2\} = P\{3 < R \leq 6\} = P\{R \leq 6\} - P\{R \leq 3\} = \frac{27}{81},$$

$$P\{X = 1\} = P\{6 < R \leq 9\} = P\{R \leq 9\} - P\{R \leq 6\} = \frac{45}{81}.$$

Note that the sum of the probabilities is equal to 1 (as it should be). ▲

1.6. Finer points ♣

Continuity of the probability measure

The following fact expresses a continuity property of probability measures as functions of events. See Exercise 1.61 for the companion statement for decreasing sequences.

Fact 1.39. Suppose we have an infinite sequence of events $A_1, A_2, A_3, \ldots$ that are nested increasing: $A_1 \subseteq A_2 \subseteq A_3 \subseteq \cdots \subseteq A_n \subseteq \cdots$. Let $A_\infty = \cup_{k=1}^{\infty} A_k$ denote the union. Then

$$\lim_{n \to \infty} P(A_n) = P(A_\infty). \tag{1.30}$$

Another way to state this is as follows: if we have increasing events then the probability of the union of all the events (the probability that at least one of them happens) is equal to the limit of the individual probabilities.

Recall from calculus that a function $f : \mathbb{R} \to \mathbb{R}$ is continuous at x if and only if for each sequence of points $x_1, x_2, x_3, \ldots$ that converge to x, we have $f(x_n) \to f(x)$ as $n \to \infty$. In a natural way increasing sets A_n converge to their union A_∞, because the difference $A_\infty \setminus A_n$ shrinks away as $n \to \infty$. Thus (1.30) is an analogue of the calculus notion of continuity.

Proof of Fact 1.39. To take advantage of the additivity axiom of probability, we break up the events A_n into disjoint pieces. For $n = 2, 3, 4, \ldots$ let $B_n = A_n \setminus A_{n-1}$.

Since each B_n is contained in A_n and disjoint from A_{n-1}, the events B_n are pairwise disjoint. We have the disjoint decomposition of A_n as

$$\begin{aligned} A_n &= A_{n-1} \cup B_n = A_{n-2} \cup B_{n-1} \cup B_n \\ &= \cdots = A_1 \cup B_2 \cup B_3 \cup \cdots \cup B_n. \end{aligned} \tag{1.31}$$

Taking a union of all the events A_n gives us the disjoint decomposition

$$\cup_{n=1}^{\infty} A_n = A_1 \cup B_2 \cup B_3 \cup B_4 \cup \cdots$$

By the additivity of probability, and by expressing the infinite series as the limit of partial sums,

$$\begin{aligned} P(\cup_{n=1}^{\infty} A_n) &= P(A_1) + P(B_2) + P(B_3) + (B_4) + \cdots \\ &= \lim_{n\to\infty} (P(A_1) + P(B_2) + \cdots + P(B_n)) = \lim_{n\to\infty} P(A_n). \end{aligned}$$

The last equality above came from the decomposition in (1.31). ▲

Measurability

The notion of an event is actually significantly deeper than we let on in this chapter. But for discrete sample spaces there is no difficulty. Usually each sample point ω has a well-defined probability and then each subset A of Ω has the probability given by the sum

$$P(A) = \sum_{\omega:\,\omega\in A} P\{\omega\}. \tag{1.32}$$

Thus every subset of a discrete sample space Ω is a legitimate event. For example, for the flip of a single coin the sample space is $\Omega = \{\mathrm{H}, \mathrm{T}\}$ and the collection of events is $\mathcal{F} = \{\varnothing, \{\mathrm{H}\}, \{\mathrm{T}\}, \{\mathrm{H}, \mathrm{T}\}\}$, which is exactly the collection of *all* subsets of Ω. The collection of all subsets of Ω is called the *power set* of Ω and is denoted by $2^{\Omega} = \{A : A \text{ is a subset of } \Omega\}$.

This all seems very straightforward. But it turns out that there can be good reasons to use smaller collections $\mathcal{F}$ of events. One reason is that a smaller $\mathcal{F}$ can be useful for modeling purposes. This is illustrated in Example 1.40 and Exercise 1.63 below. A second reason is that technical problems with uncountable sample spaces *prevent* us from taking $\mathcal{F}$ as the power set. We discuss this issue next.

Recall from Example 1.17 that for the uniform probability measure on $[0, 1]$ (1.32) fails for intervals because $P\{\omega\} = 0$ for each sample point ω. However, the technical issues go beyond this. It turns out that it is *impossible* to define consistently the uniform probability measure for all subsets of $[0, 1]$. Hence $\mathcal{F}$ must be something smaller than the power set of $[0, 1]$, but what exactly?

To put the theory on a sound footing, the axiomatic framework of Definition 1.1 is extended to impose the following requirements on the collection of events $\mathcal{F}$:

(i) the empty set $\varnothing$ is a member of $\mathcal{F}$,
(ii) if A is in $\mathcal{F}$, then A^c is also in $\mathcal{F}$,
(iii) if $A_1, A_2, A_3, \ldots$ is a sequence of events in $\mathcal{F}$, then their union $\cup_{i=1}^{\infty} A_i$ is also in $\mathcal{F}$.

Any collection $\mathcal{F}$ of sets satisfying properties (i)–(iii) is called a *σ-algebra* or a *σ-field*. The members of a σ-algebra are called *measurable sets*. The properties of a σ-algebra imply that countably many applications of the usual set operations to events is a safe way to produce new events.

To create the σ-algebra for the probability model we desire, we typically start with some nice sets that we want in $\mathcal{F}$, and let the axioms determine which other sets must also be members of $\mathcal{F}$. When Ω is $[0, 1]$, the real line, or some other interval of real numbers, the standard choice for $\mathcal{F}$ is to take the *smallest σ-algebra that contains all subintervals of* Ω. The members of this $\mathcal{F}$ are called *Borel sets*. It is impossible to give a simple description of all Borel sets. But this procedure allows us to rule out pathological sets for which probabilities cannot be assigned. Construction of probability spaces belongs in a subject called *measure theory* and lies beyond the scope of this book.

Similar issues arise with random variables. Not every function X from an uncountable Ω into $\mathbb{R}$ can be a random variable. We have to require that $\{X \in B\}$ is a member of $\mathcal{F}$ whenever B is a Borel subset of $\mathbb{R}$. This defines the notion of a *measurable function*, and it is the measurable functions that can appear as random variables. In particular, in Definition 1.33 the probability distribution $P\{X \in B\}$ of a random variable X is defined only for Borel sets $B \subseteq \mathbb{R}$.

Fortunately it turns out that all reasonable sets and functions encountered in practice are measurable. Consequently this side of the theory can be completely ignored in an undergraduate probability course. Readers who wish to pursue the full story can turn to graduate textbooks on real analysis and probability, such as [Bil95], [Fol99], and [Dur10].

Collections of events as information

Another aspect of the collection $\mathcal{F}$ of events is that it can represent *information*. This point becomes pronounced in advanced courses of probability. The next example gives a glimpse of the idea.

Example 1.40. (Continuation of Example 1.6) Consider again the experiment of rolling a pair of dice but this time we cannot tell the two dice apart. This means that we cannot separate outcomes $(1, 3)$ and $(3, 1)$ from each other. For a concrete experimental setup, imagine that I roll the dice behind a curtain and then tell you, "you got a one and a three," without specifying which die gave which number.

We can still model the situation with the same sample space $\Omega = \{(i, j) : i, j \in \{1, 2, 3, 4, 5, 6\}\}$ of ordered pairs. By a judicious choice of $\mathcal{F}$ we can forbid the model from separating $(1, 3)$ and $(3, 1)$. This is how we define the collection of events: $\mathcal{F}$ consists of

- the singletons $\{(i,i)\}$ for integers $1 \le i \le 6$,
- the sets of pairs $\{(i,j),(j,i)\}$ for $1 \le i,j \le 6$ such that $i \ne j$,
- and all unions of sets of the above two types, including the empty set.

The probability measure P is the same as before, except that its domain is now the smaller collection $\mathcal{F}$ of events described above. $(1,3)$ and $(3,1)$ are still separate points of the space Ω, but $\{(1,3)\}$ is not a member of $\mathcal{F}$ and hence does not have a probability assigned to it. The event $\{(1,3),(3,1)\}$ does have a probability assigned to it, namely $P\{(1,3),(3,1)\} = \frac{2}{36}$ as before. The point of this example is that by restricting $\mathcal{F}$ we can model the information available to the observer of the experiment, without changing Ω. ▲

Exercises

We start with some warm-up exercises arranged by section.

Section 1.1

Exercise 1.1. We roll a fair die twice. Describe a sample space Ω and a probability measure P to model this experiment. Let A be the event that the second roll is larger than the first. Find the probability $P(A)$ that the event A occurs.

Exercise 1.2. For breakfast Bob has three options: cereal, eggs or fruit. He has to choose exactly two items out of the three available.

(a) Describe the sample space of this experiment.

Hint. What are the different possible outcomes for Bob's breakfast?

(b) Let A be the event that Bob's breakfast includes cereal. Express A as a subset of the sample space.

Exercise 1.3.

(a) You flip a coin and roll a die. Describe the sample space of this experiment.
(b) Now each of 10 people flips a coin and rolls a die. Describe the sample space of this experiment. How many elements are in the sample space?
(c) In the experiment of part (b), how many outcomes are in the event where nobody rolled a five? How many outcomes are in the event where at least one person rolled a five?

Section 1.2

Exercise 1.4. Every day a kindergarten class chooses randomly one of the 50 state flags to hang on the wall, without regard to previous choices. We are interested in the flags that are chosen on Monday, Tuesday and Wednesday of next week.

(a) Describe a sample space Ω and a probability measure P to model this experiment.
(b) What is the probability that the class hangs Wisconsin's flag on Monday, Michigan's flag on Tuesday, and California's flag on Wednesday?
(c) What is the probability that Wisconsin's flag will be hung at least two of the three days?

Exercise 1.5. In one type of state lottery 5 distinct numbers are picked from $1, 2, 3, \dots, 40$ uniformly at random.

(a) Describe a sample space Ω and a probability measure P to model this experiment.
(b) What is the probability that out of the five picked numbers exactly three will be even?

Exercise 1.6. We have an urn with 3 green and 4 yellow balls. We choose 2 balls randomly without replacement. Let A be the event that we have two different colored balls in our sample.

(a) Describe a possible sample space with equally likely outcomes, and the event A in your sample space.
(b) Compute $P(A)$.

Exercise 1.7. We have an urn with 3 green and 4 yellow balls. We draw 3 balls one by one without replacement.

(a) Find the probability that the colors we see in order are green, yellow, green.
(b) Find the probability that our sample of 3 balls contains 2 green balls and 1 yellow ball.

Exercise 1.8. Suppose that a bag of scrabble tiles contains 5 Es, 4 As, 3 Ns and 2 Bs. It is my turn and I draw 4 tiles from the bag without replacement. Assume that my draw is uniformly random. Let C be the event that I got two Es, one A and one N.

(a) Compute $P(C)$ by imagining that the tiles are drawn one by one as an ordered sample.
(b) Compute $P(C)$ by imagining that the tiles are drawn all at once as an unordered sample.

Section 1.3

Exercise 1.9. We break a stick at a uniformly chosen random location. Find the probability that the shorter piece is less than 1/5th of the original.

Exercise 1.10. We roll a fair die repeatedly until we see the number four appear and then we stop. The outcome of the experiment is the number of rolls.

(a) Following Example 1.16 describe a sample space Ω and a probability measure P to model this situation.

(b) Calculate the probability that the number four never appears.

Exercise 1.11. We throw a dart at a square shaped board of side length 20 inches. Assume that the dart hits the board at a uniformly chosen random point. Find the probability that the dart is within 2 inches of the center of the board.

Section 1.4

Exercise 1.12. We roll a fair die repeatedly until we see the number four appear and then we stop.

(a) What is the probability that we need at most 3 rolls?
(b) What is the probability that we needed an even number of die rolls?

Exercise 1.13. At a certain school, 25% of the students wear a watch and 30% wear a bracelet. 60% of the students wear neither a watch nor a bracelet.

(a) One of the students is chosen at random. What is the probability that this student is wearing a watch or a bracelet?
(b) What is the probability that this student is wearing both a watch and a bracelet?

Exercise 1.14. Assume that $P(A) = 0.4$ and $P(B) = 0.7$. Making no further assumptions on A and B, show that $P(AB)$ satisfies $0.1 \le P(AB) \le 0.4$.

Exercise 1.15. An urn contains 4 balls: 1 white, 1 green and 2 red. We draw 3 balls with replacement. Find the probability that we did not see all three colors. Use two different calculations, as specified by (a) and (b) below.

(a) Define the event $W = \{\text{white ball did not appear}\}$ and similarly for G and R. Use inclusion-exclusion.
(b) Compute the probability by considering the complement of the event that we did not see all three colors.

Section 1.5

Exercise 1.16. We flip a fair coin five times. For every heads you pay me \$1 and for every tails I pay you \$1. Let X denote my net winnings at the end of five flips. Find the possible values and the probability mass function of X.

Exercise 1.17. An urn contains 4 red balls and 3 green balls. Two balls are drawn randomly.

(a) Let Z denote the number of green balls in the sample when the draws are done without replacement. Give the possible values and the probability mass function of Z.

(b) Let W denote the number of green balls in the sample when the draws are done with replacement. Give the possible values and the probability mass function of W.

Exercise 1.18. The statement

```
SOME DOGS ARE BROWN
```

has 16 letters. Choose one of the 16 letters uniformly at random. Let X denote the length of the *word* containing the chosen letter. Determine the possible values and probability mass function of X.

Exercise 1.19. You throw a dart and it lands uniformly at random on a circular dartboard of radius 6 inches. If your dart gets to within 2 inches of the center I will reward you with 5 dollars. But if your dart lands farther than 2 inches away from the center I will give you only 1 dollar. Let X denote the amount of your reward in dollars. Find the possible values and the probability mass function of X.

Further exercises

Exercise 1.20. We roll a fair die four times.

(a) Describe the sample space Ω and the probability measure P that model this experiment. To describe P, give the value $P\{\omega\}$ for each outcome $\omega \in \Omega$.
(b) Let A be the event that there are at least two fives among the four rolls. Let B be the event that there is at most one five among the four rolls. Find the probabilities $P(A)$ and $P(B)$ by finding the ratio of the number of favorable outcomes to the total, as in Fact 1.8.
(c) What is the set $A \cup B$? What equality should $P(A)$ and $P(B)$ satisfy? Check that your answers to part (b) satisfy this equality.

Exercise 1.21. Suppose an urn contains three black chips, two red chips, and two green chips. We draw three chips at random without replacement. Let A be the event that all three chips are of different color.

(a) Compute $P(A)$ by imagining that the chips are drawn one by one as an ordered sample.
(b) Compute $P(A)$ by imagining that the three chips are drawn all at once as an unordered sample.

Exercise 1.22. We pick a card uniformly at random from a standard deck of 52 cards. (If you are unfamiliar with the deck of 52 cards, see the description above Example C.19 in Appendix C.)

(a) Describe the sample space Ω and the probability measure P that model this experiment.

(b) Give an example of an event in this probability space with probability $\frac{3}{52}$.
(c) Show that there is no event in this probability space with probability $\frac{1}{5}$.

Exercise 1.23. The **Monty Hall problem** is a famous math problem loosely based on a television quiz show hosted by Monty Hall. You (the contestant) face three closed doors. There is a big prize behind one door, but nothing behind the other two. You are asked to pick one door out of the three, but you cannot yet see what is behind it. Monty opens one of the two other doors to reveal that there is nothing behind it. Then he offers you one chance to switch your choice. Should you switch?

This question has generated a great deal of confusion. An intuitively appealing quick answer goes like this. After Monty showed you a door with no prize, the prize is equally likely to be behind either one of the other two doors. Thus switching makes no difference. Is this correct?

To get clarity, let us use the simplest possible model. Let the outcome of the experiment be the door behind which the prize hides. So the sample space is $\Omega = \{1, 2, 3\}$. Assume that the prize is equally likely to be behind any one of the doors, and so we set $P\{1\} = P\{2\} = P\{3\} = 1/3$. Your decision to switch or not is not part of the model. Rather, use the model to compute the probability of winning separately for the two scenarios. We can fix the labeling of the doors so that your initial choice is door number 1. In both parts (a) and (b) below determine, for each door i, whether you win or not when the prize is behind door i.

(a) Suppose you will not switch. What is the probability that you win the prize?
(b) Suppose you *will* switch after Monty shows a door without a prize. What is the probability that you win the prize?

Exercise 1.24. Suppose the procedure of the quiz show is as described in the previous Exercise 1.23: you choose a door, Monty opens one of the two other doors with no prize, and you get to switch if you wish. Now suppose we have three positive numbers p_1, p_2, p_3 such that $p_1 + p_2 + p_3 = 1$ and the prize is behind door i with probability p_i. By labeling the doors suitably we can assume that $p_1 \geq p_2 \geq p_3$. Assume that you know the probabilities p_1, p_2, p_3 associated to each door. What is the strategy that maximizes your chances of winning the prize?

Exercise 1.25. A student lives within walking distance of 6 restaurants. Unbeknownst to her, she has exactly 1 friend at 3 of the restaurants, she has 2 friends at 2 of the restaurants, and she has no friends at one of the restaurants. Suppose that this accounts for all of her friends.

(a) Suppose that she chooses a restaurant at random. What is the probability that she has at least one friend at the restaurant?

(b) Suppose she calls one of her friends at random in order to meet up for dinner. What is the probability that she called someone at a restaurant where two of her friends are present?

Exercise 1.26. 10 men and 5 women are meeting in a conference room. Four people are chosen at random from the 15 to form a committee.

(a) What is the probability that the committee consists of 2 men and 2 women?
(b) Among the 15 is a couple, Bob and Jane. What is the probability that Bob and Jane both end up on the committee?
(c) What is the probability that Bob ends up on the committee but Jane does not?

Exercise 1.27. Suppose an urn contains seven chips labeled $1, \ldots, 7$. Three of the chips are black, two are red, and two are green. The chips are drawn randomly one at a time without replacement until the urn is empty.

(a) What is the probability that the ith draw is chip number 5?
(b) What is the probability that the ith draw is black?

Exercise 1.28. We have an urn with m green balls and n yellow balls. Two balls are drawn at random. What is the probability that the two balls have the same color?

(a) Assume that the balls are sampled without replacement.
(b) Assume that the balls are sampled with replacement.
(c) When is the answer to part (b) larger than the answer to part (a)? Justify your answer. Can you give an intuitive explanation for what the calculation tells you?

Exercise 1.29. At a political meeting there are 7 liberals and 6 conservatives. We choose five people uniformly at random to form a committee. Let A be the event that we end up with a committee consisting of more conservatives than liberals.

(a) Describe a possible sample space for this experiment, and the event A in your sample space.
(b) Compute $P(A)$.

Exercise 1.30. Eight rooks are placed randomly on a chess board. What is the probability that none of the rooks can capture any of the other rooks? Translation for those who are not familiar with chess: pick 8 unit squares at random from an 8×8 square grid. What is the probability that no two chosen squares share a row or a column?

Hint. You can think of placing the rooks either with or without order, both approaches work.

Exercise 1.31. Two cards are dealt from an ordinary deck of 52 cards. This means that two cards are sampled uniformly at random without replacement.

(a) What is the probability that both cards are aces and one of them is the ace of spades?
(b) What is the probability that at least one of the cards is an ace?

Exercise 1.32. You are dealt five cards from a standard deck of 52. Find the probability of being dealt a *full house*. (This means that you have three cards of one face value and two cards of a different face value. An example would be three queens and two fours. See Exercise C.10 in Appendix C for a description of all the poker hands.)

Exercise 1.33. You roll a fair die 5 times. What is the probability of seeing a "full house" in the sense of seeing three rolls of one type, and two rolls of another different type? Note that we do not allow for all five rolls to be of the same type.

Exercise 1.34. Pick a uniformly chosen random point inside a unit square (a square of sidelength 1) and draw a circle of radius 1/3 around the point. Find the probability that the circle lies entirely inside the square.

Exercise 1.35. Pick a uniformly chosen random point inside the triangle with vertices $(0, 0)$, $(3, 0)$ and $(0, 3)$.

(a) What is the probability that the distance of this point to the y-axis is less than 1?
(b) What is the probability that the distance of this point to the origin is more than 1?

Exercise 1.36.

(a) Let (X, Y) denote a uniformly chosen random point inside the unit square

$$[0, 1]^2 = [0, 1] \times [0, 1] = \{(x, y) : 0 \leq x, y \leq 1\}.$$

Let $0 \leq a < b \leq 1$. Find the probability $P(a < X < b)$, that is, the probability that the x-coordinate X of the chosen point lies in the interval (a, b).
(b) What is the probability $P(|X - Y| \leq 1/4)$?

Exercise 1.37. These questions pertain to Example 1.20 where Peter and Mary take turns rolling a fair die. To answer the questions, be precise about the definitions of your events and their probabilities.

(a) As in Example 1.20, suppose Peter takes the first roll. What is the probability that Mary wins and her last roll is a six?
(b) Suppose Mary takes the first roll. What is the probability that Mary wins?
(c) What is the probability that the game lasts an even number of rolls? Consider separately the case where Peter takes the first roll and the case

where Mary takes the first roll. Based on your intuition, which case should be likelier to end in an even number of rolls? Does the calculation confirm your intuition?

(d) What is the probability that the game does not go beyond k rolls? Consider separately the case where Peter takes the first roll and the case where Mary takes the first roll. Based on your intuition, in which case should the game be over faster? Does the calculation confirm your intuition?

Hint. Consider separately games that last $2j$ rolls (even number of rolls) and games that last $2j - 1$ rolls (odd number of rolls).

Exercise 1.38. Show that it is not possible to choose a uniform positive integer at random. (In other words, we cannot define a probability measure on the positive integers that can be considered uniform.)

Hint. What would be the probability of choosing a particular number?

Exercise 1.39. A restaurant offers three ways of paying the bill: cash, check, and card. At the end of the day there were 100 paid bills. Some of the bills were paid with a mix of cash, check, and card. There are 78 cash receipts, 26 card receipts and 16 checks. There were 13 bills that used cash and card, 4 that used card and check, 6 that used cash and check, and 3 that used all three methods of payment.

(a) If a bill is chosen at random, what is the probability that the bill was paid fully using only one of the methods?

(b) If two distinct bills are chosen at random, what is the probability that at least one was paid using two or more methods?

Exercise 1.40. An urn contains 1 green ball, 1 red ball, 1 yellow ball and 1 white ball. I draw 4 balls with replacement. What is the probability that there is at least one color that is repeated exactly twice?

Hint. Use inclusion-exclusion with events $G = \{\text{exactly two balls are green}\}$, $R = \{\text{exactly two balls are red}\}$, etc.

Exercise 1.41. Imagine a game of three players where exactly one player wins in the end and all players have equal chances of being the winner. The game is repeated four times. Find the probability that there is at least one person who wins no games.

Hint. Let $A_i = \{\text{person } i \text{ wins no games}\}$ and utilize the inclusion-exclusion formula.

Exercise 1.42. Suppose $P(A) > 0.8$ and $P(B) > 0.5$. Show that $P(AB) > 0.3$.

Exercise 1.43. Show that for any events $A_1, A_2, \ldots, A_n$,

$$P(A_1 \cup \cdots \cup A_n) \leq \sum_{k=1}^{n} P(A_k). \tag{1.33}$$

Hint. Obtain the case $n = 2$ from inclusion-exclusion, and more generally apply the $n = 2$ case repeatedly. Or formulate a proof by induction.

Exercise 1.44. Two fair dice are rolled. Let X be the maximum of the two numbers and Y the minimum of the two numbers on the dice.

(a) Find the possible values of X and the possible values of Y.
(b) Find the probabilities $P(X \leq k)$ for all integers k. Find the probability mass function of X.

 Hint. Noticing that $P(X = k) = P(X \leq k) - P(X \leq k - 1)$ can save you some work.
(c) Find the probability mass function of Y.

Exercise 1.45. Imagine the following game of chance. There are four dollar bills on the table. You roll a fair die repeatedly. Every time you fail to get a six, one dollar bill is removed. When you get your first six, you get to take the money that remains on the table. If the money runs out before you get a six, you have lost and the game is over. Let X be the amount of your award. Find the possible values and the probability mass function of X.

Exercise 1.46. An urn contains 3 red balls and 1 yellow ball. We draw balls from the urn one by one without replacement until we see the yellow ball. Let X denote the number of red balls we see before the yellow. Find the possible values and the probability mass function of X.

Exercise 1.47. Consider the experiment of drawing a point uniformly at random from the unit interval, as in Examples 1.17 and 1.31, with $\Omega = [0, 1]$ and P determined by $P([a, b]) = b - a$ for $[a, b] \subseteq \Omega$. Define $Z(\omega) = e^{\omega}$ for $\omega \in \Omega$. Find $P(Z \leq t)$ for all real values t.

Exercise 1.48. Consider the experiment of drawing a point uniformly at random from the unit interval $[0, 1]$. Let Y be the first digit after the decimal point of the chosen number. Explain why Y is discrete and find its probability mass function.

Exercise 1.49. An urn has $n - 3$ green balls and 3 red balls. Draw ℓ balls with replacement. Let B denote the event that a red ball is seen at least once. Find $P(B)$ using the following methods.

(a) Use inclusion-exclusion with the events $A_i = \{i\text{th draw is red}\}$.

 Hint. Use the general inclusion-exclusion formula from Fact 1.26 and the binomial theorem from Fact D.2.

(b) Decompose the event by considering the events of seeing a red ball exactly k times, with $k = 1, 2, \dots, \ell$.
(c) Compute the probability by considering the complement B^c.

Challenging problems

Exercise 1.50. We flip a fair coin repeatedly, without stopping. In Examples 1.16 and 1.22 we have seen that with probability one we will eventually see tails. Prove the following generalizations.

(a) Show that with probability one we will eventually get 5 tails in a row.
Hint. It could help to group the sequence of coin flips into groups of five.

(b) Let $\bar{a} = (a_1, a_2, \dots, a_r)$ be a fixed ordered r-tuple of heads and tails. (That is, each $a_i \in \{\mathrm{H}, \mathrm{T}\}$.) Show that with probability one the sequence $\bar{a}$ will show up eventually in the sequence of coin flips.

Exercise 1.51. A monkey sits down in front of a computer and starts hitting the keys of the keyboard randomly. Somebody left open a text editor with an empty file, so everything the monkey types is preserved in the file. Assume that each keystroke is a uniform random choice among the entire keyboard and does not depend on the previous keystrokes. Assume also that the monkey never stops. Show that with probability one the file will eventually contain your favorite novel in full. (This means that the complete text of the novel appears from some keystroke onwards.)

Assume for simplicity that each symbol (letter and punctuation mark) needed for writing the novel can be obtained by a single keystroke on the computer.

Exercise 1.52. Three married couples (6 guests altogether) attend a dinner party. They sit at a round table randomly in such a way that each outcome is equally likely. What is the probability that somebody sits next to his or her spouse?

Hint. Label the seats, the individuals, and the couples. There are $6! = 720$ seating arrangements altogether. Apply inclusion-exclusion to the events $A_i = \{i\text{th couple sit next to each other}\}$, $i = 1, 2, 3$. Count carefully the numbers of arrangements in the intersections of the A_is.

Exercise 1.53. Recall the definition of a degenerate random variable from Example 1.32. Construct an example of a random variable that is degenerate but not a constant function on Ω. Explicitly: define a sample space Ω, a probability measure P on Ω, and a random variable $X : \Omega \to \mathbb{R}$ such that X satisfies $P(X = b) = 1$ for some $b \in \mathbb{R}$ but $X(\omega)$ is not the same for all $\omega \in \Omega$.

Exercise 1.54. Verify the remark under Definition 1.33 for discrete random variables. Let X be a random variable with countably many values and define

$Q(A) = P\{X \in A\}$ for $A \subset \mathbb{R}$. Show that Q satisfies the axioms of probability. (Note that the sample space for Q is $\mathbb{R}$.)

Exercise 1.55. Show that it is not possible to choose a uniform random number from the whole real line. (In other words, we cannot define a probability on the real numbers that can be considered uniform.)

Exercise 1.56.

(a) Put k balls into n boxes at random. What is the probability that no box remains empty? (You may assume that balls and boxes are numbered and each assignment of k balls into n boxes is equally likely.)
(b) Using the result of part (a) calculate the value of

$$\sum_{j=1}^{n} (-1)^{j-1} j^k \binom{n}{j}$$

for $k \leq n$.

Exercise 1.57. We tile a 2×9 board with 9 dominos of size 2×1 so that each square is covered by exactly one domino. Dominos can be placed horizontally or vertically. Suppose that we choose a tiling randomly so that each possible tiling configuration is equally likely. The figure below shows two possible tiling configurations.

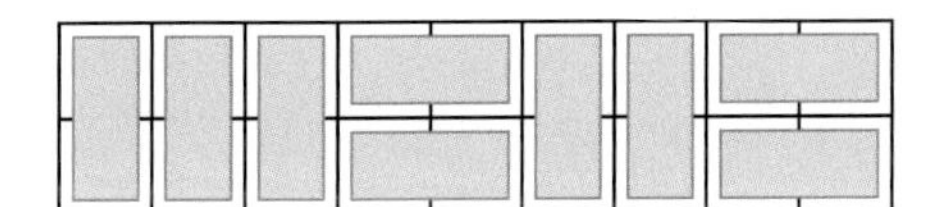
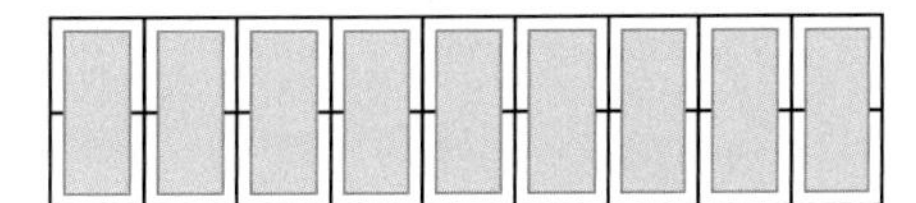

(a) What is the probability that all the dominos are placed vertically?
(b) Find the probability that there is a vertical domino in the middle of the board (at the 5th position).

Exercise 1.58. Recall the setup of Example 1.27. Find the probability that exactly ℓ people out of n receive the correct hat, and find the limit of this probability as $n \to \infty$.

Hint. Multiply (1.23) by $n!$ to get this counting identity:

$$\#(\text{arrangements where no person among } n \text{ gets the correct hat}) = n! \sum_{k=0}^{n} \frac{(-1)^k}{k!}. \tag{1.34}$$

Find the probability that individuals $i_1, i_2, \ldots, i_\ell$ and nobody else gets the correct hat. Add over all $i_1 < i_2 < \cdots < i_\ell$. For the limit you should get

$$\lim_{n\to\infty} P\{\text{exactly } \ell \text{ people out of } n \text{ receive the correct hat}\} = \frac{e^{-1}}{\ell!}.$$

This probability distribution defined by $p(\ell) = e^{-1}/\ell!$ for nonnegative integers ℓ is a particular case of the *Poisson distribution*. We meet it again in Chapter 4.

Exercise 1.59. (Buffon's needle problem) Suppose that we have an infinite grid of parallel lines on the plane, spaced one unit apart. (For example, on the xy-plane take the lines $y = n$ for all integers n.) We also have a needle of length ℓ, with $0 < \ell < 1$. We drop the needle on the grid and it lands in a random position. Show that the probability that the needle does not intersect any of the parallel lines is $2\ell/\pi$.

Hint. It is important to set up a probability space for the problem before attempting the solution. Suppose the grid consists of the horizontal lines $y = n$ for all integers n. The needle is a line segment of length ℓ. To check whether the needle intersects one of the parallel lines we need the following two quantities:

- the distance d from the center of the needle to the nearest parallel line below,
- the angle α of the needle measured from the horizontal direction.

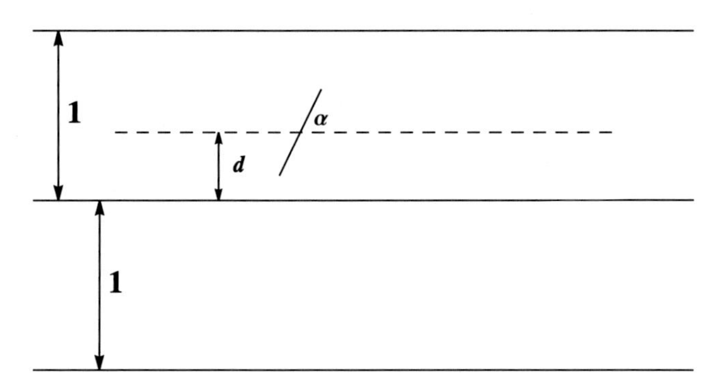

We have $d \in [0, 1)$ and $\alpha \in [0, \pi)$. We may assume (although the problem did not state it explicitly) that none of the possible pairs (d, α) is any likelier than another. This means that we can model the random position of the needle as a uniformly chosen point (d, α) from the rectangle $[0, 1) \times [0, \pi)$. To finish the proof complete the following two steps.

(a) Describe the event $A = \{\text{the needle does not intersect any of the lines}\}$ in terms of d and α. You need a little bit of trigonometry.
(b) Compute $P(A)$ as a ratio of areas.

This problem was first introduced and solved by Georges Louis Leclerc, Comte de Buffon, in 1777.

Exercise 1.60. (Continuation of Examples 1.7 and 1.40) Consider the experiment of flipping a coin three times with Ω as in Example 1.7. Suppose you learn the outcome of the first flip. What is the collection of events $\mathcal{F}_1$ that represents the information you possess? Suppose you learn the outcome of the first two flips. What is the collection of events $\mathcal{F}_2$ that represents the information you possess?

This exercise suggests that increasing collections of events $\mathcal{F}_k$, $k = 1, 2, 3, \ldots$, can represent the passage of time, as we learn the outcomes of more experiments. This idea is central for modeling random processes that unfold over time.

Exercise 1.61. Use Fact 1.39 and de Morgan's law to show that if $B_1 \supseteq B_2 \supseteq B_3 \supseteq \cdots \supseteq B_n \supseteq \cdots$ is a sequence of nested decreasing events then

$$\lim_{n \to \infty} P(B_n) = P(\cap_{k=1}^{\infty} B_k). \tag{1.35}$$

Exercise 1.62. Using Fact 1.39 to take a limit, deduce the infinite version of Exercise 1.43: for any sequence of events $\{A_k\}_{k=1}^{\infty}$,

$$P\left(\bigcup_{k=1}^{\infty} A_k\right) \leq \sum_{k=1}^{\infty} P(A_k). \tag{1.36}$$

Exercise 1.63. Recall the properties of a σ-algebra from Section 1.6. Consider the experiment of rolling a die which has sample space $\Omega = \{1, 2, 3, 4, 5, 6\}$.

(a) What is the smallest collection of events $\mathcal{F}$ satisfying the properties of a σ-algebra?

(b) Suppose that the sets $\{1, 2\}$ and $\{2, 3, 4\}$ are in the collection of events $\mathcal{F}$. Now what is the smallest collection of events $\mathcal{F}$ satisfying the properties of a σ-algebra?

2

Conditional probability and independence

Chapter 1 established the basic vocabulary for discussing randomness in mathematical terms. In this chapter we introduce independence and dependence, two complementary features of probability models. The mathematical tool for describing dependence is conditional probability. When independence is present it enables us to calculate probabilities of complicated events. Consideration of a sequence of independent trials leads to important named distributions: the Bernoulli, binomial and geometric distributions.

2.1. Conditional probability

Consider the following question. How should we update a probability model in light of additional information that constrains the outcome? Let us introduce the idea through a simple example. Consider an experiment with three outcomes $\Omega = \{1, 2, 3\}$ and probabilities $P\{1\} = \frac{1}{5}$, $P\{2\} = \frac{2}{5}$, and $P\{3\} = \frac{2}{5}$. Suppose we learn that the outcome is 1 or 2. This just means that the event $B = \{1, 2\}$ happens. How should our model change?

We will keep the same sample space $\Omega = \{1, 2, 3\}$, but define a new probability measure $\widetilde{P}$ on Ω that reflects our new information. Obviously $\widetilde{P}\{3\} = 0$ since 3 is no longer possible. Consequently $\widetilde{P}\{1\} + \widetilde{P}\{2\} = 1$. The new information should not alter the relative chances of 1 and 2, hence 2 should continue to be twice as likely as 1 under $\widetilde{P}$, as it was under P. We can satisfy all these demands by dividing the old probabilities of the remaining outcomes with the probability of B:

$$\widetilde{P}\{1\} = \frac{P\{1\}}{P(B)} = \frac{1}{3} \quad \text{and} \quad \widetilde{P}\{2\} = \frac{P\{2\}}{P(B)} = \frac{2}{3}. \tag{2.1}$$

To summarize, we restrict the old probability to the elements of the event B and renormalize the total to 1 by dividing by $P(B)$. We call this new probability a *conditional probability* and give it a special notation. It is among the most important concepts of this entire course. Figure 2.1 illustrates the next definition.

Definition 2.1. Let B be an event in the sample space Ω such that $P(B) > 0$. Then for all events A the **conditional probability of A given B** is defined as

$$P(A \mid B) = \frac{P(AB)}{P(B)}. \tag{2.2}$$

The definition of $P(A \mid B)$ does not make sense if $P(B) = 0$. In Chapter 10 we develop tools that enable us to condition even on some zero probability events.

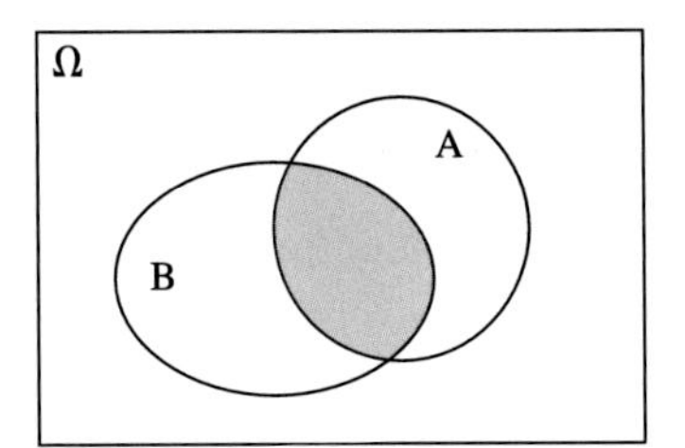

Figure 2.1. Venn diagram representation of conditioning on an event. The conditional probability of A given B is the probability of the part of A inside B (the shaded region), divided by the probability of B.

Conditioning produces a new probability measure that also satisfies the axioms of probability from Definition 1.1. We state the fact, but leave the proof as Exercise 2.77.

Fact 2.2. Let B be an event in the sample space Ω such that $P(B) > 0$. Then, as a function of the event A, the conditional probability $P(A \mid B)$ satisfies the axioms of Definition 1.1.

We illustrate Definition 2.1 with coin flips.

Example 2.3. Counting outcomes as in Example 1.7, the probability of getting 2 heads out of three coin flips is 3/8. Suppose the first coin flip is revealed to be heads. Heuristically, the probability of getting exactly two heads out of the three is now 1/2. This is because we are simply requiring the appearance of precisely one heads in the final two flips of the coin, which has a probability of 1/2.

We can rigorously verify the answer by the following conditioning calculation. As before, we denote outcomes as triples of zeros and ones where zero represents heads and one represents tails. Let

$$A = \{\text{exactly 2 heads}\} = \{(0,0,1),(0,1,0),(1,0,0)\}$$

and

$$B = \{\text{first flip is heads}\} = \{(0,0,0),(0,0,1),(0,1,0),(0,1,1)\}.$$

Then

$$P(A \mid B) = \frac{P(AB)}{P(B)} = \frac{P\{(0,0,1),(0,1,0)\}}{P\{(0,0,0),(0,0,1),(0,1,0),(0,1,1)\}} = \frac{1}{2}.$$

▲

Note that in the example above $P(A \mid B)$ is also equal to $\frac{\#AB}{\#B}$, that is, the ratio of the numbers of outcomes in the events. This is true in general *if* the sample space consists of finitely many *equally likely* outcomes.

Fact 2.4. Suppose that we have an experiment with **finitely many equally likely outcomes** and B is not the empty set. Then, for any event A

$$P(A|B) = \frac{\#AB}{\#B}.$$

This fact is immediately verified to be consistent with equation (2.2) of Definition 2.1: since $P(B) = \frac{\#B}{\#\Omega} > 0$ and $P(AB) = \frac{\#AB}{\#\Omega}$, we have $\frac{\#AB}{\#B} = \frac{P(AB)}{P(B)}$.

Example 2.5. We have an urn with 4 red and 6 green balls. We choose a sample of 3 without replacement. Find the conditional probability of having exactly 2 red balls in the sample given that there is at least one red ball in the sample.

Let A be the event that there are exactly two red balls in the sample and B that there is at least one. We need to compute $P(A \mid B) = \frac{P(AB)}{P(B)}$. We have

$$P(B) = 1 - P(\text{no red balls in the sample}) = 1 - \frac{\binom{6}{3}}{\binom{10}{3}} = \frac{5}{6},$$

$$P(AB) = P(A) = \frac{\binom{4}{2}\binom{6}{1}}{\binom{10}{3}} = \frac{3}{10}.$$

This gives $P(A \mid B) = \frac{3/10}{5/6} = \frac{9}{25}$.

We could also have computed this conditional probability using Fact 2.4 by checking that $\#B = 100$ and $\#AB = \#A = 36$ and by writing $P(A \mid B) = \frac{\#AB}{\#B}$. ▲

If we rearrange the terms in (2.2) in the definition of conditional probability we get the *multiplication rule*

$$P(AB) = P(A)P(B \mid A). \tag{2.3}$$

The multiplication rule generalizes to three events as follows:

$$P(ABC) = P(AB)P(C \mid AB) = P(A)P(B \mid A)P(C \mid AB) \tag{2.4}$$

provided all the conditional probabilities make sense. You can see (2.4) as an instance of simple algebra:

$$P(ABC) = P(A) \cdot \frac{P(AB)}{P(A)} \cdot \frac{P(ABC)}{P(AB)} = P(A)\, P(B \mid A)\, P(C \mid AB).$$

The general version can be stated as follows.

Fact 2.6. (Multiplication rule for n events) If $A_1, \dots, A_n$ are events and all the conditional probabilities below make sense then we have

$$P(A_1 A_2 \cdots A_n) = P(A_1)P(A_2 \mid A_1)P(A_3 \mid A_1 A_2) \cdots P(A_n \mid A_1 \cdots A_{n-1}). \tag{2.5}$$

Example 2.7. Suppose an urn contains 8 red and 4 white balls. Draw two balls without replacement. What is the probability that both are red?

This question can of course be answered by simple counting, but there is also a natural solution using conditional probability. Define the events

$$R_1 = \{\text{first draw is red}\} \quad \text{and} \quad R_2 = \{\text{second draw is red}\}.$$

Then

$$P(R_1 R_2) = P(R_1)P(R_2 \mid R_1) = \tfrac{8}{12} \cdot \tfrac{7}{11} = \tfrac{14}{33}.$$

$P(R_2 \mid R_1) = 7/11$ comes from observing that after the first red draw, 7 out of the remaining 11 balls are red.

We may use Fact 2.6 to generalize to any number of draws: for example,

$$\begin{aligned} &P\{\text{first two draws are red and the third and fourth draws are white}\} \\ &\qquad = P(R_1 R_2 W_3 W_4) = P(R_1)P(R_2 \mid R_1)P(W_3 \mid R_1 R_2)P(W_4 \mid R_1 R_2 W_3) \\ &\qquad = \tfrac{8}{12} \cdot \tfrac{7}{11} \cdot \tfrac{4}{10} \cdot \tfrac{3}{9} = \tfrac{28}{495}. \end{aligned}$$

▲

Next we combine the multiplication rule with the decomposition of an event into pairwise disjoint parts, in order to calculate the probability of a complicated event.

Example 2.8. We have two urns. Urn I has 2 green balls and 1 red ball. Urn II has 2 red balls and 3 yellow balls. We perform a *two-stage experiment*. First choose one of the urns with equal probability. Then sample one ball uniformly at random from the selected urn.

What is the probability that we draw a red ball?

Let {urn I} denote the event that we choose urn I, {urn II} similarly, and let {red} denote the event that the drawn ball is red. We decompose the event {red} according to the urn that the ball came from, and then use the multiplication rule:

$$\begin{aligned} P(\text{red}) &= P(\{\text{red}\} \cap \{\text{urn I}\}) + P(\{\text{red}\} \cap \{\text{urn II}\}) \\ &= P(\text{red} \mid \text{urn I})\, P(\text{urn I}) + P(\text{red} \mid \text{urn II})\, P(\text{urn II}) \\ &= \tfrac{1}{3} \cdot \tfrac{1}{2} + \tfrac{2}{5} \cdot \tfrac{1}{2} = \tfrac{11}{30}. \end{aligned}$$

What is the probability that we draw a green ball?

A green ball can be drawn only from urn I. We can use this information in two ways. One way is to set $P(\text{green} \mid \text{urn II}) = 0$ and compute as above:

$$\begin{aligned}
P(\text{green}) &= P(\{\text{green}\} \cap \{\text{urn I}\}) + P(\{\text{green}\} \cap \{\text{urn II}\}) \\
&= P(\text{green} \mid \text{urn I})\, P(\text{urn I}) + P(\text{green} \mid \text{urn II})\, P(\text{urn II}) \\
&= \tfrac{2}{3} \cdot \tfrac{1}{2} + 0 \cdot \tfrac{1}{2} = \tfrac{1}{3}.
\end{aligned}$$

Another way is to use the set containment $\{\text{green}\} \subset \{\text{urn I}\}$ and compute as follows:

$$P(\text{green}) = P(\{\text{green}\} \cap \{\text{urn I}\}) = P(\text{green} \mid \text{urn I})\, P(\text{urn I}) = \tfrac{2}{3} \cdot \tfrac{1}{2} = \tfrac{1}{3}. \qquad \blacktriangle$$

Here is the general version of the reasoning used in the example above:

$$P(A) = P(AB) + P(AB^c) = P(A \mid B)\, P(B) + P(A \mid B^c)\, P(B^c). \tag{2.6}$$

It follows from additivity of probability and the multiplication rule. The basic idea here (one that recurs throughout the book) is the decomposition of a complicated event A into disjoint pieces that are easier to deal with. Above we used the pair $\{B, B^c\}$ to split A into two pieces. $\{B, B^c\}$ is an example of a *partition.*

Definition 2.9. A finite collection of events $\{B_1, \dots, B_n\}$ is a **partition** of Ω if the sets B_i are pairwise disjoint and together they make up Ω. That is, $B_iB_j = \varnothing$ whenever $i \neq j$ and $\bigcup_{i=1}^{n} B_i = \Omega$.

Here is the extension of (2.6) to a partition with an arbitrary number of sets.

Fact 2.10. Suppose that $B_1, \dots, B_n$ is a partition of Ω with $P(B_i) > 0$ for $i = 1, \dots, n$. Then for any event A we have

$$P(A) = \sum_{i=1}^{n} P(AB_i) = \sum_{i=1}^{n} P(A|B_i)P(B_i). \tag{2.7}$$

Equation (2.7) is true for the same reason as (2.6). Namely, set algebra gives

$$A = A \cap \Omega = A \cap \left(\bigcup_{i=1}^{n} B_i \right) = \bigcup_{i=1}^{n} AB_i \tag{2.8}$$

which expresses A as a union of the pairwise disjoint events AB_i (see Figure 2.2). Hence additivity of probability gives the first equality of (2.7). Then apply the multiplication rule (2.3) to each term. Identity (2.7) is sometimes called the *law of total probability.*

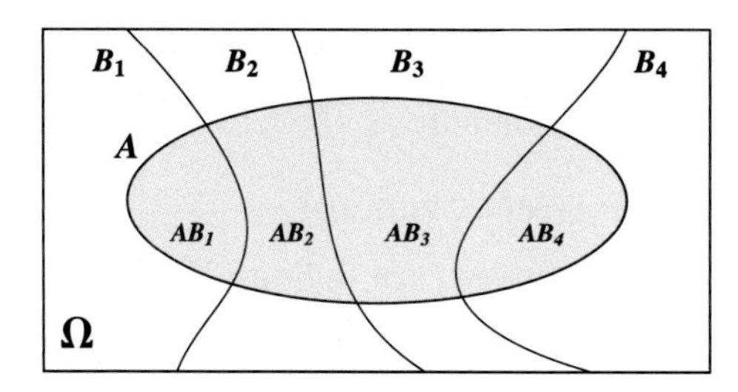

Figure 2.2. Venn diagram representation of decomposition (2.8). The events B_1, B_2, B_3, B_4 form a partition of Ω. The shaded set A is the disjoint union of AB_1, AB_2, AB_3 and AB_4.

Example 2.11. There are three types of coins in circulation. 90% of coins are fair coins that give heads and tails with equal probability. 9% of coins are moderately biased and give tails with probability $3/5$. The remaining 1% of coins are heavily biased and give tails with probability $9/10$. The type of a coin cannot be determined from its appearance. I have a randomly chosen coin in my pocket and I flip it. What is the probability I get tails?

If we knew the type of the coin the answer would be one of the probabilities given above. But the type of the coin is now random and consequently we use (2.7). The partition is given by the type of coin: denote the events by

$$F = \{\text{fair coin}\},\ M = \{\text{moderately biased coin}\} \text{ and } H = \{\text{heavily biased coin}\}.$$

The probabilities given above are $P(F) = \frac{90}{100}$, $P(M) = \frac{9}{100}$ and $P(H) = \frac{1}{100}$. Note that these probabilities add up to 1 as they should, since $\{B, M, H\}$ is a partition of the sample space of all coins.

The probabilities of tails given in the problem statement are the conditional probabilities

$$P(\text{tails} \mid F) = \tfrac{1}{2}, \quad P(\text{tails} \mid M) = \tfrac{3}{5}, \quad \text{and} \quad P(\text{tails} \mid H) = \tfrac{9}{10}.$$

Now apply (2.7):

$$\begin{aligned} P(\text{tails}) &= P(\text{tails} \mid F)P(F) + P(\text{tails} \mid M)P(M) + P(\text{tails} \mid H)P(H) \\ &= \tfrac{1}{2} \cdot \tfrac{90}{100} + \tfrac{3}{5} \cdot \tfrac{9}{100} + \tfrac{9}{10} \cdot \tfrac{1}{100} = \tfrac{513}{1000}. \end{aligned}$$ ▲

2.2. Bayes' formula

Continuing to build on conditional probability and the law of total probability, we ask the following question: what are the conditional probabilities of the competing explanations for an observed event? This leads us to Bayes' formula and the idea of reversing the order of conditioning.

Example 2.12. As in Example 2.8, we have two urns: urn I has 2 green balls and 1 red ball, while urn II has 2 red balls and 3 yellow balls. An urn is picked randomly and a ball is drawn from it. Given that the chosen ball is red, what is the probability that the ball came from urn I?

By the definition of conditional probability and the law of total probability, we compute as follows:

$$\begin{aligned} P(\text{urn I} \mid \text{red}) &= \frac{P(\{\text{red}\} \cap \{\text{urn I}\})}{P(\text{red})} \\ &= \frac{P(\text{red} \mid \text{urn I})P(\text{urn I})}{P(\text{red} \mid \text{urn I})P(\text{urn I}) + P(\text{red} \mid \text{urn II})P(\text{urn II})} \\ &= \frac{\frac{1}{3} \cdot \frac{1}{2}}{\frac{1}{3} \cdot \frac{1}{2} + \frac{2}{5} \cdot \frac{1}{2}} = \frac{5}{11}. \end{aligned}$$

Since $\frac{5}{11} < \frac{1}{2}$, urn I is less likely to be the source of the red ball than urn II. Hopefully this agrees with your intuition. ▲

Here is the general version of the formula we derived in the example.

Fact 2.13. (Bayes' formula) If $P(A), P(B), P(B^c) > 0$, then

$$P(B \mid A) = \frac{P(AB)}{P(A)} = \frac{P(A \mid B)\,P(B)}{P(A \mid B)\,P(B) + P(A \mid B^c)\,P(B^c)}. \tag{2.9}$$

To derive the formula above, start with the definition of the conditional probability $P(B \mid A)$ and then substitute the decomposition

$$P(A) = P(AB) + P(AB^c) = P(A \mid B)P(B) + P(A \mid B^c)P(B^c)$$

into the denominator.

Example 2.14. Suppose we have a medical test that detects a particular disease 96% of the time, but gives false positives 2% of the time. Assume that 0.5% of the population carries the disease. If a random person tests positive for the disease, what is the probability that they actually carry the disease?

Define the events $D = \{\text{person has the disease}\}$ and $A = \{\text{person tests positive}\}$. The problem statement gives the probabilities

$$P(A \mid D) = \tfrac{96}{100}, \quad P(A \mid D^c) = \tfrac{2}{100}, \quad \text{and} \quad P(D) = \tfrac{5}{1000}.$$

From Bayes' formula

$$P(D \mid A) = \frac{\frac{96}{100} \cdot \frac{5}{1000}}{\frac{96}{100} \cdot \frac{5}{1000} + \frac{2}{100} \cdot \frac{995}{1000}} = \frac{96}{494} \approx 0.194.$$

The number at the end often comes as a shock when you see this kind of calculation for the first time: only approximately 1 in 5 positive tests come from the disease. How does this square with the fact that medical testing is used to great benefit in treating people? The key point is that a doctor would seldom order a test on a random person. Tests are typically ordered in response to symptoms or a judgment by the doctor. The outcome of the calculation is *very sensitive* to the inputs. Suppose that, on the basis of an examination, the doctor believes that this person is quite likely, say with a 50% chance, to have the disease. So we take $P(D) = 1/2$. With this updated $P(D)$ the calculation becomes

$$P(D \mid A) = \frac{\frac{96}{100} \cdot \frac{1}{2}}{\frac{96}{100} \cdot \frac{1}{2} + \frac{2}{100} \cdot \frac{1}{2}} = \frac{96}{98} \approx 0.980.$$

Bayes' formula teaches us to look for the likeliest explanation for an observed event. In this example the disease is so rare that even if we see a positive test on a randomly selected person, an error in the test is still a likelier explanation than the disease. The tables are turned when the a priori chance of disease is raised to 50%. Now the disease is the very likely explanation of a positive test result. ▲

To generalize Bayes' formula to more than two explanatory events, let $B_1, \dots, B_n$ be a partition of the sample space Ω such that each $P(B_i) > 0$ (recall Definition 2.9). For any event A we have the decomposition

$$P(A) = \sum_{i=1}^{n} P(AB_i) = \sum_{i=1}^{n} P(A \mid B_i)P(B_i).$$

Inserting this into the denominator of the formula for the conditional probability $P(B_k|A)$ gives us the general *Bayes' formula.*

Fact 2.15. (General version of Bayes' formula) Let $B_1, \dots, B_n$ be a partition of the sample space Ω such that each $P(B_i) > 0$. Then for any event A with $P(A) > 0$, and any $k = 1, \dots, n$,

$$P(B_k \mid A) = \frac{P(AB_k)}{P(A)} = \frac{P(A \mid B_k)\,P(B_k)}{\sum_{i=1}^{n} P(A \mid B_i)\,P(B_i)}. \tag{2.10}$$

The probabilities $P(B_k)$ are often called *prior* probabilities because they correspond to our beliefs about the model before additional information is collected. Bayes' formula calculates the *posterior* probabilities $P(B_k \mid A)$, which update our model based on the occurrence of event A.

Example 2.16. Return to Example 2.11 with three types of coins: fair (F), moderately biased (M) and heavily biased (H), with probabilities of tails

$$P(\text{tails} \mid F) = \tfrac{1}{2}, \quad P(\text{tails} \mid M) = \tfrac{3}{5}, \quad \text{and} \quad P(\text{tails} \mid H) = \tfrac{9}{10}.$$

We hold a coin of unknown type. The probabilities of its type were given by

$$P(F) = \tfrac{90}{100}, \quad P(M) = \tfrac{9}{100}, \quad \text{and} \quad P(H) = \tfrac{1}{100}.$$

These are the prior probabilities. We flip the coin once and observe tails. Bayes' formula calculates our new posterior probabilities.

Here is the posterior probability for the coin being fair:

$$\begin{aligned}P(F \mid \text{tails}) &= \frac{P(\{\text{tails}\} \cap F)}{P(\text{tails})} \\ &= \frac{P(\text{tails} \mid F)P(F)}{P(\text{tails} \mid F)\, P(F) + P(\text{tails} \mid M)\, P(M) + P(\text{tails} \mid H)\, P(H)} \\ &= \frac{\frac{1}{2} \cdot \frac{90}{100}}{\frac{1}{2} \cdot \frac{90}{100} + \frac{3}{5} \cdot \frac{9}{100} + \frac{9}{10} \cdot \frac{1}{100}} = \frac{\frac{45}{100}}{\frac{513}{1000}} = \frac{450}{513} = 0.877.\end{aligned}$$

Similar calculations give the other two posterior probabilities:

$$P(M \mid \text{tails}) = \frac{P(\text{tails} \mid M)P(M)}{P(\text{tails})} = \frac{\frac{3}{5} \cdot \frac{9}{100}}{\frac{513}{1000}} = \frac{54}{513} = 0.105$$

and

$$P(H \mid \text{tails}) = \frac{P(\text{tails} \mid H)P(H)}{P(\text{tails})} = \frac{\frac{9}{10} \cdot \frac{1}{100}}{\frac{513}{1000}} = \frac{9}{513} = 0.018.$$

The calculation shows that observing tails strengthens the case for a biased coin. ▲

2.3. Independence

The conditional probability $P(A|B)$ quantifies the effect that the occurrence of an event B has on the probability of another event A. Intuitively, we would say that the event A is independent of B if this conditional probability is the same as the unconditional one: $P(A|B) = P(A)$. In this case

$$P(A) = \frac{P(AB)}{P(B)}, \qquad \text{which implies} \qquad P(AB) = P(A)P(B).$$

If $P(A) > 0$ then this last equation also implies $P(B|A) = \frac{P(AB)}{P(A)} = P(B)$. This motivates our next definition.

Definition 2.17. Two events A and B are **independent** if

$$P(AB) = P(A)P(B). \tag{2.11}$$

Independence makes sense only when A and B are events on the same sample space. One cannot ask about the independence of events from different sample spaces.

Example 2.18. Suppose that we flip a fair coin three times. Let A be the event that we have exactly one tails among the first two coin flips, B the event that we have exactly one tails among the last two coin flips and D the event that we get exactly one tails among the three coin flips. Show that A and B are independent, A and D are not independent, and B and D are also not independent.

We have $\#\Omega = 8$. The events A and B have four elements and D has three:

$$A = \{(\mathtt{H}, \mathtt{T}, \mathtt{H}), (\mathtt{H}, \mathtt{T}, \mathtt{T}), (\mathtt{T}, \mathtt{H}, \mathtt{H}), (\mathtt{T}, \mathtt{H}, \mathtt{T})\},$$
$$B = \{(\mathtt{H}, \mathtt{H}, \mathtt{T}), (\mathtt{H}, \mathtt{T}, \mathtt{H}), (\mathtt{T}, \mathtt{H}, \mathtt{T}), (\mathtt{T}, \mathtt{T}, \mathtt{H})\},$$
$$D = \{(\mathtt{H}, \mathtt{H}, \mathtt{T}), (\mathtt{H}, \mathtt{T}, \mathtt{H}), (\mathtt{T}, \mathtt{H}, \mathtt{H})\}.$$

The intersections are $AB = \{(\mathtt{H}, \mathtt{T}, \mathtt{H}), (\mathtt{T}, \mathtt{H}, \mathtt{T})\}$, $AD = \{(\mathtt{H}, \mathtt{T}, \mathtt{H}), (\mathtt{T}, \mathtt{H}, \mathtt{H})\}$, and $BD = \{(\mathtt{H}, \mathtt{H}, \mathtt{T}), (\mathtt{H}, \mathtt{T}, \mathtt{H})\}$ which gives

$$P(AB) = \tfrac{2}{8} = \tfrac{4}{8} \cdot \tfrac{4}{8} = P(A)P(B),$$
$$P(AD) = \tfrac{2}{8} \neq \tfrac{4}{8} \cdot \tfrac{3}{8} = P(A)P(D),$$
$$P(BD) = \tfrac{2}{8} \neq \tfrac{4}{8} \cdot \tfrac{3}{8} = P(B)P(D).$$

▲

Example 2.19. Suppose that we have an urn with 4 red and 7 green balls. We choose two balls with replacement. Let

$$A = \{\text{first ball is red}\} \quad \text{and} \quad B = \{\text{second ball is green}\}.$$

Is it true that A and B are independent? What if we sample without replacement?

Try to guess the answer first without actually computing it. When we sample with replacement, the contents of the urn are the same for each draw. Consequently the first draw does not influence the second. When we sample without replacement this is not the case anymore: the first draw alters the composition of the urn and thereby the probabilities. Consequently the conditional probability of the second draw should differ from the unconditional probability. Let us check this intuition by doing the math.

Sampling with replacement. We can pretend that the balls are numbered from 1 to 11 and the first four are red. Then $\#\Omega = 11^2$ (Ω is the set of all ordered pairs made up from these numbers), $\#A = 4 \cdot 11$ (all pairs where the first number is between 1 and 4), $\#B = 11 \cdot 7$ (all pairs where the second number is between 5 and 11) and $\#(A \cap B) = 4 \cdot 7$ (all pairs where the first number is between 1 and 4 and the second is between 5 and 11). Now we can check that

$$P(AB) = \frac{4 \cdot 7}{11^2} = \frac{4 \cdot 11}{11^2} \cdot \frac{11 \cdot 7}{11^2} = P(A)P(B).$$

This equality shows that the two events are independent.

Sampling without replacement. Numbering the balls as in the previous case, $\#\Omega = 11 \cdot 10 = 110$, $\#A = 4 \cdot 10 = 40$, $\#(A \cap B) = 4 \cdot 7 = 28$ and $\#B = \#(A \cap B) + \#(A^c \cap B) = 4 \cdot 7 + 7 \cdot 6 = 70$. In the last step we counted the red-green and green-green picks to get the number of ways to get green in the second draw. This gives

$$P(AB) = \frac{28}{110} \quad \text{and} \quad P(A)P(B) = \frac{40 \cdot 70}{110^2} = \frac{28}{11^2} = \frac{28}{121}.$$

Thus $P(AB) \neq P(A)P(B)$ and so A and B are not independent.

We can also see this by comparing $P(B)$ and $P(B \mid A)$: $P(B) = \frac{70}{110} = \frac{7}{11}$ while $P(B \mid A) = \frac{7}{10}$ since after conditioning on A we have 3 red and 7 green balls left.

Note that $P(B) = \frac{7}{11}$ is the same as the probability of drawing a green ball on the *first* draw. There is a symmetry going on here which we will study in Chapter 7. ▲

Fact 2.20. Suppose that A and B are independent. Then the same is true for each of these pairs: A^c and B, A and B^c, and A^c and B^c.

Proof. If A and B are independent then $P(A)P(B) = P(AB)$. To prove the independence of A^c and B start from the identity $P(B) = P(A^cB) + P(AB)$ which follows from the fact that A^cB and AB are disjoint and their union is B. Then

$$\begin{aligned} P(A^cB) &= P(B) - P(AB) = P(B) - P(A)P(B) = (1 - P(A))P(B) \\ &= P(A^c)P(B) \end{aligned}$$

which says that A^c and B are independent. The proof for the remaining two pairs follows the same way. ▲

Fact 2.20 ties in with the notion that independence has to do with information. Knowing whether A happened is exactly the same as knowing whether A^c happened. (Simply because A happens if and only if A^c does not happen, and vice versa.) Thus if knowledge about A does not alter the probability of B, neither should knowledge about A^c.

Example 2.21. Suppose that A and B are independent and $P(A) = 1/3$, $P(B) = 1/4$. Find the probability that exactly one of the two events is true.

The event in question is $AB^c \cup A^cB$. Using the disjointness of AB^c and A^cB and the fact that both intersections involve independent events we get

$$\begin{aligned} P(AB^c \cup A^cB) &= P(AB^c) + P(A^cB) = P(A)P(B^c) + P(A^c)P(B) \\ &= \tfrac{1}{3} \cdot \tfrac{3}{4} + \tfrac{2}{3} \cdot \tfrac{1}{4} = \tfrac{5}{12}. \end{aligned}$$

▲

The definition of independence of more than two events is unfortunately more complicated. These complications are forced on us because we want the definition to capture the notion that no information about events independent of an event A can change our assessment of the probability of A.

Definition 2.22. Events $A_1, \dots, A_n$ are **independent (or mutually independent)** if for every collection $A_{i_1}, \dots, A_{i_k}$, where $2 \leq k \leq n$ and $1 \leq i_1 < i_2 < \cdots < i_k \leq n$,

$$P(A_{i_1}A_{i_2} \cdots A_{i_k}) = P(A_{i_1})P(A_{i_2}) \cdots P(A_{i_k}). \tag{2.12}$$

In words, the definition states that the probability of the intersection of any choice of events from $A_1, \dots, A_n$ is equal to the product of the probabilities of the individual events. Definition 2.22 prescribes $2^n - n - 1$ equations. To illustrate the definition, the requirement for the independence of three events A, B, and C is that these four equations all hold:

$$P(AB) = P(A)P(B), \quad P(AC) = P(A)P(C), \quad P(BC) = P(B)P(C),$$
$$\text{and} \quad P(ABC) = P(A)P(B)P(C).$$

As we saw in Fact 2.20 for two events, Definition 2.22 also implies that the product property holds for any combination of the events A_i and their complements.

Fact 2.23. Suppose events $A_1, \dots, A_n$ are mutually independent. Then for every collection $A_{i_1}, \dots, A_{i_k}$, where $2 \le k \le n$ and $1 \le i_1 < i_2 < \dots < i_k \le n$, we have

$$P(A_{i_1}^* A_{i_2}^* \cdots A_{i_k}^*) = P(A_{i_1}^*)P(A_{i_2}^*) \cdots P(A_{i_k}^*) \tag{2.13}$$

where each A_i^* represents either A_i or A_i^c.

Let us illustrate again with three independent events A, B, and C. Among the identities that Fact 2.23 implies are for example these:

$$P(AB^c) = P(A)P(B^c), \quad P(A^cC^c) = P(A^c)P(C^c),$$
$$P(AB^cC) = P(A)P(B^c)P(C) \quad \text{and} \quad P(A^cB^cC^c) = P(A^c)P(B^c)P(C^c).$$

Exercise 2.60 asks you to verify these identities.

The next example illustrates that we can have $P(ABC) = P(A)P(B)P(C)$ for three events A, B and C without the events being independent.

Example 2.24. Choose a random real number uniformly from the unit interval $\Omega = [0, 1]$. (This model was introduced in Example 1.17.) Consider these events:

$$A = [\tfrac{1}{2}, 1], \quad B = [\tfrac{1}{2}, \tfrac{3}{4}], \quad C = [\tfrac{1}{16}, \tfrac{9}{16}].$$

Then $ABC = [\frac{1}{2}, \frac{9}{16}]$ and

$$P(ABC) = \tfrac{1}{16} = \tfrac{1}{2} \cdot \tfrac{1}{4} \cdot \tfrac{1}{2} = P(A)P(B)P(C).$$

However the three events are *not* mutually independent, since for example

$$P(AB) = \tfrac{1}{4} \neq \tfrac{1}{8} = P(A)P(B).$$

▲

We say that the events $A_1, \dots, A_n$ are *pairwise independent* if for any two indices $i \neq j$ the events A_i and A_j are independent. This is a weaker constraint on the events than the definition of mutual independence. The next example shows that there are events which are pairwise independent but not mutually independent.

Example 2.25. Consider Example 2.18 of the three fair coin flips and define events A and B as in the example. Let C be the event that we have exactly one tails among the first and third coin flip. Example 2.18 showed that A and B are independent. The same argument shows that B and C are independent and this is also true for A and C. Thus A, B and C are pairwise independent. However these events are not mutually independent. We have seen that $AB = \{(\mathrm{H}, \mathrm{T}, \mathrm{H}), (\mathrm{T}, \mathrm{H}, \mathrm{T})\}$. Consequently $ABC = \varnothing$ and $P(ABC) = 0$. But $P(A) = P(B) = P(C) = \frac{1}{2}$. Thus $P(ABC) \neq P(A)P(B)P(C)$.

To see an example of three independent events, for $i = 1, 2, 3$ let G_i be the event that the ith flip is tails. So for example

$$G_1 = \{(\mathrm{T}, \mathrm{H}, \mathrm{H}), (\mathrm{T}, \mathrm{H}, \mathrm{T}), (\mathrm{T}, \mathrm{T}, \mathrm{H}), (\mathrm{T}, \mathrm{T}, \mathrm{T})\}.$$

Each $P(G_i) = \frac{1}{2}$. Each pairwise intersection has two outcomes: for example, $G_1G_2 = \{(\mathrm{T}, \mathrm{T}, \mathrm{H}), (\mathrm{T}, \mathrm{T}, \mathrm{T})\}$. So $P(G_iG_j) = \frac{1}{4} = P(G_i)P(G_j)$ for $i \neq j$. Finally, for the triple intersection

$$P(G_1G_2G_3) = P\{(\mathrm{T}, \mathrm{T}, \mathrm{T})\} = \tfrac{1}{8} = (\tfrac{1}{2})^3 = P(G_1)P(G_2)P(G_3).$$

We have verified that G_1, G_2, G_3 are independent. ▲

The next example illustrates calculation of probabilities of complicated events constructed from independent events.

Example 2.26. The picture below represents an electric network. Current can flow through the top branch if switches 1 and 2 are closed, and through the lower branch if switch 3 is closed. Current can flow from A to B if current can flow either through the top branch or through the lower branch.

Assume that the switches are open or closed independently of each other, and that switch i is closed with probability p_i. Find the probability that current can flow from point A to point B.

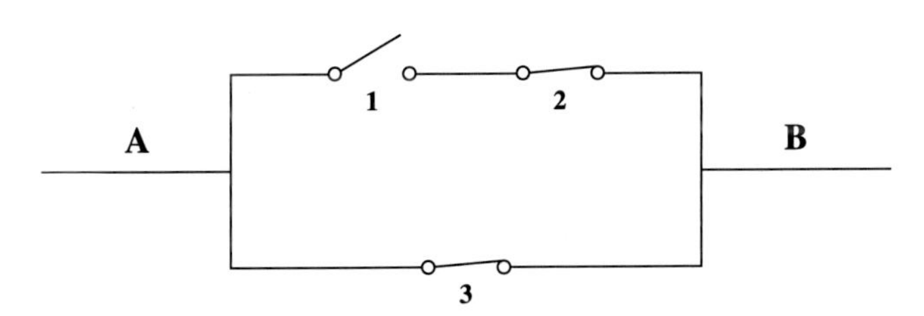

Let S_i be the event that switch i is closed, C_1 the event that current can flow through the top branch, C_2 the same for the lower branch, and finally let D be the event that current flows from A to B. Then $C_1 = S_1S_2$, $C_2 = S_3$ and

$$D = C_1 \cup C_2 = (S_1S_2) \cup S_3.$$

Using the inclusion-exclusion formula we get

$$P(D) = P(C_1) + P(C_2) - P(C_1C_2) = P(S_1S_2) + P(S_3) - P(S_1S_2S_3)$$
$$= p_1p_2 + p_3 - p_1p_2p_3.$$

In the last step we used (2.12) for the mutually independent events S_i. ▲

Independence of random variables

The definition of independence of random variables amounts to requiring that the product rule works for any intersection of events concerning the random variables.

Definition 2.27. Let $X_1, X_2, \ldots, X_n$ be random variables defined on the same probability space. Then $X_1, X_2, \ldots, X_n$ are independent if

$$P(X_1 \in B_1, X_2 \in B_2, \ldots, X_n \in B_n) = \prod_{k=1}^{n} P(X_k \in B_k) \tag{2.14}$$

for all choices of subsets $B_1, B_2, \ldots, B_n$ of the real line. ♣

This definition seems impractical because it involves thinking of all possible sets B_k. Fortunately it simplifies in cases of interest. For example, if the random variables are discrete, then it is enough to check the condition for particular values.

Fact 2.28. Discrete random variables $X_1, X_2, \ldots, X_n$ are independent if and only if

$$P(X_1 = x_1, X_2 = x_2, \ldots, X_n = x_n) = \prod_{k=1}^{n} P(X_k = x_k) \tag{2.15}$$

for all choices $x_1, x_2, \ldots, x_n$ of possible values of the random variables.

Proof. The task is to show that for discrete random variables Definition 2.27 and the condition in Fact 2.28 amount to the same thing.

If you start from Definition 2.27, then identity (2.15) is just the special case obtained by taking $B_k = \{x_k\}$ for $k = 1, \ldots, n$.

Conversely, assume that the condition in Fact 2.28 holds, and choose some sets $B_1, \ldots, B_n$. Decompose the probability and then do some algebra. In the sums below the indices $x_1, \ldots, x_n$ range over those possible values of the random variables $X_1, \ldots, X_n$ that satisfy the constraints $x_1 \in B_1, \ldots, x_n \in B_n$.

$$\begin{aligned}
&P\{X_1 \in B_1, X_2 \in B_2, \ldots, X_n \in B_n\} \\
&\quad = \sum_{x_1 \in B_1, x_2 \in B_2, \ldots, x_n \in B_n} P\{X_1 = x_1, X_2 = x_2, \ldots, X_n = x_n\} \\
&\quad = \sum_{x_1 \in B_1, x_2 \in B_2, \ldots, x_n \in B_n} P\{X_1 = x_1\} P\{X_2 = x_2\} \cdots P\{X_n = x_n\}
\end{aligned}$$

$$= \left(\sum_{x_1 \in B_1} P\{X_1 = x_1\} \right) \left(\sum_{x_2 \in B_2} P\{X_2 = x_2\} \right) \cdots \left(\sum_{x_n \in B_n} P\{X_n = x_n\} \right)$$
$$= P\{X_1 \in B_1\} \cdot P\{X_2 \in B_2\} \cdots P\{X_n \in B_n\},$$

where condition (2.15) justifies the second equality. Thereby we verified (2.14). ▲

Example 2.29. (Continuation of Example 2.25) In the setting of three fair coin flips, define random variables X_i, $i = 1, 2, 3$, by

$$X_i = \begin{cases} 0, & \text{if the } i\text{th flip is heads} \\ 1, & \text{if the } i\text{th flip is tails.} \end{cases}$$

In Example 2.25 we verified that the events $G_i = \{X_i = 1\}$ are independent. Using this we can verify the product property for X_1, X_2, X_3. For example,

$$\begin{aligned} P(X_1 = 1, X_2 = 0, X_3 = 1) &= P(G_1 G_2^c G_3) = P(G_1)P(G_2^c)P(G_3) \\ &= P(X_1 = 1)P(X_2 = 0)P(X_3 = 1), \end{aligned}$$

where we utilized Fact 2.23 in the second equality. Similar calculation verifies the product property for all combinations of possible values of the random variables X_i. Thus by Fact 2.28 X_1, X_2, X_3 are independent. ▲

Example 2.30. (Continuation of Example 2.19) Suppose that we choose one by one a sample of size k from the set $\{1, 2, \ldots, n\}$. Denote the outcomes of the successive draws by $X_1, X_2, \ldots, X_k$. Then if we sample with replacement the random variables $X_1, \ldots, X_k$ are mutually independent, while if we sample without replacement then the random variables are not mutually independent. Let us verify these claims.

Sampling with replacement. In this case $\Omega = \{1, \ldots, n\}^k$ (Cartesian product set) and $\#\Omega = n^k$. Let us check that $X_1, X_2, \ldots, X_k$ are independent. First compute, for any $x \in \{1, \ldots, n\}$,

$$P(X_j = x) = \frac{n^{k-1}}{n^k} = \frac{1}{n}$$

by counting favorable arrangements (the jth draw gives x, the other $k - 1$ are unrestricted). Let $x_1, x_2, \ldots, x_k \in \{1, \ldots, n\}$. Note that specifying the values of all $X_1, X_2, \ldots, X_k$ amounts to specifying a unique outcome from Ω:

$$P(X_1 = x_1, X_2 = x_2, \ldots, X_k = x_k) = \frac{1}{\#\Omega} = \frac{1}{n^k} = \left(\frac{1}{n}\right)^k = \prod_{j=1}^{k} P(X_j = x_j).$$

We have verified the criterion in Fact 2.28 and thereby checked that $X_1, \ldots, X_k$ are mutually independent.

Sampling without replacement. Now we must have $k \leq n$. Ω is the space of k-tuples of distinct entries from $\{1,\dots,n\}$, and so $\#\Omega = n(n-1)\cdots(n-k+1)$. Again we begin with the probability mass functions of the individual draws. For the first draw, counting favorable arrangements,

$$P(X_1 = x) = \frac{1\cdot(n-1)(n-2)\cdots(n-k+1)}{n(n-1)\cdots(n-k+1)} = \frac{1}{n}.$$

For the second draw

$$P(X_2 = x) = \frac{(n-1)\cdot 1\cdot(n-2)(n-3)\cdots(n-k+1)}{n(n-1)\cdots(n-k+1)} = \frac{1}{n}$$

because the first draw can be anything other than x ($n-1$ choices), the second draw is restricted to x (1 choice), and the third draw has $n-2$ alternatives, and so on. The general pattern is

$$P(X_j = x) = \frac{(n-1)\cdots(n-j+1)\cdot 1\cdot(n-j)\cdots(n-k+1)}{n(n-1)\cdots(n-k+1)} = \frac{1}{n}.$$

To show that independence fails, we only need to come up with one devious choice of outcomes $x_1,\dots,x_k$ where identity (2.15) fails. If we take $x_1 = \cdots = x_k = 1$, then the event $\{X_1 = 1, X_2 = 1, \dots, X_k = 1\}$ is empty. Consequently

$$P(X_1 = 1, X_2 = 1, \dots, X_k = 1) = 0 \neq \left(\frac{1}{n}\right)^k = \prod_{j=1}^{k} P(X_j = 1).$$

We have shown that for sampling without replacement, the random variables X_1, $\dots$, X_k are not independent. Note that the inequality above could also have been justified simply by noting that each probability $P(X_j = 1)$ is positive, without calculating their values.

In fact, these random variables are not even pairwise independent since $P(X_i = 1, X_j = 1) = 0$ but $P(X_i = 1)P(X_j = 1) = n^{-2} \neq 0$. ▲

2.4. Independent trials

A basic setting with independence is that of *repeated trials*. The simplest kind of trial has two outcomes, *success* and *failure*. Success comes with probability p and failure with probability $1-p$.

Let us encode a success as 1 and failure as 0. Suppose we repeat the trial n times. The sample space Ω is then the space of all n-vectors of zeros and ones:

$$\Omega = \{\omega = (s_1,\dots,s_n) : \text{each } s_i \text{ equals 0 or 1}\}. \tag{2.16}$$

We assume from the outset that the trials are independent and we build this feature into the probability measure P.

Suppose the trial is repeated $n = 6$ times. Then examples of probabilities of particular sample points are

$$P\{(1,1,1,1,1,1)\} = p \cdot p \cdot p \cdot p \cdot p \cdot p = p^6$$
$$\text{and} \quad P\{(0,1,0,0,1,0)\} = (1-p) \cdot p \cdot (1-p) \cdot (1-p) \cdot p \cdot (1-p) = p^2(1-p)^4. \tag{2.17}$$

Let us emphasize that the multiplication above does not come from a calculation. We are *defining* P, based on our *assumption* that the trials are independent.

In general, for a vector $\omega = (s_1, \dots, s_n)$ where each $s_i \in \{0, 1\}$, we would put

$$P\{\omega\} = P\{(s_1, \dots, s_n)\} = p^{\text{\# of 1s in } \omega}(1-p)^{\text{\# of 0s in } \omega}. \tag{2.18}$$

Equation (2.20) below implies that these probabilities add up to 1. We have thus constructed a probability space that models n independent trials with success probability p.

Recall that each random variable has a probability distribution. The most important probability distributions have names. Repeated independent trials give rise to three important probability distributions: Bernoulli, binomial, and geometric. We introduce them in this section.

Bernoulli distribution

The Bernoulli random variable records the result of a single trial with two possible outcomes.

Definition 2.31. Let $0 \le p \le 1$. A random variable X has the **Bernoulli distribution with success probability** p if X is $\{0, 1\}$-valued and satisfies $P(X = 1) = p$ and $P(X = 0) = 1 - p$. Abbreviate this by $X \sim \text{Ber}(p)$.

A sequence of n independent repetitions of a trial with success probability p gives rise to a sequence of n independent Bernoulli random variables with success probability p. To illustrate how to write things down with Bernoulli random variables, let X_i denote the outcome of trial i. Then probabilities such as the second probability in (2.17) can be expressed as

$$P(X_1 = 0, X_2 = 1, X_3 = X_4 = 0, X_5 = 1, X_6 = 0) = p^2(1-p)^4. \tag{2.19}$$

Binomial distribution

The binomial distribution arises from counting successes. Let S_n be the number of successes in n independent trials with success probability p. Note that if X_i denotes the outcome of trial i, then

$$S_n = X_1 + X_2 + \cdots + X_n.$$

There are $\binom{n}{k}$ vectors of length n with exactly k ones and $n - k$ zeros. From (2.18), each particular outcome with exactly k successes has probability $p^k(1-p)^{n-k}$. Thus by adding up all the possibilities,

$$P(S_n = k) = \binom{n}{k} p^k (1-p)^{n-k}.$$

This is an example of a binomial random variable, and here is the formal definition.

Definition 2.32. Let n be a positive integer and $0 \le p \le 1$. A random variable X has the **binomial distribution** with parameters n and p if the possible values of X are $\{0, 1, \dots, n\}$ and the probabilities are

$$P(X = k) = \binom{n}{k} p^k (1-p)^{n-k} \quad \text{for } k = 0, 1, \dots, n.$$

Abbreviate this by $X \sim \text{Bin}(n, p)$.

The fact that binomial probabilities add to 1 is a particular case of the binomial theorem (see Fact D.2 in Appendix D):

$$\sum_{k=0}^{n} \binom{n}{k} p^k (1-p)^{n-k} = (p + 1 - p)^n = 1. \tag{2.20}$$

Example 2.33. What is the probability that five rolls of a fair die yield two or three sixes?

The repeated trial is a roll of the die, and success means rolling a six. Let S_5 be the number of sixes that appeared in five rolls. Then $S_5 \sim \text{Bin}(5, 1/6)$.

$$\begin{aligned} P(S_5 \in \{2, 3\}) &= P(S_5 = 2) + P(S_5 = 3) \\ &= \binom{5}{2}\left(\frac{1}{6}\right)^2\left(\frac{5}{6}\right)^3 + \binom{5}{3}\left(\frac{1}{6}\right)^3\left(\frac{5}{6}\right)^2 = \frac{1500}{7776} \approx 0.193. \end{aligned}$$ ▲

Geometric distribution

With a little leap of faith we can consider *infinite sequences* of independent trials. This conceptual step is in fact necessary if we are to make sense of statements such as "when a fair die is rolled repeatedly, the long term frequency of fives is $\frac{1}{6}$." So imagine an infinite sequence of independent trials with success probability p. We shall not worry about the construction of the sample space, but this will not hinder us from doing calculations. We can still let X_j denote the outcome of the jth trial, with $X_j = 1$ if trial j is a success and $X_j = 0$ if trial j is a failure. The probability of any event concerning finitely many trials is computed exactly as before. For example, (2.19) is still true as a statement about the first six trials. Similarly, if $S_n = X_1 + \cdots + X_n$ denotes the number of successes in the first n trials, then

$$P(S_n = k) = \binom{n}{k} p^k (1-p)^{n-k}. \tag{2.21}$$

By reference to an infinite sequence of independent trials we can properly define the geometric distribution. Let N be the number of trials needed to see the first success in a sequence of independent trials with success probability p. Then for any positive integer k

$$P(N = k) = P(X_1 = 0, X_2 = 0, \ldots, X_{k-1} = 0, X_k = 1) = (1-p)^{k-1}p. \tag{2.22}$$

This random variable has the geometric distribution that we encountered in Example 1.16. Now we give its formal definition.

Definition 2.34. Let $0 < p \le 1$. A random variable X has the **geometric distribution** with success parameter p if the possible values of X are $\{1, 2, 3, \ldots\}$ and X satisfies $P(X = k) = (1-p)^{k-1}p$ for positive integers k. Abbreviate this by $X \sim \text{Geom}(p)$.

You can check that the values of the probability mass function add up to one by using the geometric series formula (D.1).

The case $p = 0$ does not really make sense for the geometric random variable because if $p = 0$ there is never a first success and $P(X = k) = 0$ for all positive integers k. Alternatively, we could take the case $p = 0$ to mean that $P(X = \infty) = 1$.

Example 2.35. What is the probability that it takes more than seven rolls of a fair die to roll a six?

Let N be the number of rolls of a fair die until the first six. Then $N \sim \text{Geom}(1/6)$.

$$P(N > 7) = \sum_{k=8}^{\infty} P(N = k) = \sum_{k=8}^{\infty} \left(\tfrac{5}{6}\right)^{k-1} \tfrac{1}{6} = \tfrac{1}{6}\left(\tfrac{5}{6}\right)^{7} \sum_{j=0}^{\infty} \left(\tfrac{5}{6}\right)^{j} = \frac{\frac{1}{6}\left(\frac{5}{6}\right)^{7}}{1 - \frac{5}{6}} = \left(\tfrac{5}{6}\right)^{7}.$$

We changed the index of summation to $j = k - 8$ and applied the geometric series formula (D.1).

Alternatively, $N > 7$ is the same as the failure of the first seven trials, which has probability $(5/6)^7$. ▲

Sometimes the geometric random variable is defined as the number of failures before the first success. This is a shifted version of our definition above. That is, if X is the number of trials needed to see the first success then the number of failures preceding the first success is $Y = X - 1$. The possible values of Y are $0, 1, 2, \ldots$ and $P(Y = k) = (1-p)^k p$.

Here is another example involving infinitely many die rolls where we use the familiar strategy of decomposing a complicated event into manageable pieces.

Example 2.36. Roll a pair of fair dice until you get either a sum of 5 or a sum of 7. What is the probability that you get 5 first?

Let A be the event that 5 comes first, and let

$$A_n = \{\text{no 5 or 7 in rolls } 1, \dots, n-1, \text{ and 5 at roll } n\}.$$

The events A_n are pairwise disjoint and their union is A. First some simple calculations:

$$P(\text{pair of dice gives 5}) = \tfrac{4}{36} \quad \text{and} \quad P(\text{pair of dice gives 7}) = \tfrac{6}{36}.$$

By the independence of the die rolls

$$P(A_n) = \left(1 - \tfrac{10}{36}\right)^{n-1} \tfrac{4}{36} = \left(\tfrac{13}{18}\right)^{n-1} \tfrac{1}{9}.$$

Consequently

$$P(A) = \sum_{n=1}^{\infty} P(A_n) = \sum_{n=1}^{\infty} \left(\tfrac{13}{18}\right)^{n-1} \tfrac{1}{9} = \frac{\frac{1}{9}}{1 - \frac{13}{18}} = \tfrac{2}{5}.$$

Note that the answer agrees with this conditional probability:

$$\tfrac{2}{5} = P(\text{pair of dice gives } 5 \,|\, \text{pair of dice gives 5 or 7}).$$

This is not a coincidence. It is a special case of a general phenomenon derived in Exercise 2.78. ▲

2.5. Further topics on sampling and independence

In this final section of the chapter, independence is extended to conditional independence and new independent events are constructed out of old ones. Sampling without replacement is revisited to add the hypergeometric distribution to our collection of named probability distributions. We introduce a probability classic called the birthday problem. Last we leave pure mathematics to illustrate the pitfalls of careless probabilistic reasoning with two cases from recent history. The subsections of this section can be read independently of each other.

Conditional independence

Recall that conditioning on an event B gives us a new probability measure $P(\cdot \,|\, B)$. If events $A_1, A_2, \dots, A_n$ are independent under this new measure $P(\cdot \,|\, B)$, we talk about conditional independence.

Definition 2.37. Let $A_1, A_2, \dots, A_n$ and B be events and $P(B) > 0$. Then $A_1, A_2, \dots, A_n$ are **conditionally independent, given** B, if the following condition holds: for any $k \in \{2, \dots, n\}$ and $1 \le i_1 < i_2 < \cdots < i_k \le n$,

$$P(A_{i_1} A_{i_2} \cdots A_{i_k} \,|\, B) = P(A_{i_1} \,|\, B)\, P(A_{i_2} \,|\, B) \cdots P(A_{i_k} \,|\, B). \tag{2.23}$$

Let us emphasize that this is not really a new definition. It is the same old Definition 2.22 of independence, but this time applied to the conditional probability measure $P(\cdot \mid B)$.

Example 2.38. Suppose 90% of coins in circulation are fair and 10% are biased coins that give tails with probability 3/5. I have a random coin and I flip it twice. Denote by A_1 the event that the first flip yields tails and by A_2 the event that the second flip yields tails. Should the two flips be independent, in other words, should we expect to have $P(A_1A_2) = P(A_1)P(A_2)$? To investigate this, we compute the probabilities with the law of total probability, as we did previously in Examples 2.8 and 2.11.

Let F denote the event that the coin is fair and B that it is biased. The information above gives us the probabilities of the two types of coins: $P(F) = \frac{9}{10}$ and $P(B) = \frac{1}{10}$, and the conditional probabilities of tails:

$$P(A_1 \mid F) = P(A_2 \mid F) = \tfrac{1}{2}, \qquad P(A_1 \mid B) = P(A_2 \mid B) = \tfrac{3}{5}.$$

Above we made the natural assumption that *for a given coin*, the probability of tails does not change between the first and second flip. Then we compute, for both $i = 1$ and 2,

$$P(A_i) = P(A_i \mid F)P(F) + P(A_i \mid B)P(B) = \tfrac{1}{2} \cdot \tfrac{9}{10} + \tfrac{3}{5} \cdot \tfrac{1}{10} = \tfrac{51}{100}.$$

To compute the probability of two tails, we need to make an assumption: the successive flips *of a given coin* are independent. This gives us the conditional independence:

$$P(A_1A_2 \mid F) = P(A_1 \mid F)\, P(A_2 \mid F) \quad \text{and} \quad P(A_1A_2 \mid B) = P(A_1 \mid B)\, P(A_2 \mid B).$$

Then we apply again the law of total probability:

$$\begin{aligned} P(A_1A_2) &= P(A_1A_2 \mid F)\, P(F) + P(A_1A_2 \mid B)\, P(B) \\ &= P(A_1 \mid F)\, P(A_2 \mid F)\, P(F) + P(A_1 \mid B)\, P(A_2 \mid B)\, P(B) \\ &= \tfrac{1}{2} \cdot \tfrac{1}{2} \cdot \tfrac{9}{10} + \tfrac{3}{5} \cdot \tfrac{3}{5} \cdot \tfrac{1}{10} = \tfrac{261}{1000}. \end{aligned}$$

Now $\frac{261}{1000} \neq \left(\frac{51}{100}\right)^2$ and we conclude that $P(A_1A_2) \neq P(A_1)P(A_2)$. In other words, A_1 and A_2 are *not* independent without the conditioning. The intuitive reason is that the first flip gives us information about the coin we hold, and thereby alters our expectations about the second flip. Exercise 2.71 develops these ideas further. ▲

Example 2.39. Example 2.14 had a medical test that detects a particular disease 96% of the time, but gives a false positive 2% of the time. It was assumed that 0.5% of the population carry the disease. We found that if a random person tests positive for the disease, the probability that he actually carries the disease is approximately 0.194.

Suppose now that after testing positive, the person is retested for the disease and this test also comes back positive. What is the probability of disease after two positive tests?

Following Example 2.14, let $D = \{\text{person has the disease}\}$, and now also $A_1 = \{\text{person tests positive on first test}\}$ and $A_2 = \{\text{person tests positive on second test}\}$. We assume that A_1 and A_2 are conditionally independent, given D. We have

$$\begin{aligned}
P(D|A_1A_2) &= \frac{P(DA_1A_2)}{P(A_1A_2)} = \frac{P(A_1A_2|D)P(D)}{P(A_1A_2|D)P(D) + P(A_1A_2|D^c)P(D^c)} \\
&= \frac{P(A_1|D)P(A_2|D)P(D)}{P(A_1|D)P(A_2|D)P(D) + P(A_1|D^c)P(A_2|D^c)P(D^c)} \\
&= \frac{(0.96)(0.96)(0.005)}{(0.96)(0.96)(0.005) + (0.02)(0.02)(0.995)} \\
&= \frac{2304}{2503} \approx 0.9205.
\end{aligned}$$

The probability of actually having the disease increased dramatically because two false positives is much less likely than the disease.

The calculation above rested on the assumption of conditional independence of the successive test results. Whether this is reasonable depends on the case at hand. If the tests are performed by the same laboratory, a systematic error in their equipment might render this assumption invalid. ▲

In the previous two examples conditional independence was built into the model by an assumption. Here is an example where we discover conditional independence.

Example 2.40. Roll two fair dice. Define two events

$$A = \{\text{first die gives 1 or 2}\} \quad \text{and} \quad B = \{\text{3 appears at least once}\}.$$

The reader should check that A and B are not independent. However, let

$$D = \{\text{the sum of the dice is 5}\} = \{(1,4),(2,3),(3,2),(4,1)\}.$$

We derive

$$\begin{aligned}
P(A \mid D) &= \frac{P(AD)}{P(D)} = \frac{P\{(1,4),(2,3)\}}{P\{(1,4),(2,3),(3,2),(4,1)\}} = \frac{2/36}{4/36} = \frac{1}{2}, \\
P(B \mid D) &= \frac{P(BD)}{P(D)} = \frac{P\{(2,3),(3,2)\}}{4/36} = \frac{2/36}{4/36} = \frac{1}{2}, \\
\text{and} \quad P(AB \mid D) &= \frac{P(ABD)}{P(D)} = \frac{P\{(2,3)\}}{4/36} = \frac{1/36}{4/36} = \frac{1}{4} = P(A \mid D)P(B \mid D).
\end{aligned}$$

The last line shows that A and B are conditionally independent, given D. In other words, conditioning can turn dependence into independence. (And also vice versa.) ▲

Independence of events constructed from independent events

Suppose we know that events $A_1, \dots, A_n$ are independent. Then we construct new events $B_1, \dots, B_k$ by applying set operations to $A_1, \dots, A_n$, but in such a manner that two different B_ℓs never use the same A_i. Then it stands to reason that the events $B_1, \dots, B_k$ are also independent. A general proof of this fact is beyond this course. Here is a small example that illustrates.

Example 2.41. Let A, B and C be independent events. We claim that the two events A and $B^c \cup C$ are independent. To show this, we use inclusion-exclusion and the fact that when events are independent, the product rule of independence continues to hold if some events are replaced with complements:

$$\begin{aligned} P\big(A \cap (B^c \cup C)\big) &= P\big((A \cap B^c) \cup (A \cap C)\big) \\ &= P(A \cap B^c) + P(A \cap C) - P(A \cap B^c \cap C) \\ &= P(A)P(B^c) + P(A)P(C) - P(A)P(B^c)P(C) \\ &= P(A)\left[P(B^c) + P(C) - P(B^c)P(C)\right] \\ &= P(A)P(B^c \cup C). \end{aligned}$$

You can imagine how tedious the proof would be for an example with more events. However, this is not inherently so, but only because we do not have the right mathematical tools at our disposal. ▲

A similar statement holds for random variables. For example, if X, Y, Z are independent random variables and we construct a new random variable U as a function of X and Y (so $U = g(X, Y)$ for some function g) then this new random variable U is independent of Z.

Hypergeometric distribution

We revisit sampling without replacement. Suppose there are two types of items, type A and type B. Let N_A be the number of type A items, N_B the number of type B items, and $N = N_A + N_B$ the total number of items. We sample n items without replacement (with $n \le N$) and denote the number of type A items in the sample by X. This is sampling without replacement with order not mattering.

The probability mass function of X can be found through the types of calculations that we saw in Example 1.15 of Section 1.2: choose a subset of size k from the set of N_A type A items, choose a subset of size $n - k$ from the set of $N - N_A$ type B items, and divide by the total number of ways of choosing a subset of size n from a set of N items. This leads to one more probability distribution with a name, the hypergeometric distribution.

Definition 2.42. Let $0 \le N_A \le N$ and $1 \le n \le N$ be integers. A random variable X has the **hypergeometric distribution** with parameters (N, N_A, n) if X takes values in the set $\{0, 1, \dots, n\}$ and has probability mass function

$$P(X = k) = \frac{\binom{N_A}{k}\binom{N-N_A}{n-k}}{\binom{N}{n}}, \qquad \text{for } k = 0, 1, \ldots, n. \tag{2.24}$$

Abbreviate this by $X \sim \text{Hypergeom}(N, N_A, n)$.

Note that some values k in $\{0, 1, \ldots, n\}$ can have probability zero. This is because X cannot be larger than N_A, and the number $n - X$ of type B items sampled cannot be larger than $N_B = N - N_A$. Hence the possible values k of X satisfy $\max(0, n - N_B) \le k \le \min(n, N_A)$. Formula (2.24) gives zero probability to the impossible values if we follow the convention that $\binom{a}{k} = 0$ for integers $k > a \ge 0$. With this convention (2.24) is valid for all $k \in \{0, 1, \ldots, n\}$.

Example 2.43. A basket contains a litter of 6 kittens, 2 males and 4 females. A neighbor comes and picks 3 kittens randomly to take home with him. Let X be the number of male kittens in the group the neighbor chose. Then $X \sim \text{Hypergeom}(6, 2, 3)$. The probability mass function of X is as follows.

$$P(X = 0) = \frac{\binom{2}{0}\binom{4}{3}}{\binom{6}{3}} = \frac{1 \cdot 4}{20} = \frac{4}{20}, \qquad P(X = 1) = \frac{\binom{2}{1}\binom{4}{2}}{\binom{6}{3}} = \frac{2 \cdot 6}{20} = \frac{12}{20},$$

$$P(X = 2) = \frac{\binom{2}{2}\binom{4}{1}}{\binom{6}{3}} = \frac{1 \cdot 4}{20} = \frac{4}{20}, \qquad P(X = 3) = \frac{\binom{2}{3}\binom{4}{0}}{\binom{6}{3}} = \frac{0 \cdot 1}{20} = 0.$$

Notice the use of the convention $\binom{2}{3} = 0$ because $3 > 2$. It agrees with the example because with only 2 male kittens available $X = 3$ cannot happen.

Contrast the experiment above with this one. The neighbor's daughter picks up a random kitten from the litter of six, pets it, and puts it back in the basket. A moment later she comes around again, picks up a random kitten, pets it, and puts it back in the basket with its siblings. She repeats this altogether three times. Let Y be the number of times she chose a male kitten to pet. This experiment is sampling *with replacement*, the individual trials are independent, so $Y \sim \text{Bin}(3, \frac{2}{6})$ and the probability mass function is

$$P(Y = k) = \binom{3}{k}\left(\frac{2}{6}\right)^k\left(\frac{4}{6}\right)^{3-k} \qquad \text{for } k = 0, 1, 2, 3. \qquad \blacktriangle$$

The birthday problem

Every student of probability should learn the *birthday problem*. You might even use it to make a bet with your friends at a party.

Example 2.44. (Birthday Problem) How large should a randomly selected group of people be to guarantee that with probably at least 1/2 there are two people with the same birthday? The modest size required often comes as a surprise. In fact, the problem is sometimes called the *birthday paradox*, even though there is nothing paradoxical about it.

To solve the problem we set up a probability model. We ignore leap years and assume that each of the possible 365 birthdays are equally likely for each person. We assume that birthdays of different people are independent. Then we can restate the problem as follows. Take a random sample of size k with replacement from the set $\{1, 2, \ldots, 365\}$. Let p_k be the probability that there is repetition in the sample. How large should k be to have $p_k > 1/2$?

The complementary event that there is no repetition is easier to handle. Let

$$A_k = \{\text{the first } k \text{ picks are all distinct}\}.$$

Then $p_k = P(A_k^c) = 1 - P(A_k)$.

To find $P(A_k)$, the ratio of the number of favorable outcomes to all outcomes gives

$$P(A_k) = \frac{365 \cdot 364 \cdots (365 - (k-1))}{365^k} = \frac{\prod_{i=0}^{k-1}(365 - i)}{365^k}. \tag{2.25}$$

By taking complements we get a formula for p_k:

$$p_k = 1 - \frac{\prod_{i=0}^{k-1}(365 - i)}{365^k}. \tag{2.26}$$

If we compute successive values of p_k from (2.26), we see an increasing sequence and eventually we come across

$$p_{22} \approx 0.4757 \qquad \text{and} \qquad p_{23} \approx 0.5073.$$

This shows that 23 is the smallest size of the group for which the probability of a match exceeds 1/2. ▲

How can we explain that the final answer is such a small number compared to 365? One way to think about it is the following. For any particular pair of people, there is a probability of 1/365 that they share a birthday. Thus, if there are approximately 365 pairs of people, there should be a good probability of success. For a group of size k, there are $\binom{k}{2} = \frac{k(k-1)}{2}$ pairs. Hence we should expect that k should satisfy $\frac{k(k-1)}{2} \approx 365$. The following approximation will shed more light on why this guess is reasonable.

For small x, $1 - x \approx e^{-x}$. (This is true because the Taylor expansion begins $e^x = 1 + x + x^2/2 + \cdots$) With this approximation

$$\begin{aligned} p_k &= 1 - \prod_{i=0}^{k-1} \frac{365 - i}{365} = 1 - \prod_{i=0}^{k-1}\left(1 - \frac{i}{365}\right) \\ &\approx 1 - \prod_{i=0}^{k-1} e^{-\frac{i}{365}} = 1 - e^{-\sum_{i=0}^{k-1}\frac{i}{365}} = 1 - e^{-\frac{k(k-1)}{2 \cdot 365}} \approx 1 - e^{-\frac{k^2}{2 \cdot 365}}. \end{aligned}$$

Above we summed an arithmetic series (see (D.9) in Appendix D) and used the approximation $k(k-1) \approx k^2$. Using this new approximate value for p_k and solving $p_k \geq 1/2$ we get

$$1 - e^{-\frac{k^2}{2\cdot 365}} \geq 1/2 \qquad \text{if and only if} \qquad k \geq \sqrt{2 \cdot 365 \cdot \ln 2} \approx 22.49.$$

This also gives 23 as the smallest integer k with $p_k \geq 1/2$. Of course, the computation we just did is not fully rigorous because of the approximations. But the last line clearly shows that the result should be on the order of $\sqrt{365}$ and not 365.

Repeating the same computation on an alien planet where a year is N days long (with N not too small), the solution of the birthday problem would be close to $\sqrt{2N \ln 2}$.

There are plenty of anecdotes related to the birthday problem. We recount two here. The first story is from *Lady Luck: The Theory of Probability* by Warren Weaver.[1]

> In World War II, I mentioned these facts at a dinner attended by a group of high-ranking officers of the Army and Navy. Most of them thought it incredible that there was an even chance with only 22 or 23 persons. Noticing that there were exactly 22 at the table, someone proposed a test. We got all the way around the table without a duplicate birthday. At which point the waitress remarked: "Excuse me. But I am the 23rd person in the room, and my birthday is May 17, just like the general over there." I admit that this story is almost too good to be true (for, after all, the test should succeed only half the time when the odds are even); but you can take my word for it.

The second story is about Johnny Carson, the former host of the Tonight Show. It is a good example of a common misinterpretation of the problem.[2] Apparently, Johnny Carson had heard about the birthday problem and its solution, and decided to test it on the show's audience one evening. However, instead of looking for a match anywhere in the crowd, he asked for the birthday of one particular guest and then checked if anybody in the audience shared the same birthday. It turned out that there was no match even though the size of the audience was roughly 500. Johnny took this as confirmation that the solution of the birthday problem is incorrect.

Where did Johnny Carson go wrong? He performed the following, very different experiment. Fix a date (a number between 1 and 365) and let r_k be the probability

---

[1] Warren Weaver, *Lady Luck: The Theory of Probability*, Dover Publications, New York, 1982, page 135.

[2] We learned about this anecdote from Steven Strogatz's post at the New York Times website (It's my Birthday Too, Yeah. October 1, 2012).

that among randomly chosen k individuals someone's birthday falls on that date. Use the complement and $1 - x \approx e^{-x}$ again to get

$$\begin{aligned} r_k &= 1 - P\{\text{among } k \text{ individuals nobody has the given date as birthday}\} \\ &= 1 - \frac{364^k}{365^k} = 1 - \left(1 - \frac{1}{365}\right)^k \approx 1 - e^{-\frac{k}{365}}. \end{aligned}$$

Even with k as large as 500 we get $r_{500} \approx 0.75$ which is large, but not overwhelmingly close to 1.

Reasoning with uncertainty in the real world ♦

Humans are not naturally good at reasoning with uncertainty. This is one lesson from the Nobel Prize winning investigations into human decision-making by Daniel Kahneman and Amos Tversky [Kah11]. Below we describe two cases of faulty reasoning with serious real-world consequences.

Example 2.45. The Sally Clark case is a famous wrongful conviction in England. Sally Clark's two boys died as infants without obvious causes. In 1998 she was charged with murder. At the trial an expert witness made the following calculation. Population statistics indicated that there is about a 1 in 8500 chance of an unexplained infant death in a family like the Clarks. Hence the chance of *two* deaths is $(1/8500)^2$, roughly 1 in 72 million. This number was presented as an indication of how unlikely Clark's innocence was. The jury convicted her.

Much went wrong in the process and the reader is referred to [SC13] for the history. Two errors relevant to our topic happened: (i) the probabilities of the two deaths were multiplied as if these were independent random events, and (ii) the resulting number was misinterpreted as the probability of Clark's innocence. Let us consider these two points in turn.

(i) The assumption of independence that led to the 1 in 72 million probability can readily fail due to unknown genetic or environmental factors. Here is a simple illustration of how that can happen.

Suppose a disease appears in 0.1% of the population. Suppose further that this disease comes from a genetic mutation passed from father to son with probability 0.5 and that a carrier of the mutation develops the disease with probability 0.5. What is the probability that both sons of a particular family have the disease? If the disease strikes completely at random, the answer is 0.001^2, or 1 in a million. However, the illness of the first son implies that the father carries the mutation. Hence the conditional probability that the second son also falls ill is $0.5 \cdot 0.5 = 0.25$. Thus the correct answer is $0.001 \cdot 0.25$, which is 1 in 4000, a much larger probability than 1 in a million.

(ii) The second error is the interpretation of the 1 in 72 million figure. Even if it were the correct number, it is not the probability of Sally Clark's innocence. This mistake is known as *prosecutor's fallacy*. The mathematically correct reasoning proceeds with Bayes' rule: by comparing the relative likelihood of two competing explanations for the deaths, namely Sally Clark's guilt and a medical explanation. This involves the prior probability of Clark's guilt, which in turn depends crucially on the strength of the other evidence against her.

To put a different view on this point, consider a lottery. The probability of getting 6 numbers out of 40 exactly right is $\binom{40}{6}^{-1}$, about 1 in 3.8 million, extremely unlikely. Yet there are plenty of lottery winners, and we do not automatically suspect cheating just because of the low odds. In a large enough population even a low probability event is likely to happen to *somebody*. ▲

Example 2.46. The financial crisis of 2008 wiped away trillions of dollars of wealth and cost the jobs and homes of millions of people. The crisis originated with mortgages. We present a brief explanation to highlight an independence assumption that failed when circumstances changed.

Mortgages are loans that banks make to people for buying homes. Investment banks buy up large numbers of mortgages and bundle them into financial assets called *mortgage-backed securities*. These securities are sold to investors. As the original borrowers pay interest and principal on their loans, the owner of the security receives income. The investor might be for example a pension fund that invests contributions from current workers and pays out pensions to retirees.

Some homeowners *default*, that is, fail to pay back their loans. That obviously hurts the owner of the security. But defaults were assumed random unrelated rare events. Under this assumption only a tiny proportion of mortgages default and the mortgage-backed security is a safe investment. Yet the danger in this assumption is evident: if a single event brings down many mortgages, these securities are much riskier than originally thought.

Developments in the early 2000s undermined this assumption of safety. A *housing bubble* drove home values high and enticed homeowners to borrow money against their houses. To increase business, banks lowered their standards for granting loans. Investors bought and sold mortgage-backed securities and other derivative securities related to mortgages without fully understanding the risks involved.

Eventually home values fell and interest rates rose. Borrowers began to default on their loans. A cycle of contagion ensued. Banks loaded with bad mortgages lost money and could no longer extend credit to businesses. Businesses suffered and people lost jobs. People out of jobs could not pay back their loans. Hence more mortgages defaulted and more mortgage-backed securities declined in value. *The assumption of independent and rare defaults was no longer valid because events in the overall economy were causing large numbers of mortgages to default.* Ultimately the crisis led to the worst global economic downturn since the 1930s.

Many actors made mistakes in this process, both in industry and in government. Complex mathematical models were used to analyze mortgage-backed securities. Decision-makers are often not the people who understand the limitations of these models. Warnings from experts were not heeded because there were large profits to be made. ▲

2.6. Finer points ♣

Independence

Note the following technical detail about Definition 2.27. As we saw in Section 1.6, $P(X \in B)$ is in general defined only for Borel subsets B of $\mathbb{R}$. Hence, in the precise version of Definition 2.27, the sets $B_1, B_2, \dots, B_n$ must be Borel sets.

Sample space for infinitely many trials

The n-fold Cartesian product space (2.16) that serves as the sample space for n trials generalizes naturally to the *sequence space*

$$\Omega = \{\omega = (s_1, s_2, s_3, \dots) : \text{each } s_i \text{ equals 0 or 1}\}.$$

A sample point ω is now an infinite sequence of 0s and 1s. Entry s_i of the sequence represents the outcome of the ith trial. The Bernoulli random variables X_i are defined as functions on Ω by $X_i(\omega) = s_i$.

In the finite n case we could define the probability measure P by simply giving formula (2.18). For infinite sequences this will not work because each sequence ω has either infinitely many 0s or infinitely many 1s or both, and hence the right-hand side of (2.18) is zero for each ω. From this you can already surmise that the sequence space is an example of an uncountable space. (Recall here the discussion on countably infinite and uncountable sample spaces at the end of Section 1.3 and on measurability in Section 1.6.)

It is possible to define a probability measure P on the sequence space under which the random variables $X_i(\omega) = s_i$ are independent Bernoulli random variables with success probability p. This rests on *Kolmogorov's extension theorem* which is taken up at the beginning of a measure-theoretic course in probability.

For fair coin flips there is an attractive alternative construction. We start by choosing a random point uniformly from $\Omega = (0, 1]$. The probability that this point lies in a specific subinterval of $(0, 1]$ is the length of the interval. (This is exactly as in Example 1.17 with the omission of the point 0.) Each real number $\omega \in (0, 1]$ has a binary expansion $\omega = \sum_{i=1}^{\infty} a_i 2^{-i}$ where each $a_i \in \{0, 1\}$. Numbers of the form $\frac{k}{2^n}$ have two representations, for example $\frac{1}{2} = \frac{0}{2} + \sum_{i=2}^{\infty} \frac{1}{2^i}$. For these numbers we pick the representation with infinitely many ones. Then each $\omega \in (0, 1]$ has a unique binary expansion.

Fixing a_k restricts ω to a union of 2^{k-1} disjoint intervals of length 2^{-k}, of total length $\frac{1}{2}$. For example, $a_2 = 1$ if and only if $\omega \in (\frac{1}{4}, \frac{1}{2}] \cup (\frac{3}{4}, 1]$. Furthermore, fixing the first n digits $a_1, \dots, a_n$ restricts ω to an interval of length 2^{-n}. Take these digits as the random variables: for each $\omega \in (0, 1]$ let $X_k(\omega)$ be the kth binary digit of ω. That is, $X_k(\omega) = a_k$ for $\omega = \sum_{i=1}^{\infty} a_i 2^{-i}$. Then the previous two observations imply that each X_k is a Ber(1/2) random variable and $P(X_1 = a_1, \dots, X_n = a_n) = 2^{-n}$ for all choices of $a_i \in \{0, 1\}$. This implies that the $X_1, X_2, X_3, \dots$ are mutually independent Ber(1/2) random variables.

Exercises

We start with some warm-up exercises arranged by section.

Section 2.1

Exercise 2.1. We roll two dice. Find the conditional probability that at least one of the numbers is even, given that the sum is 8.

Exercise 2.2. A fair coin is flipped three times. What is the probability that the second flip is tails, given that there is at most one tails among the three flips?

Exercise 2.3. What is the probability that a randomly chosen number between 1 and 100 is divisible by 3, given that the number has at least one digit equal to 5?

Exercise 2.4. We have two urns. The first urn contains two balls labeled 1 and 2. The second urn contains three balls labeled 3, 4 and 5. We choose one of the urns at random (with equal probability) and then sample one ball (uniformly at random) from the chosen urn. What is the probability that we picked the ball labeled 5?

Exercise 2.5. We have two urns. The first urn contains three balls labeled 1, 2 and 3. The second urn contains four balls labeled 2, 3, 4 and 5. We choose one of the urns randomly, so that the probability of choosing the first one is $1/5$ and the probability of choosing the second is $4/5$. Then we sample one ball (uniformly at random) from the chosen urn. What is the probability that we picked a ball labeled 2?

Exercise 2.6. When Alice spends the day with the babysitter, there is a 0.6 probability that she turns on the TV and watches a show. Her little sister Betty cannot turn the TV on by herself. But once the TV is on, Betty watches with probability 0.8. Tomorrow the girls spend the day with the babysitter.

(a) What is the probability that both Alice and Betty watch TV tomorrow?
(b) What is the probability that Betty watches TV tomorrow?
(c) What is the probability that only Alice watches TV tomorrow?

Hint. Define events precisely and use the product rule and the law of total probability.

Exercise 2.7.

(a) Use the definition of conditional probability and additivity of probability to show that $P(A^c \mid B) = 1 - P(A \mid B)$.
(b) Suppose $P(A \mid B) = 0.6$ and $P(B) = 0.5$. Find $P(A^c B)$.

Exercise 2.8. We shuffle a deck of cards and deal three cards (without replacement). Find the probability that the first card is a queen, the second is a king and the third is an ace.

Section 2.2

Exercise 2.9. We return to the setup of Exercise 2.5. Suppose that ball 3 was chosen. What is the probability that it came from the second urn?

Exercise 2.10. I have a bag with 3 fair dice. One is 4-sided, one is 6-sided, and one is 12-sided. I reach into the bag, pick one die at random and roll it. The outcome of the roll is 4. What is the probability that I pulled out the 6-sided die?

Exercise 2.11. The Acme Insurance company has two types of customers, careful and reckless. A careful customer has an accident during the year with probability 0.01. A reckless customer has an accident during the year with probability 0.04. 80% of the customers are careful and 20% of the customers are reckless. Suppose a randomly chosen customer has an accident this year. What is the probability that this customer is one of the careful customers?

Section 2.3

Exercise 2.12. We choose a number from the set $\{1, 2, 3, \ldots, 100\}$ uniformly at random and denote this number by X. For each of the following choices decide whether the two events in question are independent or not.

(a) $A = \{X \text{ is even}\}$, $B = \{X \text{ is divisible by } 5\}$.
(b) $C = \{X \text{ has two digits}\}$, $D = \{X \text{ is divisible by } 3\}$.
(c) $E = \{X \text{ is a prime}\}$, $F = \{X \text{ has a digit } 5\}$. Note that 1 is not a prime number.

Exercise 2.13. Suppose that $P(A) = 1/3$, $P(B) = 1/3$ and $P(AB^c) = 2/9$. Decide whether A and B are independent or not.

Exercise 2.14. Let A and B be two disjoint events. Under what condition are they independent?

Exercise 2.15. Every morning Ramona misses her bus with probability $\frac{1}{10}$, independently of other mornings. What is the probability that next week she catches her bus on Monday, Tuesday and Thursday, but misses her bus on Wednesday and Friday?

Exercise 2.16. We flip a fair coin three times. For $i = 1, 2, 3$, let A_i be the event that among the first i coin flips we have an odd number of heads. Check whether the events A_1, A_2, A_3 are independent or not.

Exercise 2.17. Suppose that the events A, B and C are mutually independent with $P(A) = 1/2$, $P(B) = 1/3$ and $P(C) = 1/4$. Compute $P(AB \cup C)$.

Exercise 2.18. We choose a number from the set $\{10, 11, 12, \dots, 99\}$ uniformly at random.

(a) Let X be the first digit and Y the second digit of the chosen number. Show that X and Y are independent random variables.
(b) Let X be the first digit of the chosen number and Z the sum of the two digits. Show that X and Z are not independent.

Exercise 2.19. We have an urn with balls labeled $1, \dots, 7$. Two balls are drawn. Let X_1 be the number of the first ball drawn and X_2 the number of the second ball drawn. By counting favorable outcomes, compute the probabilities $P(X_1 = 4)$, $P(X_2 = 5)$, and $P(X_1 = 4, X_2 = 5)$ in cases (a) and (b) below.

(a) The balls are drawn *with* replacement.
(b) The balls are drawn *without* replacement.
(c) Does the answer to either (a) or (b) *prove* something about the independence of the random variables X_1 and X_2?

Section 2.4

Exercise 2.20. A fair die is rolled repeatedly. Use precise notation of probabilities of events and random variables for the solutions to the questions below.

(a) Write down a precise sum expression for the probability that the first five rolls give a three at most two times.
(b) Calculate the probability that the first three does not appear before the fifth roll.
(c) Calculate the probability that the first three appears before the twentieth roll but not before the fifth roll.

Exercise 2.21. Jane must get at least three of the four problems on the exam correct to get an A. She has been able to do 80% of the problems on old exams, so she assumes that the probability she gets any problem correct is 0.8. She also assumes that the results on different problems are independent.

(a) What is the probability she gets an A?
(b) If she gets the first problem correct, what is the probability she gets an A?

Exercise 2.22. Ann and Bill play rock-paper-scissors. Each has a strategy of choosing uniformly at random out of the three possibilities every round (independently of the other player and the previous choices).

(a) What is the probability that Ann wins the first round? (Remember that the round could end in a tie.)
(b) What is the probability that Ann's first win happens in the fourth round?
(c) What is the probability that Ann's first win comes after the fourth round?

Exercise 2.23. The probability that there is no accident at a certain busy intersection is 95% on any given day, independently of the other days.

(a) Find the probability that there will be no accidents at this intersection during the next 7 days.
(b) Find the probability that next September there will be exactly 2 days with accidents.
(c) Today was accident free. Find the probability that there is no accident during the next 4 days, but there is at least one by the end of the 10th day.

Section 2.5

Exercise 2.24. A team of three is chosen randomly from an office with 2 men and 4 women. Let X be the number of women on the team.

(a) Identify the probability distribution of X by name.
(b) Give the probability mass function of X.

Exercise 2.25. I have a bag with 3 fair dice: a 4-sided die, a 6-sided die, and a 12-sided die. I reach into the bag, pick one die at random and roll it *twice*. The first roll is a 3 and the second roll is a 4. What is the probability that I pulled out the 6-sided die?

Hint. The rolls of a given die are independent.

Exercise 2.26. Suppose events A, B, C, D are mutually independent. Show that events AB and CD are independent. Justify each step from the definition of mutual independence.

Exercise 2.27. We have two urns. Urn I has 1 green and 2 red balls. Urn II has 2 green and 1 yellow ball.

(a) Pick an urn uniformly at random, and then sample one ball from this urn. What is the probability that the ball is green?
(b) After sampling the ball in part (a) and recording its color, put it back into the same urn. Then repeat the entire experiment: choose one of the urns uniformly at random and sample one ball from this urn. What is the probability that we picked a green ball in both the first and the second experiment?
(c) Pick an urn uniformly at random, and then sample two balls with replacement from this same urn. What is the probability that both picks are green?
(d) Sample one ball from each urn. What is the probability that both picks are green?

Exercise 2.28. We play a card game where we receive 13 cards at the beginning out of the deck of 52. We play 50 games one evening. For each of the following random variables identify the name and the parameters of the distribution.

(a) The number of aces I get in the first game.
(b) The number of games in which I receive at least one ace during the evening.
(c) The number of games in which all my cards are from the same suit.
(d) The number of spades I receive in the 5th game.

Further exercises

Exercise 2.29. There is a softball game at the company picnic. It is Uncle Bob's turn at bat. His probability of hitting a single is 0.35, of hitting a double is 0.25, of hitting a triple is 0.1, and of not making it on base is 0.3. Once on base, his probability of scoring after hitting a single is 0.2, after a double is 0.3, and after a triple is 0.4. What is the probability that Uncle Bob will be able to score in this turn?

Exercise 2.30. Assume that $\frac{1}{3}$ of all twins are identical twins. You learn that Miranda is expecting twins, but you have no other information.

(a) Find the probability that Miranda will have two girls.
(b) You learn that Miranda gave birth to two girls. What is the probability that the girls are identical twins?

Explain any assumptions you make.

Exercise 2.31. Suppose a family has two children of different ages. We assume that all combinations of boys and girls are equally likely.

(a) Formulate precisely the sample space and probability measure that describes the genders of the two children in the order in which they are born.
(b) Suppose we learn that there is a girl in the family. (Precisely: we learn that there is at least one girl.) What is the probability that the other child is a boy?
(c) Suppose we see the parents with a girl, and the parents tell us that this is their younger child. What is the probability that the older child we have not yet seen is a boy?

Exercise 2.32. Suppose a family has three children of different ages. We assume that all combinations of boys and girls are equally likely.

(a) Formulate precisely the sample space and probability measure that describes the genders of the three children in the order in which they are born.
(b) Suppose we see the parents with two girls. Assuming we have no other information beyond that at least two of the children are girls, what is the probability that the child we have not yet seen is a boy?

(c) Suppose we see the parents with two girls, and the parents tell us that these are the two youngest children. What is the probability that the oldest child we have not yet seen is a boy?

Exercise 2.33. Suppose that we have 5 urns and for each $k = 1, 2, 3, 4, 5$, urn k contains k red and $10 - k$ green balls. We choose an urn randomly and then draw a ball from it.

(a) What is the probability that we draw a red ball?
(b) Find the conditional probability that we chose urn k, given that we drew a red ball.

Exercise 2.34. You play the following game against your friend. You have two urns and three balls. One of the balls is marked. You get to place the balls in the two urns any way you please, including leaving one urn empty. Your friend will choose one urn at random and then draw a ball from that urn. (If he chose an empty urn, there is no ball.) His goal is to draw the marked ball.

(a) How would you arrange the balls in the urns to *minimize* his chances of drawing the marked ball?
(b) How would your friend arrange the balls in the urns to *maximize* his chances of drawing the marked ball?
(c) Repeat (a) and (b) for the case of n balls with one marked ball.

Exercise 2.35. We shuffle a deck of cards and deal two cards. Find the probability that the first card is a queen and the second is a spade.

Exercise 2.36. A bag contains three kinds of dice: seven 4-sided dice, three 6-sided dice, and two 12-sided dice. A die is drawn from the bag and then rolled, producing a number. For example, the 12-sided die could be chosen and rolled, producing the number 10. Assume that each die is equally likely to be drawn from the bag.

(a) What is the probability that the roll gave a six?
(b) What is the probability that a 6-sided die was chosen, given that the roll gave a six?

Exercise 2.37. An urn contains one 6-sided die, two 8-sided dice, three 10-sided dice, and four 20-sided dice. One die is chosen at random and then rolled.

(a) What is the probability that the roll gave a five?
(b) What is the probability that the die rolled was the 20-sided die, given that the outcome of the roll was seven?

Exercise 2.38. We choose one of the words in the following sentence uniformly at random and then choose one of the letters of that word, again uniformly at random:

SOME DOGS ARE BROWN

(a) Find the probability that the chosen letter is R.
(b) Let X denote the length of the chosen *word*. Determine the probability mass function of X.
(c) For each possible value k of X determine the conditional probability $P(X = k \mid X > 3)$.
(d) Determine the conditional probability $P(\text{the chosen letter is R} \mid X > 3)$.

Hint. The decomposition idea works just as well for conditional probabilities: if $\{B_1, \dots, B_n\}$ is a partition of Ω, then

$$P(A \mid D) = \sum_{k=1}^{n} P(AB_k \mid D).$$

(e) Given that the chosen letter is R, what is the probability that the chosen word was BROWN?

Exercise 2.39. We choose one of the words in the following sentence uniformly at random and then choose one of the letters of that word, again uniformly at random:

THE QUICK BROWN FOX JUMPED OVER THE GATE

(a) Find the probability that the chosen letter is O.
(b) Let X denote the length of the chosen word. Determine the probability mass function of X.

Exercise 2.40. Incoming students at a certain school take a mathematics placement exam. The possible scores are 1, 2, 3, and 4. From past experience, the school knows that if a particular student's score is $x \in \{1, 2, 3, 4\}$, then the student will become a mathematics major with probability $\frac{x-1}{x+3}$. Suppose that the incoming class had the following scores: 10% of the students scored a 1, 20% scored a 2, 60% scored a 3, and 10% scored a 4.

(a) What is the probability that a randomly selected student from the incoming class will become a mathematics major?
(b) Suppose a randomly selected student from the incoming class turns out to be a mathematics major. What is the probability that he or she scored a 4 on the placement exam?

Exercise 2.41. Two factories I and II produce phones for brand ABC. Factory I produces 60% of all ABC phones, and factory II produces 40%. 10% of phones produced by factory I are defective, and 20% of those produced by factory II are defective. You purchase a brand ABC phone, and assume this phone is randomly chosen from all ABC phones. Suppose the phone is *not defective*. What is the probability that it came from factory II?

Exercise 2.42. Urn A contains 2 red and 4 white balls, and urn B contains 1 red and 1 white ball. A ball is randomly chosen from urn A and put into urn B, and a ball is then chosen from urn B. What is the conditional probability that the transferred ball was white given that a white ball is selected from urn B?

Exercise 2.43. We have an urn with 3 green balls and 2 yellow balls. We pick a sample of two *without replacement* and put these two balls in a second urn that was previously empty. Next we sample two balls from the second urn *with replacement.*

(a) What is the probability that the first sample had two balls of the same color?
(b) What is the probability that the second sample had two balls of the same color?
(c) Given that the two balls chosen from the second urn have the same color, what is the probability that the second urn contains two balls of the same color?

Exercise 2.44. We have two bins. The first bin has 6 blue marbles and 4 yellow marbles. The second bin has 3 blue marbles and 4 yellow marbles. We choose a bin at random and then draw a marble from that bin.

(a) If the marble we select is yellow, what is the probability that we chose the first bin?
(b) Now suppose we put the yellow marble from (a) back in the bin it was drawn from and then draw a marble from the same bin. This marble is also yellow. What is the probability we chose the first bin?

Exercise 2.45. Suppose that 10% of 7-year-olds in Madison, Wisconsin, like the Chicago Bears, 75% like the Green Bay Packers, and the remaining 15% like some other team. (These are professional football teams.) Suppose that the probability of a 7-year-old fan of the Chicago Bears going to a game in a given year is 0.01, the probability of a 7-year-old Green Bay Packers fan going to a game in a given year is 0.05, and the probability of a 7-year-old fan of a different team going to a game in a given year is 0.005. A 7 year old is selected randomly from Madison.

(a) What is the probability that the selected 7 year old goes to a professional football game next year?
(b) If the selected 7 year old does go to a game in the coming year, what is the probability that this 7 year old was a fan of the Packers?

Exercise 2.46. Consider three boxes with numbered balls in them. Box A contains six balls numbered $1, \dots, 6$. Box B contains twelve balls numbered $1, \dots, 12$. Finally, box C contains four balls numbered $1, \dots, 4$. One ball is selected from each box uniformly at random.

(a) What is the probability that the ball chosen from box A is labeled 1 if exactly two balls numbered 1 were selected?
(b) What is the probability that the ball chosen from box B is 12 if the arithmetic mean of the three balls selected is exactly 7?

Exercise 2.47. A medical trial of 80 patients is testing a new drug to treat a certain condition. This drug is expected to be effective for each patient with probability p, independently of the other patients. You personally have two friends in this trial. Given that the trial is a success for 55 patients, what is the probability that it was successful for both of your two friends?

Exercise 2.48. A crime has been committed in a town of 100,000 inhabitants. The police are looking for a single perpetrator, believed to live in town. DNA evidence is found on the crime scene. Kevin's DNA matches the DNA recovered from the crime scene. According to DNA experts, the probability that a random person's DNA matches the crime scene DNA is 1 in 10,000. Before the DNA evidence, Kevin was no more likely to be the guilty person than any other person in town. What is the probability that Kevin is guilty after the DNA evidence appeared?

Hint. Reason as in Example 2.14.

Exercise 2.49. Let X be a discrete random variable with possible values $\{0, 1, 2, \dots\}$ and the following probability mass function: $P(X = 0) = \frac{4}{5}$ and for $k \in \{1, 2, 3, \dots\}$

$$P(X = k) = \tfrac{1}{10} \cdot \left(\tfrac{2}{3}\right)^k.$$

(a) Verify that the above is a probability mass function.
(b) For $k \in \{1, 2, \dots\}$, find $P(X \geq k \,|\, X \geq 1)$.

Exercise 2.50. (Prisoner's paradox) Three prisoners A, B and C have been sentenced to die tomorrow. The king has chosen one of the three uniformly at random to be pardoned tomorrow, while the two unlucky ones head for the gallows. The guard already knows who is to be pardoned. Prisoner A begs the guard to name someone other than A himself who will be executed. He cajoles, "Even if you tell me which one of B and C will be executed, I will not have gained any knowledge because I know already that at least one of them will die tomorrow." The guard is persuaded and reveals that B is to die tomorrow.

(a) After receiving this information, does A still have probability $\frac{1}{3}$ of being pardoned?
(b) Prisoner A whispers his new information to prisoner C. Prisoner C learned conditional probability before turning to a life of crime and is now hopeful. What is his new probability of being pardoned?

Hint. Use events $A = \{A \text{ is pardoned}\}$, similarly for B and C, and $D = \{\text{the guard names } B \text{ as one to die}\}$. Compute $P(A|D)$ and $P(C|D)$. Your answer will depend on the quantity $p = P(D|A) =$ the probability that the guard names B if both B and C are to die. Interpret in particular the special cases $p = 0$, 1 and $\frac{1}{2}$.

Exercise 2.51. Three events A, B and C satisfy the following: A and B are independent, C is a subset of B, C is disjoint from A, $P(A) = 1/2$, $P(B) = 1/4$ and $P(C) = 1/10$. Compute $P(A \cup B \cup C)$.

Exercise 2.52. Suppose that $P(A) = 0.3$, $P(B) = 0.2$, and $P(C) = 0.1$. Further, $P(A \cup B) = 0.44$, $P(A^c C) = 0.07$, $P(BC) = 0.02$, and $P(A \cup B \cup C) = 0.496$. Decide whether A, B, and C are mutually independent.

Exercise 2.53. Suppose $P(A) = 0.3$ and $P(B) = 0.6$.

(a) If A and B are disjoint, what is $P(A \cup B)$?
(b) If A and B are independent, what is $P(A \cup B)$?

Exercise 2.54. Let A and B be events with these properties: $0 < P(B) < 1$ and $P(A \mid B) = P(A \mid B^c) = \frac{1}{3}$.

(a) Is it possible to calculate $P(A)$ from this information? Either declare that it is not possible, or find the value of $P(A)$.
(b) Are A and B independent, not independent, or is it impossible to determine?

Exercise 2.55. Peter and Mary take turns throwing one dart at the dartboard. Peter hits the bullseye with probability p and Mary hits the bullseye with probability r. Whoever hits the bullseye first wins. Suppose Peter throws the first dart.

(a) What is the probability that Mary wins?
(b) Let X be the number of times Mary throws a dart in the game.
Find the possible values and the probability mass function of X. Check that the function you give sums to 1. Is this a familiar named distribution?
(c) Find the conditional probability $P(X = k|\text{Mary wins})$, for all possible values k of X. Is this a familiar named distribution?

Exercise 2.56. Show that if $P(A) = 0$ or $P(A) = 1$ then any event B is independent of A.

Exercise 2.57. We have a system that has two independent components. Both components must function in order for the system to function. The first component has 8 independent elements that each work with probability 0.95. If at least 6 of the elements are working then the first component will function. The second component has 4 independent elements that each work with probability 0.90. If at least 3 of the elements are working then the second component will function.

(a) What is the probability that the system functions?
(b) Suppose the system is not functioning. Given that information, what is the probability that the second component is not functioning?

Exercise 2.58. Suppose that a person's birthday is a uniformly random choice from the 365 days of a year (leap years are ignored), and one person's birthday is independent of the birthdays of other people. Alex, Betty and Conlin are comparing birthdays. Define these three events:

$$A = \{\text{Alex and Betty have the same birthday}\}$$
$$B = \{\text{Betty and Conlin have the same birthday}\}$$
$$C = \{\text{Conlin and Alex have the same birthday}\}.$$

(a) Are events A, B and C pairwise independent? (See the definition of pairwise independence on page 54 and Example 2.25 for illustration.)
(b) Are events A, B and C independent?

Exercise 2.59. Two towns are connected by a daily bus and by a daily train that go through a valley. On any given day one or more of the following three mutually independent events may happen: (i) the bus breaks down with probability $2/10$, (ii) the train breaks down with probability $1/10$, and (iii) a storm closes down the valley and cuts off both the bus and the train with probability $1/20$. What is the probability that travel is possible between the two towns tomorrow?

Exercise 2.60. Assume that A, B and C are mutually independent events according to Definition 2.22. Verify the identities below, using the definition of independence, set operations and rules of probability.

(a) $P(AB^c) = P(A)P(B^c)$
(b) $P(A^cC^c) = P(A^c)P(C^c)$
(c) $P(AB^cC) = P(A)P(B^c)P(C)$
(d) $P(A^cB^cC^c) = P(A^c)P(B^c)P(C^c)$

Exercise 2.61. Suppose an urn has 3 green balls and 4 red balls.

(a) Nine draws are made with replacement. Let X be the number of times a green ball appeared. Identify by name the probability distribution of X. Find the probabilities $P(X \geq 1)$ and $P(X \leq 5)$.
(b) Draws with replacement are made until the first green ball appears. Let N be the number of draws that were needed. Identify by name the probability distribution of N. Find the probability $P(N \leq 9)$.
(c) Compare $P(X \geq 1)$ and $P(N \leq 9)$. Is there a reason these should be the same?

Exercise 2.62. We have a bin with 9 blue marbles and 4 gold marbles. Draw 3 marbles from the bin without replacement and record their colors. Put the marbles back in the bin. Perform this procedure 20 times. Let X be the number of times that the three draws resulted in exactly 3 blue marbles. Find the probability mass function of X and identify it by name.

Exercise 2.63. You play a game where you first choose a positive integer n and then flip a fair coin n times. You win a prize if you get exactly 2 heads. How should you choose n to maximize your chances of winning? What is the probability of winning with an optimal choice of n? There are two equally good choices for the best number of flips. Find both.

Hint. Let $f_n = P(n \text{ flips yield exactly 2 heads})$. Derive precise conditions for $f_n < f_{n+1}$ and $f_n > f_{n+1}$.

Exercise 2.64. On a test there are 20 true-or-false questions. For each problem the student

- knows the answer with probability p,
- thinks he knows the answer, but is wrong with probability q,
- is aware of the fact that he does not know the answer with probability r.

Assume that these alternatives happen independently for each question, and that $p + q + r = 1$. If the student does not know the answer he chooses true or false with equal probability. What is the probability that he will get the correct answer for at least 19 questions out of 20?

Exercise 2.65. Three dice are rolled. What is the conditional probability that at least one lands on a 4 given that all three land on different numbers?

Exercise 2.66. You are given a fair die. You must decide ahead of time how many times to roll. If you get exactly 2 sixes, you get a prize. How many rolls should you take to maximize your chances and what are the chances of winning? There are two equally good choices for the best number of rolls.

Exercise 2.67. Show that if $X \sim \text{Geom}(p)$ then

$$P(X = n + k \mid X > n) = P(X = k) \qquad \text{for } n, k \geq 1. \tag{2.27}$$

This could be called the *memoryless property* of the geometric distribution, because it states the following. Given that there are no successes in the first n trials, the probability that the first success comes at trial $n + k$ is the same as the probability that a freshly started sequence of trials yields the first success at trial k. In other words, the earlier n failures are forgotten. On an intuitive level this is a self-evident consequence of the independence of trials.

Exercise 2.68. Suppose that X is a random variable whose possible values are $\{1, 2, 3, \dots\}$ and X satisfies $0 < P(X = 1) < 1$ and the memoryless property (2.27). Show that X must be a geometric random variable for some parameter value p. In other words, the geometric distribution is the *unique* discrete distribution on the positive integers with the memoryless property.

Hint. Use $P(X = k) = P(X = k + 1 \mid X > 1)$ repeatedly.

Exercise 2.69. In the context of Examples 2.11 and 2.16 where we have three types of coins in circulation, suppose I flip the same random coin twice. Denote by A_1 the event that the first flip yields tails and by A_2 the event that the second flip yields tails. Example 2.11 calculated $P(A_1) = P(A_2) = \frac{513}{1000}$. Find $P(A_1 A_2)$ and decide whether A_1 and A_2 are independent or not.

Hint. Use conditional independence.

Exercise 2.70. Flip a coin three times. Assume the probability of tails is p and that successive flips are independent. Let A be the event that we have exactly one tails among the first two coin flips and B the event that we have exactly one tails among the last two coin flips. For which values of p are events A and B independent? (This generalizes Example 2.18.)

Exercise 2.71. As in Example 2.38, assume that 90% of the coins in circulation are fair, and the remaining 10% are biased coins that give tails with probability 3/5. I hold a randomly chosen coin and begin to flip it.

(a) After one flip that results in tails, what is the probability that the coin I hold is a biased coin? After two flips that both give tails? After n flips that all come out tails?
(b) After how many straight tails can we say that with 90% probability the coin I hold is biased?
(c) After n straight tails, what is the probability that the next flip is also tails?
(d) Suppose we have flipped a very large number of times (think number of flips n tending to infinity), and each time gotten tails. What are the chances that the next flip again yields tails?

Exercise 2.72. Describe the probability model for a sequence of n independent trials where trial i has success probability p_i, for $i = 1, 2, \ldots, n$. That is, describe the sample space Ω and define the probability measure P for elements of Ω.

Exercise 2.73. The population of a small town is 500. 20% of the population is blond. One morning a visitor enters the local pancake place where she finds 14 local customers (and nobody else).

(a) What is the probability that the visitor finds exactly 10 customers who are blond?
(b) What is the probability that the visitor finds at most 2 blond customers?

Hint. Assume that the 14 customers are chosen uniformly at random from the population of the town.

Exercise 2.74. Suppose that 1% of the employees of a certain company use illegal drugs. This company performs random drug tests that return positive results 99% of the time if the person is a drug user. However, it also has a 2% false positive rate. The results of the drug test are known to be independent from test to test for a given person.

(a) Steve, an employee at the company, has a positive test. What is the probability he is a drug user?
(b) Knowing he failed his first test, what is the probability that Steve will fail his next drug test?
(c) Steve just failed his second drug test. Now what is the probability he is a drug user?

Exercise 2.75. Two factories I and II produce phones for brand ABC. Factory I produces 60% of all ABC phones, and factory II produces 40%. 10% of phones produced by factory I are defective, and 20% of those produced by factory II are defective. You know that the store where you buy your phones is supplied by *one* of the factories, but you do not know which one. You buy two phones, and *both* are defective. What is the probability that the store is supplied by factory II?

Hint. Make an appropriate assumption of conditional independence.

Exercise 2.76. Consider again Examples 2.14 and 2.39 of a medical test that detects a particular disease 96% of the time, but gives a false positive 2% of the time. This exercise gives an alternative solution to Example 2.39.

Let D be the event that a random person has the disease. In Example 2.14 we assumed that $P(D) = 0.005$ and derived $P(D|A_1) = 0.194$ for the probability of disease after one positive test. Using this new probability as the prior in equation (2.9), compute the posterior probability that this person has the disease if a second test comes back positive. Compare your solution with that of Example 2.39.

Exercise 2.77. Suppose $P(B) > 0$. Prove that, as a function of the event A, the conditional probability $P(A \mid B)$ satisfies the axioms of Definition 1.1. Explicitly, this means checking that $0 \le P(A \mid B) \le 1$, $P(\Omega \mid B) = 1$, $P(\varnothing \mid B) = 0$, and $P(\cup_i A_i \mid B) = \sum_i P(A_i \mid B)$ for a pairwise disjoint sequence $\{A_i\}$.

Exercise 2.78. Let A and B be two disjoint events for a random experiment. Perform independent repetitions of the experiment until either A or B happens. Show that

$$P(A \text{ happens before } B) = P(A \mid A \cup B).$$

Hint. Decompose as in Example 2.36 and sum up a geometric series.

Exercise 2.79. Find the probability that a randomly chosen group of 23 people emulates Warren Weaver's anecdote from page 68. In other words, find the probability that the following happens. In a randomly chosen group of 22 people there are no matching birthdays, but a randomly chosen 23rd person has a birthday that matches the birthday of one in the original group of 22.

Exercise 2.80. Find the probability that among 7 randomly chosen guests at a party no two share a birth month.

Exercise 2.81. A Martian year is 669 Martian days long. Solve the birthday problem for Mars. That is, for each n find the probability that among n randomly chosen Martians there are at least two with the same birthday. Estimate the value of n where this probability becomes larger than 90%.

Challenging problems

Exercise 2.82. There are N applicants for an open job. Assume that the applicants can be ranked from best to worst. The interviewer sees the N applicants sequentially, one after the other, in a random order. She must accept or reject each applicant on the spot without a chance to go back and reverse her decisions. She uses the following strategy (call it the *k-strategy*): she lets k applicants walk away and after that she chooses the first one who is better than all the previous applicants. If there is no such applicant then she has to hire the last applicant. She chooses the parameter k ahead of time.

(a) What is the probability that with the k-strategy she hires the best applicant?
(b) For a given N, let k^* denote the value of k that gives the highest probability of hiring the best applicant. Express k^* in terms of N.
(c) Find the limit of k^*/N as $N \to \infty$.

Exercise 2.83. Initially we have 1 red and 1 green ball in a box. At each step we choose a ball randomly from the box, look at its color and then return it in the box *with an additional ball of the same color*. This way there will be 3 balls in the box after the first step, 4 balls after the second step, and so on. Let X denote the number of red balls in the box after the 1000th step. Find the probability mass function of X.

Exercise 2.84. (Monty Hall problem with Bayes' rule) With conditional probability we can give a more sophisticated analysis of the Monty Hall problem. Recall the rules of the game from Exercise 1.23. Suppose again that your initial choice is door 1. We need now one extra bit of information. If the prize is behind door 1, Monty has to open either door 2 or door 3. Let $0 \leq p \leq 1$ be the probability that Monty opens door 2 if the prize is behind door 1.

(a) Assuming that you will not switch your choice, compute the probability of winning, conditioned on Monty opening door 2. Do the same conditioned on Monty opening door 3.

(b) Assume that you will switch once Monty has opened a door. Compute the probability of winning, conditioned on Monty opening door 2, and then conditioned on Monty opening door 3. Check that, regardless of the value of p, and regardless of which door you see Monty open, deciding to switch is always at least as good as deciding not to switch.

Hint. Use events $A_i = \{\text{prize is behind door } i\}$, $M_j = \{\text{Monty opens door } j\}$, and $W = \{\text{win}\}$. Probabilities $P(A_i)$ and $P(M_j|A_i)$ come from the assumptions and rules of the game. $P(A_i|M_j)$ can be derived with Bayes' rule. $P(W|M_j)$ is one of the $P(A_i|M_j)$, depending on the scenario. Observe how the mathematical structure of the problem is similar to that of the prisoner's paradox (Exercise 2.50).

(c) Use the law of total probability to calculate $P(W)$ under both cases (a) and (b) to confirm that the results of this exercise are consistent with those you found in Exercise 1.23.

Exercise 2.85. (Gambler's ruin with a fair coin) You play the following simple game of chance. A fair coin is flipped. If it comes up heads, you win a dollar. If it comes up tails, you lose a dollar. Suppose you start with N dollars in your pocket. You play repeatedly until you either reach M dollars or lose all your money, whichever comes first. M and N are fixed positive integers such that $0 < N < M$.

(a) Show that with probability one the game ends, in other words, that the amount of money in your pocket will eventually hit 0 or M.
(b) What is the probability that the game ends with M dollars in your pocket?

Hint. You can condition on the outcome of the first coin flip.

Exercise 2.86. (Gambler's ruin with a biased coin) Solve both parts of Exercise 2.85 assuming that the probability of heads is $p \neq 1/2$. In other words, the coin in the game is biased.

Exercise 2.87. Read the construction of infinitely many fair coin flips at the end of Section 2.6. Find a construction that produces infinitely many independent Bernoulli random variables with the success probability p specified below.

(a) $p = \frac{1}{4}$.
(b) $p = \frac{1}{3}$.

Exercise 2.88. We have a coin with an unknown probability $p \in (0, 1)$ for heads. Use it to produce a fair coin. You are allowed to flip the coin as many times as you wish.

3

Random variables

We have already encountered several particular random variables. Now we begin a more general treatment of probability distributions, expectation, and variance. The last section of the chapter introduces the centrally important normal, or Gaussian, random variable.

3.1. Probability distributions of random variables

Recall from Definition 1.28 in Section 1.5 that a random variable is a real-valued function on a sample space. The important probabilistic information about a random variable X is contained in its *probability distribution*, which means a description of the probabilities $P(X \in B)$ for subsets B of the real line (Definition 1.33). In this section we discuss three concrete ways of achieving this description: the probability mass function that we have already encountered, the probability density function, and the cumulative distribution function.

Probability mass function

We review some main points from Section 1.5. A random variable X is *discrete* if there is a finite or countably infinite set $A = \{k_1, k_2, k_3, \dots\}$ so that $P(X \in A) = 1$. We say that k is a possible value of X if $P(X = k) > 0$.

The probability distribution of a discrete random variable is entirely determined by its probability mass function (p.m.f.) $p(k) = P(X = k)$. The probability mass function is a function from the set of possible values of X into $[0, 1]$. If we wish to label the probability mass function with the random variable we write $p_X(k)$. The values of a probability mass function must sum to 1 on account of the axioms of probability:

$$\sum_k p_X(k) = \sum_k P(X = k) = 1, \tag{3.1}$$

where the sum is over all possible values k of X.

Recall from (1.28) that we compute the probability $P(X \in B)$ for a subset $B \subset \mathbb{R}$ by

$$P(X \in B) = \sum_{k \in B} p_X(k), \tag{3.2}$$

where the sum runs over those possible values k of X that lie in B.

It is sometimes useful to extend the domain of the probability mass function to values k that are not possible for X. For these values we naturally set $p_X(k) = 0$. This does not affect any calculations.

The probability mass function can be represented as a bar chart (see Figure 3.1). If $p_X(k) > 0$ then the bar chart has a column centered at k with height $p_X(k)$.

Probability density function

We have seen many examples of discrete random variables, including some of the most important ones: Bernoulli, binomial and geometric. The next definition introduces another major class of random variables. This class is studied with the tools of calculus.

Definition 3.1. Let X be a random variable. If a function f satisfies

$$P(X \leq b) = \int_{-\infty}^{b} f(x)\,dx \tag{3.3}$$

for all real values b, then f is the **probability density function** (p.d.f.) of X.

In plain English the definition requires that the probability that the value of X lies in the interval $(-\infty, b]$ equals the area under the graph of f from $-\infty$ to b. An important (and somewhat surprising) technical fact is that if f satisfies Definition 3.1 then

$$P(X \in B) = \int_{B} f(x)\,dx \tag{3.4}$$

for *any* subset B of the real line for which integration makes sense. We prefer (3.3) for the definition because this ties in with the cumulative distribution function (to be introduced below) and (3.3) is easier to check than a more general condition such as (3.4).

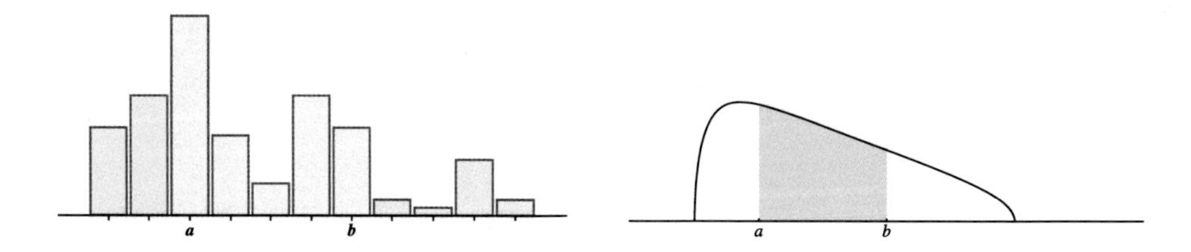

Figure 3.1. On the left is a graphical representation of the probability mass function p_X of a discrete random variable X. The column at position k has height $p_X(k)$. The sum of the heights of the blue bars equals $P(a \leq X \leq b)$. On the right is the graph of the probability density function f_Y of a continuous random variable Y. The area of the shaded region below the graph of f_Y and above the x-axis equals $P(a \leq Y \leq b)$.

The set B in (3.4) can be any interval, bounded or unbounded, or any collection of intervals. Examples include

$$P(a \le X \le b) = \int_a^b f(x)\,dx \quad \text{and} \quad P(X > a) = \int_a^\infty f(x)\,dx \tag{3.5}$$

for any real $a \le b$; see Figure 3.1. Note especially the following case.

Fact 3.2. If a random variable X has density function f then point values have probability zero:

$$P(X = c) = \int_c^c f(x)\,dx = 0 \qquad \text{for any real c.} \tag{3.6}$$

It follows that a random variable with a density function is not discrete. It also follows that probabilities of intervals are not changed by including or excluding endpoints. For example, in (3.5) we have also $P(a < X \le b) = P(a < X < b) = \int_a^b f(x)\,dx$.

Which functions qualify as probability density functions? Since probabilities are always nonnegative and since $P(-\infty < X < \infty) = 1$, a density function f must satisfy

$$f(x) \ge 0 \text{ for all } x \in \mathbb{R} \text{ and } \int_{-\infty}^{\infty} f(x)\,dx = 1. \tag{3.7}$$

Any function that satisfies condition (3.7) will be called a probability density function.

Example 3.3. Below are three functions f_1, f_2 and f_3, with some constants $a > 0$ and $b, c \in \mathbb{R}$. Which ones are probability density functions?

$$f_1(x) = \begin{cases} \frac{1}{x^2}, & x \ge 1 \\ 0, & x < 1, \end{cases} \qquad f_2(x) = \begin{cases} b\sqrt{a^2 - x^2}, & |x| \le a \\ 0, & |x| > a, \end{cases}$$

$$\text{and} \qquad f_3(x) = \begin{cases} c \sin x, & x \in [0, 2\pi] \\ 0, & x \notin [0, 2\pi]. \end{cases}$$

We show that f_1 is a probability density function, f_2 can be a probability density function if the constants are chosen correctly, but f_3 cannot be a probability density function for any choice of c.

The function f_1 is a probability density function because it is nonnegative and integrates to one:

$$\int_{-\infty}^{\infty} f_1(x)\,dx = \int_1^\infty \frac{1}{x^2}\,dx = \left(-\frac{1}{x}\right)\Bigg|_{x=1}^{x=\infty} = 1.$$

The function f_2 is nonnegative if $b \geq 0$. Let us take $a > 0$ as given and see what value of b makes f_2 a probability density function:

$$1 = \int_{-\infty}^{\infty} f_2(x)\, dx = b \int_{-a}^{a} \sqrt{a^2 - x^2}\, dx = b \cdot \tfrac{1}{2}\pi a^2.$$

Hence f_2 is a probability density function if $b = 2/(\pi a^2)$, otherwise not. (To evaluate the integral, note that it is half the area of a disk of radius a.) The probability density function f_2 is called *Wigner's semicircle law* after the physicist Eugene Wigner. It is important in the subject called random matrix theory.

If c is nonzero, then f_3 takes both positive and negative values. For $c = 0$ the function f_3 is identically zero. Hence no choice of c can turn f_3 into a probability density function. ▲

Remark 3.4. You may wonder whether a random variable X can have two different density functions. The answer is essentially no. Section 3.6 contains further comments on this point. ♣ ▲

Random variables that have a density function are sometimes called *continuous* random variables. The term is not ideal because it is not necessarily the case that the random variable is continuous as a function on the sample space Ω. We will use it nevertheless as the terminology is commonly accepted.

In Example 1.17 we encountered a uniform distribution. Let us now define it precisely and more generally.

Definition 3.5. Let $[a, b]$ be a bounded interval on the real line. A random variable X has the *uniform distribution on the interval* $[a, b]$ if X has density function

$$f(x) = \begin{cases} \frac{1}{b-a}, & \text{if } x \in [a, b] \\ 0, & \text{if } x \notin [a, b]. \end{cases} \tag{3.8}$$

Abbreviate this by $X \sim \text{Unif}[a, b]$.

If $X \sim \text{Unif}[a, b]$ and $[c, d] \subset [a, b]$, then

$$P(c \leq X \leq d) = \int_c^d \frac{1}{b-a}\, dx = \frac{d-c}{b-a}, \tag{3.9}$$

the ratio of the lengths of the intervals $[c, d]$ and $[a, b]$. In particular, this verifies that Definition 3.5 is consistent with Example 1.17 for the case $[a, b] = [0, 1]$.

Recall again that by (3.6), individual points make no difference to any probability calculation with a density function. Hence in Definition 3.5 we can drop one or both endpoints a and b if we so prefer, and define a uniform random variable on the half-open interval $(a, b]$ or on the open interval (a, b). It makes no difference to the probability calculation because in any case $P(X = a \text{ or } X = b) = 0$.

Example 3.6. Let Y be a uniform random variable on $[-2, 5]$. Find the probability that its absolute value is at least 1.

The random variable Y takes values in $[-2, 5]$ so its absolute value is at least 1 if and only if $Y \in [-2, -1] \cup [1, 5]$. The density function of Y is $f(x) = \frac{1}{5-(-2)} = \frac{1}{7}$ on $[-2, 5]$ and $f(x) = 0$ otherwise. Thus

$$\begin{aligned} P(|Y| \geq 1) &= P(Y \in [-2, -1] \cup [1, 5]) \\ &= P(-2 \leq Y \leq -1) + P(1 \leq Y \leq 5) \\ &= \frac{-1-(-2)}{7} + \frac{5-1}{7} = \frac{1}{7} + \frac{4}{7} = \frac{5}{7}, \end{aligned}$$

where we used (3.9). ▲

Infinitesimal method

It is important to keep straight that *the value $f(x)$ of a density function is not a probability.* A density function f gives probabilities of sets by integration, as in (3.4) and (3.5) above. However, multiplied by the length of a tiny interval, it approximates the probability of the interval.

> **Fact 3.7.** Suppose that random variable X has density function f that is continuous at the point a. Then for small $\varepsilon > 0$
>
> $$P(a < X < a + \varepsilon) \approx f(a) \cdot \varepsilon.$$

The fact above comes from the limit

$$\lim_{\varepsilon \to 0} \frac{P(a < X < a + \varepsilon)}{\varepsilon} = \lim_{\varepsilon \to 0} \frac{1}{\varepsilon} \int_a^{a+\varepsilon} f(x)\, dx = f(a). \tag{3.10}$$

The last limit is true because in a small interval around a point a the continuous function f is close to the constant $f(a)$. See Figure 3.2.

A limit like (3.10) also justifies these similar approximations:

$$P(a - \varepsilon < X < a + \varepsilon) \approx f(a) \cdot 2\varepsilon \quad \text{and} \quad P(a - \varepsilon < X < a) \approx f(a) \cdot \varepsilon. \tag{3.11}$$

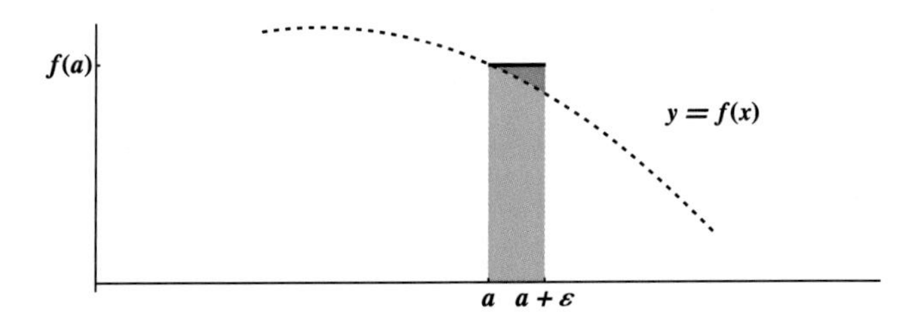

Figure 3.2. Illustration of Fact 3.7. The dotted curve is the probability density function $y = f(x)$ and the solid line is the constant function $y = f(a)$. The area below the dotted curve is the probability $P(a < X < a + \varepsilon) = \int_a^{a+\varepsilon} f(x)\, dx$ while the area below the solid line is $\varepsilon f(a)$. Fact 3.7 states that these two areas are approximately equal for very small ε.

Example 3.8. Suppose the density function of X is $f(x) = 3x^2$ for $0 < x < 1$ and $f(x) = 0$ elsewhere. Compare the precise value of $P(0.50 < X < 0.51)$ with its approximation by Fact 3.7.

The precise probability is

$$P(0.50 < X < 0.51) = \int_{0.50}^{0.51} 3x^2\, dx = 0.51^3 - 0.50^3 = 0.007651.$$

The approximation is

$$P(0.50 < X < 0.51) \approx f(0.50) \cdot 0.01 = 3 \cdot 0.50^2 \cdot 0.01 = 0.0075.$$

The relative error of the approximation is 2%. ▲

The approximations in Fact 3.7 and equation (3.11) assume that we already know the density function f. These approximations can also be turned around to *identify* f by finding an expression for the probability. We shall call this the *infinitesimal method* for finding f. It is illustrated in the next example.

Example 3.9. We revisit Example 1.18 where a dart is thrown at a board of radius 9 inches. We assume the dart hits the board at a location chosen uniformly at random. Let R be the distance of the dart from the center of the board. In Example 3.19 in the next section we will show that R is a continuous random variable. Here we use the infinitesimal method to find f_R, the density function of R.

Fix a number t in $(0, 9)$. We compute the probability density function at t utilizing the approximation $f_R(t) \approx \varepsilon^{-1} P(t < R < t + \varepsilon)$ from Fact 3.7. Take ε positive and small enough so that $t + \varepsilon < 9$. Then $\{t < R < t + \varepsilon\}$ is the event that the dart lands in the annulus (or ring) bounded by the two circles of radii t and $t + \varepsilon$. (See Figure 3.3.) By subtracting the area of the smaller disk from the area of the larger disk, we calculate the area of the annulus as

$$\pi (t + \varepsilon)^2 - \pi t^2 = 2\pi t\varepsilon + \pi \varepsilon^2.$$

Since the probability is the ratio of areas,

$$P(t < R < t + \varepsilon) = \frac{2\pi t\varepsilon + \pi \varepsilon^2}{\pi 9^2} = \frac{2t\varepsilon}{9^2} + \frac{\varepsilon^2}{9^2}.$$

Consequently

$$f_R(t) \approx \varepsilon^{-1} P(t < R < t + \varepsilon) = \frac{2t}{9^2} + \frac{\varepsilon}{9^2}.$$

This approximation becomes precise in the limit $\varepsilon \to 0$. So we can write the identity

$$f_R(t) = \lim_{\varepsilon \to 0} \varepsilon^{-1} P(t < R < t + \varepsilon) = \frac{2t}{9^2} \qquad \text{for } t \in (0, 9).$$

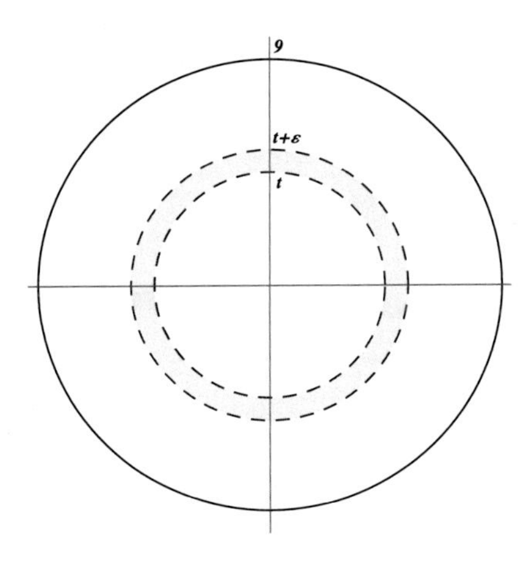

Figure 3.3. The shaded area is the annulus bounded by the circles with radii t and $t + \varepsilon$. The fact that the area of this annulus is close to $2\pi t\varepsilon$ for small ε should not be surprising: this is just the circumference of the smaller circle times the width of the annulus.

We complete the description of f_R by (i) observing that $f_R = 0$ outside $[0, 9]$ because by definition R lies in $[0, 9]$, and by (ii) noting that the two values $f_R(0)$ and $f_R(9)$ can be assigned arbitrarily. You can check that f_R is indeed a density function by noting that it takes only nonnegative values and that it integrates to 1. ▲

3.2. Cumulative distribution function

Probability mass functions are defined only for discrete random variables and density functions are defined only for continuous random variables. By contrast, the cumulative distribution function gives a way to describe the probability distribution of *any* random variable, including those that do not fall into the discrete or continuous categories.

Definition 3.10. The cumulative distribution function (c.d.f.) of a random variable X is defined by

$$F(s) = P(X \le s) \quad \text{for all } s \in \mathbb{R}. \tag{3.12}$$

It is very important to be mindful of the convention that the inequality is $\le$ in equation (3.12). The cumulative distribution function gives probabilities of left-open right-closed intervals of the form $(a, b]$:

$$P(a < X \le b) = P(X \le b) - P(X \le a) = F(b) - F(a).$$

Knowing these probabilities is enough to determine the distribution of X completely. But proving this fact is beyond the scope of this book.

Cumulative distribution function of a discrete random variable

The connection between the probability mass function and the cumulative distribution function of a discrete random variable is

$$F(s) = P(X \le s) = \sum_{k:k\le s} P(X = k), \tag{3.13}$$

where the sum extends over those possible values k of X that are less than or equal to s. This is an application of (3.2) with the set $B = (-\infty, s]$.

Example 3.11. Let $X \sim \text{Bin}(2, \frac{2}{3})$ (recall Definition 2.32). Find the cumulative distribution function of X.

We first need the probability mass function of X:

$$P(X = 0) = \left(\tfrac{1}{3}\right)^2 = \tfrac{1}{9}, \quad P(X = 1) = 2 \cdot \tfrac{2}{3} \cdot \tfrac{1}{3} = \tfrac{4}{9}, \quad \text{and} \quad P(X = 2) = \left(\tfrac{2}{3}\right)^2 = \tfrac{4}{9}.$$

To use equation (3.13) for a particular value of s, we collect those values of X that are at most s, and add their probabilities. To take an example, for $s = \frac{3}{2}$ we have

$$F\left(\tfrac{3}{2}\right) = P\left(X \le \tfrac{3}{2}\right) = P(X = 0) + P(X = 1) = \tfrac{5}{9}.$$

In the same manner we deduce $F(s)$ for each real number s. For $s < 0$ we have $F(s) = P(X \le s) = 0$ since X takes no negative values. For $0 \le s < 1$,

$$F(s) = P(X \le s) = P(X = 0) = \tfrac{1}{9}.$$

For $1 \le s < 2$,

$$F(s) = P(X \le s) = P(X = 0) + P(X = 1) = \tfrac{5}{9}.$$

Finally, for $s \ge 2$,

$$F(s) = P(X \le s) = P(X = 0) + P(X = 1) + P(X = 2) = 1.$$

Collecting these cases yields the function

$$F(s) = \begin{cases} 0, & s < 0 \\ \frac{1}{9}, & 0 \le s < 1 \\ \frac{5}{9}, & 1 \le s < 2 \\ 1, & s \ge 2. \end{cases} \tag{3.14}$$

Figure 3.4 illustrates F. ▲

Note the following features that distinguish the cumulative distribution function F of a discrete random variable X.

(i) The graph of F jumps exactly at the possible values of X and the graph is constant between the jumps. A function of this type is called a *step function* or a *piecewise constant* function.

(ii) The probability of a value of X equals the size of the jump of F. For example, in (3.14), F jumps from $\frac{1}{9}$ to $\frac{5}{9}$ at $s = 1$, and $P(X = 1) = \frac{4}{9} = \frac{5}{9} - \frac{1}{9}$.

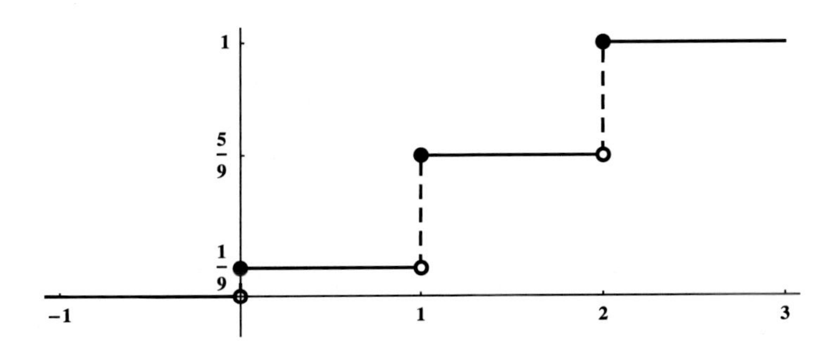

Figure 3.4. The c.d.f. of a random variable with distribution Bin(2, 2/3).

Warning. We observed earlier that in a formula for a density function such as (3.8) it makes no difference how the function is defined at the endpoints a and b of the interval. The reason is that this choice does not change any probability. By contrast, for a cumulative distribution function it is *essential* to get the inequalities right. If we changed formula (3.14) at $x = 1$ to the value $F(1) = \frac{1}{9}$, then F would no longer be the cumulative distribution function of X.

Cumulative distribution function of a continuous random variable

For a continuous random variable with density function f, the connection between the cumulative distribution function F and the density function f is

$$F(s) = P(X \le s) = \int_{-\infty}^{s} f(x)\,dx. \tag{3.15}$$

The second equality comes from Definition 3.1 of the probability density function.

Example 3.12. Let X be a Unif[1, 3] random variable. Find the cumulative distribution function of X.

The density function for X is

$$f(x) = \begin{cases} 0, & x < 1 \\ \frac{1}{2}, & 1 \le x \le 3 \\ 0, & x > 3. \end{cases}$$

The cumulative distribution function is computed from (3.15). Since the density function $f(x)$ is 0 for $x < 1$ we get $F(s) = 0$ for any $s < 1$. If $s \ge 3$ then

$$F(s) = \int_{-\infty}^{s} f(x)\,dx = \int_{1}^{3} \tfrac{1}{2}\,dx = 1.$$

If $1 \le s < 3$ then we only have to worry about the integral between 1 and s:

$$F(s) = \int_{-\infty}^{s} f(x)\,dx = \int_{1}^{s} \tfrac{1}{2}\,dx = \frac{s-1}{2}.$$

To summarize, the cumulative distribution function of the Unif[1, 3] random variable is given by

$$F(s) = \begin{cases} 0 & \text{if } s < 1 \\ \frac{s-1}{2} & \text{if } 1 \le s \le 3 \\ 1 & \text{if } s > 3 \end{cases}$$

and is illustrated in Figure 3.5. The contrast with Figure 3.4 is evident: the cumulative distribution function of a continuous random variable is a continuous function. ▲

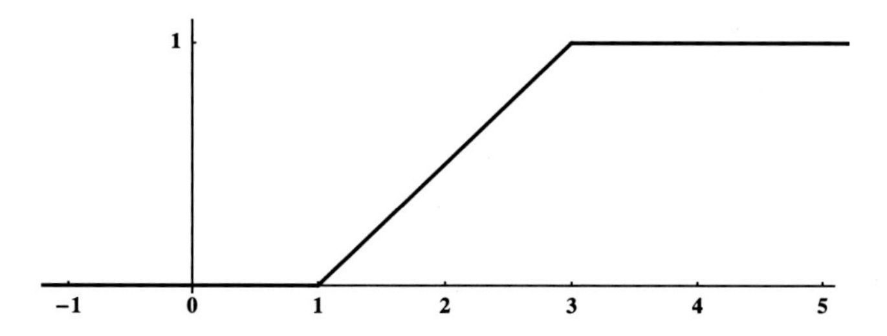

Figure 3.5. The cumulative distribution function of a Unif [1, 3] random variable

Finding the p.m.f. or the p.d.f. from the c.d.f.

The previous two examples showed how to utilize the probability mass function or the probability density function of a random variable to find its cumulative distribution function. Now we demonstrate the opposite: how to find the probability mass function or the probability density function from the cumulative distribution function. But first we must address how to tell from the cumulative distribution function whether the random variable is discrete or continuous.

Fact 3.13. Let the random variable X have cumulative distribution function F.

(a) Suppose F is piecewise constant. Then X is a discrete random variable. The possible values of X are the locations where F has jumps, and if x is such a point, then $P(X = x)$ equals the magnitude of the jump of F at x.

(b) Suppose F is continuous and the derivative $F'(x)$ exists everywhere on the real line, except possibly at finitely many points. Then X is a continuous random variable and $f(x) = F'(x)$ is the density function of X. If F is not differentiable at x, then the value $f(x)$ can be set arbitrarily. ♣

We will not give a proof of Fact 3.13. Part (a) can be checked with the ideas of Example 3.11. Exercise 3.76 asks you to prove a simple case. Justification of part (b) is discussed in Section 3.6. The statement of Fact 3.13 above does not cover all possible cases, but it is general enough for most practical needs. Next we apply Fact 3.13 to two examples.

Example 3.14. Consider a random variable X with cumulative distribution function

$$F(x) = \begin{cases} 0, & x < 1 \\ \frac{1}{7}, & 1 \le x < 2 \\ \frac{4}{7}, & 2 \le x < 5 \\ \frac{5}{7}, & 5 \le x < 8 \\ 1, & x > 8. \end{cases}$$

Since F is a step function X is discrete, and the possible values of X are the points where F jumps: $\{1, 2, 5, 8\}$. To figure out the probability mass function we find the size of the jump at each of these points.

The function F jumps from 0 to $\frac{1}{7}$ at $x = 1$. Thus, $P(X = 1) = \frac{1}{7}$. The jump in F at $x = 2$ is $\frac{3}{7}$ (as F changes from $\frac{1}{7}$ to $\frac{4}{7}$), thus $P(X = 2) = \frac{3}{7}$. Similarly, the jump at $x = 5$ is $\frac{1}{7}$, and at $x = 8$ is $\frac{2}{7}$. Hence, $P(X = 5) = \frac{1}{7}$ and $P(X = 8) = \frac{2}{7}$.

We conclude that X is a discrete random variable whose possible values are $\{1, 2, 5, 8\}$ and whose probability mass function is

$$p(1) = \tfrac{1}{7}, \quad p(2) = \tfrac{3}{7}, \quad p(5) = \tfrac{1}{7} \quad \text{and} \quad p(8) = \tfrac{2}{7}.$$

As a test, we could repeat the steps of Example 3.11 to verify that if X has this probability mass function, then its cumulative distribution function is F. ▲

Next we consider a continuous example.

Example 3.15. Suppose the cumulative distribution function F of Figure 3.5 is given:

$$F(s) = \begin{cases} 0, & \text{if } s < 1 \\ \frac{s-1}{2}, & \text{if } 1 \le s \le 3 \\ 1, & \text{if } s > 3. \end{cases}$$

Our task is to find the density function f.

Note that F is a continuous function that is differentiable everywhere except at $s = 1$ and $s = 3$. Differentiation of F gives

$$F'(s) = \begin{cases} 0, & \text{if } s < 1 \text{ or } s > 3 \\ \frac{1}{2}, & \text{if } 1 < s < 3. \end{cases}$$

By Fact 3.13(b), for the density function we can take

$$f(s) = \begin{cases} 0, & \text{if } s \le 1 \text{ or } s \ge 3 \\ \frac{1}{2}, & \text{if } 1 < s < 3. \end{cases}$$

Since F is not differentiable at $s = 1$ or $s = 3$, we can set the values $f(1)$ and $f(3)$ to be whatever we want. We chose zero above. ▲

General properties of cumulative distribution functions ♦

The examples above suggest that all cumulative distribution functions share certain properties. These properties can also be used to characterize all possible cumulative distribution functions. ♣

Fact 3.16. Every cumulative distribution function F has the following properties.

(i) Monotonicity: if $s < t$ then $F(s) \le F(t)$.
(ii) Right continuity: for each $t \in \mathbb{R}$, $F(t) = \lim_{s \to t^+} F(s)$ where $s \to t^+$ means that s approaches t from the right.
(iii) $\lim_{t \to -\infty} F(t) = 0$ and $\lim_{t \to \infty} F(t) = 1$.

Monotonicity (i) is a consequence of the monotonicity of probability, according to which a larger event has a larger probability: for $s < t$

$$F(s) = P(X \le s) \le P(X \le t) = F(t).$$

You should observe properties (ii) and (iii) in the c.d.f.s in Figures 3.4 and 3.5. Precise justification of properties (ii) and (iii) requires use of the continuity properties of probability given in Fact 1.39 and Exercise 1.61.

By definition, the cumulative distribution function F of a random variable X gives the probabilities $F(a) = P(X \le a)$. A probability of the type $P(X < a)$ can also be expressed in terms of F, but this requires a limit.

Fact 3.17. Let X be a random variable with cumulative distribution function F. Then for any $a \in \mathbb{R}$

$$P(X < a) = \lim_{s \to a^-} F(s). \tag{3.16}$$

Note that in (3.16) we take the limit of F from the left at a. The left limit at a is usually denoted by $F(a-)$. The identity (3.16) can be proved using Fact 1.39. Exercise 3.77 gives an outline of the proof.

A useful consequence is the following formula for point probabilities:

$$\begin{aligned} P(X = a) &= P(X \le a) - P(X < a) = F(a) - F(a-) \\ &= \text{size of jump in } F \text{ at point } a. \end{aligned} \tag{3.17}$$

We already used this feature for discrete random variables in Examples 3.11 and 3.14 and in Fact 3.13. However, (3.17) actually holds for *all* random variables.

Further examples

We finish this section with three more examples.

Example 3.18. Consider the density functions from Example 3.3, with a constant $a > 0$:

$$f_1(x) = \begin{cases} \frac{1}{x^2}, & x \geq 1 \\ 0, & x < 1 \end{cases} \quad \text{and} \quad f_2(x) = \begin{cases} \frac{2}{\pi a^2}\sqrt{a^2 - x^2}, & |x| \leq a \\ 0, & |x| > a. \end{cases}$$

Let us derive their cumulative distribution functions F_1 and F_2.

$$F_1(s) = \int_{-\infty}^{s} f_1(x)\,dx = \begin{cases} 0, & s < 1 \\ \int_1^s \frac{1}{x^2}\,dx = 1 - \frac{1}{s}, & s \geq 1. \end{cases}$$

Figure 3.6 shows the graph of the cumulative distribution function F_1.

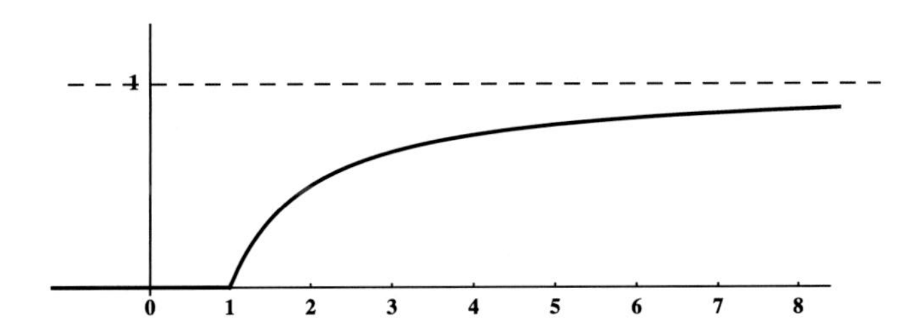

Figure 3.6. The graph of the cumulative distribution function F_1. Note that $F_1(s)$ gets closer and closer to 1 as $s \to \infty$, but never reaches it.

For F_2, first find a useful antiderivative by integration by parts:

$$\int \sqrt{1 - x^2}\,dx = \frac{1}{2}\arcsin x + \frac{x}{2}\sqrt{1 - x^2}.$$

Now for $-a \leq s \leq a$, after a change of variable $x = ay$,

$$\begin{aligned} F_2(s) &= \int_{-a}^{s} \frac{2}{\pi a^2}\sqrt{a^2 - x^2}\,dx = \int_{-1}^{s/a} \frac{2}{\pi}\sqrt{1 - y^2}\,dy \\ &= \frac{1}{\pi}\arcsin\frac{s}{a} + \frac{s}{\pi a}\sqrt{1 - \left(\frac{s}{a}\right)^2} + \frac{1}{2}. \end{aligned}$$

In particular, notice that the formula gives the values $F(-a) = 0$ and $F(a) = 1$, as it should. The full formula for the cumulative distribution function is

$$F_2(s) = \begin{cases} 0, & s < -a \\ \frac{1}{\pi}\arcsin\frac{s}{a} + \frac{s}{\pi a}\sqrt{1 - (\frac{s}{a})^2} + \frac{1}{2}, & -a \leq s < a \\ 1, & s \geq a. \end{cases}$$

▲

In the next example, we must figure out the cumulative distribution function and the density function from the experiment, instead of having one of them given to us.

Example 3.19. Consider a disk with radius r_0 as our dartboard and assume the dart lands on a uniformly random point on the disk. (This example continues the theme of Examples 1.18 and 3.9.) Let R denote the random distance of the dart from the center of the disk. Show that R is a continuous random variable and find the density function f_R.

Let us use the fact that probability is proportional to area to find the cumulative distribution function $F_R(t) = P(R \le t)$ and then differentiate to find the density. For $0 \le t \le r_0$,

$$\begin{aligned} F_R(t) &= P(R \le t) = P\{\text{the dart landed on the disk of radius } t\} \\ &= \frac{\text{area of disk of radius } t}{\text{area of disk of radius } r_0} = \frac{\pi t^2}{\pi r_0^2} = \frac{t^2}{r_0^2}. \end{aligned}$$

The cumulative distribution function is then

$$F_R(t) = \begin{cases} 0, & t < 0 \\ t^2/r_0^2, & 0 \le t \le r_0 \\ 1, & t > r_0. \end{cases}$$

This is a continuous function that is differentiable everywhere apart from $t = r_0$ (where the slope from the left is $2/r_0$ but the slope from the right is zero). Thus by Fact 3.13(b) we take the density function to be

$$f_R(t) = \begin{cases} 2t/r_0^2, & 0 \le t < r_0 \\ 0, & t < 0 \text{ or } t \ge r_0. \end{cases} \tag{3.18}$$

▲

Not all random variables are either discrete or continuous. The next example illustrates a random variable that has both a density function on an interval and a discrete part consisting of some probability assigned to a single point.

Example 3.20. Carter has an insurance policy on his car with a \$500 deductible. This means that if he gets into an accident he will personally pay for 100% of the repairs up to \$500, with the insurance company paying the rest. For example, if the repairs cost \$215, then Carter pays the whole amount. However, if the repairs cost \$832, then Carter pays \$500 and the remaining \$332 is covered by the insurance company.

Suppose that the cost of repairs for the next accident is uniformly distributed between \$100 and \$1500. Let X denote the amount Carter will have to pay. Find the cumulative distribution function of X.

Let $Y \sim \text{Unif}[100, 1500]$ denote the total cost of repairs. Because of the deductible rule, if $100 \le Y \le 500$, then $X = Y$. However, if $500 < Y \le 1500$ then $X = 500$. Thus, X can only take values between 100 and 500. This means that $F_X(x) = 0$ for $x < 0$ and $F_X(x) = 1$ for $x \ge 500$.

If $100 \le x < 500$ then the event $\{X \le x\}$ is exactly the same as $\{Y \le x\}$. Thus, in this case

$$F_X(x) = P(X \le x) = P(Y \le x) = \frac{x - 100}{1400}.$$

In the last step we used the fact that Y is uniform on $[100, 1500]$. Collecting the pieces gives the following formula, illustrated in Figure 3.7:

$$F_X(x) = \begin{cases} 0, & \text{if } x < 100 \\ \frac{x-100}{1400}, & \text{if } 100 \le x < 500 \\ 1, & \text{if } x \ge 500. \end{cases}$$

Note that the cumulative distribution function has a jump at $x = 500$: the limit from the left at this point is

$$F_X(500-) = \lim_{x \to 500^-} F_X(x) = \lim_{x \to 500^-} \frac{x - 100}{1400} = \frac{500 - 100}{1400} = \frac{2}{7},$$

while $F_X(500) = 1$. Thus $P(X = 500) = F_X(500) - F_X(500-) = 1 - \frac{2}{7} = \frac{5}{7}$. This can also be seen from the fact that

$$P(X = 500) = P(Y \ge 500) = \frac{1500 - 500}{1400} = \frac{5}{7}.$$

Since $P(X = 500) > 0$, the distribution of X cannot be continuous. On the other hand, since $P(X = 500) < 1$ and all other point probabilities are zero, X cannot be discrete either. ▲

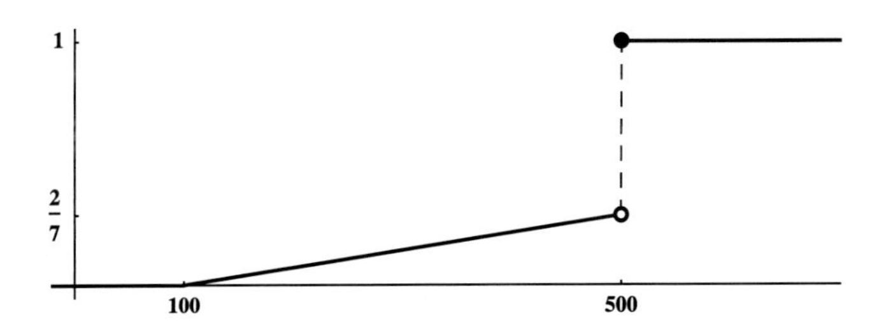

Figure 3.7. The graph of the cumulative distribution function F_X of Example 3.20.

3.3. Expectation

The previous sections discussed various ways to describe the probability distribution of a random variable. Next we turn to key quantities that represent useful partial information about a random variable, such as its expectation (also called the mean), variance, and moments.

Expectation of a discrete random variable

The next definition and its continuous counterpart (Definition 3.26) present one of the most important concepts in the field of probability.

Definition 3.21. The expectation or mean of a discrete random variable X is defined by

$$E(X) = \sum_k k\, P(X = k) \tag{3.19}$$

where the sum ranges over all the possible values k of X. ♣

The expectation is also called the *first moment*, and another conventional symbol for it is $\mu = E(X)$.

The expectation is the weighted average of the possible outcomes, where the weights are given by the probabilities. If you imagine these weights arranged on a number line at the positions of the values of the random variable, then the expectation is the center of mass of this collection of weights. See Figure 3.8 for an illustration.

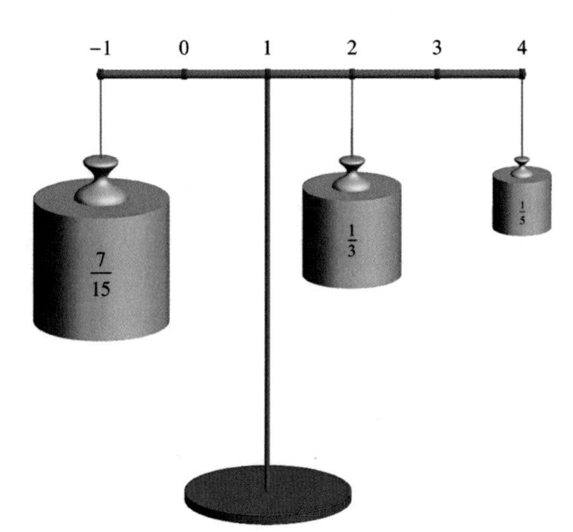

Figure 3.8. The expectation as the center of mass. If X takes the values $-1, 2, 4$ with probabilities $p_X(-1) = \frac{7}{15}$, $p_X(2) = \frac{1}{3}$, $p_X(4) = \frac{1}{5}$ then the expectation is $E[X] = -1 \cdot \frac{7}{15} + 2 \cdot \frac{1}{3} + 4 \cdot \frac{1}{5} = 1$. Hence the scale in the picture is perfectly balanced.

Example 3.22. Let Z denote the number from the roll of a fair die. Find $E(Z)$.

Since $P(Z = k) = \frac{1}{6}$ for $k \in \{1, \dots, 6\}$, from the definition we get

$$E[Z] = 1 \cdot \frac{1}{6} + 2 \cdot \frac{1}{6} + \cdots + 6 \cdot \frac{1}{6} = \frac{1}{6}(1 + 2 + \cdots + 6) = \frac{7}{2}.$$ ▲

Example 3.23. (Mean of a binomial random variable) Let $X \sim \text{Bin}(n, p)$. Recall from Definition 2.32 that the probability mass function is

$$P(X = k) = \binom{n}{k} p^k (1 - p)^{n-k} \text{ for } k = 0, 1, \dots, n.$$

The calculation of the mean of X relies on several useful computational tricks that we detail below:

$$\begin{aligned}
E(X) &= \sum_{k=0}^{n} kP(X=k) = \sum_{k=1}^{n} k \frac{n!}{k!(n-k)!} p^k (1-p)^{n-k} && \text{(drop the } k=0 \text{ term)} \\
&= \sum_{k=1}^{n} \frac{n!}{(k-1)!(n-k)!} p^k (1-p)^{n-k} && \text{(cancel } k \text{ from } k!) \\
&= np \sum_{k=1}^{n} \frac{(n-1)!}{(k-1)!(n-k)!} p^{k-1} (1-p)^{n-k} && \text{(bring } np \text{ outside the sum)} \\
&= np \sum_{j=0}^{n-1} \frac{(n-1)!}{j!(n-1-j)!} p^j (1-p)^{n-1-j} && \text{(introduce } j=k-1) \\
&= np \sum_{j=0}^{n-1} \binom{n-1}{j} p^j (1-p)^{n-1-j} \\
&= np(p+(1-p))^{n-1} && \text{(binomial theorem)} \\
&= np.
\end{aligned}$$

That the sum on the third to last row equals 1 can also be explained by pointing out that the probabilities of the $\text{Bin}(n-1,p)$ p.m.f. must sum to 1. ▲

Example 3.24. (Mean of a Bernoulli random variable) Let $0 \le p \le 1$ and $X \sim \text{Ber}(p)$. Recall from Definition 2.31 that $P(X=1) = p$ and $P(X=0) = 1-p$. The expectation is

$$E[X] = 1 \cdot P(X=1) + 0 \cdot P(X=0) = p.$$ ▲

A particular case of a Bernoulli random variable is an indicator random variable associated to an event. Let A be an event on a sample space Ω. Its *indicator random variable* I_A is defined for $\omega \in \Omega$ by

$$I_A(\omega) = \begin{cases} 1, & \omega \in A \\ 0, & \omega \notin A. \end{cases} \tag{3.20}$$

In plain English, I_A records whether the event A happens or not. Since the event $\{I_A = 1\}$ is the same as the event A, $P(I_A = 1) = P(A)$ and we see that I_A is a Bernoulli random variable with parameter $P(A)$. Its mean is then $E(I_A) = P(A)$. Thus every probability of an event is actually also an expectation, namely the expectation of the corresponding indicator random variable. We shall find later in Section 8.1 that collections of well-chosen indicator random variables can be used to compute complicated expectations.

Example 3.25. (Mean of a geometric random variable) Let $0 < p < 1$ and $X \sim \text{Geom}(p)$. Recall from Definition 2.34 that the p.m.f. of X is $P(X=k) = pq^{k-1}$ for $k = 1, 2, 3, \ldots$ where we set $q = 1-p$. Then

$$E[X] = \sum_{k=1}^{\infty} kpq^{k-1} = \frac{1}{p}. \tag{3.21}$$

There are several ways to do this calculation. You may recall from calculus that a power series can be differentiated term by term inside its radius of convergence. Let $|t| < 1$. Then

$$\sum_{k=1}^{\infty} kt^{k-1} = \sum_{k=0}^{\infty} \frac{d}{dt}(t^k) = \frac{d}{dt}\left(\sum_{k=0}^{\infty} t^k\right) = \frac{d}{dt}\left(\frac{1}{1-t}\right) = \frac{1}{(1-t)^2}. \tag{3.22}$$

(In the beginning the $k = 0$ term can be added in because it is zero.) Using (3.22) with $t = q$ we get

$$E[X] = \sum_{k=1}^{\infty} kpq^{k-1} = p\sum_{k=1}^{\infty} kq^{k-1} = p\frac{1}{(1-q)^2} = \frac{p}{p^2} = \frac{1}{p}.$$

Exercises 3.51 and 3.54 offer alternative ways of calculating the mean of the geometric random variable. ▲

Expectation of a continuous random variable

We just saw that in the discrete case the expectation is the weighted average of the possible outcomes, with the weights given by the probabilities. In the continuous case averaging is naturally done via integrals. The weighting is given by the density function.

Definition 3.26. The **expectation** or **mean** of a continuous random variable X with density function f is

$$E[X] = \int_{-\infty}^{\infty} x\, f(x)\, dx. \tag{3.23}$$

An alternative symbol is $\mu = E[X]$.

Compare (3.19) and (3.23). It is a useful rule of thumb that when passing from discrete to continuous random variables, sums are changed to integrals and probability mass functions to density functions. In (3.23) it is very important to keep separate the random variable (upper case X on the left) and the integration variable (lower case x on the right).

Example 3.27. (Mean of a uniform random variable) Let $X \sim$ Unif $[a, b]$. Then

$$E[X] = \int_{-\infty}^{\infty} x\, f(x)\, dx = \int_a^b x\frac{1}{b-a}\, dx = \frac{b^2 - a^2}{2(b-a)} = \frac{a+b}{2}.$$

If your calculus is rusty, fill in carefully the steps above. The calculation begins with the general formula (3.23). Then substitute in the density from (3.8) without forgetting the integration limits. The rest is calculus and some algebra to simplify the answer. ▲

Example 3.28. (Continuation of Example 3.19) Our dartboard is a disk of radius r_0 and the dart lands uniformly at random on the disk. Let R be the distance of the dart from the center of the disk. Find $E(R)$, the expected distance of the dart from the center of the disk.

We found the density function in Example 3.19:

$$f_R(t) = \begin{cases} 2t/r_0^2, & 0 \le t \le r_0 \\ 0, & t < 0 \text{ or } t > r_0. \end{cases}$$

The mean is

$$E(R) = \int_{-\infty}^{\infty} t\, f_R(t)\, dt = \int_0^{r_0} t \cdot \frac{2t}{r_0^2}\, dt = \tfrac{2}{3} r_0. \qquad \blacktriangle$$

Infinite and nonexistent expectations

Before proceeding further, we must take care of one technical point. Once we go beyond discrete random variables with finitely many values, infinite expectations can arise and it can even happen that no value for the expectation can be given at all. The examples below illustrate this.

Example 3.29. (Discrete example of an infinite expectation) Consider the following gamble. You flip a fair coin. If it comes up heads, you win 2 dollars and the game is over. If it comes up tails, your prize is doubled and then you flip again. Continue in this same manner: every tails doubles the prize, and once you flip the first heads the game is over and you collect the money. Let Y denote the prize you win. Find $E(Y)$.

The description of the game gives $P(Y = 2^n) = 2^{-n}$ for positive integers n. Then

$$E(Y) = \sum_{n=1}^{\infty} 2^n 2^{-n} = \sum_{n=1}^{\infty} 1 = 1 + 1 + 1 + \cdots = \infty.$$

The game is sure to end (recall the discussions in Examples 1.16 and 1.22) and consequently the prize Y is finite with probability one. Yet the expected value $E(Y)$ is infinite. This result may not agree with your initial intuition but it does not violate any mathematical rules.

This famous example is known as the *St. Petersburg paradox*. It was first considered by Nicolas Bernoulli at the beginning of the eighteenth century. ▲

Example 3.30. (Continuous example of an infinite expectation) Let X have density function

$$f(x) = \begin{cases} x^{-2}, & x \ge 1 \\ 0, & x < 1. \end{cases}$$

By the rules by which an improper integral is evaluated,

$$E(X) = \int_{-\infty}^{\infty} x f(x)\, dx = \int_1^{\infty} x \frac{1}{x^2}\, dx = \lim_{b\to\infty} \int_1^b \frac{1}{x}\, dx = \lim_{b\to\infty} \ln b = \infty. \qquad \blacktriangle$$

In the next example we cannot give the expectation any value whatsoever, finite or infinite.

Example 3.31. (Example of an undefined expectation) You and I flip a fair coin until we see the first heads. Let n denote the number of flips needed. If n is odd you pay me 2^n dollars, while if n is even I pay you 2^n dollars. Let X denote my net reward. (The reasoning of Example 1.16 or 1.22 assures us that X is a finite number with probability one.) Can I calculate my expected net reward $E[X]$?

The rules of the game specify that

$$P(X = 2^n) = 2^{-n} \text{ for odd } n \geq 1 \quad \text{and} \quad P(X = -2^n) = 2^{-n} \text{ for even } n \geq 1.$$

An attempt to calculate (3.19) leads to

$$2^1 \cdot 2^{-1} + (-2^2) \cdot 2^{-2} + 2^3 \cdot 2^{-3} + (-2^4) \cdot 2^{-4} + \cdots = 1 - 1 + 1 - 1 + \cdots$$

This series does not have a finite or infinite limit, and therefore the expectation $E[X]$ does not exist. ▲

An expectation $E(X)$ is *well defined* if it has a definite value, either a finite number or positive or negative infinity.

Expectation of a function of a random variable

Taking a function of an existing random variable creates a new random variable. That is, if X is a random variable on Ω, and g is a real-valued function defined on the range of X, then the composition $g(X)$ is a new random variable. Its value at sample point $\omega \in \Omega$ is $g(X(\omega))$. The more familiar notation for composition of functions is $g \circ X$, but in probability we prefer the notation $g(X)$.

In this section we are interested in the expectation $E[g(X)]$ and we begin with an example.

Example 3.32. In Example 1.30 a roll of a fair die determined the winnings (or loss) W of a player as follows:

$$W = \begin{cases} -1, & \text{if the roll is 1, 2, or 3} \\ 1, & \text{if the roll is 4} \\ 3, & \text{if the roll is 5 or 6.} \end{cases}$$

Let X denote the outcome of the die roll. The connection between X and W can be expressed as $W = g(X)$ where the function g is defined by $g(1) = g(2) = g(3) = -1$, $g(4) = 1$, and $g(5) = g(6) = 3$.

Using the probabilities for W derived earlier in (1.26), we can compute the mean of W:

$$\begin{aligned} E[g(X)] = E[W] &= (-1) \cdot P(W = -1) + 1 \cdot P(W = 1) + 3 \cdot P(W = 3) \\ &= (-1) \cdot \tfrac{1}{2} + 1 \cdot \tfrac{1}{6} + 3 \cdot \tfrac{1}{3} = \tfrac{2}{3}. \end{aligned}$$

There is an alternative way to think about this expectation. We can compute $P(W = a)$ for a given a by finding all the values k for which $g(k) = a$ and adding up the probabilities $P(X = k)$:

$$\begin{aligned}P(W = -1) &= P(X = 1) + P(X = 2) + P(X = 3),\\ P(W = 1) &= P(X = 4),\\ P(W = 3) &= P(X = 5) + P(X = 6).\end{aligned}$$

By substituting these values in the previous computation of $E(W)$ we get

$$\begin{aligned}E[g(X)] = E[W] &= (-1)\cdot P(W = -1) + 1\cdot P(W = 1) + 3\cdot P(W = 3)\\ &= (-1)\cdot (P(X = 1) + P(X = 2) + P(X = 3))\\ &\quad + 1\cdot P(X = 4) + 3\cdot (P(X = 5) + P(X = 6))\\ &= g(1)P(X = 1) + g(2)P(X = 2) + g(3)P(X = 3)\\ &\quad + g(4)P(X = 4) + g(5)P(X = 5) + g(6)P(X = 6) = \tfrac{2}{3}.\end{aligned}$$

The last line shows that we can obtain the expected value of $g(X)$ as

$$E[g(X)] = \sum_{k=1}^{6} g(k)P(X = k).$$

▲

The formula for $E[g(X)]$ discovered in the previous example is valid in general for discrete random variables. There is also a natural counterpart for continuous random variables. Both are stated in the next fact.

Fact 3.33. Let g be a real-valued function defined on the range of a random variable X. If X is a discrete random variable then

$$E[g(X)] = \sum_{k} g(k)P(X = k) \tag{3.24}$$

while if X is a continuous random variable with density function f then

$$E[g(X)] = \int_{-\infty}^{\infty} g(x)f(x)\,dx. \tag{3.25}$$

Proof. The justification of formula (3.24) is a generalized version of the calculation done in Example 3.32. The key is that the event $\{g(X) = y\}$ is the disjoint union of the events $\{X = k\}$ over those values k that satisfy $g(k) = y$:

$$
\begin{aligned}
E[g(X)] &= \sum_y y\,P(g(X)=y) = \sum_y y \sum_{k:g(k)=y} P(X=k) \\
&= \sum_y \sum_{k:g(k)=y} y\,P(X=k) = \sum_y \sum_{k:g(k)=y} g(k)P(X=k) \\
&= \sum_k g(k)P(X=k).
\end{aligned} \tag{3.26}
$$

In the second equality above we split the event $\{g(X)=y\}$ into the disjoint union of events $\{X=k\}$ over k with $g(k)=y$. In the third equality we move y inside the inner sum and then notice that in the inner sum $y=g(k)$. In the final step we merge all the terms to get a sum over all possible values k of X.

The proof of formula (3.25) for the continuous case goes beyond this course. So we take the formula for granted. ▲

Example 3.34. A stick of length ℓ is broken at a uniformly chosen random location. What is the expected length of the longer piece?

Let the interval $[0,\ell]$ represent the stick and let $X \sim \text{Unif}[0,\ell]$ be the position where the stick is broken. X has density $f(x)=1/\ell$ on $[0,\ell]$. Let $g(x)$ denote the length of the longer piece when the stick is broken at x, that is,

$$
g(x) = \begin{cases} \ell - x, & 0 \le x < \ell/2 \\ x, & \ell/2 \le x \le \ell. \end{cases}
$$

Then

$$
E[g(X)] = \int_{-\infty}^{\infty} g(x)f(x)\,dx = \int_0^{\ell/2} \frac{\ell - x}{\ell}\,dx + \int_{\ell/2}^{\ell} \frac{x}{\ell}\,dx = \frac{3\ell}{4}.
$$

▲

The special case $g(x)=x^n$ is common enough to have its own name.

Fact 3.35. The nth **moment** of the random variable X is the expectation $E(X^n)$. In the discrete case the nth moment is calculated by

$$
E(X^n) = \sum_k k^n P(X=k). \tag{3.27}
$$

If X has density function f its nth moment is given by

$$
E(X^n) = \int_{-\infty}^{\infty} x^n f(x)\,dx. \tag{3.28}
$$

The second moment, $E(X^2)$, is also called the *mean square*.

Example 3.36. Let $c > 0$ and let U be a uniform random variable on the interval $[0,c]$. Find the nth moment of U for all positive integers n.

The density function of U is

$$f(x) = \begin{cases} \frac{1}{c}, & \text{if } x \in [0, c] \\ 0, & \text{otherwise.} \end{cases}$$

Thus the nth moment of U is

$$E[U^n] = \int_{-\infty}^{\infty} x^n f(x)\, dx = \int_0^c x^n \frac{1}{c}\, dx = \frac{c^n}{n+1}. \qquad \blacktriangle$$

Example 3.37. Let Y be a uniformly chosen random integer from $\{0, 1, 2, \dots, m\}$. Find the first and second moment of Y.

The probability mass function of Y is $p_Y(k) = \frac{1}{m+1}$ for $0 \le k \le m$. Thus

$$E[Y] = \sum_{k=0}^{m} k \tfrac{1}{m+1} = \frac{1}{m+1} \sum_{k=1}^{m} k = \frac{1}{m+1} \frac{m(m+1)}{2} = \frac{m}{2}.$$

The sum was evaluated using identity (D.6) from Appendix D. Then using equation (3.27) in Fact 3.35,

$$E[Y^2] = \sum_{k=0}^{m} k^2 \tfrac{1}{m+1} = \frac{1}{m+1} \sum_{k=1}^{m} k^2 = \frac{1}{m+1} \frac{m(m+1)(2m+1)}{6} = \frac{m(2m+1)}{6},$$

where we first dropped the $k = 0$ term and then used identity (D.7) from Appendix D. ▲

Example 3.38. (Revisiting Example 3.20) Recall from Example 3.20 the case of Carter and the \$500 deductible. Find the expected amount that Carter pays for his next accident.

As before, let X denote the amount Carter has to pay, and Y the total cost of the repairs. We checked in Example 3.20 that X is *neither discrete nor continuous*. Our situation seems hopeless: we have not seen any method for computing the expectation of such a random variable. What saves us is that X is a simple function of Y, and Y is continuous. Indeed, $X = g(Y)$ where

$$g(y) = \min(y, 500) = \begin{cases} y, & \text{if } y < 500 \\ 500, & \text{if } y \ge 500. \end{cases}$$

Here $\min(a, b)$ is the minimum of the numbers a and b.

Since Y is uniform on $[100, 1500]$, its probability density function is

$$f_Y(y) = \begin{cases} \frac{1}{1400}, & \text{if } 100 \le y \le 1500 \\ 0, & \text{otherwise.} \end{cases}$$

From this we can compute

$$E[X] = E[g(Y)] = \int_{-\infty}^{\infty} g(y) f_Y(y)dy = \int_{100}^{1500} g(y)\frac{1}{1400}\,dy$$
$$= \int_{100}^{500} y\frac{1}{1400}\,dy + \int_{500}^{1500} 500 \cdot \frac{1}{1400}\,dy$$
$$= \frac{500^2 - 100^2}{2 \cdot 1400} + \frac{500 \cdot (1500 - 500)}{1400} = \frac{3100}{7} \approx 442.86.$$

The expectation of the amount that Carter has to cover is \$442.86. ▲

Median and quantiles ♦

When a random variable has rare, abnormally large values, its expectation may be a bad indicator of where the center of the distribution lies. The median provides an alternative measure.

> **Definition 3.39.** The **median** of a random variable X is any real value m that satisfies
>
> $$P(X \geq m) \geq \tfrac{1}{2} \quad \text{and} \quad P(X \leq m) \geq \tfrac{1}{2}.$$

Since at least half the probability is on both $\{X \leq m\}$ and $\{X \geq m\}$, the median represents in some sense the midpoint of the distribution. The median does not have to be unique. Our first example contrasts the mean and the median.

Example 3.40. Let X be uniformly distributed on the set $\{-100, 1, 2, 3, \ldots, 9\}$, so X has probability mass function

$$p_X(-100) = \tfrac{1}{10} \quad \text{and} \quad p_X(k) = \tfrac{1}{10} \quad \text{for } k = 1, 2, 3, \ldots, 9.$$

Then $E(X) = -5.5$ while the median is any number $m \in [4, 5]$. The median closely reflects the fact that 90% of the probability is in the interval $[1, 9]$, while the mean is dominated by the less likely value -100. ▲

When the cumulative distribution function is continuous and strictly increasing on the range of the random variable, the median is unique, as illustrated by the next example.

Example 3.41. We revisit the dartboard example from Examples 3.9, 3.19, and 3.28. We throw a dart uniformly at random at a dartboard with a radius of 9 inches. Let R be the distance between our dart's position and the center of the dartboard. We have seen that the cumulative distribution function of R is

$$F_R(t) = \begin{cases} t^2/9^2, & 0 \leq t \leq 9 \\ 0, & t < 0 \text{ or } t > 9. \end{cases} \tag{3.29}$$

The equation $F_R(m) = \frac{1}{2}$ has a unique solution:

$$\frac{m^2}{9^2} = \frac{1}{2} \quad \text{implies} \quad m = \frac{9}{\sqrt{2}} \approx 6.36,$$

which is the median. Note that the mean was computed in Example 3.28 as $E(R) = \frac{2}{3} \cdot 9 = 6$. ▲

The concept of the median generalizes by considering fractions other than $\frac{1}{2}$.

Definition 3.42. For $0 < p < 1$, the pth **quantile** of a random variable X is any real value x satisfying

$$P(X \geq x) \geq 1 - p \quad \text{and} \quad P(X \leq x) \geq p. \tag{3.30}$$

Thus the median is the 0.5th quantile. The quantiles corresponding to $p = 0.25$ and $p = 0.75$ are called the *first* and *third quartiles.*

Example 3.43. Continuing Example 3.41, let the cumulative distribution function of R satisfy (3.29). Find the third quartile of X.

Equation $P(R \leq x) = \frac{3}{4}$ is the same as $\frac{x^2}{9^2} = \frac{3}{4}$, which implies $x = \frac{9 \cdot \sqrt{3}}{2} \approx 7.79$. ▲

3.4. Variance

The variance measures how much the random variable fluctuates around its mean.

Definition 3.44. Let X be a random variable with mean μ. The **variance** of X is defined by

$$\text{Var}(X) = E\left[(X - \mu)^2\right]. \tag{3.31}$$

An alternative symbol is $\sigma^2 = \text{Var}(X)$.

The square root of the variance is called the *standard deviation* $\text{SD}(X) = \sigma = \sqrt{\text{Var}(X)}$. It carries the same information as the variance.

The reader may wonder why not use $E[|X - \mu|]$ to measure the fluctuations. After all, it seems simpler without the square. However, we will discover later that the variance as defined with the square has many useful properties that would be lost if we attempted to work with $E[|X - \mu|]$.

As with the mean, the definition of variance splits into two recipes for calculation.

Fact 3.45. Let X be a random variable with mean μ. Then

$$\operatorname{Var}(X) = \sum_k (k-\mu)^2 P(X=k) \qquad \text{if } X \text{ is discrete} \tag{3.32}$$

and

$$\operatorname{Var}(X) = \int_{-\infty}^{\infty} (x-\mu)^2 f(x)\,dx \qquad \text{if } X \text{ has density function } f. \tag{3.33}$$

According to (3.31) the variance of X is the expectation of the random variable $(X-\mu)^2$. Formulas (3.32) and (3.33) above are the specializations of (3.24) and (3.25) to $g(x) = (x-\mu)^2$.

The next example illustrates how larger deviations from the mean create a larger variance.

Example 3.46. Consider two investment opportunities. One yields a profit of \$1 or a loss of \$1, each with probability $\frac{1}{2}$. The other one yields a profit of \$100 or a loss of \$100, also each with probability $\frac{1}{2}$. Let X denote the random outcome of the former and Y the latter. We have the probabilities

$$P(X=1) = P(X=-1) = \tfrac{1}{2} \quad \text{and} \quad P(Y=100) = P(Y=-100) = \tfrac{1}{2}.$$

The expected return of both investments is $E(X) = E(Y) = 0$ so the mean does not distinguish between the two. Yet they are obviously quite different since Y offers a much larger upside and a much larger risk than X. The variances reveal the difference:

$$\operatorname{Var}(X) = E\left[(X - E[X])^2\right] = (1-0)^2 \cdot \tfrac{1}{2} + (-1-0)^2 \cdot \tfrac{1}{2} = 1$$
$$\operatorname{Var}(Y) = E\left[(Y - E[Y])^2\right] = (100-0)^2 \cdot \tfrac{1}{2} + (-100-0)^2 \cdot \tfrac{1}{2} = 10{,}000.$$

The variance of Y is much larger than the variance of X because the random oscillations of Y around the mean are much larger than those of X. ▲

The next example calculates the variance of a Bernoulli random variable.

Example 3.47. (Variance of a Bernoulli and indicator random variable) Let $0 \le p \le 1$. Recall that $X \sim \operatorname{Ber}(p)$ has probability mass function $P(X=1) = p$ and $P(X=0) = 1-p$ and expectation $E(X) = p$. Hence its variance is

$$\begin{aligned}\operatorname{Var}(X) &= E\left[(X-p)^2\right] = (1-p)^2 \cdot P(X=1) + (0-p)^2 \cdot P(X=0) \\ &= (1-p)^2 p + p^2 \cdot (1-p) = p(1-p).\end{aligned}$$

In particular, for an indicator random variable as defined by (3.20), we have $\operatorname{Var}(I_A) = P(A)P(A^c)$. ▲

Expanding the squares in Fact 3.45 leads to another useful formula for calculating the variance. We illustrate with the discrete case:

$$\begin{aligned}\operatorname{Var}(X) &= \sum_k (k^2 - 2\mu k + \mu^2)P(X = k) \\ &= \sum_k k^2 P(X = k) - 2\mu \sum_k kP(X = k) + \mu^2 \sum_k P(X = k) \\ &= E(X^2) - 2\mu^2 + \mu^2 = E(X^2) - \mu^2 = E(X^2) - (E[X])^2.\end{aligned}$$

The resulting formula is actually valid for *all* random variables. We state it as the next fact.

Fact 3.48. (Alternative formula for the variance)

$$\operatorname{Var}(X) = E(X^2) - (E[X])^2. \tag{3.34}$$

Armed with Fact 3.48, let us compute the variances of three named distributions.

Example 3.49. (Variance of a binomial random variable) Let $X \sim \text{Bin}(n, p)$. In Example 3.23 we calculated $E(X) = np$. A similar computation gives $E(X^2) = n(n-1)p^2 + np$. Here is the derivation.

$$\begin{aligned}E(X^2) &= \sum_{k=1}^{n} k^2 \frac{n!}{k!(n-k)!} p^k (1-p)^{n-k} \\ &= \sum_{k=1}^{n} k \frac{n!}{(k-1)!(n-k)!} p^k (1-p)^{n-k} \qquad \text{(cancel one } k \text{ factor)} \\ &= \sum_{k=2}^{n} (k-1) \frac{n!}{(k-1)!(n-k)!} p^k (1-p)^{n-k} + \sum_{k=1}^{n} \frac{n!}{(k-1)!(n-k)!} p^k (1-p)^{n-k} \\ &= \sum_{k=2}^{n} \frac{n!}{(k-2)!(n-k)!} p^k (1-p)^{n-k} + \sum_{k=1}^{n} \frac{n!}{(k-1)!(n-k)!} p^k (1-p)^{n-k} \\ &= n(n-1)p^2 + np.\end{aligned}$$

In the third equality we split $k = (k-1) + 1$, while in the fourth we canceled $k - 1$. We leave verification of the last step above as an exercise to the reader. It can be done as in Example 3.23, by introducing new summation indices and recognizing the binomial theorem.

From Fact 3.48 we deduce

$$\operatorname{Var}(X) = E(X^2) - (E[X])^2 = n(n-1)p^2 + np - n^2p^2 = np(1-p).$$

Example 3.47 showed that $p(1-p)$ is the variance of a single Bernoulli random variable of a trial with success probability p. Here we just saw that the variance of the number of successes out of n such independent trials is n times $p(1-p)$. This is not a coincidence. We shall understand what is going on in Section 8.2. ▲

Example 3.50. (Variance of a uniform random variable) Let $X \sim \text{Unif}[a, b]$. In Example 3.27 we found $E[X] = \frac{a+b}{2}$. Another integration and some algebraic simplification gives

$$E(X^2) = \int_{-\infty}^{\infty} x^2 f(x)\,dx = \frac{1}{b-a}\int_a^b x^2\,dx = \frac{b^3-a^3}{3(b-a)} = \frac{1}{3}(b^2+ba+a^2).$$

Then by Fact 3.48

$$\text{Var}(X) = E(X^2) - (E[X])^2 = \frac{b^2+ab+a^2}{3} - \frac{(b+a)^2}{4} = \frac{(b-a)^2}{12}.$$ ▲

Example 3.51. (Variance of a geometric random variable) Let $X \sim \text{Geom}(p)$ and $q = 1-p$. In Example 3.25 we found $E[X] = 1/p$. The formula for the mean square is

$$E(X^2) = \frac{1+q}{p^2}. \tag{3.35}$$

We leave this calculation as Exercise 3.56 with hints. Then from (3.34) we get

$$\text{Var}(X) = E(X^2) - (E[X])^2 = \frac{1+q}{p^2} - \frac{1}{p^2} = \frac{q}{p^2} = \frac{1-p}{p^2}.$$ ▲

The mean and variance of a linear function of a random variable appear often enough to merit a separate statement. A *linear* function g is of the form $g(x) = ax+b$. Such functions are also called *affine*.

Fact 3.52. Let X be a random variable and a and b real numbers. Then

$$E(aX+b) = aE(X)+b \tag{3.36}$$

and

$$\text{Var}(aX+b) = a^2\,\text{Var}(X) \tag{3.37}$$

provided the mean and variance are well defined.

Proof. Identity (3.36) is a consequence of the properties of sums, integrals and probabilities and Fact 3.33. Here is the derivation for X that has density function f:

$$\begin{aligned} E(aX+b) &= \int_{-\infty}^{\infty} (ax+b)f(x)\,dx = a\int_{-\infty}^{\infty} xf(x)\,dx + b\int_{-\infty}^{\infty} f(x)\,dx \\ &= aE(X)+b. \end{aligned}$$

For (3.37) it is convenient to simplify things inside the expectation:

$$\begin{aligned}\operatorname{Var}(aX+b) &= E\left[(aX+b-E(aX+b))^2\right] = E\left[(aX+b-aE(X)-b)^2\right] \\ &= E\left[a^2(X-E(X))^2\right] = a^2E\left[(X-E(X))^2\right] = a^2\operatorname{Var}(X).\end{aligned}$$

Note that the second and fourth equalities make use of (3.36). ▲

You can understand formula (3.37) as follows. Since the variance measures fluctuation around the mean, shifting by b has no influence because the mean is shifted too. And since the variance is the expectation of squared deviations, the constant a must come out as a square.

Example 3.53. Let $Z \sim \text{Bin}(10, 1/5)$. Find the expectation and variance of $3Z+2$.

From Examples 3.23 and 3.49 we know that

$$E(Z) = 10 \cdot \tfrac{1}{5} = 2 \quad \text{and} \quad \operatorname{Var}(Z) = 10 \cdot \tfrac{1}{5} \cdot \tfrac{4}{5} = \tfrac{8}{5}.$$

Using Fact 3.52 we get

$$E(3Z+2) = 3E(Z)+2 = 8 \quad \text{and} \quad \operatorname{Var}(3Z+2) = 9\operatorname{Var}(Z) = \tfrac{72}{5}.$$ ▲

The linearity property (3.36) generalizes to sums of higher moments. In particular, for any choice of constants $a_0, \dots, a_n$, and as long as the moments are finite, we have

$$E\left[\sum_{k=0}^{n} a_k X^k\right] = \sum_{k=0}^{n} a_k E(X^k). \tag{3.38}$$

For example, if the expectations below are finite,

$$E[3X^2 - 7X + 2] = 3E[X^2] - 7E[X] + 2.$$

We will see in Fact 8.1 that linearity of expectations is even more general than what is presented in (3.38).

We end this section with the fact that a random variable has zero variance if and only if it is degenerate in the sense of Example 1.32.

Fact 3.54. For a random variable X, $\operatorname{Var}(X) = 0$ if and only if $P(X=a)=1$ for some real value a.

Proof. One direction is straightforward. If $P(X = a) = 1$, then X is discrete, $E[X] = a$ and

$$\operatorname{Var}(X) = E[(X-a)^2] = (a-a)^2 P(X=a) = 0.$$

Our tools allow us to prove the other implication under the additional assumption that X is discrete. So, suppose $\mu = E(X)$ and

$$0 = \mathrm{Var}(X) = \sum_k (k - \mu)^2 P(X = k).$$

This is a sum of nonnegative terms. Hence it vanishes if and only if each individual term vanishes. For each term we have

$$(k - \mu)^2 P(X = k) = 0 \quad \text{if and only if} \quad k = \mu \;\text{ or }\; P(X = k) = 0.$$

Thus the only k-value with $P(X = k) > 0$ is $k = \mu$, and hence $P(X = \mu) = 1$. ▲

Summary of properties of random variables

The following table summarizes the features of discrete and continuous random variables that we have discussed in this chapter. However, there are random variables that do not possess either a probability mass function or a density function, as illustrated in Example 3.20. The entries in the table that go across both columns are in fact valid for *all* random variables.

<table>
<tr><th colspan="2">Properties of Random Variables</th></tr>
<tr><td>Discrete</td><td>Continuous</td></tr>
<tr><td>Probability mass function
$p_X(k) = P(X = k)$</td><td>Probability density function
$f_X(x)$</td></tr>
<tr><td>$P(X \in B) = \sum_{k:k \in B} p_X(k)$</td><td>$P(X \in B) = \int_B f_X(x)dx$</td></tr>
<tr><td colspan="2">Cumulative distribution function
$F_X(a) = P(X \le a)$</td></tr>
<tr><td>$F_X(a) = \sum_{k:k \le a} p_X(k)$
F_X is a step function.</td><td>$F_X(a) = \int_{-\infty}^{a} f(x)\,dx$
F_X is a continuous function.</td></tr>
<tr><td colspan="2">$P(X < a) = \lim_{t \to a^-} F(t) = F(a-)$
$P(X = a) = F(a) - \lim_{t \to a^-} F(t) = F(a) - F(a-)$</td></tr>
<tr><td>$E(X) = \sum_k k\,p_X(k)$</td><td>$E(X) = \int_{-\infty}^{\infty} x f(x)\,dx$</td></tr>
<tr><td colspan="2">$E(aX + b) = aE[X] + b$</td></tr>
<tr><td>$E[g(X)] = \sum_k g(k)\,p_X(k)$</td><td>$E[g(X)] = \int_{-\infty}^{\infty} g(x) f(x)\,dx$</td></tr>
<tr><td colspan="2">$\mathrm{Var}(X) = E\left[(X - E[X])^2\right] = E[X^2] - (E[X])^2$</td></tr>
<tr><td colspan="2">$\mathrm{Var}(aX + b) = a^2\,\mathrm{Var}(X)$</td></tr>
</table>

3.5. Gaussian distribution

The Gaussian (or normal) random variable plays a central role in the theory of probability.

Definition 3.55. A random variable Z has the **standard normal distribution** (also called **standard Gaussian distribution**) if Z has density function

$$\varphi(x) = \frac{1}{\sqrt{2\pi}} e^{-x^2/2} \tag{3.39}$$

on the real line. Abbreviate this by $Z \sim \mathcal{N}(0, 1)$.

Figure 3.9 shows the plot of the probability density of the standard normal distribution, the familiar bell shaped curve.

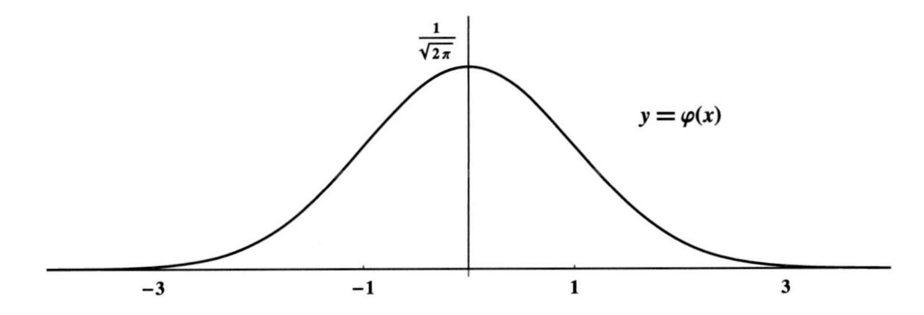

Figure 3.9. The probability density function φ of the standard normal distribution.

The standard normal distribution is so important that it has its own notation. Instead of the generic f and F for the density and cumulative distribution function we write φ for the standard normal density and

$$\Phi(x) = \frac{1}{\sqrt{2\pi}} \int_{-\infty}^{x} e^{-s^2/2}\, ds, \qquad x \in \mathbb{R}, \tag{3.40}$$

for the standard normal cumulative distribution function (see Figure 3.10). Working with the normal distribution is complicated by the fact that there is no explicit antiderivative for φ. It is not even obvious that φ integrates to 1 as a density function must. We begin by checking this property.

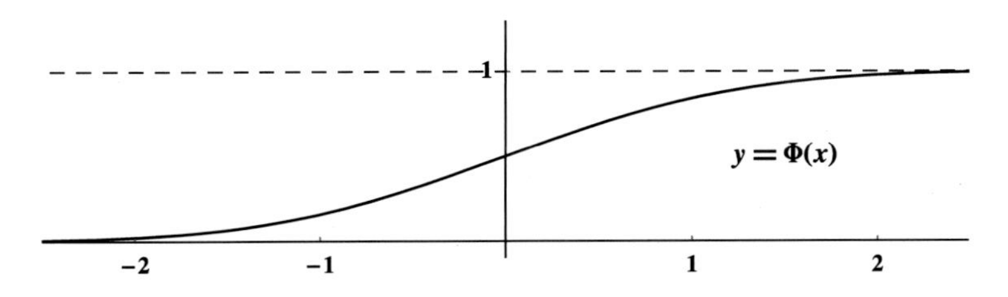

Figure 3.10. The cumulative distribution function Φ of the standard normal distribution. It satisfies $0 < \Phi(x) < 1$ for all x.

Fact 3.56.

$$\int_{-\infty}^{\infty} e^{-s^2/2}\, ds = \sqrt{2\pi}.$$

Proof. The trick here is to compute the square of the integral as a double integral and switch to polar coordinates.

$$\begin{aligned}\left(\int_{-\infty}^{\infty} e^{-s^2/2}\, ds\right)^2 &= \int_{-\infty}^{\infty} e^{-x^2/2}\, dx \cdot \int_{-\infty}^{\infty} e^{-y^2/2}\, dy \\ &= \int_{-\infty}^{\infty}\int_{-\infty}^{\infty} e^{-x^2/2-y^2/2}\, dx\, dy = \int_{0}^{2\pi}\int_{0}^{\infty} e^{-r^2/2} r\, dr\, d\theta \\ &= \int_{0}^{2\pi} \left(-e^{-r^2/2}\, \Big|_{r=0}^{r=\infty}\right) d\theta = \int_{0}^{2\pi} d\theta = 2\pi.\end{aligned}$$

▲

Since we do not have a simple explicit formula that gives numerical values for $\Phi(x)$, we have to turn to a table such as the one given in Appendix E. This table gives the values of $\Phi(x)$ for $0 \le x \le 3.49$ accurate to four decimal digits. For larger x, $\Phi(x)$ will be closer than 0.0002 to 1. For negative values we use symmetry (see Figure 3.11). Since $\varphi(x) = \varphi(-x)$ we have

$$\Phi(-x) = \frac{1}{\sqrt{2\pi}} \int_{-\infty}^{-x} e^{-s^2/2}\, ds = \frac{1}{\sqrt{2\pi}} \int_{x}^{\infty} e^{-s^2/2}\, ds = 1 - \Phi(x). \tag{3.41}$$

For example, to find $\Phi(-1.7)$, look up $\Phi(1.7) \approx 0.9554$ in the table and use $\Phi(-1.7) = 1 - \Phi(1.7) \approx 0.0446$.

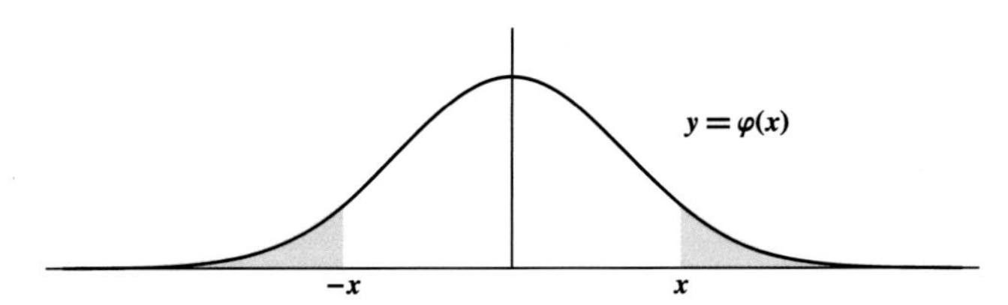

Figure 3.11. The symmetry of the function φ. The shaded blue area equals $\Phi(-x)$, the shaded red area equals $1 - \Phi(x)$, and the two are equal.

Example 3.57. Let $Z \sim \mathcal{N}(0, 1)$. Find the numerical value of $P(-1 \le Z \le 1.5)$.

By the properties of density and c.d.f.,

$$\begin{aligned}P(-1 \le Z \le 1.5) &= \int_{-1}^{1.5} \varphi(s) ds = \int_{-\infty}^{1.5} \varphi(s) ds - \int_{-\infty}^{-1} \varphi(s) ds \\ &= \Phi(1.5) - \Phi(-1) = \Phi(1.5) - (1 - \Phi(1)) \\ &\approx 0.9332 - (1 - 0.8413) = 0.7745.\end{aligned}$$

The second to last step used the table in Appendix E. Note that the answer is just an approximation of the probability because the values in the table are themselves decimal approximations of the exact values. ▲

Example 3.58. Find $z > 0$ so that a standard normal random variable Z has approximately $2/3$ probability of being in the interval $(-z, z)$.

We are asking for $z > 0$ such that $P(-z < Z < z) = 2/3$. First turn the question into a question about $\Phi(x)$ for positive x:

$$P(-z < Z < z) = \Phi(z) - \Phi(-z) = \Phi(z) - (1 - \Phi(z)) = 2\Phi(z) - 1.$$

Thus we need z for which $\Phi(z) = \frac{1}{2}(1 + \frac{2}{3}) = \frac{5}{6} \approx 0.833$. From the table in Appendix E we find $\Phi(0.96) = 0.8315$ and $\Phi(0.97) = 0.8340$. So $z = 0.97$ is a good approximation. We could use linear interpolation to get a slightly better approximation, but this is not important here. ▲

Once we know that φ is a well-defined probability density, we can compute its mean and variance.

Fact 3.59. Let $Z \sim \mathcal{N}(0, 1)$. Then $E(Z) = 0$ and $\operatorname{Var}(Z) = E(Z^2) = 1$.

Proof. By definition

$$E(Z) = \frac{1}{\sqrt{2\pi}} \int_{-\infty}^{\infty} x e^{-x^2/2}\, dx.$$

The first point is that this improper integral converges so the integral is well defined. This is tested by integrating the absolute value of the integrand:

$$\int_{-\infty}^{\infty} |x|\, e^{-x^2/2}\, dx = 2 \int_{0}^{\infty} x e^{-x^2/2}\, dx = 2.$$

We used the symmetry of the integrated function and the fact that the last integral was already evaluated above in the proof of Fact 3.56. Then the value $E(Z) = 0$ follows from a general fact about integration: if f is an odd function, which means that $f(-x) = -f(x)$, then

$$\int_{-a}^{a} f(x)\, dx = 0$$

for any $a \in [0, \infty]$, as long as the integral is well defined. Applying this to the odd function $f(x) = xe^{-x^2/2}$ gives $E(Z) = 0$.

For the mean square we integrate by parts:

$$\begin{aligned} E(Z^2) &= \frac{1}{\sqrt{2\pi}} \int_{-\infty}^{\infty} x^2 e^{-x^2/2}\, dx = -\frac{1}{\sqrt{2\pi}} \int_{-\infty}^{\infty} x \cdot (-xe^{-x^2/2})\, dx \\ &= -\frac{1}{\sqrt{2\pi}} \left\{ xe^{-x^2/2} \Big|_{x=-\infty}^{x=\infty} - \int_{-\infty}^{\infty} e^{-x^2/2}\, dx \right\} \\ &= \int_{-\infty}^{\infty} \frac{1}{\sqrt{2\pi}} e^{-x^2/2}\, dx = 1. \end{aligned}$$

In the second to last equality we used the limit $\lim_{x\to\pm\infty} xe^{-x^2/2} = 0$, while in the last equality we integrated the standard normal density function over the entire real line. ▲

Fact 3.59 gives the meaning of the parameters 0 and 1 in $Z \sim \mathcal{N}(0,1)$: a standard normal random variable has mean 0 and variance 1.

The general family of normal distributions is obtained by linear (or affine) transformations of Z. Let μ be real, $\sigma > 0$, and

$$X = \sigma Z + \mu.$$

By Fact 3.52, $E(X) = \mu$ and $\mathrm{Var}(X) = \sigma^2$. We can express the cumulative distribution function of X in terms of Φ:

$$F(x) = P(X \le x) = P(\sigma Z + \mu \le x) = P(Z \le \tfrac{x-\mu}{\sigma}) = \Phi\left(\tfrac{x-\mu}{\sigma}\right). \tag{3.42}$$

The probability density function of X can now be obtained by differentiating F:

$$f(x) = F'(x) = \frac{d}{dx}\left[\Phi\left(\tfrac{x-\mu}{\sigma}\right)\right] = \frac{1}{\sigma}\varphi\left(\tfrac{x-\mu}{\sigma}\right) = \frac{1}{\sqrt{2\pi\sigma^2}} e^{-\frac{(x-\mu)^2}{2\sigma^2}}. \tag{3.43}$$

We used the chain rule and the fact that $\Phi'(x) = \varphi(x)$.

We record this below as the definition of the general normal random variable. We can call the parameter μ the mean and σ^2 the variance since we checked above that these are the mean and variance of X.

Definition 3.60. Let μ be real and $\sigma > 0$. A random variable X has the *normal distribution with mean μ and variance σ^2* if X has density function

$$f(x) = \frac{1}{\sqrt{2\pi\sigma^2}} e^{-\frac{(x-\mu)^2}{2\sigma^2}} \tag{3.44}$$

on the real line. Abbreviate this by $X \sim \mathcal{N}(\mu, \sigma^2)$.

The argument above can be used to check the following useful fact. Exercise 3.70 asks you to verify this.

Fact 3.61. Let μ be real, $\sigma > 0$, and suppose $X \sim \mathcal{N}(\mu, \sigma^2)$. Let $a \neq 0$, b real, and $Y = aX + b$. Then $Y \sim \mathcal{N}(a\mu + b, a^2\sigma^2)$, that is, Y is normally distributed with mean $a\mu + b$ and variance $a^2\sigma^2$.

In particular, $Z = \frac{X-\mu}{\sigma}$ is a standard normal random variable.

The last part of the fact above is the key to evaluating probabilities of general normal random variables.

Example 3.62. Suppose $X \sim \mathcal{N}(-3, 4)$. Find the probability $P(X \le -1.7)$.

Since $Z = \frac{X-(-3)}{2} \sim \mathcal{N}(0, 1)$ we have

$$P(X \le -1.7) = P\left(\frac{X-(-3)}{2} \le \frac{-1.7-(-3)}{2}\right) = P(Z \le 0.65)$$
$$= \Phi(0.65) \approx 0.7422,$$

with the last equality from the table in Appendix E. ▲

Example 3.63. Suppose $X \sim \mathcal{N}(\mu, \sigma^2)$. What is the probability that the observed value of X deviates from μ by more than 2σ? With $Z = \frac{X-\mu}{\sigma}$ (a standard normal) we have

$$\begin{aligned} P(|X-\mu| > 2\sigma) &= P(X < \mu - 2\sigma) + P(X > \mu + 2\sigma) \\ &= P\left(\frac{X-\mu}{\sigma} < -2\right) + P\left(\frac{X-\mu}{\sigma} > 2\right) \\ &= P(Z < -2) + P(Z > 2) = 2\,(1 - P(Z \le 2)) \\ &= 2\,(1 - \Phi(2)) \approx 2(1 - 0.9772) = 0.0456. \end{aligned}$$

This can be remembered as a rule of thumb: a normal random variable is within two standard deviations of its mean with probability over 95%. ▲

3.6. Finer points ♣

Density function

With density functions we continue to sweep measure-theoretic details under the rug. Which functions can be density functions? Basically any function that can be legitimately integrated so that it can represent probabilities as in (3.3) and that satisfies (3.7). Rigorously speaking this leads us back to the class of measurable functions on $\mathbb{R}$. What functions should the reader think about? Any function that can be integrated with calculus techniques. In this book there are no density functions wilder than piecewise continuous functions, so we remain safely within the purview of calculus.

The reader may have realized a uniqueness problem with the density function. Let $c \in (a, b)$ and alter formula (3.8) by writing

$$\tilde{f}(x) = \begin{cases} \frac{1}{b-a}, & \text{if } x \in [a, c) \cup (c, b] \\ 0, & \text{if } x \notin [a, b] \text{ or } x = c. \end{cases} \tag{3.45}$$

This density $\tilde{f}$ defines exactly the same probability distribution as the original f in (3.8). This is because the value of any integral of f is not changed by altering f at a single point. By the same token, integrals are not changed by altering the function at finitely many points. Hence many actual functions can serve as "the density function." This does not create any confusion for calculations.

A logically satisfactory resolution of this lack of uniqueness is reached in measure theory. Two functions are said to be *equal almost everywhere* if their integrals over all intervals agree. Whenever two probability density functions are equal almost everywhere, they define the same probability distribution.

Range, support, and possible values of a random variable

As a function, a random variable X has a *range* which is the set $\{X(\omega) : \omega \in \Omega\}$. The set of possible values of a discrete random variable is a subset of its range, but these are not necessarily equal. The following example illustrates this idea.

Example 3.64. Let the sample space be $\Omega = \{a, b, c\}$ with probabilities $P\{a\} = 0$, $P\{b\} = \frac{1}{3}$, and $P\{c\} = \frac{2}{3}$. Define the function X on Ω by $X(a) = 1$, $X(b) = 2$ and $X(c) = 3$. Then the range of X is $\{1, 2, 3\}$. The possible values of X are $\{2, 3\}$ because $P(X = 1) = 0$ while $P(X = 2)$ and $P(X = 3)$ are positive. ▲

The next example illustrates why the range cannot be used to define a discrete random variable.

Example 3.65. Let the sample space Ω be the real line. Define the probability measure P on Ω by setting $P\{k\} = 1/3$ for $k = 1, 2, 3$, and $P(A) = 0$ for all sets A that are disjoint from $\{1, 2, 3\}$. Let Y be the identity function: $Y(\omega) = \omega$ for all real ω. Then Y is a random variable with probabilities $P(Y = k) = 1/3$ for $k = 1, 2, 3$. Clearly Y is a discrete random variable with possible values $\{1, 2, 3\}$. However, the range of Y as a function on Ω is the entire real line. ▲

To avoid these pathologies, it is useful to consider a slightly different concept, the *support* of the distribution of the random variable. The support of the random variable X is defined as the smallest closed set B such that $P(X \in B) = 1$. When the support of the random variable X is finite, or countably infinite with no accumulation points, then X is discrete and the support coincides with the set of possible values. In the two examples above the supports are $\{2, 3\}$ for X and $\{1, 2, 3\}$ for Y.

Continuity of a random variable

Fact 3.13(b) states that if the cumulative distribution function F of a random variable X is continuous on all of $\mathbb{R}$ and differentiable at all but finitely many points, then the derivative $f = F'$ is the density function of X. In fact, it is enough to assume that any bounded interval contains only finitely many points of nondifferentiability. In other words, the set of points of nondifferentiability of F can be countably infinite, as long as this set has no accumulation points.

If the derivative f is assumed continuous on those intervals where it is defined, Fact 3.13(b) follows from the fundamental theorem of calculus because then $F(b) - F(a) = \int_a^b f(s)\, ds$ whenever F is differentiable on (a, b). Without the continuity assumption on f the result follows by an application of Theorem 7.21 in [Rud87].

Properties of the cumulative distribution function

Properties (i)–(iii) of Fact 3.16 are both necessary and sufficient for F to be a cumulative distribution function. We indicate how the sufficiency is proved. Suppose F satisfies Fact 3.16(i)–(iii). To construct a random variable X whose cumulative distribution function is F, take the real line $\mathbb{R}$ as the sample space with probability measure P defined by $P\{(a,b]\} = F(b) - F(a)$ for all $a < b$. It is a nontrivial fact that such a probability measure exists, see [Dur10]. Let X be the identity function: $X(s) = s$ for real s. Then directly from the definitions: $P(X \le s) = P\{(-\infty, s]\} = F(s)$.

Expectation

A shortcoming of a probability course at the elementary level is that we cannot give a definition of the expectation that is independent of the particular computational recipes for discrete and continuous random variables. More advanced analytical tools can treat the expectations of all random variables in a unified manner. For example, the *Stieltjes integral* allows us to express both expectation formulas (3.19) and (3.23) as

$$E(X) = \int_{-\infty}^{\infty} x \, dF(x)$$

where F is the cumulative distribution function of X. With measure theory $E(X)$ can be defined as the *Lebesgue integral* of X with respect to a probability measure P on an abstract sample space Ω.

A condition that guarantees that the discrete expectation (3.19) is a well-defined finite number is the following:

$$\sum_k |k| \, P(X = k) < \infty. \tag{3.46}$$

Analogously, the expectation in (3.23) is a well-defined finite number if

$$\int_{-\infty}^{\infty} |x| f(x) \, dx < \infty.$$

A random variable that satisfies either of these two conditions is called *absolutely integrable.*

Moments

Just as expectations can be infinite or undefined, the same holds for moments. The following fact is useful. If $1 \le k < \ell$ are integers and $E[X^\ell]$ is finite, then $E[X^k]$ is also finite. This follows from Hölder's inequality (see [Dur10]).

Quantiles

An alternative definition of the pth quantile that also appears in practice is the following:

$$Q(p) = \min\{x \in \mathbb{R} : F(x) \ge p\} \qquad \text{for } 0 < p < 1.$$

The function Q is termed the **quantile function.** The median can then be defined as $Q(1/2)$. In contrast with the median of Definition 3.39, $Q(1/2)$ is a *uniquely* determined number for all distributions. For example, $Q(1/2) = 4$ in Example 3.40.

Exercises

We start with some warm-up exercises arranged by section.

Section 3.1

Exercise 3.1. Let X have possible values $\{1, 2, 3, 4, 5\}$ and probability mass function

x	1	2	3	4	5
$p_X(x)$	1/7	1/14	3/14	2/7	2/7

(a) Calculate $P(X \le 3)$.
(b) Calculate $P(X < 3)$.
(c) Calculate $P(X < 4.12 \mid X > 1.638)$.

Exercise 3.2. Suppose the random variable X has possible values $\{1, 2, 3, 4, 5, 6\}$ and probability mass function of the form $p(k) = ck$.

(a) Find c.
(b) Find the probability that X is odd.

Exercise 3.3. Let X be a continuous random variable with density function

$$f(x) = \begin{cases} 3e^{-3x}, & x > 0 \\ 0, & \text{else.} \end{cases}$$

(a) Verify that f is a density function.
(b) Calculate $P(-1 < X < 1)$.
(c) Calculate $P(X < 5)$.
(d) Calculate $P(2 < X < 4 \mid X < 5)$.

Exercise 3.4. Let $X \sim \text{Unif}[4, 10]$.

(a) Calculate $P(X < 6)$.
(b) Calculate $P(|X - 7| > 1)$.
(c) For $4 \le t \le 6$, calculate $P(X < t \mid X < 6)$.

Section 3.2

Exercise 3.5. Suppose that the discrete random variable X has cumulative distribution function given by

$$F(x) = \begin{cases} 0, & \text{if } x < 1 \\ 1/3, & \text{if } 1 \le x < \frac{4}{3} \\ 1/2, & \text{if } \frac{4}{3} \le x < \frac{3}{2} \\ 3/4, & \text{if } \frac{3}{2} \le x < \frac{9}{5} \\ 1, & \text{if } x \ge \frac{9}{5}. \end{cases}$$

Find the possible values and the probability mass function X.

Exercise 3.6. Find the cumulative distribution function of the random variable X from both Exercise 3.1 and 3.3.

Exercise 3.7. Suppose that the continuous random variable X has cumulative distribution function given by

$$F(x) = \begin{cases} 0, & \text{if } x < \sqrt{2} \\ x^2 - 2, & \text{if } \sqrt{2} \le x < \sqrt{3} \\ 1, & \text{if } \sqrt{3} \le x. \end{cases}$$

(a) Find the smallest interval $[a, b]$ such that of $P(a \le X \le b) = 1$.
(b) Find $P(X = 1.6)$.
(c) Find $P(1 \le X \le \frac{3}{2})$.
(d) Find the probability density function of X.

Section 3.3

Exercise 3.8. Let X be the random variable from Exercise 3.1.

(a) Compute the mean of X.
(b) Compute $E[\,|X - 2|\,]$.

Exercise 3.9. Let X be the random variable from Exercise 3.3.

(a) Find the mean of X.
(b) Compute $E[e^{2X}]$.

Exercise 3.10. Let X have probability mass function

$$P(X = -1) = \tfrac{1}{2}, \quad P(X = 0) = \tfrac{1}{3}, \quad \text{and} \quad P(X = 1) = \tfrac{1}{6}.$$

Calculate $E[|X|]$ using the approaches in (a) and (b) below.

(a) First find the probability mass function of the random variable $Y = |X|$ and using that compute $E[|X|]$.
(b) Apply formula (3.24) with $g(x) = |x|$.

Exercise 3.11. Let Y be a random variable with density function $f(x) = \frac{2}{3}x$ for $x \in [1, 2]$ and $f(x) = 0$ otherwise. Compute $E[(Y-1)^2]$.

Exercise 3.12. Suppose that X is a random variable taking values in $\{1, 2, 3, \dots\}$ with probability mass function

$$p_X(n) = \frac{6}{\pi^2} \cdot \frac{1}{n^2}.$$

Show that $E[X] = \infty$.

Hint. See Example D.5.

Exercise 3.13. Compute the following.

(a) The median of the random variable X from both Exercise 3.1 and 3.3.
(b) The 0.9th quantile of the random variable X from Exercise 3.3.

Section 3.4

Exercise 3.14. Find the variance of the random variable X from both Exercise 3.1 and 3.3.

Exercise 3.15. Suppose that the random variable X has expected value $E[X] = 3$ and variance $\text{Var}(X) = 4$. Compute the following quantities.

(a) $E[3X + 2]$
(b) $E[X^2]$
(c) $E[(2X + 3)^2]$
(d) $\text{Var}(4X - 2)$

Exercise 3.16. Let Z have the following density function

$$f_Z(z) = \begin{cases} \frac{1}{7}, & 1 \le z \le 2 \\ \frac{3}{7}, & 5 \le z \le 7 \\ 0, & \text{otherwise.} \end{cases}$$

Compute both $E[Z]$ and $\text{Var}(Z)$.

Section 3.5

Exercise 3.17. Let X be a normal random variable with mean $\mu = -2$ and variance $\sigma^2 = 7$. Find the following probabilities using the table in Appendix E.

(a) $P(X > 3.5)$
(b) $P(-2.1 < X < -1.9)$
(c) $P(X < 2)$
(d) $P(X < -10)$
(e) $P(X > 4)$

Exercise 3.18. Let X be a normal random variable with mean 3 and variance 4.

(a) Find the probability $P(2 < X < 6)$.
(b) Find the value c such that $P(X > c) = 0.33$.
(c) Find $E(X^2)$.

Hint. You can integrate with the density function, but it is quicker to relate $E(X^2)$ to the mean and variance.

Further exercises

Exercise 3.19. Let $Z \sim \text{Bin}(10, \frac{1}{3})$. Find the value of its cumulative distribution function at 2 and at 8.

Exercise 3.20. Let $c > 0$ and $X \sim \text{Unif}[0, c]$. Show that the random variable $Y = c - X$ has the same cumulative distribution function as X and hence also the same density function.

Exercise 3.21. A fair coin is flipped twice. Let X be the number of heads observed.
(a) Give both the possible values and probability mass function for X.
(b) Find $P(X \geq 1)$ and $P(X > 1)$.
(c) Compute $E[X]$ and $\text{Var}(X)$.

Exercise 3.22. A fair coin is flipped three times. Let X be the number of heads observed.

(a) Give both the possible values and probability mass function of X.
(b) Find $P(X \geq 1)$ and $P(X > 1)$.
(c) Compute $E[X]$ and $\text{Var}(X)$.

Exercise 3.23. Ten thousand people each buy a lottery ticket. Each lottery ticket costs \$1. 100 people are chosen as winners. Of those 100 people, 80 will win \$2, 19 will win \$100, and one lucky winner will win \$7000. Let X denote the profit (profit = winnings − cost) of a randomly chosen player of this game.

(a) Give both the possible values and probability mass function for X.
(b) Find $P(X \geq 100)$.
(c) Compute $E[X]$ and $\text{Var}(X)$.

Exercise 3.24. Suppose X has a discrete distribution with probability mass function given by

x	1	2	3
$p_X(x)$	1/7	2/7	4/7

(a) What is $P(X \geq 2)$?
(b) What is $E\left(\frac{1}{1+X}\right)$?

Exercise 3.25. In each of the following cases find all values of b for which the given function is a probability density function.
(a)

$$f(x) = \begin{cases} x^2 - b, & \text{if } 1 \leq x \leq 3 \\ 0, & \text{otherwise.} \end{cases}$$

(b)

$$h(x) = \begin{cases} \cos x, & \text{if } -b \leq x \leq b \\ 0, & \text{otherwise.} \end{cases}$$

Exercise 3.26. Suppose that X is a discrete random variable with possible values $\{1, 2, \dots\}$, and probability mass function

$$p_X(k) = \frac{c}{k(k+1)}$$

with some constant $c > 0$.

(a) What is the value of c?
Hint. $\frac{1}{k(k+1)} = \frac{1}{k} - \frac{1}{k+1}$.
(b) Find $E(X)$.
Hint. Example D.5 could be helpful.

Exercise 3.27. Let X have probability mass function

x	1	2	3	4
$p_X(x)$	2/5	1/5	1/5	1/5

(a) Calculate $P(X \geq 2)$.
(b) Calculate $P(X \leq 3 | X \geq 2)$.
(c) Calculate $\text{Var}(X)$.

Exercise 3.28. There are five closed boxes on a table. Three of the boxes have nice prizes inside. The other two do not. You open boxes one at a time until you find a prize. Let X be the number of boxes you open.

(a) Find the probability mass function of X.
(b) Find $E[X]$.
(c) Find $\text{Var}(X)$.
(d) Suppose the prize inside each of the three good boxes is \$100, but each empty box you open costs you \$100. What is your expected gain or loss in this game?

Hint. Express the gain or loss as a function of X.

Exercise 3.29. In a school there are four kindergarten classes, with 21, 24, 17 and 28 students. We choose one of the 90 kindergarten students randomly, and denote the number of students in her class by X. Find the expectation and variance of X.

Exercise 3.30. When a certain basketball player takes his first shot in a game he succeeds with probability 1/2. If he misses his first shot, he loses confidence and his second shot will go in with probability 1/3. If he misses his first 2 shots then his third shot will go in with probability 1/4. His success probability goes down further to 1/5 after he misses his first 3 shots. If he misses his first 4 shots then the coach will remove him from the game. Assume that the player keeps shooting until he succeeds or he is removed from the game. Let X denote the *number of shots he misses* until his first success or until he is removed from the game.

(a) Calculate the probability mass function of X.
(b) Compute the expected value of X.

Exercise 3.31. Suppose a random variable X has density function

$$f(x) = \begin{cases} cx^{-4}, & \text{if } x \geq 1 \\ 0, & \text{else,} \end{cases}$$

where c is a constant.

(a) What must be the value of c?
(b) Find $P(0.5 < X < 1)$.
(c) Find $P(0.5 < X < 2)$.
(d) Find $P(2 < X < 4)$.
(e) Find the cumulative distribution function $F_X(x)$.
(f) Find $E(X)$ and $\text{Var}(X)$.
(g) Find $E[5X^2 + 3X]$.
(h) Find $E[X^n]$ for *all* integers n. Your answer will be a formula that contains n.

Exercise 3.32. Let X be a continuous random variable with density function

$$f_X(x) = \begin{cases} \frac{1}{2}x^{-3/2}, & 1 < x < \infty \\ 0, & \text{else.} \end{cases}$$

(a) Find $P(X > 10)$.
(b) Find the cumulative distribution function F_X of X.
(c) Find $E[X]$.
(d) Find $E[X^{1/4}]$.

Exercise 3.33. Let X be a random variable with density function

$$f(x) = \begin{cases} \frac{1}{4}, & 1 < x < 2 \\ c, & 3 < x < 5 \\ 0, & \text{otherwiese,} \end{cases}$$

where c is a constant.

(a) What is the value of c?
(b) Find $P(\frac{3}{2} < X < 4)$.
(c) Find $E[X]$.

Exercise 3.34. Let X be a random variable with probability mass function

$$P(X = 1) = \tfrac{1}{2}, \quad P(X = 2) = \tfrac{1}{3} \quad \text{and} \quad P(X = 5) = \tfrac{1}{6}.$$

(a) Find a function g such that $E[g(X)] = \frac{1}{3}\ln 2 + \frac{1}{6}\ln 5$. Your answer should give at least the values $g(k)$ for all possible values k of X, but you can also specify g on a larger set if possible.
(b) Let t be some real number. Find a function g such that

$$E[g(X)] = \tfrac{1}{2}e^{t} + \tfrac{2}{3}e^{2t} + \tfrac{5}{6}e^{5t}.$$

(c) Find a function g such that $E[g(X)] = 2$.

Exercise 3.35. Suppose random variable X has probability mass function

$$P(X = -2) = \tfrac{1}{16}, \quad P(X = 0) = \tfrac{53}{64}, \quad P(X = 4) = \tfrac{7}{64}.$$

Compute the fourth moment $E(X^4)$.

Exercise 3.36. Suppose random variable X has density function

$$f(x) = \begin{cases} \dfrac{2}{x^2}, & 1 \le x \le 2 \\ 0, & x < 1 \text{ or } x > 2. \end{cases}$$

Compute the fourth moment $E(X^4)$.

Exercise 3.37. Suppose random variable X has a cumulative distribution function

$$F(x) = \begin{cases} \frac{x}{1+x}, & x \ge 0 \\ 0, & x < 0. \end{cases}$$

(a) Find the probability density function of X.
(b) Calculate $P\{2 < X < 3\}$.
(c) Calculate $E[(1 + X)^2 e^{-2X}]$.

Exercise 3.38. Let the random variable Z have probability density function

$$f_Z(z) = \begin{cases} \frac{5}{2}z^4, & -1 \le z \le 1 \\ 0, & \text{otherwise.} \end{cases}$$

(a) Calculate $E[Z]$.
(b) Calculate $P(0 < Z < \frac{1}{2})$.
(c) Calculate $P(Z < \frac{1}{2} \mid Z > 0)$.
(d) Calculate all the moments $E[Z^n]$ for $n = 1, 2, 3, \ldots$ Your answer will be a formula that contains n.

Exercise 3.39. Parts (a) and (b) below ask for an example of a random variable X whose cumulative distribution function $F(x)$ satisfies $F(1) = 1/3$, $F(2) = 3/4$, and $F(3) = 1$.

(a) Make X discrete and give its probability mass function.
(b) Make X continuous and give its probability density function.

Exercise 3.40. Give an example of a discrete or continuous random variable X (by giving the p.m.f. or p.d.f.) whose cumulative distribution function $F(x)$ satisfies $F(n) = 1 - \frac{1}{n}$ for each positive integer n.

Exercise 3.41. We produce a random real number X through the following two-stage experiment. First roll a fair die to get an outcome Y in the set $\{1, 2, \ldots, 6\}$. Then, if $Y = k$, choose X uniformly from the interval $(0, k]$. Find the cumulative distribution function $F(s)$ and the probability density function $f(s)$ of X for $3 < s < 4$.

Exercise 3.42. Consider the trapezoid D with corners $(0, 0)$, $(1, 0)$, $(1, 1)$ and $(0, 2)$. We choose a point (X, Y) uniformly at random from D.

(a) Find the cumulative distribution function of F_X of X and F_Y of Y.
(b) Find the probability density function of f_X of X and f_Y of Y.

Exercise 3.43. Choose a point uniformly at random in a unit square (square of side length 1). Let X be the distance from the point chosen to the nearest edge of the square.

(a) Find the cumulative distribution function of X.
(b) Find the probability density function of X.

Exercise 3.44. Choose a point uniformly at random in that half of the unit disk that is to the right of the y-axis. In other words, choose a point uniformly at random from the set $\{(x, y) : x > 0, x^2 + y^2 < 1\}$. Let S be the slope of the line through the chosen point and the origin.

(a) Find the cumulative distribution function of S.
(b) Find the probability density function of S.

The density function you find should be $f(x) = \frac{1}{\pi} \cdot \frac{1}{1+x^2}$ for $-\infty < x < \infty$. This is the density function of the *Cauchy distribution*.

Exercise 3.45. Choose a point uniformly at random inside the unit square with corners at $(0, 0)$, $(1, 0)$, $(0, 1)$ and $(1, 1)$. Let S be the slope of the line through the chosen point and the origin.

(a) Find the cumulative distribution function of S.
(b) Find the probability density function of S.

Exercise 3.46. A stick of length ℓ is broken at a uniformly chosen random location. We denote the length of the smaller piece by X.

(a) Find the cumulative distribution function of X.
(b) Find the probability density function of X.

Exercise 3.47. Choose a point uniformly at random from the triangle with vertices $(0, 0)$, $(30, 0)$, and $(30, 20)$. Let (X, Y) be the coordinates of the chosen point.

(a) Find the cumulative distribution function of X.
(b) Use part (a) to find the density function of X.
(c) Use part (b) to find the expectation of X.

Exercise 3.48. Pick a random point uniformly inside the triangle with vertices $(0, 0)$, $(2, 0)$ and $(0, 1)$. Compute the expectation of the distance of this point to the y-axis.

Exercise 3.49. A random point is picked uniformly inside the triangle on the plane with vertices $(0, 0)$, $(0, 2)$ and $(4, 1)$. Let X be the x-coordinate of this random point. Find the cumulative distribution function F and density function f of X using the following two approaches. Examples 3.19 and 3.9 give guidance.

(a) Find F from the connection between probability and area, and then differentiate to find f.
(b) Find f with the infinitesimal method illustrated by Example 3.9, and then integrate to find F.

Exercise 3.50. This exercise continues the application of the infinitesimal method in Example 3.9.

(a) Find the density function f_R of Example 3.9 with the approximation $f_R(t) \approx \varepsilon^{-1} P(t - \varepsilon < R < t)$.
(b) Find the density function f_R of Example 3.9 with the approximation $f_R(t) \approx (2\varepsilon)^{-1} P(t - \varepsilon < R < t + \varepsilon)$.

Exercise 3.51. (Another way to compute the mean of $X \sim \text{Geom}(p)$) Since $ka = \sum_{j=1}^{k} a$ for any quantity a, we can start the computation of the mean of $X \sim \text{Geom}(p)$ as follows:

$$E(X) = \sum_{k=1}^{\infty} k(1-p)^{k-1} p = \sum_{k=1}^{\infty} \sum_{j=1}^{k} (1-p)^{k-1} p.$$

Now switch the order of summation. This means that you take the j-summation outside and put the k-summation inside. But be careful with the limits of the sums. After this you get to the answer by evaluating two geometric series.

Exercise 3.52. The idea in Exercise 3.51 generalizes to give a new formula for the expected value of any nonnegative integer-valued random variable. Show that if the random variable X takes only nonnegative integers as its values then

$$E(X) = \sum_{k=1}^{\infty} P(X \geq k).$$

This holds even when $E(X) = \infty$, in which case the sum on the right-hand side is infinite.

Hint. Write $P(X \geq k)$ as $\sum_{i=k}^{\infty} P(X = i)$ in the sum, and then switch the order of the two summations.

Exercise 3.53. The random variable X has the following probability mass function:

$$P(X = 0) = \tfrac{3}{4}, \qquad P(X = k) = \tfrac{1}{2} \cdot \left(\tfrac{1}{3}\right)^k \quad \text{for integers } k \geq 1.$$

(a) Compute $E[X]$.
(b) Compute $\text{Var}[X]$.

Exercise 3.54.

(a) Let $X \sim \text{Geom}(p)$. Compute $P(X \geq k)$ for all positive integers k.
(b) Use Exercise 3.52 and part (a) to give a new proof that the expected value of a $\text{Geom}(p)$ random variable is p^{-1}.

Exercise 3.55. (Exercise 2.55 continued) Peter and Mary take turns throwing one dart at the dartboard. Peter hits the bullseye with probability p and Mary hits the bullseye with probability r. Whoever hits the bullseye first wins. Suppose Peter throws the first dart and let Y denote the number of darts thrown in the game. Find $E[Y]$. (Your answer in the $p = r$ case should be familiar to you.)

Exercise 3.56. (Mean square of a geometric random variable) Let $X \sim \text{Geom}(p)$ and $q = 1 - p$. Derive the formula

$$E(X^2) = \frac{1+q}{p^2}.$$

Hint. To compute this conveniently, start with $E(X^2) = E(X) + E[X(X-1)]$. Then note that in the geometric case $E[X(X-1)]$ can be computed with the derivative technique of Example 3.25 (see equation (3.22)), but with the *second* derivative because $(x^k)'' = k(k-1)x^{k-2}$ for $k \geq 2$.

Exercise 3.57. Let $X \sim \text{Geom}(p)$. Find the expected value of $\frac{1}{X}$.

Exercise 3.58. Let $X \sim \text{Binom}(n, p)$. Find the expected value of $\frac{1}{1+X}$.

Exercise 3.59. Recall Example 1.38 of the dartboard with the concentric circles. We throw a dart uniformly on the board. Let R be the distance of the dart to the center and X the number of points we get for the throw.

(a) Find a function $g(r)$ so that $X = g(R)$.
(b) Find the expectation $E[X]$ by using the p.m.f. of X computed in Example 1.38.
(c) Find the expectation of $E[X] = E[g(R)]$ using the probability density function of R.
Hint. Example 3.19 will be helpful for finding the density function of R.

Exercise 3.60. Show that if all the expectations below are finite, then

$$E[u(X) + v(X)] = E[u(X)] + E[v(X)]$$

(a) assuming that X is discrete
(b) assuming that X has density function f.

Exercise 3.61. Let the value of your car be M dollars. (M can be, for example, 2000 but it is easier to write mathematics with M than with long numbers.) When you get into an accident, the amount of damage to your car is X dollars, where X is a random variable with probability density function

$$f(x) = \begin{cases} 2(M - x)/M^2, & 0 \le x \le M \\ 0, & x < 0 \text{ or } x > M. \end{cases}$$

You have insurance with an $M/2$ dollar deductible. This means that if $X \le M/2$ then you pay the full amount of damages, while if $X > M/2$ then you pay $M/2$ dollars and the insurance company covers the remaining $X - M/2$ dollars. Let Y be the amount you pay for the damage.

(a) Find the cumulative distribution function of X.
(b) Give a formula that expresses Y as a function of X.
(c) Find the probabilities $P(Y \le y)$ for $y < M/2$, $P(Y = M/2)$, and $P(Y \le y)$ for $y > M/2$. Based on these, write a formula for and sketch the graph of the cumulative distribution function F_Y of the random variable Y.
(d) Find the probability $P(Y < M/2)$. How can this probability be found from the c.d.f. F_Y?
(e) Does Y qualify either as a discrete or as a continuous random variable?
Hint. When you write down probabilities, be careful with $\le$ and $<$ when it matters. If you need inspiration, look at Example 3.20.

Exercise 3.62. A little boy plays outside in the yard. On his own he would come back inside at a random time uniformly distributed on the interval [0, 1]. (Let us take the units to be hours.) However, if the boy is not back inside in 45 minutes, his mother brings him in. Let X be the time when the boy comes back inside.

(a) Find the cumulative distribution function F of X.
(b) Find the mean $E(X)$.
(c) Find the variance $\mathrm{Var}(X)$.
Hint. You should see something analogous in Examples 3.20 and 3.38.

Exercise 3.63. A random variable X is *symmetric* if X has the same probability distribution as $-X$. In the discrete case symmetry means that $P(X = k) = P(X = -k)$ for all possible values k. In the continuous case it means that the density function satisfies $f(x) = f(-x)$ for all x.

Assume that X is symmetric and $E[X]$ is finite. Show that $E[X] = 0$ in the

(a) discrete case
(b) continuous case.

Exercise 3.64. Give an example of a nonnegative discrete random variable for which $E[X] < \infty$ and $E[X^2] = \infty$. Next, give an example of a nonnegative continuous random variable for which $E[X] < \infty$ and $E[X^2] = \infty$.

Exercise 3.65. Suppose that X is a random variable with mean 2 and variance 3.

(a) Compute $\mathrm{Var}(2X + 1)$.
(b) Compute $E[(3X - 4)^2]$.

Exercise 3.66. Let X be a normal random variable with mean 8 and variance 3. Find the value of α such that $P(X > \alpha) = 0.15$.

Exercise 3.67. Let $Z \sim \mathcal{N}(0, 1)$ and $X \sim \mathcal{N}(\mu, \sigma^2)$. This means that Z is a standard normal random variable with mean 0 and variance 1, while X is a normal random variable with mean μ and variance σ^2.

(a) Calculate $E(Z^3)$ (this is the *third moment* of Z).
(b) Calculate $E(X^3)$.
Hint. Do not integrate with the density function of X unless you love messy integration. Instead use the fact that X can be represented as $X = \sigma Z + \mu$ and expand the cube inside the expectation.

Exercise 3.68. Let $Z \sim \mathcal{N}(0, 1)$ and $X \sim \mathcal{N}(\mu, \sigma^2)$.

(a) Calculate $E[Z^4]$.
(b) Calculate $E[X^4]$.

Exercise 3.69. Find a general formula for all the moments $E(Z^n)$, $n \geq 1$, for a standard normal random variable Z.

Exercise 3.70. Let $X \sim \mathcal{N}(\mu, \sigma^2)$ and $Y = aX + b$. By adapting the calculation from (3.42) and (3.43), show that $Y \sim \mathcal{N}(a\mu + b, a^2\sigma^2)$.

Exercise 3.71. My bus is scheduled to depart at noon. However, in reality the departure time varies randomly, with average departure time 12 o'clock noon and a standard deviation of 6 minutes. Assume the departure time is normally distributed. If I get to the bus stop 5 minutes past noon, what is the probability that the bus has not yet departed?

Exercise 3.72. In a lumberjack competition a contestant is blindfolded and spun around 9 times. The blindfolded contestant then tries to hit the target point in the middle of a horizontal log with an axe. The contestant receives

- 15 points if his hit is within 3 cm of the target,
- 10 points if his hit is between 3 cm and 10 cm off the target,
- 5 points if his hit is between 10 cm and 20 cm off the target, and
- zero points if his hit is 20 cm or more away from the target (and someone may lose a finger!).

Let Y record the position of the hit, so that $Y = y > 0$ corresponds to missing the target point to the right by y cm and $Y = -y < 0$ corresponds to missing the target to the left by y cm. Assume that Y is normally distributed with mean $\mu = 0$ and variance 100 cm^2. Find the expected number of points that the contestant wins.

Exercise 3.73. Let Y have density function

$$f_Y(x) = \begin{cases} \frac{1}{2}x^{-2}, & x \le -1 \text{ or } x \ge 1 \\ 0, & -1 < x < 1. \end{cases}$$

Is it possible to give $E[Y]$ a meaningful value?

Exercise 3.74. Let k be a positive integer. Give an example of a nonnegative random variable X for which $E[X^k] < \infty$ but $E[X^{k+1}] = \infty$.

Challenging problems

Exercise 3.75. Let $X \sim \text{Bin}(n, p)$. Find the probability that X is even. (Your answer should be a simple expression.)

Exercise 3.76. This exercise asks you to prove part (a) of Fact 3.13 in the case where the piecewise constant F has finitely many jumps. Let F be the cumulative distribution function of the random variable X. Suppose F is piecewise constant with finitely many jumps at the locations $s_1 < \cdots < s_n$. Precisely, this means that $0 < F(s_1) < \cdots < F(s_n) = 1$, $F(x) = 0$ for $x < s_1$, $F(x) = F(s_i)$ for $s_i \le x < s_{i+1}$ as $i = 1, \dots, n-1$, and $F(x) = 1$ for $x \ge s_n$.

Show that X is a discrete random variable with possible values $\{s_1, \dots, s_n\}$ and probability mass function $P(X = s_i) = F(s_i) - F(s_i-)$ for $1 \le i \le n$.

Exercise 3.77. Prove Fact 3.17 using the following outline. Suppose that $s_1 < s_2 < s_3 < \dots$ is an increasing sequence with $\lim_{n\to\infty} s_n = a$. Consider the events $A_n = \{X \le s_n\}$.

(i) Show that $\bigcup_n A_n = \{X < a\}$.
(ii) Use Fact 1.39 to prove that $\lim_{n\to\infty} P(A_n) = P(\bigcup_n A_n)$.

Exercise 3.78. In Example 1.27, find the expected number of people who receive the correct hat.

Exercise 3.79. In a town there are on average 2.3 children in a family and a randomly chosen child has on average 1.6 siblings. Determine the variance of the number of children in a randomly chosen family.

4

Approximations of the binomial distribution

A central feature that makes probability theory such an exciting subject is that out of the mess of large scale randomness, striking patterns and regularity appear. In this chapter we encounter two such phenomena that arise from a sequence of independent trials.

Figure 4.1 shows that the probability mass function of the Bin(n, p) distribution can be very close to the bell curve of the normal distribution, when viewed in the right scale. In the situations depicted, the mean and variance of the binomial are growing large. This is a particular case of the *central limit theorem.*

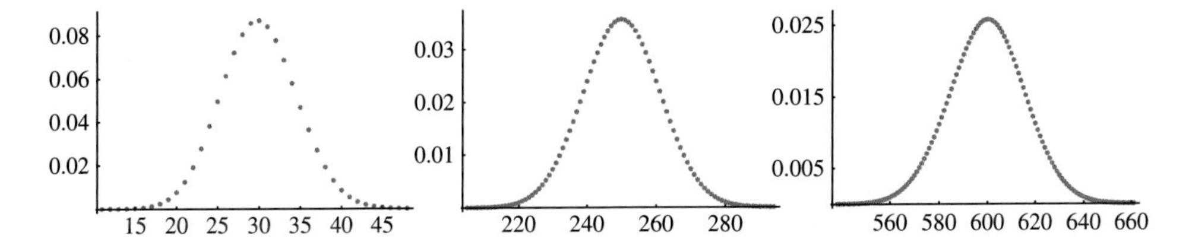

Figure 4.1. The probability mass function of the Bin(n, p) distribution with parameter values $(100, 0.3)$, $(500, 0.5)$, and $(1000, 0.6)$. In each case the graph shows the function on the interval $(np - 4\sqrt{np(1-p)}, np + 4\sqrt{np(1-p)})$ and the scale on the y-axis changes also.

Figure 4.2 looks at the probability mass function of the Bin(n, p) distribution when p is very small, so that successes are rare. In each case in Figure 4.2 the mean number of successes is 2, and the graphs are virtually indistinguishable

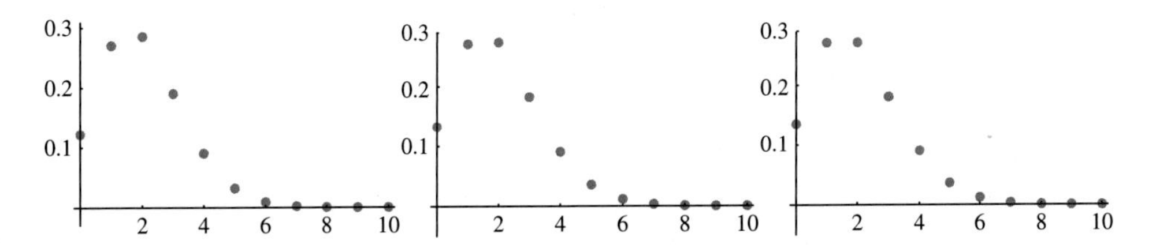

Figure 4.2. The probability mass function of the Bin(n, p) distribution with parameter values $(20, 0.1)$, $(100, 0.02)$, and $(1000, 0.002)$. In each case the graph shows the function for values $\{0, 1, \dots, 10\}$. The scales on the x and y axes are the same in all three graphs.

from the probability mass function of the Poisson(2) distribution, which we define later in this chapter. This is an instance of the *law of rare events*.

In the sections that follow we study the mathematics behind the approximations revealed by these graphs and practice using these approximations in examples.

4.1. Normal approximation

This section is among the most significant ones in the text, and one whose message should stay with you long after reading this book. The idea here is foundational to just about all human activity that makes use of statistical evidence.

We begin with the statement of the central limit theorem (CLT) for the binomial distribution and provide some examples of its application. At the end of the section we give the main idea of a derivation of the theorem that relies on calculus.

The CLT for the binomial and examples

As shown in the third graph of Figure 4.1, the probability mass function of the Bin(1000, 0.6) distribution resembles a normal density function. But which normal density function? Let us try matching its mean μ and variance σ^2 with those of Bin(1000, 0.6), and set $\mu = 1000 \cdot 0.6 = 600$ and $\sigma^2 = 1000 \cdot 0.6 \cdot 0.4 = 240$. Figure 4.3 shows how strikingly the Bin(1000, 0.6) probability mass function matches the $\mathcal{N}(600, 240)$ density function.

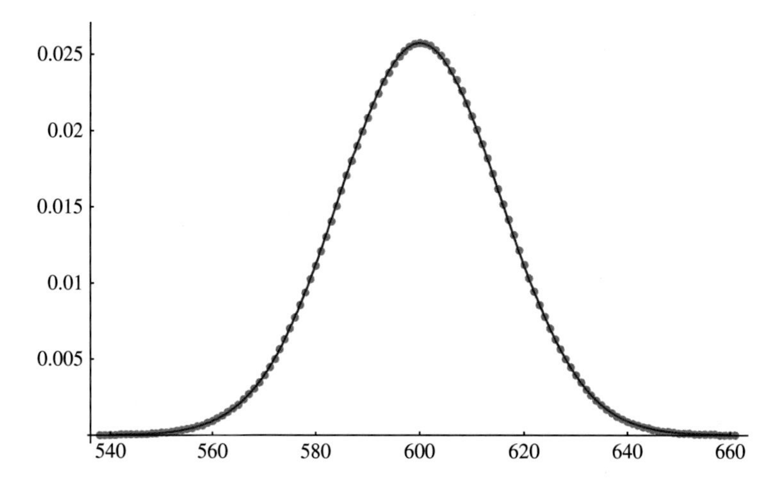

Figure 4.3. The bullets represent the probability mass function of Bin(1000, 0.6) and the continuous curve is the density function of $\mathcal{N}(600, 240)$ in the interval [540, 660].

With this example in mind it is reasonable to suspect that, in general, the probability mass function of $S_n \sim \text{Bin}(n, p)$ should approximate the density function of $X \sim \mathcal{N}(np, np(1-p))$ as n becomes large. However, the growing mean np and variance $np(1-p)$ are inconvenient for comparison of S_n with X. Hence we *standardize* the random variables to fix the mean at 0 and the variance at 1. This is achieved by subtracting the mean and by dividing by the standard deviation, which gives the standard normal random variable $Z = \frac{X-np}{\sqrt{np(1-p)}}$ and the standardized binomial $\frac{S_n-np}{\sqrt{np(1-p)}}$. By Fact 3.52 we have $E(Z) = 0$ and $\text{Var}(Z) = 1$ and also

$$E\left[\frac{S_n - np}{\sqrt{np(1-p)}}\right] = \frac{E[S_n] - np}{\sqrt{np(1-p)}} = 0 \quad \text{and} \quad \text{Var}\left(\frac{S_n - np}{\sqrt{np(1-p)}}\right) = \frac{\text{Var}(S_n)}{np(1-p)} = 1.$$

We expect now that $\frac{S_n-np}{\sqrt{np(1-p)}}$ approximates a standard normal random variable. This is exactly what the central limit theorem tells us.

Theorem 4.1. (Central limit theorem for binomial random variables) *Let $0 < p < 1$ be fixed and suppose that $S_n \sim \text{Bin}(n, p)$. Then for any fixed $-\infty \le a \le b \le \infty$ we have the limit*

$$\lim_{n\to\infty} P\left(a \le \frac{S_n - np}{\sqrt{np(1-p)}} \le b\right) = \int_a^b \frac{1}{\sqrt{2\pi}} e^{-\frac{x^2}{2}}\, dx. \tag{4.1}$$

Another name for this theorem is the *de Moivre–Laplace theorem*, and it goes back to the early 1700s. Note that the theorem holds also when $a = -\infty$ or $b = \infty$. The case $a = -\infty$ gives the limit of cumulative distribution functions:

$$\lim_{n\to\infty} P\left(\frac{S_n - np}{\sqrt{np(1-p)}} \le b\right) = \Phi(b) = \int_{-\infty}^b \frac{1}{\sqrt{2\pi}} e^{-\frac{x^2}{2}}\, dx. \tag{4.2}$$

The central limit theorem is one of the important *limit theorems* of probability theory. The CLT is an example of a *limit in distribution* because it is the probability distribution that converges. Namely, the probability distribution of $\frac{S_n-np}{\sqrt{np(1-p)}}$ converges to the standard normal distribution.

A limit theorem can be used to give a simple estimate to a complicated quantity. When n is large, calculating with a binomial distribution can be impractical. We can use limit (4.1) to give a quick approximation. This works well enough in many cases and is of enormous practical significance. We state this approximation in the next fact, together with an often cited criterion for deciding when its use is appropriate.

Normal approximation of the binomial distribution. Suppose that $S_n \sim \text{Bin}(n, p)$ with n large and p not too close to 0 or 1. Then

$$P\left(a \le \frac{S_n - np}{\sqrt{np(1-p)}} \le b\right) \quad \text{is close to} \quad \Phi(b) - \Phi(a). \tag{4.3}$$

As a rule of thumb, the approximation is good if $np(1-p) > 10$.

The mathematically oriented reader may be troubled by the imprecision of approximation (4.3). Rigorous error bounds for this approximation exist in the literature, and such a bound is stated in Section 4.7. For the purpose of this course, it is best to turn to examples to get a feeling for how the approximation works.

The next example shows the basic pattern that the normal approximation takes.

Example 4.2. A fair coin is flipped 10,000 times. Estimate the probability that the number of heads is between 4850 and 5100.

Let S denote the number of heads observed. Then $S \sim \text{Bin}(10{,}000, \frac{1}{2})$ with mean $E(S) = 10{,}000 \cdot \frac{1}{2} = 5000$ and variance $\text{Var}(S) = 10{,}000 \cdot \frac{1}{2} \cdot \frac{1}{2} = 2500$. The probability in question is $P(4850 \le S \le 5100)$. Centering with the mean and dividing by the standard deviation transforms the probability as follows:

$$\begin{aligned} P(4850 \le S \le 5100) &= P\left(\frac{4850 - 5000}{\sqrt{2500}} \le \frac{S - 5000}{\sqrt{2500}} \le \frac{5100 - 5000}{\sqrt{2500}}\right) \\ &= P\left(-3 \le \frac{S - 5000}{50} \le 2\right). \end{aligned}$$

So far everything is precise. Now we continue with approximation (4.3):

$$\begin{aligned} P\left(-3 \le \frac{S - 5000}{50} \le 2\right) &\approx \Phi(2) - \Phi(-3) \\ &\approx 0.9772 - (1 - 0.9987) = 0.9759. \end{aligned}$$

We actually made two approximations. First we approximated the binomial with the normal distribution, and then we approximated the cumulative distribution function of the standard normal using the table in the appendix.

What was the benefit of the approximation? The exact probability is

$$P(4850 \le S \le 5100) = \sum_{k=4850}^{5100} \binom{10,000}{k} 2^{-10,000}.$$

With modern computational tools it is easy to evaluate this expression. It is approximately 0.9765. Our approximate value 0.9759 is within 0.06% of the truth. The normal approximation is a quick way to get a number that is often good enough. ▲

Sometimes an example needs to be processed a little to cast it in a form where the normal approximation to a binomial applies.

Example 4.3. Suppose a game piece moves along a board according to the following rule: if a roll of a die gives 1 or 2 then take two steps ahead, otherwise take three steps. Approximate the probability that after 120 rolls the piece has moved more than 315 steps.

Let X_n denote the number of steps taken up to and including the nth die roll. X_n is not a binomial random variable, but it can be easily related to one. If S_n is the total number of ones and twos among the first n rolls then $X_n = 2S_n + 3(n - S_n) = 3n - S_n$. Using this to rewrite the probability, we need to estimate

$$P(X_{120} > 315) = P(360 - S_{120} > 315) = P(S_{120} < 45).$$

We may now apply the normal approximation to $S_{120} \sim \text{Bin}(120, \frac{1}{3})$. The mean and variance are $E(S_{120}) = 40$ and $\text{Var}(S_{120}) = 120 \cdot \frac{1}{3} \cdot \frac{2}{3} = \frac{80}{3}$. Thus

$$P(S_{120} < 45) = P\left(\frac{S_{120} - 40}{\sqrt{80/3}} < \frac{5}{\sqrt{80/3}}\right) \approx \Phi\left(\frac{5}{\sqrt{80/3}}\right) \approx 0.834. \qquad \blacktriangle$$

Example 4.4. (The three sigma rule) The probability that a standard normal random variable lies in $[-3, 3]$ is $\Phi(3) - \Phi(-3) \approx 0.9974$. By the normal approximation of a binomial, this tells us that with probability larger than 0.997 a $\text{Bin}(n, p)$ random variable is within $3\sqrt{np(1-p)}$ of its mean np, provided n is large enough and p is not too close to 0 or 1. This is sometimes called the *three sigma rule* because sigma σ stands for the standard deviation $\sqrt{np(1-p)}$. ▲

Continuity correction

If $S_n \sim \text{Bin}(n, p)$ then it can take only integer values. Thus, if k_1, k_2 are integers then

$$P(k_1 \le S_n \le k_2) = P(k_1 - 1/2 \le S_n \le k_2 + 1/2)$$

since the interval $[k_1, k_2]$ contains exactly the same integers as $[k_1 - 1/2, k_2 + 1/2]$. It turns out that if we apply the normal approximation to the modified interval $[k_1 - 1/2, k_2 + 1/2]$ we get a slightly better approximation of the exact binomial probability. In other words

$$\Phi\left(\frac{k_2 + 1/2 - np}{\sqrt{np(1-p)}}\right) - \Phi\left(\frac{k_1 - 1/2 - np}{\sqrt{np(1-p)}}\right)$$

gives a better estimate for $P(k_1 \le S_n \le k_2)$ than the formula without the 1/2 terms.

Switching from $[k_1, k_2]$ to $[k_1 - 1/2, k_2 + 1/2]$ is called the *continuity correction.* It is especially important if k_1, k_2 are close to each other or $np(1-p)$ is not very large.

Example 4.5. Roll a fair die 720 times. Estimate the probability that we have exactly 113 sixes.

Denote the number of sixes by S. This is a $\text{Bin}(720, \frac{1}{6})$ distributed random variable with mean 120 and variance $720 \cdot \frac{1}{6} \cdot \frac{5}{6} = 100$. To estimate $P(S = 113)$ we write it as

$$P(S = 113) = P(112.5 \le S \le 113.5).$$

The normal approximation now gives

$$P(S = 113) \approx \Phi\left(\frac{113.5 - 120}{10}\right) - \Phi\left(\frac{112.5 - 120}{10}\right)$$
$$= \Phi(-0.65) - \Phi(-0.75) \approx (1 - 0.7422) - (1 - 0.7734) = 0.0312.$$

The exact probability is

$$P(S = 113) = \binom{720}{113}\frac{5^{607}}{6^{720}} \approx 0.0318.$$

Our estimate is within 2% of the actual value, which is an excellent approximation. Note that a mindless normal approximation without the continuity correction gives an absurd estimate, namely zero:

$$P(S = 113) = P\left(\frac{113 - 120}{10} \le \frac{S - 120}{10} \le \frac{113 - 120}{10}\right)$$
$$= P(-0.7 \le \tfrac{S-120}{10} \le -0.7) \approx \Phi(-0.7) - \Phi(-0.7) = 0.$$

In this example $np(1 - p) = 100$ so we are well within range of the rule of thumb given in the approximation (4.3). ▲

Example 4.6. (Continuation of Example 4.2) When the number of trials is large, the continuity correction does not make much of a difference. We illustrate this by applying the continuity correction to Example 4.2. A fair coin is flipped 10,000 times and we want the probability that we observe between 4850 and 5100 heads.

$$\begin{aligned}
P(4850 \le S \le 5100) &= P(4849.5 \le S \le 5100.5) \\
&= P\left(\frac{4849.5 - 5000}{\sqrt{2500}} \le \frac{S - 5000}{\sqrt{2500}} \le \frac{5100.5 - 5000}{\sqrt{2500}}\right) \\
&= P\left(-3.01 \le \frac{S - 5000}{50} \le 2.01\right) \\
&\approx \Phi(2.01) - \Phi(-3.01) \\
&\approx 0.9778 - (1 - 0.9987) = 0.9765.
\end{aligned}$$

Now the estimate agrees with the exact value up to at least four decimals. ▲

A partial proof of the CLT for the binomial ♦

This section sketches a proof of Theorem 4.1 with calculus, without going through all the details.

For ease of notation set $q = 1 - p$. Begin by writing the probability on the left-hand side of (4.1) in terms of the binomial probability mass function:

$$P\left(a \le \frac{S_n - np}{\sqrt{npq}} \le b\right) = P\left(np + a\sqrt{npq} \le S_n \le np + b\sqrt{npq}\right)$$
$$= \sum_{np+a\sqrt{npq} \le k \le np+b\sqrt{npq}} \frac{n!}{(n-k)!k!} p^k q^{n-k}, \tag{4.4}$$

where the sum is over integers k between $np + a\sqrt{npq}$ and $np + b\sqrt{npq}$.

The rest of the proof relies on the following two steps. (i) Approximation of the factorial functions in the terms above with an expression that is easier to analyze. (ii) Interpretation of the resulting sum as a Riemann approximation of the integral on the right-hand side of (4.1).

To approximate the factorial function we use Stirling's formula. The symbol $\sim$ in the statement has a precise technical meaning, namely

$$a_n \sim b_n \quad \text{means that} \quad \frac{a_n}{b_n} \to 1 \quad \text{as } n \to \infty. \tag{4.5}$$

Fact 4.7. (Stirling's formula) As $n \to \infty$, $n! \sim n^n e^{-n} \sqrt{2\pi n}$.

For a proof of Stirling's formula see Exercises D.13 and 4.57.

An application of Stirling's formula to the factorials in (4.4) and some manipulation yields

$$\frac{n!}{(n-k)!k!} p^k q^{n-k} \sim \frac{1}{\sqrt{2\pi\, npq}} \exp\left\{-\frac{(k-np)^2}{2npq}\right\}.$$

From this follows the approximate equality

$$P\left(a \le \frac{S_n - np}{\sqrt{npq}} \le b\right) \approx \sum_{np+a\sqrt{npq} \le k \le np+b\sqrt{npq}} \frac{1}{\sqrt{2\pi\, npq}} \exp\left\{-\frac{(k-np)^2}{2npq}\right\}. \tag{4.6}$$

The sum on the right can be cast in the familiar form of a Riemann sum over the interval $[a, b]$ by defining partition points

$$x_k = \frac{k - np}{\sqrt{npq}}.$$

The summation limits $np + a\sqrt{npq} \le k \le np + b\sqrt{npq}$ specify exactly $a \le x_k \le b$, and the length of the subintervals is $\Delta x = \frac{1}{\sqrt{npq}}$. The approximation above becomes

$$P\left(a \le \frac{S_n - np}{\sqrt{npq}} \le b\right) \approx \sum_{k:\, a \le x_k \le b} \frac{1}{\sqrt{2\pi}} e^{-x_k^2/2} \Delta x$$
$$\underset{n\to\infty}{\longrightarrow} \int_a^b \frac{1}{\sqrt{2\pi}} e^{-x^2/2}\, dx.$$

The last step comes from the Riemann sum approximation of the integral. Making this proof fully rigorous requires an error estimate for the approximation in (4.6), which in turn needs a more quantitative version of Stirling's formula.

4.2. Law of large numbers

We probably feel confident that if a fair coin is flipped forever, the frequency of tails comes closer and closer to $1/2$. This is a mathematically true fact. It is an instance of the *law of large numbers* (LLN), another limit theorem. At this point in the course we can use the central limit theorem to derive a version of the LLN for independent trials.

Fix $0 < p < 1$ and let $X_1, X_2, X_3, \ldots$ be the outcomes of independent repetitions of a trial with success probability p. Then for each j we have $P(X_j = 1) = p$ and $P(X_j = 0) = 1 - p$. Let $S_n = X_1 + \cdots + X_n$ be the number of successes in the first n trials.

Theorem 4.8. (Law of large numbers for binomial random variables) *For any fixed $\varepsilon > 0$ we have*

$$\lim_{n\to\infty} P\left(\left|\frac{S_n}{n} - p\right| < \varepsilon\right) = 1. \tag{4.7}$$

Limit (4.7) says that no matter how small an interval $(p - \varepsilon, p + \varepsilon)$ you put around p, as n becomes large the *observed frequency of successes* $\frac{S_n}{n}$ will lie inside this interval with overwhelming probability. Here is the law of large numbers stated for a specific situation.

Example 4.9. Let S_n denote the number of sixes in n rolls of a fair die. Then S_n/n is the observed frequency of sixes in n rolls. Let $\varepsilon = 0.0001$. Then we have the limit

$$\lim_{n\to\infty} P\left(\left|\frac{S_n}{n} - \frac{1}{6}\right| < 0.0001\right) = 1.$$

The limit says that, as the number of rolls grows, deviations of S_n/n from $1/6$ by more than 0.0001 become extremely unlikely. ▲

Example 4.10. Show that the probability that fair coin flips yield 51% or more tails converges to zero as the number of flips tends to infinity.

Let S_n denote the number of tails in n fair coin flips.

$$\begin{aligned} P(\text{at least 51\% tails in } n \text{ flips}) &= P\left(\frac{S_n}{n} \geq 0.51\right) = P\left(\frac{S_n}{n} - 0.5 \geq 0.01\right) \\ &\leq P\left(\left|\frac{S_n}{n} - 0.5\right| \geq 0.01\right) \to 0 \quad \text{as } n \to \infty. \end{aligned}$$

The inequality above comes from the monotonicity of probability: the event $\{\frac{S_n}{n} - 0.5 \geq 0.01\}$ lies inside the larger event $\{\,|\frac{S_n}{n} - 0.5| \geq 0.01\}$. The limit comes from taking complements in (4.7) with $p = 0.5$ and $\varepsilon = 0.01$. ▲

Logically the central limit theorem comes after the law of large numbers because the central limit theorem describes the error in the law of large numbers. This is evident if we express the central limit theorem as the approximation

$$\frac{S_n}{n} = p + \frac{\sigma}{\sqrt{n}} \cdot \frac{S_n - np}{\sigma\sqrt{n}} \approx p + \frac{\sigma}{\sqrt{n}} Z \tag{4.8}$$

where $\sigma = \sqrt{p(1-p)}$ is the standard deviation of X_j and Z is a standard normal random variable. The approximation $\approx$ in (4.8) should be taken to mean that the probability distributions of the left-hand and right-hand sides are close for large n. The first equality in (4.8) decomposes the observed frequency $\frac{S_n}{n}$ into a sum of its mean p and a random error. For large n this random error is approximately normally distributed with standard deviation $\sigma/\sqrt{n}$.

In Chapter 9 we extend both the LLN and the CLT to more general random variables, and there we present the LLN first. Here we give a derivation of the law of large numbers from the central limit theorem.

Proof of Theorem 4.8. Fix the value of $\varepsilon > 0$. Let C be a large constant. Consider n large enough so that $\varepsilon\sqrt{n}/\sigma > C$. Then

$$\begin{aligned} P\left(\left|\frac{S_n}{n} - p\right| < \varepsilon\right) &= P(-n\varepsilon < S_n - np < n\varepsilon) \\ &= P\left(-\frac{\varepsilon\sqrt{n}}{\sigma} < \frac{S_n - np}{\sigma\sqrt{n}} < \frac{\varepsilon\sqrt{n}}{\sigma}\right) \\ &\geq P\left(-C < \frac{S_n - np}{\sigma\sqrt{n}} < C\right). \end{aligned}$$

By the CLT this last probability converges as $n \to \infty$:

$$P\left(-C < \frac{S_n - np}{\sigma\sqrt{n}} < C\right) \longrightarrow \int_{-C}^{C} \frac{1}{\sqrt{2\pi}} e^{-x^2/2}\,dx.$$

The integral on the right can be taken as close to 1 as we please, by increasing C. Thus $P\left(\left|\frac{S_n}{n} - p\right| < \varepsilon\right)$ will eventually be larger than any number smaller than 1, once n is large enough. As a probability it cannot be above 1, hence it has to converge to 1, and thereby we have proved statement (4.7). ▲

4.3. Applications of the normal approximation

The following applications form the basis of a number of statistical methods.

Confidence intervals

Suppose we have a biased coin and we do not know the true probability p that it lands on heads. How can we estimate p? The law of large numbers suggests a natural approach: flip the coin a large number n times, count the number S_n of heads, and take the observed frequency $\widehat{p} = \frac{S_n}{n}$ as the estimate for p. In practice we cannot flip the coin forever to get an accurate estimate. Can we estimate the error of the approximation for a finite n?

Let us see if the normal approximation can say anything about this problem. We estimate the probability that the error $|\widehat{p}-p|$ is bounded by some small margin of error ε. First rearrange the inequality inside the probability and standardize:

$$\begin{aligned} P(|\widehat{p} - p| < \varepsilon) &= P\left(\left|\tfrac{S_n}{n} - p\right| < \varepsilon\right) = P(-n\varepsilon < S_n - np < n\varepsilon) \\ &= P\left(-\frac{\varepsilon\sqrt{n}}{\sqrt{p(1-p)}} < \frac{S_n-np}{\sqrt{np(1-p)}} < \frac{\varepsilon\sqrt{n}}{\sqrt{p(1-p)}}\right). \end{aligned}$$

Up to this point we have used only algebra. Now comes the normal approximation:

$$\begin{aligned} P\left(-\frac{\varepsilon\sqrt{n}}{\sqrt{p(1-p)}} < \frac{S_n-np}{\sqrt{np(1-p)}} < \frac{\varepsilon\sqrt{n}}{\sqrt{p(1-p)}}\right) &\approx \Phi\left(\frac{\varepsilon\sqrt{n}}{\sqrt{p(1-p)}}\right) - \Phi\left(-\frac{\varepsilon\sqrt{n}}{\sqrt{p(1-p)}}\right) \\ &= 2\Phi\left(\frac{\varepsilon\sqrt{n}}{\sqrt{p(1-p)}}\right) - 1. \end{aligned}$$

We do not know the value of p (that is the whole point!) so how can we evaluate the last term? We can get a lower bound for it that works for all p. The maximum value of $p(1-p)$ is $1/4$ which is achieved at $p = 1/2$. (Check this with calculus.) Consequently $\frac{\varepsilon\sqrt{n}}{\sqrt{p(1-p)}} \ge 2\varepsilon\sqrt{n}$. Since the function Φ is increasing,

$$2\Phi\left(\frac{\varepsilon\sqrt{n}}{\sqrt{p(1-p)}}\right) - 1 \;\ge\; 2\Phi(2\varepsilon\sqrt{n}) - 1.$$

Combining the steps above gives the following lower bound on the probability, valid for all p, with the usual caveats about the validity of the normal approximation:

$$P(\,|\widehat{p} - p| < \varepsilon) \;\ge\; 2\Phi\left(2\varepsilon\sqrt{n}\right) - 1. \tag{4.9}$$

Inequality (4.9) can be used to answer several types of questions. First, we might want to know how many trials are needed to reach a prescribed level of confidence in the estimate.

Example 4.11. How many times should we flip a coin with unknown success probability p so that the estimate $\widehat{p} = S_n/n$ is within 0.05 of the true p, with probability at least 0.99?

Using (4.9), to ensure a lower bound of at least 0.99 we need

$$P(\,|\widehat{p} - p| < \varepsilon) \ge 2\Phi\left(2\varepsilon\sqrt{n}\right) - 1 \ge 0.99.$$

This last inequality is satisfied if

$$\Phi\left(2\varepsilon\sqrt{n}\right) \geq 0.995.$$

From the table for Φ the last inequality is equivalent to

$$2\varepsilon\sqrt{n} \geq 2.58 \quad \text{which is the same as} \quad n \geq \frac{2.58^2}{4\varepsilon^2} = \frac{2.58^2}{4 \cdot 0.05^2} \approx 665.64.$$

In the last step we used the value $\varepsilon = 0.05$ that was given in the problem statement.

The conclusion is that if we flip the coin 666 times, the estimate $\widehat{p}$ is within 0.05 of the real p, with probability at least 0.99. ▲

Another task is to find the confidence interval around $\widehat{p}$ that captures the true p, with a given (high) probability. The $100r\,\%$ *confidence interval* for the unknown success probability p is given by $(\widehat{p} - \varepsilon, \widehat{p} + \varepsilon)$ where ε is chosen to satisfy $P(|\widehat{p} - p| < \varepsilon) \geq r$. In other words, the random interval $(\widehat{p} - \varepsilon, \widehat{p} + \varepsilon)$ contains the true p with probability at least r.

Example 4.12. We repeat a trial 1000 times and observe 450 successes. Find the 95% confidence interval for the true success probability p.

This time n is given and we look for ε such that $P(|\widehat{p} - p| < \varepsilon) \geq 0.95$. From (4.9) we need to solve the inequality $2\Phi\left(2\varepsilon\sqrt{n}\right) - 1 \geq 0.95$ for ε. First simplify and then turn to the Φ table:

$$\Phi\left(2\varepsilon\sqrt{n}\right) \geq 0.975 \quad \Longleftrightarrow \quad 2\varepsilon\sqrt{n} \geq 1.96 \quad \Longleftrightarrow \quad \varepsilon \geq \frac{1.96}{2\sqrt{1000}} \approx 0.031.$$

Thus if $n = 1000$, then with probability at least 0.95 the random quantity $\widehat{p}$ satisfies $|\widehat{p} - p| < 0.031$. If our observed ratio is $\widehat{p} = \frac{450}{1000} = 0.45$, we say that the *95% confidence interval* for the true success probability p is $(0.45 - 0.031, 0.45 + 0.031) = (0.419, 0.481)$. ▲

Note carefully the terminology used in the example above. Once the experiment has been performed and 450 successes observed, $\widehat{p} = \frac{450}{1000}$ is no longer random. The true p is also not random since it is just a fixed parameter. Thus we can no longer say that "the true p lies in the interval $(\widehat{p} - 0.031, \widehat{p} + 0.031) = (0.419, 0.481)$ with probability 0.95." That is why we say instead that $(\widehat{p} - 0.031, \widehat{p} + 0.031) = (0.419, 0.481)$ is the *95% confidence interval* for the true p.

Remark 4.13. (Maximum likelihood estimator) The use of $\widehat{p} = S_n/n$ as the estimate of the unknown success probability p was justified above by the law of large numbers. Here is an alternative justification. Once the outcome $S_n = k$ has been observed, we can use the value of the probability mass function of S_n to compare how likely the outcome k is under different parameter values p. We call it the *likelihood function* $L(p) = P(S_n = k) = \binom{n}{k} p^k (1-p)^{n-k}$, which is a function of p with fixed k and n.

The value $\widehat{p}$ that maximizes $L(p)$ is the *maximum likelihood estimator* of p. This is the value of p that gives the outcome k the highest probability. The reader can do the calculus to check that $L(p)$ is maximized uniquely by the value $\widehat{p} = k/n$. Thus the maximum likelihood estimator of p is the same $\widehat{p} = S_n/n$ for which we derived confidence intervals above. ▲

Polling

Polling means estimating public opinion by interviewing a sample of people. This leads naturally to confidence intervals. We discuss below some mathematical issues that arise. Creating polls that are genuinely representative of the larger population is a very difficult practical problem which we do not address at all.

Example 4.14. Suppose that the fraction of a population who like broccoli is p. We wish to estimate p. In principle we could record the preferences of every individual, but this would be slow and expensive. Instead we take a random sample: we choose randomly n individuals, ask each of them whether they like broccoli or not, and estimate p with the ratio $\widehat{p}$ of those who said yes. We would like to quantify the accuracy of this estimate.

When we take a poll we are actually sampling *without replacement* because we do not ask the same individual twice. Recall that it is sampling *with replacement* that leads to independent trials and a binomially distributed number of successes. Consequently the number of people who said yes to broccoli is not exactly $\text{Bin}(n, p)$ but a hypergeometric distributed random variable (see Definition 2.42). So strictly speaking, the approach developed above for estimating p for independent trials is not valid now.

However, if the sample size n is small compared to the size of the population, even if we sampled with replacement the chances of asking the same person twice would be small. Consequently sampling with and without replacement are very close to each other. (We discuss this point in more detail in Remark 4.16 below.) With this justification we can pretend that the sample of the poll was taken with replacement and thereby results in a $\text{Bin}(n, p)$ random variable. Then we can use the techniques developed above for the binomial.

Continuing with the example, suppose we interviewed 100 people and 20 of them liked broccoli. Thus our estimate is $\widehat{p} = \frac{20}{100} = 0.20$. Let us find the 90% confidence interval for the true p.

With $n = 100$ and a desired confidence level of 0.90, we seek ε such that

$$P(|\widehat{p} - p| < \varepsilon) \geq 0.90.$$

By inequality (4.9) this can be achieved by making sure ε satisfies

$$2\Phi\left(2\varepsilon\sqrt{n}\right) - 1 \geq 0.90 \iff \Phi\left(2\varepsilon\sqrt{n}\right) \geq 0.95$$
$$\iff 2\varepsilon\sqrt{n} \geq 1.645 \iff 20\varepsilon \geq 1.645 \iff \varepsilon \geq 0.082.$$

The value 1.645 in the calculation above was chosen because, according to the table, $\Phi(1.64) = 0.9495$ and $\Phi(1.65) = 0.9505$. Thus the 90% confidence interval for p is $(0.20 - 0.082, 0.20 + 0.082) = (0.118, 0.282)$.

Suppose the broccoli growers who commissioned the poll come back and tell you that the error 0.082 is too large. How many more people do you need to interview to reduce the margin of error down to 0.05, still achieving the same 90% confidence level? This time we take $\varepsilon = 0.05$ and solve for n:

$$\begin{aligned} &2\Phi\left(2\varepsilon\sqrt{n}\right) - 1 \geq 0.90 \iff \Phi\left(2\varepsilon\sqrt{n}\right) \geq 0.95 \\ \iff\ &2\varepsilon\sqrt{n} \geq 1.645 \iff 2 \cdot 0.05\sqrt{n} \geq 1.645 \iff n \geq 16.45^2 \approx 270.6. \end{aligned}$$

Thus to reach margin of error 0.05 with confidence level 90% requires 271 trials. In other words, after interviewing 100 people, another 171 interviews are needed. ▲

Remark 4.15. (Confidence levels in political polls) During election seasons we are bombarded with news of the following kind: "The latest poll shows that 44% of voters favor candidate Honestman, with a margin of error of 3 percentage points." This report gives the confidence interval of the unknown fraction p that favor Honestman, namely $(0.44-0.03, 0.44+0.03) = (0.41, 0.47)$. The level of confidence used to produce the estimate is usually omitted from news reports. This is no doubt partly due to a desire to avoid confusing technicalities. It is also a fairly common convention to set the confidence level at 95%, so it does not need to be stated explicitly. ▲

The remark below explains why the normal approximation is justified in sampling without replacement from a large population, as for example in the polling application of Example 4.14.

Remark 4.16. (Binomial limit of the hypergeometric) Let a set of N items consist of N_A type A items and $N_B = N - N_A$ type B items. Take a sample of size n *without replacement* and let X denote the number of type A items in the sample. Recalling the notation $(a)_b = a \cdot (a-1) \cdots (a-b+1)$, rewrite the hypergeometric probability mass function (2.24) of X as

$$\begin{aligned} P(X = k) &= \frac{\binom{N_A}{k}\binom{N-N_A}{n-k}}{\binom{N}{n}} = \frac{\frac{(N_A)_k}{k!} \cdot \frac{(N-N_A)_{n-k}}{(n-k)!}}{\frac{(N)_n}{n!}} \\ &= \binom{n}{k} \frac{(N_A)_k\,(N-N_A)_{n-k}}{(N)_n}. \end{aligned} \tag{4.10}$$

Next fix n and k and let N and N_A go to infinity in such a way that $\frac{N_A}{N}$ (the ratio of type A items) converges to p:

$$\frac{(N_A)_k (N-N_A)_{n-k}}{(N)_n} = \frac{N_A(N_A-1)\cdots(N_A-k+1)}{N(N-1)\cdots(N-k+1)}$$
$$\times \frac{(N-N_A)(N-N_A-1)\cdots(N-N_A-n+k+1)}{(N-k)(N-k-1)\cdots(N-n+1)}$$
$$= \left(\frac{N_A}{N}\right)^k \cdot \prod_{i=1}^{k} \frac{(1-\frac{i-1}{N_A})}{(1-\frac{i-1}{N})} \cdot \left(1-\frac{N_A}{N}\right)^{n-k} \cdot \prod_{i=k+1}^{n} \frac{(1-\frac{i-k-1}{N-N_A})}{(1-\frac{i-1}{N})}$$
$$\longrightarrow p^k(1-p)^{n-k}.$$

Thus $P(X = k)$ converges to $\binom{n}{k}p^k(1-p)^{n-k}$. This is another limit theorem: the probability mass function of a Hypergeom(N, N_A, n) random variable converges to the probability mass function of a Bin(n, p) random variable as $N \to \infty$ and $N_A/N \to p$.

The hypergeometric-to-binomial limit explains why we can use the normal approximation for sampling without replacement for a reasonably large population (as in Example 4.14). In that case the hypergeometric distribution is close to the binomial distribution, which we can approximate with the normal. ▲

Random walk ♦

Imagine a person who repeatedly flips a coin that gives heads with probability $0 < p < 1$. Every time she flips heads, she takes a step to the right. Every time she flips tails, she takes a step to the left. This simple model lies at the heart of probability theory.

To turn the informal description into mathematics, let $X_1, X_2, X_3, \dots$ be independent random variables such that each X_j satisfies $P(X_j = 1) = p$ and $P(X_j = -1) = 1 - p$. Define the sums $S_0 = 0$ and $S_n = X_1 + \cdots + X_n$ for $n \geq 1$. The random variable X_j is the jth step of the walker and S_n her position after n time steps. The random sequence $S_0, S_1, S_2, \dots$ is the *simple random walk.*

If $p = \frac{1}{2}$ then the walk is equally likely to go right and left and S_n is called a *symmetric* simple random walk. (See Figure 4.4 for one realization.) If $p \neq \frac{1}{2}$ it is an *asymmetric* simple random walk.

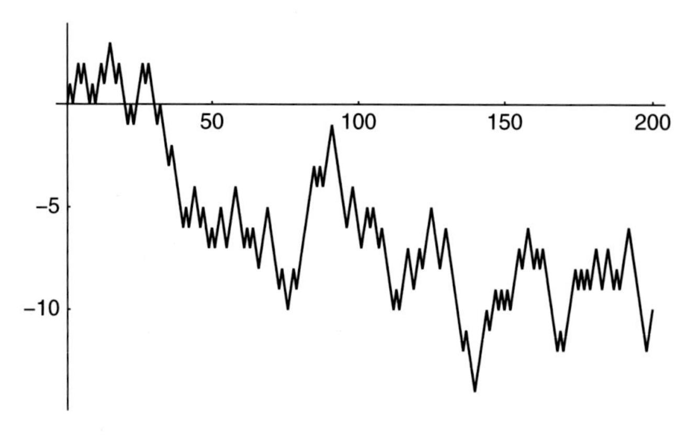

Figure 4.4. Simulation of a symmetric random walk S_n. The horizontal axis is indexed by n.

Except for replacing the values $\{0, 1\}$ of the trials with $\{\pm 1\}$, random walk is just like the independent trials of Section 2.4. Consequently the probability distribution of S_n can be readily derived (Exercise 4.29).

The central limit theorem tell us what the distribution of S_n looks like after a long time. Let T_n count the number of times the coin came up heads in the first n flips. Then $S_n = T_n - (n - T_n) = 2T_n - n$. Since $T_n \sim \text{Bin}(n, p)$, we get the mean $E(S_n) = 2E(T_n) - n = n(2p-1)$ and the variance $\text{Var}(S_n) = 4\,\text{Var}(T_n) = 4np(1-p)$. Now the CLT gives us the limit

$$P\left(\frac{S_n - E[S_n]}{\sqrt{\text{Var}[S_n]}} \le b\right) = P\left(\frac{T_n - np}{\sqrt{np(1-p)}} \le b\right) \to \frac{1}{\sqrt{2\pi}} \int_{-\infty}^{b} e^{-x^2/2}\, dx.$$

In qualitative terms, after a long time the random walk is roughly normally distributed, and its fluctuations around its mean have order of magnitude $\sqrt{n}$.

We can also approximate the probability mass function of S_n. For example, consider the probability $P(S_{2n} = 0)$ that after $2n$ steps the walker is back to her original starting position. Since $S_{2n} = 0$ happens if the walker stepped to the right exactly n times, we have

$$P(S_{2n} = 0) = \binom{2n}{n} p^n (1-p)^{2n-n}.$$

Assuming $p = 1/2$, and applying Stirling's approximation (Fact 4.7),

$$P(S_{2n} = 0) = \frac{(2n)!}{n!n!}\left(\frac{1}{2}\right)^{2n} \sim \frac{1}{\sqrt{\pi n}},$$

where the final approximation holds for large n. Note the consistency of the large and small scale approximations: by the CLT most of the probability mass of S_n is distributed over order $\sqrt{n}$ locations, so we would expect the probabilities of individual points within this range to decay at rate $1/\sqrt{n}$.

Random walk is an example of a *stochastic process* which is the general term for a random evolution in time. This book introduces two important stochastic processes: random walk and the Poisson process (Section 4.6). Further study of stochastic processes is a natural sequel to an introductory probability course.

4.4. Poisson approximation

This section introduces the Poisson distribution and shows how it arises from repeated trials. The limit in Theorem 4.19 below explains the phenomenon in Figure 4.2 on page 141. We begin with the definition.

Definition 4.17. Let $\lambda > 0$. A random variable X has the *Poisson distribution with parameter* λ if X is nonnegative integer valued and has the probability mass function

$$P(X = k) = e^{-\lambda}\frac{\lambda^k}{k!} \qquad \text{for } k \in \{0, 1, 2, \dots\}. \tag{4.11}$$

Abbreviate this by $X \sim \text{Poisson}(\lambda)$.

The fact that formula (4.11) defines a probability mass function follows from the Taylor series of the exponential function:

$$e^{\lambda} = \sum_{k=0}^{\infty} \frac{\lambda^k}{k!}, \qquad \text{which yields} \qquad 1 = \sum_{k=0}^{\infty} e^{-\lambda}\frac{\lambda^k}{k!}.$$

Let us compute the mean and variance of a Poisson(λ) random variable.

Fact 4.18. Let $X \sim \text{Poisson}(\lambda)$. Then $E[X] = \lambda$ and $\text{Var}(X) = \lambda$.

Proof. We have

$$E[X] = \sum_{k=0}^{\infty} k\, e^{-\lambda}\frac{\lambda^k}{k!} = \sum_{k=1}^{\infty} e^{-\lambda}\frac{\lambda^k}{(k-1)!} = \lambda \sum_{k=1}^{\infty} e^{-\lambda}\frac{\lambda^{k-1}}{(k-1)!} = \lambda \sum_{j=0}^{\infty} e^{-\lambda}\frac{\lambda^j}{j!} = \lambda.$$

In the second equality we dropped the $k = 0$ term and canceled a factor of k. In the next to last equality we changed the summation index to $j = k - 1$. Finally, we summed up the Poisson(λ) probability mass function to 1.

To calculate the variance, we require $E[X^2]$. Due to the factorial in the denominator, it is convenient instead to compute $E[X(X-1)]$ and then use the equality $E[X^2] = E[X(X-1)] + E[X]$, which is justified by (3.38). Note that now both the terms $k = 0$ and $k = 1$ vanish.

$$E[X(X-1)] = \sum_{k=0}^{\infty} k(k-1)\, e^{-\lambda}\frac{\lambda^k}{k!} = \sum_{k=2}^{\infty} e^{-\lambda}\frac{\lambda^k}{(k-2)!} = \lambda^2 \sum_{j=0}^{\infty} e^{-\lambda}\frac{\lambda^j}{j!} = \lambda^2.$$

Hence,

$$\begin{aligned} \text{Var}(X) &= E[X^2] - (E[X])^2 = E[X(X-1)] + E[X] - (E[X])^2 \\ &= \lambda^2 + \lambda - \lambda^2 = \lambda. \end{aligned}$$

▲

Having acquainted ourselves with the Poisson distribution, we turn to the Poisson limit of the binomial. This limit is sometimes called the *law of rare events*. Informally speaking, this principle says that if successes are very rare in a sequence of independent trials, then the number of successes is well approximated by a Poisson random variable.

Theorem 4.19. *Let $\lambda > 0$ and consider positive integers n for which $\lambda/n < 1$. Let $S_n \sim$ Bin$(n, \lambda/n)$. Then*

$$\lim_{n\to\infty} P(S_n = k) = e^{-\lambda}\frac{\lambda^k}{k!} \quad \text{for all} \quad k \in \{0, 1, 2, \ldots\}. \tag{4.12}$$

In plain English, Theorem 4.19 says that if S_n counts the number of successes in n independent trials and the mean $E[S_n] = \lambda$ does not change with n, then as $n \to \infty$ the distribution of S_n approaches the Poisson(λ) distribution. This is our second important example of a *limit in distribution*. The central limit theorem (Theorem 4.1) was the first.

Theorem 4.19 explains the similarity of the three graphs in Figure 4.2. In each of the three cases $np = 2$ and hence all three binomial probability mass functions are close to the Poisson(2) probability mass function.

The Poisson limit of the binomial is elementary enough that we can give a complete proof.

Proof. We rearrange the binomial probability cleverly and then simply observe where the different pieces converge as $n \to \infty$:

$$\begin{aligned}
P(S_n = k) &= \binom{n}{k}\left(\frac{\lambda}{n}\right)^k\left(1-\frac{\lambda}{n}\right)^{n-k} \\
&= \frac{n(n-1)\cdots(n-k+1)}{k!}\,\frac{\lambda^k}{n^k}\left(1-\frac{\lambda}{n}\right)^n\frac{1}{\left(1-\frac{\lambda}{n}\right)^k} \\
&= \frac{\lambda^k}{k!}\left(1-\frac{\lambda}{n}\right)^n\left[1\cdot\left(1-\frac{1}{n}\right)\cdot\left(1-\frac{2}{n}\right)\cdots\left(1-\frac{k-1}{n}\right)\right]\frac{1}{\left(1-\frac{\lambda}{n}\right)^k} \\
&\underset{n\to\infty}{\longrightarrow} \frac{\lambda^k}{k!}\cdot e^{-\lambda}\cdot 1\cdot 1 = e^{-\lambda}\frac{\lambda^k}{k!}.
\end{aligned}$$

The limit is straightforward since k is kept fixed as we take $n \to \infty$. But you need to remember the limit $(1 + x/n)^n \to e^x$ from (A.1) in Appendix A. ▲

Poisson approximation of counts of rare events

Theorem 4.19 states that the distribution Bin$(n, \lambda/n)$ gets closer and closer to Poisson(λ) as $n \to \infty$. But what if we want a statement for a fixed n? We would want to quantify the error in the approximation of the binomial with the Poisson. The following theorem does exactly that.

Theorem 4.20. *Let $X \sim$ Bin(n, p) and $Y \sim$ Poisson(np). Then for any subset $A \subseteq \{0, 1, 2, \ldots\}$, we have*

$$|\,P(X \in A) - P(Y \in A)\,| \le np^2. \tag{4.13}$$

Proof of this theorem can be found, for example, in [Dur10].

Theorem 4.20 states that if we replace a $\text{Bin}(n, p)$ random variable with a $\text{Poisson}(np)$ random variable in any event, then the probability of the event changes by at most np^2. Thus if np^2 is small then the binomial random variable can be accurately approximated with a Poisson random variable with the *same* expected value.

Theorem 4.20 suggests Poisson as a good model whenever one counts rare occurrences in a large number of trials that can be assumed approximately independent. For example, the number of customers arriving in a store during a fixed time period can be modeled by a Poisson distribution if it is realistic to imagine that there is a large number of potential customers each of whom chooses to enter the store with a small probability and independently of each other. On similar grounds the Poisson distribution is used to model the number of phone calls to a call center, the number of accidents that occur in some particular circumstance, or the number of typographical errorrs on a printed page.

The framework for when the Poisson approximation can be useful is described in the following statement.

Poisson approximation for counting rare events. Assume that the random variable X counts the occurrences of rare events that are not strongly dependent on each other. Then the distribution of X can be approximated with a $\text{Poisson}(\lambda)$ distribution for $\lambda = E[X]$. That is,

$$P(X = k) \text{ is close to } e^{-\lambda}\frac{\lambda^k}{k!} \quad \text{for} \quad k \in \{0, 1, 2, \dots\}. \tag{4.14}$$

Note that this is not a mathematically rigorous statement: we did not define what rare events or strong dependence mean. However, it can still be used to model many real-world phenomena, as the next two examples show.

Example 4.21. Suppose a factory experiences on average 3 accidents each month. What is the probability that during a particular month there are exactly 2 accidents?

We cannot answer the question without making an assumption about the probability distribution of the number X of accidents during a month. If the factory has a large number of workers each of whom has a small chance of having an accident independently of the others, then by approximation (4.14) the Poisson model may be appropriate. Since we know the mean to be 3 the correct choice is the Poisson with parameter 3. By assuming $X \sim \text{Poisson}(3)$ we can calculate $P(X = 2) = e^{-3}3^2/2! \approx 0.224$. ▲

Example 4.22. The proofreader of an undergraduate probability textbook noticed that a randomly chosen page of the book has no typos with probability 0.9.

Estimate the probability that a randomly chosen page contains exactly two typos.

Denote the number of typos on a randomly chosen page by X. There are a number of words and formulas on a given page, and each one contains a typo with a small probability. Thus it is reasonable to believe that the distribution of X is close to a Poisson(λ) distribution. According to approximation (4.14) the parameter λ should be chosen as $E[X]$, but this was not given in the problem. However, we know that $P(X = 0) \approx 0.9$. Since $P(X = 0)$ can be approximated by $e^{-\lambda}$ (the probability that a Poisson(λ) is equal to 0), we get $0.9 \approx e^{-\lambda}$ and $\lambda \approx -\ln 0.9 \approx 0.105$. We can now estimate $P(X = 2)$ by the Poisson approximation:

$$P(X = 2) \approx e^{-\lambda}\frac{\lambda^2}{2} \approx e^{-0.105}\frac{(0.105)^2}{2} \approx 0.005. \qquad \blacktriangle$$

The next example shows that the Poisson limit for rare events can take place even though trials are not exactly independent, as long as some approximate independence holds.

Example 4.23. Consider again Example 1.27 of mixed-up hats with n guests. Let X denote the number of guests who receive the correct hat. How can we approximate the distribution of X?

The probability that a given guest receives the correct hat is $\frac{1}{n}$, which is small if n is large. We have seen in Example 1.27 that the occurrences of successes (that a guest gets his/her own hat) are not independent, because we are sampling without replacement. However, dependence between two (or any finite number of) successes will get weaker and weaker as n increases. Hence it is plausible to approximate X with a Poisson(λ) random variable Y. Example 1.27 showed that $P(X = 0)$ converges to e^{-1} which means that $\lambda = 1$ is the appropriate choice.

This means that the distribution of the number of people who receive the correct hat should converge to a Poisson(1) distribution as $n \to \infty$. In other words,

$$\lim_{n\to\infty} P(X = \ell) = \frac{e^{-1}}{\ell!}, \qquad \text{for any } \ell \in \{0, 1, 2, \dots\}. \tag{4.15}$$

Exercise 1.58 provides an outline for the rigorous proof of (4.15). $\blacktriangle$

Comparison of the normal and Poisson approximations of the binomial

In particular cases the reader may wonder which approximation to use for a binomial: the normal approximation or the Poisson approximation. Approximation (4.3) and Theorem 4.20 give some guidance: when $np(1-p) > 10$ the normal approximation should be pretty safe as long as a and b are not too close together, while if np^2 is small then the Poisson approximation will work well. The next two examples compare the two approximations. You can see that the normal and Poisson approximations are quite different and it is usually fairly evident which one to apply.

Example 4.24. Let $X \sim \text{Bin}(10, \frac{1}{10})$. Compare the Poisson and normal approximations of the probability $P(X \leq 1)$. The exact value is

$$P(X \leq 1) = P(X = 0) + P(X = 1) = (\tfrac{9}{10})^{10} + 10 \cdot \tfrac{1}{10} \cdot (\tfrac{9}{10})^9 \approx 0.7361.$$

Now $E[X] = np = 10 \cdot \frac{1}{10} = 1$ so the correct Poisson approximation to use is $Y \sim \text{Poisson}(1)$. We have $np^2 = 10 \cdot (\frac{1}{10})^2 = 0.1$ so by Theorem 4.20 the Poisson approximation will be within 0.1 of the true probability. In fact, it is even better than that:

$$P(Y \leq 1) = P(Y = 0) + P(Y = 1) = e^{-1} + e^{-1} \approx 0.7358,$$

which is within 0.001 of the exact value.

Next we see how the normal approximation performs. Since the number of trials is small, we employ the continuity correction (see page 145) to enhance the accuracy of the normal approximation. With $E[X] = 1$ and $\text{Var}(X) = 10 \cdot \frac{1}{10} \cdot \frac{9}{10} = \frac{9}{10}$ normal approximation gives

$$\begin{aligned} P(X \leq 1) &= P\left(X \leq \tfrac{3}{2}\right) = P\left(\frac{X - 1}{\sqrt{9/10}} \leq \frac{\frac{3}{2} - 1}{\sqrt{9/10}}\right) \\ &\approx P\left(\frac{X - 1}{\sqrt{9/10}} \leq 0.53\right) \approx \Phi(0.53) \approx 0.7019. \end{aligned}$$

The answer is clearly not as accurate as the Poisson approximation. Since $np(1 - p) = 9/10$ is far below 10, the suboptimal performance of the normal approximation is not a surprise. Note that without the continuity correction the normal approximation gives 0.5 (do the calculation as practice) which is an awful estimate of the exact probability! ▲

The next example reverses the roles: the Poisson approximation performs poorly while the normal approximation does well.

Example 4.25. Estimate the probability that 40 flips of a fair coin give exactly 20 tails.

Let S be the number of tails among 40 coin flips. Then $S \sim \text{Bin}(40, \frac{1}{2})$ with mean $\mu = 20$ and standard deviation $\sigma = \sqrt{40 \cdot (1/2) \cdot (1/2)} = \sqrt{10}$. A calculation with a computer gives the exact probability as

$$P(S = 20) = \binom{40}{20} 2^{-40} \approx 0.1254.$$

The normal approximation with the continuity correction gives

$$\begin{aligned} P(S = 20) &= P(19.5 \leq S \leq 20.5) = P\left(\frac{19.5 - 20}{\sqrt{10}} \leq \frac{S - 20}{\sqrt{10}} \leq \frac{20.5 - 20}{\sqrt{10}}\right) \\ &\approx P\left(-0.16 \leq \frac{S - 20}{\sqrt{10}} \leq 0.16\right) \approx \Phi(0.16) - \Phi(-0.16) \\ &= 2\Phi(0.16) - 1 \approx 0.1272. \end{aligned}$$

The fit is pretty good.

For the Poisson approximation we compare S to $Y \sim \text{Poisson}(20)$. A calculation using a computer gives

$$P(Y = 20) = \frac{e^{-20} 20^{20}}{20!} \approx 0.089.$$

The Poisson is badly off the mark. The reason is simply that we are not in the regime of the law of rare events, because successes with probability 1/2 are not rare! We have $np^2 = 40 \cdot \frac{1}{4} = 10$ and so the rigorous error bound in Theorem 4.20 is larger than one. This bad error bound already suggests that Poisson is not the right approximation here. ▲

4.5. Exponential distribution

The geometric distribution is a discrete probability distribution that models waiting times, for example, the first time in a sequence of coin flips that tails appears. This section develops a continuous counterpart, for modeling waiting times such as the first arrival of a customer at a post office. We define this distribution below. A derivation from repeated trials is given at the end of the section.

Definition 4.26. Let $0 < \lambda < \infty$. A random variable X has the **exponential distribution** with parameter λ if X has density function

$$f(x) = \begin{cases} \lambda e^{-\lambda x}, & x \geq 0 \\ 0, & x < 0 \end{cases} \tag{4.16}$$

on the real line. Abbreviate this by $X \sim \text{Exp}(\lambda)$. The $\text{Exp}(\lambda)$ distribution is also called the *exponential distribution with rate* λ.

By integrating the density function we find the cumulative distribution function of the $\text{Exp}(\lambda)$ distribution:

$$F(t) = \int_{-\infty}^{t} f(x)dx = \int_{0}^{t} \lambda e^{-\lambda x}\, dx = 1 - e^{-\lambda t} \qquad \text{for } t \geq 0 \tag{4.17}$$

while $F(t) = 0$ for $t < 0$. By letting $t \to \infty$ in (4.17) we see that $\int_{-\infty}^{\infty} f(x)dx = 1$ so f is indeed a probability density function. It is often useful to keep in mind that an $\text{Exp}(\lambda)$ random variable X can also be characterized by its tail probability: for $t \geq 0$, $P(X > t) = 1 - P(X \leq t) = e^{-\lambda t}$.

To find the mean and variance of $X \sim \text{Exp}(\lambda)$, evaluate the first two moments by integration by parts:

$$E(X) = \int_{0}^{\infty} x \lambda e^{-\lambda x} dx = \frac{1}{\lambda} \quad \text{and} \quad E(X^2) = \int_{0}^{\infty} x^2 \lambda e^{-\lambda x} dx = \frac{2}{\lambda^2}. \tag{4.18}$$

Then

$$\text{Var}(X) = E(X^2) - (E[X])^2 = \frac{1}{\lambda^2}.$$

Example 4.27. Let $X \sim \text{Exp}(1/2)$. Find $P(X > \frac{7}{2})$. Also, find the median of X, that is, the value m for which $P(X \leq m) = \frac{1}{2}$.

A direct calculation shows that $P(X > \frac{7}{2}) = \int_{7/2}^{\infty} \frac{1}{2} e^{-x/2} dx = e^{-7/4}$. For $m > 0$ we have $P(X \leq m) = \int_0^m \frac{1}{2} e^{-x/2} dx = 1 - e^{-m/2}$. Thus if m is the median then it solves the equality $1 - e^{-m/2} = \frac{1}{2}$, which yields $m = 2 \ln 2$. ▲

Example 4.28. Let $X \sim \text{Exp}(\lambda)$ and $Z = X/2$. Find $P(Z > t)$ for $t > 0$.

We have

$$P(Z > t) = P(X/2 > t) = P(X > 2t) = e^{-\lambda \cdot 2t} = e^{-2\lambda t}.$$

This is exactly the tail probability of an $\text{Exp}(2\lambda)$ distribution, and so we can conclude that $X/2 \sim \text{Exp}(2\lambda)$.

Example 4.29. Suppose the length of a phone call, in minutes, is well modeled by an exponential random variable with mean 10 minutes. What is the probability that a call takes more than 8 minutes? Between 8 and 22 minutes?

Let X be the length of the call. Then $X \sim \text{Exp}(1/10)$. We have

$$P(X > 8) = e^{-8/10} \approx 0.4493 \quad \text{and}$$
$$P(8 < X < 22) = P(X > 8) - P(X \geq 22) = e^{-8/10} - e^{-22/10} \approx 0.3385.$$ ▲

Example 4.30. Consider a particular protein found in human cells. Let T denote the time, in seconds, required until one of these proteins is degraded (destroyed). A common modeling assumption in biomolecular chemistry is that $T \sim \text{Exp}(\lambda n)$ for some $\lambda > 0$, where n is the number of copies of the protein in the cell. Suppose that $\lambda = 2$. How many proteins have to be present so that $P(T > 10^{-2}) < \frac{1}{10}$?

We wish to find n so that

$$\tfrac{1}{10} > P(T > 10^{-2}) = \int_{10^{-2}}^{\infty} 2ne^{-2nx} dx = e^{-2n/100}.$$

Solving yields $n > \frac{100}{2} \ln(10) \approx 115.1$. Thus, at least $n = 116$ proteins have to be present. ▲

We state and prove next the so-called *memoryless property* of the exponential distribution. We explored this earlier for the geometric distribution in Exercises 2.67 and 2.68.

Fact 4.31. (Memoryless property of the exponential distribution) Suppose that $X \sim \text{Exp}(\lambda)$. Then for any $s, t > 0$ we have

$$P(X > t + s \mid X > t) = P(X > s). \tag{4.19}$$

To make the meaning of the memoryless property concrete, imagine that the lifetime of some machine can be modeled by an exponential random variable. Then (4.19) says the following: given that the machine has functioned for t time units, the conditional probability that it works for another s time units is the same as the unconditional probability that the original machine functions for s time units. In other words, regardless of how long the machine has been operational, the distribution of the remaining lifetime is the same as the distribution of the original (unconditioned) lifetime. As long as the machine functions, it behaves as if it were brand new.

Proof of Fact 4.31. From the definition of conditional probability we have

$$\begin{aligned} P(X > t + s \mid X > t) &= \frac{P(X > t+s,\, X > t)}{P(X > t)} = \frac{P(X > t+s)}{P(X > t)} \\ &= \frac{e^{-\lambda(t+s)}}{e^{-\lambda t}} = e^{-\lambda s} = P(X > s). \end{aligned}$$

We used the fact that the intersection of $\{X > t + s\}$ and $\{X > t\}$ is $\{X > t + s\}$ and also that $P(X > t) = e^{-\lambda t}$. ▲

Example 4.32. Animals of the forest need to cross a remote highway. From experience they know that, from the moment someone arrives at the roadside, the time till the next car is an exponential random variable with expected value 30 minutes. The turtle needs 10 minutes to cross the road. (a) What is the probability that the turtle can cross the road safely? (b) Now suppose that when the turtle arrives at the roadside, the fox tells her that he has been there already 5 minutes without seeing a car go by. What is the probability now that the turtle can cross safely?

Let X denote the arrival of the next car in minutes, measured from the moment when the turtle arrived on the side of the road. Then $X \sim \text{Exp}(\lambda)$, and since $\frac{1}{\lambda} = E[X] = 30$, the parameter is $\lambda = \frac{1}{30}$. For part (a) we need $P(X > 10) = e^{-10\lambda} = e^{-\frac{1}{3}} \approx 0.7165$. For part (b) we let X be the arrival of the next car in minutes, measured from the moment when the fox arrived. We condition on the information given by the fox, so the desired probability is $P(X > 5 + 10 \mid X > 5)$. By the memoryless property (4.19), $P(X > 15 \mid X > 5) = P(X > 10)$. Thus the information that the turtle received does not change the probability of her successful crossing. ▲

Other than the exponential, there is no distribution with a continuous density function on $[0, \infty)$ that satisfies the memoryless property. See Section 4.7 for a proof of this fact.

As the last item of this section, we show that the exponential distribution also arises from independent trials under a suitable limit. More precisely, the exponential distribution is the limit of a scaled geometric distribution.

Derivation of the exponential distribution ♦

We model the time when the first customer arrives at a post office in a discrete time scale. Suppose that there exists a constant $\lambda > 0$ such that the probability that at least one customer arrives during a time interval of length $1/n$ minutes is approximately λ/n, for very large n.

For each large n we record a discretized random arrival time T_n as follows: for $k = 1, 2, 3, \ldots$, if the first customer arrives during time interval $[(k-1)/n, k/n)$, we set $T_n = k/n$. Assuming that arrivals in disjoint time intervals are independent, we can find the probability mass function

$$P\left(T_n = \tfrac{k}{n}\right) = \left(1 - \tfrac{\lambda}{n}\right)^{k-1} \tfrac{\lambda}{n} \qquad \text{for } k \geq 1. \tag{4.20}$$

This is a familiar p.m.f. and the statement above says that $nT_n \sim \mathrm{Geom}(\lambda/n)$.

We take the $n \to \infty$ limit of the distribution of the random arrival time T_n in the next theorem. It is convenient to take the limit of tail probabilities.

Theorem 4.33. *Fix $\lambda > 0$. Consider n large enough so that $\lambda/n < 1$. Suppose that for each large enough n, the random variable T_n satisfies $nT_n \sim \mathrm{Geom}(\lambda/n)$. Then*

$$\lim_{n\to\infty} P(T_n > t) = e^{-\lambda t} \qquad \textit{for all nonnegative real } t. \tag{4.21}$$

Proof. We use the floor notation: for any real x,

$$\lfloor x \rfloor = \max\{m \in \mathbb{Z} : m \leq x\} \tag{4.22}$$

is the largest integer less than or equal to x. Note that $\lfloor \cdot \rfloor$ is the same as rounding down. For example, $\lfloor \pi \rfloor = 3$ and $\lfloor 100 \rfloor = 100$. Note also that an integer k satisfies $k > x$ if and only if $k \geq \lfloor x \rfloor + 1$. (For example, $4 > 3.14$ and $4 \geq 3 + 1$.) We compute the tail probability of T_n for $t \geq 0$:

$$\begin{aligned} P(T_n > t) &= P(nT_n > nt) = \sum_{k:k>nt} \left(1 - \tfrac{\lambda}{n}\right)^{k-1} \tfrac{\lambda}{n} = \sum_{k=\lfloor nt \rfloor+1}^{\infty} \left(1 - \tfrac{\lambda}{n}\right)^{k-1} \tfrac{\lambda}{n} \\ &= \left(1 - \tfrac{\lambda}{n}\right)^{\lfloor nt \rfloor} = \left(1 - \tfrac{\lambda}{n}\right)^{nt} \left(1 - \tfrac{\lambda}{n}\right)^{\lfloor nt \rfloor - nt} = \left(1 - \tfrac{\lambda t}{nt}\right)^{nt} \frac{1}{\left(1 - \tfrac{\lambda}{n}\right)^{nt - \lfloor nt \rfloor}}. \end{aligned} \tag{4.23}$$

Since $0 \leq nt - \lfloor nt \rfloor \leq 1$ for all n and t, we see that $\left(1 - \tfrac{\lambda}{n}\right)^{nt - \lfloor nt \rfloor} \to 1$ as $n \to \infty$. The expression $\left(1 - \tfrac{\lambda t}{nt}\right)^{nt}$, and hence the desired probability $P(T_n > t)$, converges to $e^{-\lambda t}$ by the limit (A.1). ▲

Let us rewrite limit (4.21) as a limit of cumulative distribution functions. Let F_{T_n} be the c.d.f. of the random variable T_n. Recall that the c.d.f. of the

Exp(λ) distribution is $F(t) = 1 - e^{-\lambda t}$ for $t \geq 0$. By taking complements, (4.21) gives

$$\lim_{n\to\infty} F_{T_n}(t) = \lim_{n\to\infty} (1 - P(T_n > t)) = 1 - e^{-\lambda t} = F(t)$$

for all $t \geq 0$. This is another example of a limit in distribution. The precise statement for what we have proved here is that the random variable T_n converges in distribution to an Exp(λ) random variable.

4.6. Poisson process ♦

In Section 4.4 we introduced the Poisson distribution and showed how it is used to count events (such as earthquakes, typos and accidents). In this section we introduce the *Poisson process* to model occurrences of events as they happen in time. At the end of the section a new named distribution, the gamma distribution, arises.

Suppose we want to model the times of shark attacks on a particular beach. Figure 4.5 below shows a timeline and the x mark the times at which shark attacks occurred. The particular realization shown in the figure has two shark attacks in the time interval $[0, s]$ and five in the time interval $[s, t]$.

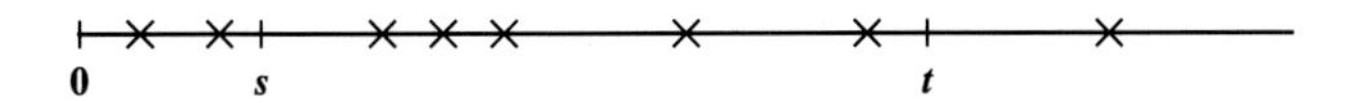

Figure 4.5. Timeline of shark attacks.

In a given time interval the number of shark attacks can be modeled with a Poisson random variable. If we assume that the average rate of attacks is constant in time, the mean of the Poisson random variable should be proportional to the length of the time interval. A useful simplifying assumption is that the numbers of occurrences in nonoverlapping time intervals are independent. (Two intervals are nonoverlapping if they share at most an endpoint.) The independence assumption would be precisely true if the Poisson random variables were approximate counts coming from underlying independent Bernoulli random variables.

The listed assumptions are enough to define the Poisson process. Below we write $|I|$ for the length of an interval I. For example, $|I| = b - a$ for $I = [a, b]$.

Definition 4.34. The Poisson process with intensity, or rate, $\lambda > 0$ is a collection of random points on the positive half-line $[0, \infty)$ with the following properties.

- The points are distinct (that is, there cannot be more than one point at any given position on $[0, \infty)$).

- The number of points in a bounded interval $I \subset [0, \infty)$, which is denoted by $N(I)$, has Poisson distribution with parameter $\lambda \cdot |I|$. For example, if $I = [a, b]$ then $N(I) \sim \text{Poisson}(\lambda(b - a))$.
- If $I_1, I_2, \dots, I_n$ are nonoverlapping intervals in $[0, \infty)$ then the random variables $N(I_1), N(I_2), \dots, N(I_n)$ are mutually independent.

The object introduced in Definition 4.34 can also be called the *Poisson point process*. Then the name Poisson process is reserved for the random function of time, $N_t = N([0, t])$, that counts the number of occurrences by time t. For simplicity we ignore this distinction.

These three simple conditions allow us to compute a number of interesting probabilities related to Poisson processes.

Example 4.35. Suppose that customers arrive in a certain store according to a Poisson process with intensity 5/hr. Suppose that the store is open between 9 AM and 6 PM.

(a) Find the probability that no customer comes to the store within one hour of its opening.
(b) Find the probability that we have two customers between 9 AM and 10 AM, three customers between 10 AM and 10:30 AM and five customers between 2 PM and 3:30 PM.
(c) Find the conditional probability that three customers arrive between 10 AM and 10:30 AM, given that 12 customers arrive between 10 AM and 12 PM.

For part (a) note that the number of customers between 9 AM and 10 AM (denoted by $N([9, 10])$) is Poisson with parameter 5. Thus

$$P(\text{no customers between 9 AM and 10 AM}) = P(N([9, 10]) = 0) = e^{-5} \approx 0.00674.$$

For part (b) we look at the number of customers in the intervals $[9, 10]$, $[10, 10.5]$ and $[14, 15.5]$. These are nonoverlapping intervals, so $N([9, 10]), N([10, 10.5])$ and $N([14, 15.5])$ are independent Poisson random variables with parameters $5, 5 \cdot \frac{1}{2}$, and $5 \cdot \frac{3}{2}$. This gives us

$$\begin{aligned} P(N([9, 10]) &= 2, N([10, 10.5]) = 3, \text{ and } N([14, 15.5]) = 5) \\ &= P(N([9, 10]) = 2)P(N([10, 10.5]) = 3)P(N([14, 15.5]) = 5) \\ &= \frac{5^2}{2!}e^{-5} \cdot \frac{(5/2)^3}{3!}e^{-5/2} \cdot \frac{(15/2)^5}{5!}e^{-15/2} \\ &= \frac{5^{10}3^5}{2!3!5!2^8}e^{-15} \approx 0.00197. \end{aligned}$$

Finally, for part (c) we have

$$\begin{aligned}
P(N([10,10.5])=3\,|\,N([10,12])=12) &= \frac{P(N([10,10.5])=3, N([10,12])=12)}{P(N([10,12])=12)} \\
&= \frac{P(N([10,10.5])=3, N([10.5,12])=9)}{P(N([10,12])=12)} \\
&= \frac{P(N([10,10.5])=3)P(N([10.5,12])=9)}{P(N([10,12])=12)} \\
&= \frac{(5/2)^3}{3!}e^{-5/2}\cdot\frac{(15/2)^9}{9!}e^{-15/2}\cdot\left(e^{-10}\frac{10^{12}}{12!}\right)^{-1} \\
&= \binom{12}{3}\left(\frac{1}{4}\right)^3\left(\frac{3}{4}\right)^9 \approx 0.258.
\end{aligned}$$

▲

Example 4.36. Suppose that phone calls arrive to an office according to a Poisson process with intensity λ. What is the distribution of the time until the first phone call? How about the second one? What about the nth one?

Assume we start the process at time zero. Let us denote the time of the first call by T_1. Then for any $t \geq 0$

$$P(T_1 > t) = P(\text{no calls in } [0,t]) = P(N([0,t]) = 0) = e^{-\lambda t},$$

because $N([0,t]) \sim \text{Poisson}(\lambda t)$. But this means that

$$P(T_1 \leq t) = 1 - e^{-\lambda t}, \qquad t \geq 0.$$

Since T_1 is by definition nonnegative, we have $P(T_1 \leq t) = 0$ for $t < 0$. This means that T_1 has the same cumulative distribution function as an $\text{Exp}(\lambda)$ random variable, and thus $T_1 \sim \text{Exp}(\lambda)$.

Now denote the time of the second call by T_2. We try a similar strategy to identify its distribution. The starting point is the fact that the second call is larger than t exactly if there is at most one call in $[0,t]$. This gives, for $t \geq 0$:

$$\begin{aligned}
&P(T_2 > t) = P(N([0,t]) \leq 1) = \lambda t e^{-\lambda t} + e^{-\lambda t}, \\
\text{and}\quad &P(T_2 \leq t) = 1 - (\lambda t e^{-\lambda t} + e^{-\lambda t}).
\end{aligned}$$

This gives the cumulative distribution function of T_2, and if we differentiate it then we get its density function:

$$f_{T_2}(t) = \frac{d}{dt}\left(1 - (\lambda t e^{-\lambda t} + e^{-\lambda t})\right) = \lambda^2 t e^{-\lambda t}, \qquad \text{for } t \geq 0. \tag{4.24}$$

The density function is zero for $t < 0$.

In fact, we can use exactly the same argument to find the distribution of the time T_n of the nth phone call. For $t \geq 0$ the event $\{T_n > t\}$ is the same as the event of having at most $n-1$ phone calls in $[0,t]$. Using the probability mass function of the Poisson distribution we get

$$P(T_n \le t) = 1 - P(T_n > t) = 1 - P(N([0,t]) \le n-1)$$
$$= 1 - \sum_{i=0}^{n-1} P(N([0,t]) = i) = 1 - \sum_{i=0}^{n-1} \frac{(\lambda t)^i}{i!} e^{-\lambda t}.$$

Next we differentiate to get the density function for $t > 0$:

$$f_{T_n}(t) = \frac{d}{dt}\left(1 - \sum_{i=0}^{n-1} \frac{(\lambda t)^i}{i!} e^{-\lambda t}\right) = \lambda e^{-\lambda t} - \sum_{i=1}^{n-1}\left(\frac{i\lambda^i t^{i-1}}{i!} e^{-\lambda t} - \lambda \frac{(\lambda t)^i}{i!} e^{-\lambda t}\right)$$
$$= -\sum_{i=1}^{n-1} \frac{\lambda^i t^{i-1}}{(i-1)!} e^{-\lambda t} + \sum_{i=0}^{n-1} \frac{\lambda^{i+1} t^i}{i!} e^{-\lambda t} = \frac{\lambda^n t^{n-1}}{(n-1)!} e^{-\lambda t}. \tag{4.25}$$

For the last equality, note that the terms of the two sums cancel each other one by one and leave only the last term of the second sum. ▲

The probability density function discovered in the previous example belongs to a named family of distributions. To describe the distribution, we introduce the *gamma function*

$$\Gamma(r) = \int_0^\infty x^{r-1} e^{-x}\, dx, \qquad \text{for } r > 0. \tag{4.26}$$

The gamma function generalizes the factorial function: if n is a positive integer then $\Gamma(n) = (n-1)!$. (Exercise 4.52 asks you to check this.)

Definition 4.37. Let $r, \lambda > 0$. A random variable X has the **gamma distribution** with parameters (r, λ) if X is nonnegative and has probability density function

$$f_X(x) = \frac{\lambda^r x^{r-1}}{\Gamma(r)} e^{-\lambda x} \qquad \text{for } x \ge 0, \tag{4.27}$$

with $f_X(x) = 0$ for $x < 0$. We abbreviate this $X \sim \text{Gamma}(r, \lambda)$.

See Figure 4.6 for examples of gamma density functions.

The density function $f_{T_n}(t)$ found in (4.25) is seen to be the Gamma(n, λ) density function. The exponential distribution is a special case of the gamma distribution: Exp(λ) is the same distribution as Gamma$(1, \lambda)$.

The probability density function of a random variable should always integrate to one. Let us check that this is true for the density given in (4.27). If $\lambda = 1$ then this follows from the definition of the gamma function. Otherwise we can evaluate the integral using the change of variable $y = \lambda x$:

$$\int_0^\infty \frac{\lambda^r x^{r-1}}{\Gamma(r)} e^{-\lambda x}\, dx = \int_0^\infty \frac{\lambda^r (y/\lambda)^{r-1}}{\Gamma(r)} e^{-y} \frac{1}{\lambda}\, dy = \int_0^\infty \frac{y^{r-1}}{\Gamma(r)} e^{-y}\, dy = 1.$$

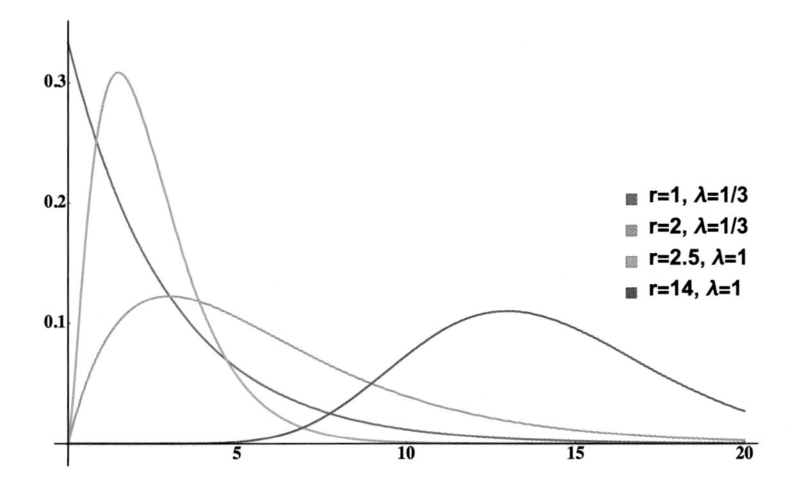

Figure 4.6. Gamma density functions for various choices of r and λ.

4.7. Finer points ♣

Error bound in the normal approximation

How can we be sure that the normal approximation to a binomial distribution is good enough? For this we need rigorous bounds on the error. The following theorem can be found in [Dur10].

Theorem 4.38. *Let $S_n \sim \text{Bin}(n, p)$. Then for all values of x we have*

$$\left| P\left(\frac{S_n - np}{\sqrt{np(1-p)}} \leq x \right) - \Phi(x) \right| \leq \frac{3}{\sqrt{np(1-p)}}.$$

The constant 3 on the right is not the best possible known value. The inequality shows how the bound on the error gets better as $np(1-p)$ increases.

Weak versus strong law of large numbers

Technically speaking Theorem 4.8 is a *weak law of large numbers* because the statement says that the probability that S_n/n is close to p converges to 1. A *strong law of large numbers* would state that the observed random frequency S_n/n itself converges to p. The distinction is that (4.7) does not completely prevent deviations larger than ε from happening as n grows. It merely says that such deviations become less and less likely. In Section 9.5 we present the strong law of large numbers.

The memoryless property

It is interesting to know that among the positive distributions that are regular enough (for example, whose density function is continuous on $[0, \infty)$) only the exponential distribution has the memoryless property. We prove this fact here.

Assume that X is a positive random variable with a density function $f(x)$ that is continuous on $[0, \infty)$ and satisfies (4.19). Since

$$P(X > t + s | X > t) = \frac{P(X > t + s)}{P(X > t)} = \frac{1 - F_X(t + s)}{1 - F_X(t)},$$

and $P(X > s) = 1 - F_X(s)$, the memoryless property

$$P(X > t + s | X > t) = P(X > s)$$

implies

$$\frac{1 - F_X(t + s)}{1 - F_X(t)} = 1 - F_X(s)$$

for any $t, s > 0$. Using the notation $G(t) = 1 - F_X(t)$ this gives

$$G(t + s) = G(t)G(s), \qquad \text{for all } s, t > 0. \tag{4.28}$$

Since f is continuous on $[0, \infty)$, we have $G'(t) = \frac{d}{dt}(1 - F_X(t)) = -f(t)$ for any $t > 0$. Differentiating both sides of (4.28) in s and then letting $s \to 0^+$ we get

$$-f(t + s) = -G(t)f(s) \quad \underset{s \to 0^+}{\Longrightarrow} \quad -f(t) = -G(t)f(0).$$

We used the continuity of f at 0 in the last step. Using $G'(t) = -f(t)$ the last identity gives the differential equation

$$G'(t) = -G(t)f(0), \qquad t > 0.$$

We also know that $G(0) = P(X > 0) = 1$. This is a differential equation which we can solve. Integrating $-f(0) = \frac{G'(t)}{G(t)}$ from 0 to $t > 0$ gives

$$-f(0)t = \int_0^t \frac{G'(s)}{G(s)} ds = \ln G(t) - \ln G(0) = \ln G(t)$$

which implies $G(t) = 1 - F_X(t) = e^{-f(0)t}$. This shows that X has to be exponential with parameter $f(0)$. This also means that $f(0) > 0$, otherwise we would not get a proper cumulative distribution function for $F_X(t) = 1 - G(t)$.

The functional equation (4.28) is called the multiplicative Cauchy equation. It can be proved that with rather weak assumptions on G (such as continuity at zero or boundedness near zero) the only solutions are of the form $G(s) = e^{as}$. Surprisingly, if we do not assume anything about G then there are other solutions too, although these cannot be described with nice formulas.

Spatial Poisson processes

Definition 4.34 of the Poisson process generalizes to higher dimensional Euclidean spaces. Thus Poisson processes can be used to model events that happen on the two-dimensional plane, in space, or even in space and time when we include time as an additional dimension. An example could be the locations (and times) of earthquakes in a region.

The definition is basically the same as before, but instead of measuring the size of a set with length, we use the measure appropriate to the space we work in. Let the integer d be the dimension of the space. For a subset $A \subseteq \mathbb{R}^d$, let $|A|$ denote its *generalized volume*. By this we mean length if $d = 1$, area if $d = 2$, ordinary volume if $d = 3$, and for $d > 3$ the natural generalization that gives a d-dimensional rectangle $A = [a_1, b_1] \times [a_2, b_2] \times \cdots \times [a_d, b_d]$ the volume $|A| = \prod_{i=1}^{d}(b_i - a_i)$.

Definition 4.39. The Poisson process with intensity $\lambda > 0$ on $\mathbb{R}^d$ is a collection of random points in $\mathbb{R}^d$ with the following properties:

- The points are all distinct. That is, there cannot be more than one point at any particular location.
- The number of points in a set A is denoted by $N(A)$. This random variable has Poisson distribution with parameter $\lambda\,|A|$, for sets A such that the volume $|A|$ is well defined and finite.
- If $A_1, A_2, \dots, A_n$ are nonoverlapping sets of finite volume then the random variables $N(A_1), N(A_2), \dots, N(A_n)$ are mutually independent.

There are further generalizations. The definition above is the *homogeneous* Poisson process because the average density of points does not vary in space. *Inhomogeneous* Poisson processes can be defined by replacing the intensity constant λ with an intensity function $\lambda(x)$. These models are part of the subject of stochastic processes.

Exercises

We start with some warm-up exercises arranged by section.

Section 4.1

Exercise 4.1. In a high school there are 1200 students. Estimate the probability that more than 130 students were born in January under each of the following assumptions. You do not have to use the continuity correction.

(a) Months are equally likely to contain birthdays.
(b) Days are equally likely to be birthdays.

Exercise 4.2. The probability of getting a single pair in a poker hand of 5 cards is approximately 0.42. Find the approximate probability that out of 1000 poker hands there will be at least 450 with a single pair.

Exercise 4.3. Approximate the probability that out of 300 die rolls we get exactly 100 numbers that are multiples of 3.

Hint. You will need the continuity correction for this.

Exercise 4.4. Liz is standing on the real number line at position 0. She rolls a die repeatedly. If the roll is 1 or 2, she takes *one* step to the right (in the positive direction). If the roll is 3, 4, 5 or 6, she takes *two* steps to the right. Let X_n be Liz's position after n rolls of the die. Estimate the probability that X_{90} is at least 160.

Section 4.2

Exercise 4.5. Consider the setup of Exercise 4.4. Find the limits below and explain your answer.

(a) Find $\lim_{n\to\infty} P(X_n > 1.6n)$.
(b) Find $\lim_{n\to\infty} P(X_n > 1.7n)$.

Section 4.3

Exercise 4.6. A pollster would like to estimate the fraction p of people in a population who intend to vote for a particular candidate. How large must a random sample be in order to be at least 95% certain that the fraction $\widehat{p}$ of positive answers in the sample is within 0.02 of the true p?

Exercise 4.7. A political interest group wants to determine what fraction $p \in (0, 1)$ of the population intends to vote for candidate A in the next election. 1000 randomly chosen individuals are polled. 457 of these indicate that they intend to vote for candidate A. Find the 95% confidence interval for the true fraction p.

Exercise 4.8. In a million rolls of a biased die the number 6 shows up 180,000 times. Find a 99.9% confidence interval for the unknown probability that the die rolls 6.

Section 4.4

Exercise 4.9. Let $X \sim \text{Poisson}(10)$.

(a) Find $P(X \geq 7)$.
(b) Find $P(X \leq 13 \,|\, X \geq 7)$.

Exercise 4.10. A hockey player scores at least one goal in roughly half of his games. How would you estimate the percentage of games where he scores a hat-trick (three goals)?

Exercise 4.11. On the first 300 pages of a book, you notice that there are, on average, 6 typos per page. What is the probability that there will be at least 4 typos on page 301? State clearly the assumptions you are making.

Section 4.5

Exercise 4.12. Let $T \sim \text{Exp}(\lambda)$. Compute $E[T^3]$.

Hint. Use integration by parts and (4.18).

Exercise 4.13. Let $T \sim \text{Exp}(1/3)$.

(a) Find $P(T > 3)$.
(b) Find $P(1 \leq T < 8)$.
(c) Find $P(T > 4 \mid T > 1)$.

Exercise 4.14. The lifetime of a lightbulb can be modeled with an exponential random variable with an expected lifetime of 1000 days.

(a) Find the probability that the lightbulb will function for more than 2000 days.
(b) Find the probability that the lightbulb will function for more than 2000 days, given that it is still functional after 500 days.

Section 4.6

Exercise 4.15. Suppose that a class of students is star-gazing on top of the local mathematics building from the hours of 11 PM through 3 AM. Suppose further that meteors arrive (i.e. they are seen) according to a Poisson process with intensity $\lambda = 4$ per hour. Find the following.

(a) The probability that the students see more than 2 meteors in the first hour.
(b) The probability they see zero meteors in the first hour, but at least 10 meteors in the final three hours (midnight to 3 AM).
(c) Given that there were 13 meteors seen all night, what is the probability there were no meteors seen in the first hour?

Further exercises

Exercise 4.16. We choose 500 numbers uniformly at random from the interval $[1.5, 4.8]$.

(a) Approximate the probability of the event that less than 65 of the numbers start with the digit 1.
(b) Approximate the probability of the event that more than 160 of the numbers start with the digit 3.

Exercise 4.17. We roll a pair of dice 10,000 times. Estimate the probability that the number of times we get *snake eyes* (two ones) is between 280 and 300.

Exercise 4.18. We are throwing darts on a disk shaped board of radius 5. We assume that the position of the dart is a uniformly chosen point in the disk.

The board has a disk shaped bullseye with radius 1. Suppose that we throw a dart 2000 times at the board. Estimate the probability that we hit the bullseye at least 100 times.

Exercise 4.19. Assume 20% of a population prefers cereal A to cereal B. I go and interview 100 randomly chosen individuals. Use the normal approximation to estimate the probability that at least 25 people in my sample prefer cereal A to cereal B. (Whether you want to use the continuity correction is up to you.)

Exercise 4.20. You flip a fair coin 10,000 times. Approximate the probability that the difference between the number of heads and number of tails is at most 100.

Exercise 4.21. In a game you win \$10 with probability $\frac{1}{20}$ and lose \$1 with probability $\frac{19}{20}$. Approximate the probability that you lost less than \$100 after the first 200 games. How will this probability change after 300 games?

Exercise 4.22. Initially a game piece is at a starting position of a game board. We repeatedly flip a fair coin and move the game piece one step left or right on the board, depending on whether the coin shows heads or tails. Approximate the probability that after 400 flips the game piece is no more than 10 steps away from the starting position. Assume that the game board is infinite so you do not have to worry about hitting the edge of the board.

Exercise 4.23. Suppose that the distribution of the lifetime of a car battery, produced by a certain car company, is well approximated by a normal distribution with a mean of 1.2×10^3 hours and variance 10^4. What is the approximate probability that a batch of 100 car batteries will contain at least 20 whose lifetimes are less than 1100 hours?

Exercise 4.24. We roll a fair die repeatedly and keep track of the observed frequencies of the outcomes $1, 2, \ldots, 6$.

(a) Show that the probability of seeing at least 17% fours converges to zero as the number of rolls tends to infinity.
(b) Let A_n be the event that after n rolls, the frequencies of *all* six outcomes are between 16% and 17%. Show that for a large enough number of rolls, the probability of A_n is at least 0.999.

Hint. Examples 4.9 and 4.10 point the way. Additionally, in part (b) use unions or intersections.

Exercise 4.25. 81 randomly chosen individuals were interviewed to estimate the unknown fraction $p \in (0, 1)$ of the population that prefers cereal to bagels for breakfast. The resulting estimate is $\widehat{p}$. With what level of confidence can we state that the true p lies in the interval $(\widehat{p} - 0.05, \widehat{p} + 0.05)$?

Exercise 4.26. 100 randomly chosen individuals were interviewed to estimate the unknown fraction $p \in (0, 1)$ of the population that prefers whole milk to skim milk. The resulting estimate is $\widehat{p}$. With what level of confidence can we state that the true p lies in the interval $(\widehat{p} - 0.1, \widehat{p} + 0.1)$?

Exercise 4.27. A marketing firm wants to determine what proportion $p \in (0, 1)$ of targeted customers prefer strawberries to blueberries. They poll n randomly chosen customers and discover that X of them prefer strawberries. How large should n be in order to know with at least 0.9 certainty that the true p is within 0.1 of the estimate X/n?

Exercise 4.28. For which values of p will the p.m.f. of the Bin(n, p) distribution have its maximum at n?

Exercise 4.29. Let $S_n = X_1 + \cdots + X_n$ be a simple random walk with step distribution $P(X_j = 1) = p$ and $P(X_j = -1) = 1 - p$. Derive the probability $P(S_n = k)$ for all values of n and k.

Exercise 4.30. Show that the probability mass function of a Poisson(λ) random variable is first increasing and then decreasing, similarly to the binomial distribution. For what range of values is the p.m.f. increasing?

Hint. Consider the ratio of subsequent values of the probability mass function.

Exercise 4.31. Let $X \sim$ Poisson(μ). Compute $E\left[\frac{1}{1+X}\right]$.

Exercise 4.32. The *nth factorial moment* of a random variable Y is defined as the expectation $E[Y(Y-1)\cdots(Y-n+1)]$.

(a) Suppose that $Y \sim$ Poisson(μ). Find an exact expression for the nth factorial moment of Y.
Hint. Use a similar argument as the one used in the proof of Fact 4.18.

(b) Compute the third moment $E(Y^3)$ of $Y \sim$ Poisson(μ).

Exercise 4.33. In a call center the number of received calls in a day can be modeled by a Poisson random variable. We know that on average about 0.5% of the time the call center receives no calls at all. What is the average number of calls per day?

Exercise 4.34. A taxi company has a large fleet of cars. On average, there are 3 accidents each week. What is the probability that at most 2 accidents happen next week? Make some reasonable assumption in order to be able to answer the question. Simplify your answer as much as possible.

Exercise 4.35. Every morning Jack flips a fair coin ten times. He does this for an entire year. Let X denote the number of days when all the flips come out the same way (all heads or all tails).

(a) Give the precise expression for the probability $P(X > 1)$.
(b) Apply either the normal or the Poisson approximation to give a simple estimate for $P(X > 1)$. Explain your choice of approximation.

Exercise 4.36. How many randomly chosen guests should I invite to my party so that the probability of having a guest with the same birthday as mine is at least 2/3?

Exercise 4.37. In low-scoring team sports (e.g. soccer, hockey) the number of goals per game can often be approximated by a Poisson random variable. (Can you explain why?) In the 2014–2015 season 8.16% of the games in the English Premier League ended in a scoreless tie (this means that there were no goals in the game). How would you estimate the percentage of games where exactly one goal was scored?

Exercise 4.38. Check Fact 4.20 with $n = 1$ for sets A of the form $\{k\}$. Note that in that case X is just a Bernoulli random variable.

Exercise 4.39. A "wheat cent" is a one-cent coin ($0.01) produced in the United States between 1909 and 1956. The name comes from a picture of wheat on the back of the coin. Assume that 1 out of every 350 pennies in circulation is a wheat cent and that wheat cents are uniformly distributed among all pennies.

Cassandra the coin collector goes to the bank and withdraws 4 dollars' worth of pennies (in other words, 400 pennies).

(a) Write an expression for the exact probability that Cassandra finds at least 2 wheat cents among her 400 pennies.
(b) Use either the Poisson or normal approximation, whichever is appropriate, to estimate the probability that Cassandra finds at least 2 wheat cents among her 400 pennies.

Exercise 4.40. We have an urn with 10 balls numbered from 1 to 10. We choose a sample of 111 with replacement. Approximate the probability of the event that the number one appears at most 3 times in the sample. Use both the normal and the Poisson approximation, and compare the results with the exact probability 0.00327556.

Exercise 4.41. We roll a die 72 times. Approximate the probability of getting exactly 3 sixes with both the normal and the Poisson approximation and compare the results with the exact probability 0.000949681.

Exercise 4.42. On average 20% of the gadgets produced by a factory are mildly defective. I buy a box of 100 gadgets. Assume this is a random sample from the production of the factory. Let A be the event that less than 15 gadgets in the random sample of 100 are mildly defective.

(a) Give an exact expression for $P(A)$, without attempting to evaluate it.
(b) Use either the normal or the Poisson approximation, whichever is appropriate, to give an approximation of $P(A)$.

Exercise 4.43. Suppose 10% of households earn over 80,000 dollars a year, and 0.25% of households earn over 450,000. A random sample of 400 households has been chosen. In this sample, let X be the number of households that earn over 80,000, and let Y be the number of households that earn over 450,000. Use the normal and Poisson approximation, whichever is appropriate in either case, to find the simplest estimates you can for the probabilities $P(X \geq 48)$ and $P(Y \geq 2)$.

Exercise 4.44. Suppose that 50% of all watches produced by a certain factory are defective (the other 50% are fine). A store buys a box with 400 watches produced by this factory. Assume this is a random sample from the factory.

(a) Write an expression for the exact probability that at least 215 of the 400 watches are defective.
(b) Approximate the probability, using either the Poisson or normal approximation, whichever is appropriate, that at least 215 of the 400 watches are defective.

Exercise 4.45. Estimate the probability that out of 10,000 poker hands (of 5 cards) we will see no four of a kinds. Use either the normal or the Poisson approximation, whichever is appropriate. Justify your choice of approximation.

Exercise 4.46. Jessica flips a fair coin 5 times every morning, for 30 days straight. Let X be the number of mornings over these 30 days on which all 5 flips are tails. Use either the normal or the Poisson approximation, whichever is more appropriate, to give an estimate for the probability $P(X = 2)$. Justify your choice of approximation.

Exercise 4.47. Each day John performs the following experiment: he flips a coin repeatedly until he gets tails and counts the number of coin flips needed.

(a) Approximate the probability that in a year there are at least 3 days when he needed more than 10 coin flips.
(b) Approximate the probability that in a year there are more than 50 days when he needed exactly 3 coin flips.

Exercise 4.48. Let $X \sim \text{Exp}(2)$. Find a real number $a < 1$ so that the events $\{X \in [0, 1]\}$ and $\{X \in [a, 2]\}$ are independent.

Exercise 4.49. Suppose that you own a store that sells a particular stove for \$1000. You purchase the stoves from the distributor for \$800 each. You believe that this stove has a lifetime which can be faithfully modeled as an exponential random variable with a parameter of $\lambda = 1/10$, where the units of time are years. You would like to offer the following extended warranty on this stove: if the stove breaks within r years, you will replace the stove completely (at a cost of \$800 to you). If the stove lasts longer than r years, the extended warranty pays nothing. Let \$$C$ be the cost you will charge the consumer for this extended warranty. For what pairs of numbers (C, r) will the expected profit you get from this warranty be zero? What do you think are reasonable choices for C and r? Why?

Exercise 4.50. Suppose an alarm clock has been set to ring after T hours, where $T \sim \text{Exp}(1/3)$. Suppose further that your friend has been staring at the clock for exactly 7 hours and can confirm that it has not yet rung. At this point, your friend wants to know when the clock will finally ring. Calculate her conditional probability that she needs to wait at least 3 more hours, given that she has already waited 7 hours. More generally, calculate the conditional probability that she needs to wait at least x more hours, given that she has already waited 7 hours.

Exercise 4.51. Suppose customer arrivals at a post office are modeled by a Poisson process N with intensity $\lambda > 0$. Let T_1 be the time of the first arrival. Let $t > 0$. Suppose we learn that by time t there has been precisely one arrival, in other words, that $N_t = 1$. What is the distribution of T_1 under this new information? In other words, find the conditional probability $P(T_1 \leq s \mid N_t = 1)$ for all $s \geq 0$.

Hint. You should see a familiar distribution.

Exercise 4.52. Recall Definition (4.26) of the gamma function $\Gamma(r)$.

(a) Show the following identity:

$$\Gamma(r + 1) = r\Gamma(r), \qquad \text{for } r > 0. \tag{4.29}$$

(b) Use this to show that if n is a positive integer then $\Gamma(n) = (n - 1)!$.

Exercise 4.53. Let $X \sim \text{Gamma}(r, \lambda)$. Find the mean and variance of X.

Hint. Equations (4.26) and (4.29) could be used to simplify your answer.

Challenging problems

Exercise 4.54. The point of this exercise is to check that it is immaterial whether the inequality is $<$ or $\leq$ in the law of large numbers of Theorem 4.8. Consider these two statements:

$$\text{for any fixed } \varepsilon > 0, \quad \lim_{n\to\infty} P\left(\left|\tfrac{S_n}{n} - p\right| \leq \varepsilon\right) = 1, \tag{4.30}$$

and

$$\text{for any fixed } \varepsilon > 0, \quad \lim_{n\to\infty} P\left(\left|\frac{S_n}{n} - p\right| < \varepsilon\right) = 1. \tag{4.31}$$

(a) Show that (4.30) implies (4.31).
(b) Show that (4.31) implies (4.30).

Exercise 4.55. Let F_n be the cumulative distribution function of the discrete random variable X_n with possible values $\{1, 2, \ldots, n\}$ and uniform probability mass function $P(X = k) = \frac{1}{n}$ for $k = 1, \ldots, n$. Find $\lim_{n\to\infty} F_n(t)$ for all real t. Interpret this limit as a limit in distribution for the random variables X_n.

Exercise 4.56. Suppose that $X \sim \text{Poisson}(\lambda)$. Find the probability $P(X \text{ is even})$. Your answer should not be an infinite series.

Exercise 4.57. Use Exercise D.13 and the proof outline for the CLT given at the end of Section 4.1 to prove that $\lim_{n\to\infty} \dfrac{n!}{n^{n+1/2}e^{-n}} = \sqrt{2\pi}$.

5

Transforms and transformations

The theme of this chapter is functions. The first section introduces the moment generating function which is an example of a *transform* of a random variable. It is a tool for working with random variables that sometimes can be more convenient than probability mass functions or density functions. The second section of the chapter shows how to derive the probability distribution of a function, that is, a *transformation*, of a random variable.

5.1. Moment generating function

Up to now we have described distributions of random variables with probability mass functions, probability density functions, and cumulative distribution functions. The moment generating function (m.g.f.) offers an alternative way to characterize the distribution. Furthermore, as the name suggests, it can also be used to compute moments of a random variable.

Definition 5.1. The **moment generating function** of a random variable X is defined by $M(t) = E(e^{tX})$. It is a function of the real variable t.

The moment generating function is analogous to Fourier and Laplace transforms commonly used in engineering and applied mathematics. As with other notation, we write $M_X(t)$ if we wish to distinguish the random variable X. We begin with two examples, first a discrete and then a continuous random variable.

Example 5.2. Let X be a discrete random variable with probability mass function

$$P(X = -1) = \tfrac{1}{3}, \quad P(X = 4) = \tfrac{1}{6}, \quad \text{and} \quad P(X = 9) = \tfrac{1}{2}.$$

Find the moment generating function of X.

The calculation is an application of formula (3.24) for the expectation of $g(X)$ with the function $g(x) = e^{tx}$. The function g contains a parameter t that can vary:

$$M_X(t) = E[e^{tX}] = \sum_k e^{tk} P(X = k) = \tfrac{1}{3}e^{-t} + \tfrac{1}{6}e^{4t} + \tfrac{1}{2}e^{9t}. \qquad \blacktriangle$$

Example 5.3. Let X be a continuous random variable with probability density function

$$f(x) = \begin{cases} \dfrac{e^x}{e-1}, & \text{if } 0 < x < 1 \\ 0, & \text{otherwise.} \end{cases}$$

Find the moment generating function of X.

In the continuous case we apply formula (3.25) for the expectation of e^{tX}:

$$M_X(t) = E[e^{tX}] = \int_{-\infty}^{\infty} e^{tx} f(x)\, dx = \int_0^1 e^{tx} \frac{e^x}{e-1}\, dx = \frac{1}{e-1} \int_0^1 e^{(t+1)x}\, dx.$$

The integration splits into two cases. If $t = -1$ then

$$M_X(t) = \frac{1}{e-1} \int_0^1 dx = \frac{1}{e-1}.$$

In the case $t \neq -1$ we have

$$M_X(t) = \frac{1}{e-1} \cdot \frac{e^{(t+1)x}}{t+1} \Big|_{x=0}^{x=1} = \frac{e^{t+1} - 1}{(e-1)(t+1)}.$$

To summarize, the moment generating function is given by the two-case formula

$$M_X(t) = \begin{cases} \frac{1}{e-1}, & \text{for } t = -1 \\ \frac{e^{t+1}-1}{(e-1)(t+1)}, & \text{for } t \neq -1. \end{cases}$$ ▲

Notice from the definition and from the examples that the moment generating function is *not random*, but it is a function of the variable t. Since $e^0 = 1$, we have $M(0) = E[e^{0 \cdot X}] = E[1] = 1$ for all random variables. You can see this in the examples above. Next we calculate moment generating functions of some familiar named distributions.

Example 5.4. (Moment generating function of the Poisson distribution) Let $X \sim \text{Poisson}(\lambda)$. The calculation is an application of formula (3.24) with the Poisson probability mass function:

$$E(e^{tX}) = \sum_{k=0}^{\infty} e^{tk} \frac{e^{-\lambda}\lambda^k}{k!} = e^{-\lambda} \sum_{k=0}^{\infty} \frac{(e^t\lambda)^k}{k!} = e^{-\lambda} \cdot e^{e^t\lambda} = e^{\lambda(e^t-1)},$$

where we used the series expansion of the exponential function (see (D.3)). Thus for $X \sim \text{Poisson}(\lambda)$, we have $M_X(t) = e^{\lambda(e^t-1)}$ for all real t. ▲

Example 5.5. (Moment generating function of the normal distribution) Let $Z \sim \mathcal{N}(0, 1)$. Now the computation uses formula (3.25). To evaluate the integral we complete the square in the exponential:

$$E(e^{tZ}) = \frac{1}{\sqrt{2\pi}} \int_{-\infty}^{\infty} e^{tx} e^{-x^2/2}\, dx = \frac{1}{\sqrt{2\pi}} \int_{-\infty}^{\infty} e^{-\frac{1}{2}x^2 + tx - \frac{1}{2}t^2 + \frac{1}{2}t^2}\, dx$$

$$= e^{t^2/2} \frac{1}{\sqrt{2\pi}} \int_{-\infty}^{\infty} e^{-\frac{1}{2}(x-t)^2}\, dx = e^{t^2/2}.$$

Notice how we immediately know the value of the last integral because we recognize the integrand as the density function of the $\mathcal{N}(t, 1)$ distribution. We deduced that $M_Z(t) = e^{t^2/2}$ for $Z \sim \mathcal{N}(0, 1)$.

To get the m.g.f. of a general normal random variable $X \sim \mathcal{N}(\mu, \sigma^2)$, write $X = \sigma Z + \mu$ and deduce

$$E(e^{tX}) = E(e^{t(\sigma Z+\mu)}) = e^{t\mu} E(e^{t\sigma Z}) = e^{t\mu} \cdot e^{t^2\sigma^2/2} = e^{\mu t+\sigma^2 t^2/2}.$$ ▲

Example 5.6. (Moment generating function of the exponential distribution) Let $X \sim \text{Exp}(\lambda)$. Then

$$E(e^{tX}) = \int_0^{\infty} e^{tx} \cdot \lambda e^{-\lambda x}\, dx = \lambda \int_0^{\infty} e^{(t-\lambda)x}\, dx = \lambda \lim_{b\to\infty} \int_0^{b} e^{(t-\lambda)x}\, dx.$$

This is an improper integral whose value depends on t, so let us do it carefully. If $t = \lambda$,

$$E(e^{tX}) = \lambda \lim_{b\to\infty} \int_0^{b} 1\, dx = \lambda \lim_{b\to\infty} b = \infty.$$

Suppose $t \neq \lambda$.

$$E(e^{tX}) = \lambda \lim_{b\to\infty} \left(\frac{e^{(t-\lambda)x}}{t-\lambda}\Bigg|_{x=0}^{x=b} \right) = \lambda \lim_{b\to\infty} \frac{e^{(t-\lambda)b} - 1}{t-\lambda} = \begin{cases} \infty, & \text{if } t > \lambda \\ \dfrac{\lambda}{\lambda - t}, & \text{if } t < \lambda. \end{cases}$$

To summarize, the m.g.f. of the $\text{Exp}(\lambda)$ distribution is

$$M(t) = \begin{cases} \infty, & \text{if } t \geq \lambda \\ \dfrac{\lambda}{\lambda - t}, & \text{if } t < \lambda. \end{cases}$$ ▲

The exponential example above shows that $M(t)$ can be infinite for some portion of t-values. In fact, it can happen that all values except $M(0) = 1$ are infinite. Such an example appears in Exercise 5.42.

Calculation of moments with the moment generating function

The moment generating functions that we see in this course are finite around the origin and can be differentiated there. This gives a way of finding moments.

Let $M(t) = E[e^{tX}]$ and consider the following calculation:

$$M'(t) = \frac{d}{dt} E[e^{tX}] = E\left[\frac{d}{dt} e^{tX}\right] = E[Xe^{tX}].$$

Substituting $t = 0$ gives the formula $M'(0) = E[X]$. The fact that we can move the differentiation inside the expectation is not self-evident, but we will not discuss the justification here. However, in the case that X takes only finitely many values this step is straightforward because the derivative of a sum is the sum of the derivatives. We have

$$M'(t) = \frac{d}{dt}\sum_k e^{kt}P\{X=k\} = \sum_k \frac{d}{dt}e^{kt}P\{X=k\}$$
$$= \sum_k ke^{kt}P\{X=k\} = E[Xe^{tX}]. \tag{5.1}$$

Returning to the general case, we can continue to differentiate as many times as we please by taking the derivative inside the expectation. Write $M^{(n)}$ for the nth derivative of the function M:

$$M^{(n)}(t) = \frac{d^n}{dt^n}E[e^{tX}] = E\left[\frac{d^n}{dt^n}e^{tX}\right] = E[X^n e^{tX}].$$

Taking $t = 0$ gives the following formula.

Fact 5.7. When the moment generating function $M(t)$ of a random variable X is finite in an interval around the origin, the moments of X are given by

$$E(X^n) = M^{(n)}(0).$$

To see the truth of Fact 5.7 for $n = 1$ for a random variable with finitely many values, just observe that setting $t = 0$ in (5.1) gives the identity

$$M'(0) = \sum_k kP\{X=k\} = E[X].$$

Example 5.8. (Moments of the Bernoulli distribution) Let X be Bernoulli with parameter p. Then $M_X(t) = (1-p) + pe^t$. Therefore, $M_X^{(n)}(t) = pe^t$ for all $n \geq 1$. Thus, $E[X^n] = M_X^{(n)}(0) = p$. This is not surprising since X takes the values 0 and 1 and hence $X^n = X$. ▲

Example 5.9. (Moments of the exponential distribution) Let $X \sim \text{Exp}(\lambda)$. From Example 5.6 its m.g.f. is

$$M(t) = \begin{cases} \dfrac{\lambda}{\lambda - t}, & \text{if } t < \lambda \\ \infty, & \text{if } t \geq \lambda. \end{cases}$$

Since $\lambda > 0$, we can differentiate around the origin and find

$$M'(t) = \lambda(\lambda-t)^{-2}, \quad M''(t) = 2\lambda(\lambda-t)^{-3}, \ldots, \quad M^{(n)}(t) = n!\lambda(\lambda-t)^{-n-1}.$$

From this $E(X^n) = M^{(n)}(0) = n!\lambda^{-n}$ for positive integers n. ▲

For the next example, we recall that the general form of the Taylor expansion of a function f around the origin is

$$f(t) = \sum_{n=0}^{\infty} \frac{f^{(n)}(0)}{n!} t^n. \tag{5.2}$$

Example 5.10. (Moments of the standard normal) Let $Z \sim \mathcal{N}(0, 1)$. From Example 5.5 we have $M_Z(t) = e^{t^2/2}$. Instead of differentiating this repeatedly, we can find the derivatives directly from the Taylor expansion.

Using the series for the exponential function we get

$$M_Z(t) = e^{t^2/2} = \sum_{k=0}^{\infty} \frac{\left(\frac{1}{2}t^2\right)^k}{k!} = \sum_{k=0}^{\infty} \frac{1}{2^k k!} t^{2k}. \tag{5.3}$$

Comparing this with

$$M_Z(t) = \sum_{n=0}^{\infty} \frac{M_Z^{(n)}(0)}{n!} t^n,$$

we can identify the values $M_Z^{(n)}(0)$ by matching the coefficients of the powers of t. Since only even powers t^{2k} appear in (5.3), we conclude that the coefficients of odd powers are zero, while

$$\frac{M_Z^{(2k)}(0)}{(2k)!} = \frac{1}{2^k k!}.$$

This gives us

$$E(Z^n) = M_Z^{(n)}(0) = \begin{cases} 0, & \text{if } n \text{ is odd} \\ \frac{(2k)!}{2^k k!}, & \text{if } n = 2k \text{ is even.} \end{cases}$$

The expression for even moments simplifies by splitting $(2k)!$ into even and odd factors:

$$(2k)! = \left(\prod_{i=1}^{k}(2i)\right) \cdot \left(\prod_{j=1}^{k}(2j-1)\right) = 2^k k! \cdot (2k-1)!!$$

where the *double factorial* is

$$n!! = \begin{cases} n(n-2)(n-4)\cdots 1, & n > 0 \text{ is odd} \\ n(n-2)(n-4)\cdots 2, & n > 0 \text{ is even.} \end{cases}$$

We can summarize the moments of Z as follows:

$$E(Z^n) = \begin{cases} 0, & \text{if } n \text{ is odd} \\ (n-1)!!, & \text{if } n \text{ is even.} \end{cases}$$

The fact that the odd moments are zero also follows from the fact that the density function $\varphi(x)$ of Z is an even function. ▲

We turn to the second topic of the section, namely that the moment generating function characterizes the distribution of a random variable. In order to make this statement precise, we discuss briefly what is meant by equality in distribution of random variables.

Equality in distribution

Consider the following example of three distinct random variables defined on entirely different sample spaces, but still essentially the same in terms of probabilities.

Example 5.11. Define three random variables from three different experiments:

$$X = \begin{cases} 1, & \text{if a fair coin flip is heads} \\ 0, & \text{if a fair coin flip is tails,} \end{cases} \qquad Y = \begin{cases} 1, & \text{if a roll of a fair die is even} \\ 0, & \text{if a roll of a fair die is odd,} \end{cases}$$

$$\text{and} \qquad Z = \begin{cases} 1, & \text{if a card dealt from a deck is spades or clubs} \\ 0, & \text{if a card dealt from a deck is hearts or diamonds.} \end{cases}$$

The random variables X, Y and Z have the same probability mass function $p(1) = p(0) = 1/2$. Thus any question about probabilities would have the same answer for each of the three random variables. ▲

The similarity illustrated in the example above is captured by the next definition.

Definition 5.12. Random variables X and Y are **equal in distribution** if $P(X \in B) = P(Y \in B)$ for all subsets B of $\mathbb{R}$. This is abbreviated by $X \stackrel{d}{=} Y$. ♣

Note in particular that the definition allows for the possibility that X and Y are defined on different sample spaces. In the context of Example 5.11 we can state that $X \stackrel{d}{=} Y \stackrel{d}{=} Z$.

Here are some illustrations of Definition 5.12.

If two discrete random variables X and Y have the same probability mass function, then $X \stackrel{d}{=} Y$. Similarly if two continuous random variables X and Y have the same density function then $X \stackrel{d}{=} Y$.

Conversely, suppose we knew that $X \stackrel{d}{=} Y$ and that X is discrete. Then in Definition 5.12 let the set $B = \{x\}$ for possible values x of X. This gives $P(X = x) = P(Y = x)$ for all such x, and consequently $\sum_x P(Y = x) = \sum_x P(X = x) = 1$. The conclusion is that Y has the same possible values as X, hence is discrete, and also has the same probability mass function as X. Furthermore, for expectations of any function g of X and Y,

$$E[g(X)] = \sum_x g(x)P(X = x) = \sum_x g(x)P(Y = x) = E[g(Y)].$$

Similar statements hold if X is continuous. $X \stackrel{d}{=} Y$ implies that Y is also continuous, X and Y have the same density function, and $E[g(X)] = E[g(Y)]$ for all functions g for which the expectations are well defined.

Here is one more example in terms of random variables that we have encountered earlier.

Example 5.13. In Example 2.30 we discovered that if $X_1, \dots, X_k$ denote successive draws with or without replacement from the set $\{1, 2, \dots, n\}$ (with $k \le n$) then $P(X_j = x) = 1/n$ for each $j = 1, \dots, k$ and $x \in \{1, 2, \dots, n\}$. Thus each X_j obeys the probability mass function $p(x) = 1/n$ and consequently all the random variables X_j in both sampling types are equal in distribution. In particular, the distributions of the individual random variables X_j do not reveal the distinction between sampling with and without replacement. ▲

Identification of distributions with moment generating functions

We can now state precisely how the moment generating function characterizes a distribution.

Fact 5.14. Let X and Y be two random variables with moment generating functions $M_X(t) = E(e^{tX})$ and $M_Y(t) = E(e^{tY})$. Suppose there exists $\delta > 0$ such that for $t \in (-\delta, \delta)$, $M_X(t) = M_Y(t)$ and these are finite numbers. Then X and Y are equal in distribution.

In the case when the random variable X takes only a finite number of values the m.g.f. takes a particularly simple form:

$$M_X(t) = E(e^{tX}) = \sum_k e^{kt} P(X = k). \tag{5.4}$$

When the moment generating function is in this form, we can read off directly the values of X and their probabilities.

Example 5.15. Suppose that X has moment generating function

$$M_X(t) = \tfrac{1}{5}e^{-17t} + \tfrac{1}{4} + \tfrac{11}{20}e^{2t}. \tag{5.5}$$

Then the possible values of X are $\{-17, 0, 2\}$ and

$$P(X = -17) = \tfrac{1}{5}, \quad P(X = 0) = \tfrac{1}{4}, \quad P(X = 2) = \tfrac{11}{20}. \tag{5.6}$$

Let us emphasize the logic. X must have the distribution given in (5.6) because (i) this probability mass function gives the moment generating function (5.5) and

(ii) by Fact 5.14 no other distribution can give this same moment generating function. ▲

Example 5.16. Find the distribution of Y if its moment generating function is

$$M_Y(t) = e^{17(e^t-1)}.$$

We recognize the moment generating function as that of a Poisson with parameter 17. ▲

5.2. Distribution of a function of a random variable

If X is a random variable (in other words, a function on the sample space) and g is a function on the range of X then $Y = g(X)$ is also a random variable. In Section 3.3 we learned how to calculate the expectation $E[g(X)]$ (recall in particular Fact 3.33). Now we tackle the larger problem of finding the distribution of $g(X)$ from the given distribution of X. The discrete and continuous cases are treated separately.

Discrete case

We begin with an example.

Example 5.17. Suppose the possible values of X are $\{-1, 0, 1, 2\}$ with $P\{X = k\} = \frac{1}{4}$ for each $k \in \{-1, 0, 1, 2\}$. Let $Y = X^2$. Then the possible values of Y are $\{0, 1, 4\}$. The probability mass function of Y is computed from the probability mass function of X:

$$\begin{aligned} P\{Y=0\} &= P\{X^2=0\} = P\{X=0\} = \tfrac{1}{4} \\ P\{Y=4\} &= P\{X^2=4\} = P\{X=-2 \text{ or } X=2\} \\ &= P\{X=2\} + P\{X=-2\} = \tfrac{1}{4} + 0 = \tfrac{1}{4} \\ P\{Y=1\} &= P\{X^2=1\} = P\{X=1 \text{ or } X=-1\} \\ &= P\{X=1\} + P\{X=-1\} = \tfrac{1}{4} + \tfrac{1}{4} = \tfrac{1}{2}. \end{aligned}$$

▲

The example above points the way to a general formulation. If X is a discrete random variable then the same is true of $Y = g(X)$. If we know the probability mass function of X then we can compute the probability mass function of Y from the formula

$$p_Y(\ell) = P(Y=\ell) = P(g(X)=\ell) = P\left(\bigcup_{k:g(k)=\ell} \{X=k\} \right) = \sum_{k:g(k)=\ell} p_X(k). \tag{5.7}$$

Note that if there are no k-values for which $g(k) = \ell$, the last sum is zero by definition. If g happens to be one-to-one then (5.7) simplifies to

$$p_Y(\ell) = p_X(g^{-1}(\ell)) \tag{5.8}$$

for the values ℓ in the range of g.

Example 5.18. Suppose that X takes values in the set $\{-1, 0, 1, 2\}$ with

$$P\{X = -1\} = \tfrac{1}{10}, \; P\{X = 0\} = \tfrac{2}{10}, \; P\{X = 1\} = \tfrac{3}{10}, \; P\{X = 2\} = \tfrac{4}{10}.$$

Let $Y = 2X^3$. Note that $g(x) = 2x^3$ is one-to-one. The possible values of Y are $\{-2, 0, 2, 16\}$, and by (5.8) the probability mass function is

$$P\{Y = -2\} = \tfrac{1}{10}, \; P\{Y = 0\} = \tfrac{2}{10}, \; P\{Y = 2\} = \tfrac{3}{10}, \; P\{Y = 16\} = \tfrac{4}{10}. \qquad \blacktriangle$$

Continuous case

Now suppose that X has density function f_X, and consider the random variable $Y = g(X)$ for some function g. What can we say about the distribution of Y? Unless we make assumptions on g the answer can be anything. Y can be discrete (for an extreme case take the constant function $g(x) = 0$), Y can be continuous (the simplest example $g(x) = x$ gives $Y = X$), and Y can be neither (see Example 3.38)!

In our examples and exercises g will be differentiable with a nonzero derivative at all but finitely many points. In other words, there can be finitely many points x where either $g'(x)$ does not exist or $g'(x) = 0$. In this case $Y = g(X)$ is continuous and we are able to compute the density function of Y. We present two methods.

(i) In many cases the easiest approach is to derive the cumulative distribution function F_Y of $Y = g(X)$ and differentiate to find the density function: $f_Y(y) = F'_Y(y)$.
(ii) There is also a general formula for f_Y that is useful to both theory and examples.

The c.d.f. approach is illustrated by the next examples. Then we develop the general formula. For the problems discussed in this text the c.d.f. approach is often more useful.

Example 5.19. Let $U \sim \text{Unif}(0, 1)$. For $\lambda > 0$, let $g(x) = -\frac{1}{\lambda} \ln(1 - x)$. What is the distribution of $Y = g(U)$?

If $x \in (0, 1)$, then $g(x) = -\lambda^{-1} \ln(1 - x) > 0$. Thus the random variable Y takes only positive values. So for $t \le 0$, we have $P(Y \le t) = 0$. For $t > 0$

$$\begin{aligned} P(Y \le t) &= P(-\lambda^{-1} \ln(1 - U) \le t) = P(\ln(1 - U) \ge -\lambda t) \\ &= P(1 - U \ge e^{-\lambda t}) = P(U \le 1 - e^{-\lambda t}) = 1 - e^{-\lambda t}. \end{aligned}$$

We recognize this cumulative distribution function and conclude that Y has exponential distribution with parameter λ. Example 5.30 below illuminates why this particular function g makes $g(U)$ into an exponential random variable. $\blacktriangle$

Example 5.20. Let $Z \sim \mathcal{N}(0, 1)$ and $Y = Z^2$. Find the cumulative distribution function F_Y and density function f_Y of Y.

Recall that the standard normal Z has cumulative distribution function $\Phi(x)$ and density function $\varphi(x) = \Phi'(x) = e^{-x^2/2}/\sqrt{2\pi}$.

For $y < 0$, $F_Y(y) = P(Y \le y) = 0$ because $Y = Z^2 \ge 0$. Consequently $f_Y(y) = 0$ for $y < 0$.

For $y \ge 0$,

$$\begin{aligned} F_Y(y) &= P(Y \le y) = P(Z^2 \le y) = P(-\sqrt{y} \le Z \le \sqrt{y}) = \Phi(\sqrt{y}) - \Phi(-\sqrt{y}) \\ &= 2\Phi(\sqrt{y}) - 1. \end{aligned}$$

Next, by the chain rule for $y > 0$,

$$f_Y(y) = F_Y'(y) = 2\varphi(\sqrt{y})\frac{1}{2\sqrt{y}} = \frac{1}{\sqrt{2\pi y}}e^{-y/2}. \tag{5.9}$$

The derivations above do not tell us what the value $f_Y(0)$ should be since F_Y is not differentiable at 0. We can define $f_Y(0)$ to be whatever we want since the value of f_Y at a single point does not affect the value of any integral. The full answer is

$$f_Y(y) = \begin{cases} 0, & y \le 0 \\ \frac{1}{\sqrt{2\pi y}}e^{-y/2}, & y > 0. \end{cases}$$

▲

Remark 5.21. (Gamma distribution) Definition 4.37 introduced the gamma distribution with parameters (r, λ), defined by the density function

$$f(x) = \frac{1}{\Gamma(r)}\lambda^r x^{r-1} e^{-\lambda x} \quad \text{for } x > 0.$$

Take $r = \lambda = \frac{1}{2}$ and consider the particular case of the Gamma$(\frac{1}{2}, \frac{1}{2})$ density function

$$\tilde{f}(y) = \frac{1}{\Gamma(1/2)}\left(\tfrac{1}{2}\right)^{1/2} y^{-1/2} e^{-y/2}.$$

This matches the power and exponential in the density function f_Y in (5.9). Might f_Y and $\tilde{f}$ be the same density function? In fact they must agree because the value of the constant factor in a density function is uniquely determined by the requirement that it integrate to 1.

Note that $f_Y(y) = \frac{\Gamma(1/2)}{\sqrt{\pi}}\tilde{f}(y)$ and use the fact that both are probability density functions:

$$1 = \int_0^\infty f_Y(y)\,dy = \int_0^\infty \frac{\Gamma(1/2)}{\sqrt{\pi}}\tilde{f}(y)\,dy = \frac{\Gamma(1/2)}{\sqrt{\pi}}\int_0^\infty \tilde{f}(y)\,dy = \frac{\Gamma(1/2)}{\sqrt{\pi}}.$$

This shows that $\Gamma(1/2) = \sqrt{\pi}$ and we conclude that the random variable $Y = Z^2$ from Example 5.20 is Gamma$(1/2, 1/2)$ distributed. ▲

Example 5.22. (Transformation by a linear, or affine, function) Suppose X has density function f_X and cumulative distribution function F_X. Fix real numbers $a \ne 0$ and b and let $Y = aX + b$. Find the density function f_Y of Y.

Depending upon the sign of a, we get

$$F_Y(y) = P(Y \le y) = P(aX + b \le y)$$
$$= \begin{cases} P\left(X \le \frac{y-b}{a}\right) = F_X\left(\frac{y-b}{a}\right), & a > 0 \\ P\left(X \ge \frac{y-b}{a}\right) = 1 - F_X\left(\frac{y-b}{a}\right), & a < 0. \end{cases}$$

Inside the probability above $\ge$ can be replaced by $>$ since we have assumed that X has a density function.

For those points y such that F_X is differentiable at $\frac{y-b}{a}$, we can deduce the density function of Y by differentiating:

$$f_Y(y) = F'_Y(y) = \begin{cases} \frac{d}{dy} F_X\left(\frac{y-b}{a}\right) = \frac{1}{a} f_X\left(\frac{y-b}{a}\right), & a > 0 \\ \frac{d}{dy}\left(1 - F_X\left(\frac{y-b}{a}\right)\right) = -\frac{1}{a} f_X\left(\frac{y-b}{a}\right), & a < 0 \end{cases}$$
$$= \frac{1}{|a|} f_X\left(\frac{y-b}{a}\right). \tag{5.10}$$

▲

Example 5.23. (Linear transformation of the normal distribution) Let us apply the previous example to the normal distribution to give an alternative proof of Fact 3.61. Suppose that $X \sim \mathcal{N}(\mu, \sigma^2)$ and $a \ne 0$ and b are fixed numbers. Find the distribution of $Y = aX + b$.

We have $f_X(x) = \frac{1}{\sqrt{2\pi\sigma^2}} e^{-\frac{(x-\mu)^2}{2\sigma^2}}$. Then by the previous example

$$f_Y(y) = \frac{1}{|a|} f_X\left(\frac{y-b}{a}\right) = \frac{1}{|a|\sqrt{2\pi\sigma^2}} e^{-\frac{\left(\frac{y-b}{a} - \mu\right)^2}{2\sigma^2}} = \frac{1}{\sqrt{2\pi a^2 \sigma^2}} e^{-\frac{(y - a\mu - b)^2}{2\sigma^2 a^2}}.$$

The result is the density function of $\mathcal{N}(a\mu + b, a^2\sigma^2)$. Thus a linear transformation of a normal random variable is also normal, and the parameters are given by the appropriate transformations of the mean and variance. ▲

We turn to develop a general formula for the density function of $Y = g(X)$. First consider the situation where the differentiable function g is one-to-one and increasing. Then the inverse function g^{-1} exists. As demonstrated in Example 5.19 above, it is often easier to understand how the c.d.f. transforms:

$$F_Y(y) = P(Y \le y) = P(g(X) \le y) = P(X \le g^{-1}(y)) = F_X(g^{-1}(y)).$$

Here we used the definition of the c.d.f. and also the fact that g is increasing, which gives $\{g(X) \le y\} = \{X \le g^{-1}(y)\}$. The above computation is valid if y is in the range of the function g, which ensures that $g^{-1}(y)$ exists.

If we have the c.d.f. of Y then we can get the density by differentiating it:

$$f_Y(y) = \frac{d}{dy} F_Y(y) = \frac{d}{dy} F_X(g^{-1}(y)). \tag{5.11}$$

In the last expression we need to differentiate the composition of F_X and g^{-1}. In order to apply the chain rule we need the derivatives of these functions:

$$\frac{d}{dx}F_X(x) = f_X(x), \qquad \frac{d}{dy}g^{-1}(y) = \frac{1}{g'(g^{-1}(y))}.$$

If you do not remember the formula for the derivative of g^{-1} then you can derive it by differentiating both sides of the identity $g(g^{-1}(y)) = y$.

Using these formulas in (5.11) we get that

$$f_Y(y) = \frac{d}{dy}F_X(g^{-1}(y)) = f_X(g^{-1}(y))\frac{1}{g'(g^{-1}(y))}.$$

This was a valid computation if $g^{-1}(y)$ exists and $g'(g^{-1}(y)) \neq 0$. Since we assumed that g is increasing and differentiable, its range is a finite or infinite interval. If y is not in the range of g then y must be below or above the range, which means that $F_Y(y) = P(g(X) \leq y)$ is either zero or one. This shows that $f_Y(y) = 0$ whenever y is not in the range of g. This is natural since all the values of Y lie in the range of g.

If we assume that g^{-1} exists and g is decreasing then we can repeat the argument from above, noting that now $\{g(X) \leq y\} = \{X \geq g^{-1}(y)\}$. This gives

$$F_Y(y) = P(Y \leq y) = P(g(X) \leq y) = P(X \geq g^{-1}(y)) = 1 - F_X(g^{-1}(y))$$

and

$$f_Y(y) = \frac{d}{dy}F_Y(y) = \frac{d}{dy}(1 - F_X(g^{-1}(y)) = -f_X(g^{-1}(y))\frac{1}{g'(g^{-1}(y))}. \tag{5.12}$$

Because g' in (5.12) is a negative function, we can write (5.11) and (5.12) in a single formula in the following way. In (5.10) we already derived this for the special case of a linear transformation.

Fact 5.24. Suppose that random variable X has density function f_X and the function g is differentiable, one-to-one, and its derivative is zero only at finitely many points. Then the density function f_Y of $Y = g(X)$ is given by

$$f_Y(y) = f_X(g^{-1}(y))\frac{1}{|g'(g^{-1}(y))|} \tag{5.13}$$

for points y such that $g^{-1}(y)$ exists and $g'(g^{-1}(y)) \neq 0$. For other points $f_Y(y) = 0$.

Remark 5.25. In the statement above we made the seemingly ad hoc choice $f_Y(y) = 0$ for values of y with $g'(g^{-1}(y)) = 0$. This choice does not affect the distribution of Y since g' can be zero at only finitely many points. For more on this, see Section 3.6. ▲

Example 5.26. Suppose that $X \sim \mathcal{N}(1,2)$. Find the density of $Y = X^3$.

Since the function $g(x) = x^3$ is differentiable and one-to-one we can use Fact 5.24. We have $g'(x) = 3x^2$ and $g^{-1}(x) = \sqrt[3]{x}$. The density of X is $f_X(x) = \frac{1}{\sqrt{4\pi}} e^{-\frac{(x-1)^2}{4}}$. For $y \neq 0$, formula (5.13) gives

$$f_Y(y) = f_X(\sqrt[3]{y}) \frac{1}{3(\sqrt[3]{y})^2} = \frac{1}{6\sqrt{\pi}} y^{-2/3} e^{-\frac{(\sqrt[3]{y}-1)^2}{4}}.$$

We can also solve this problem with the c.d.f. method. For $y \in \mathbb{R}$

$$P\{Y \le y\} = P\{X^3 \le y\} = P\{X \le y^{1/3}\} = F_X(y^{1/3}).$$

Differentiation gives the same density function as we found above:

$$f_Y(y) = \frac{d}{dy} F_X(y^{1/3}) = f_X(y^{1/3}) \frac{1}{3} y^{-2/3} = \frac{1}{3\sqrt{4\pi}} y^{-2/3} e^{-(y^{1/3}-1)^2/4}.$$ ▲

Formula (5.13) is not correct if the function g is not one-to-one. There is a modification that works for many-to-one functions.

Fact 5.27. Suppose that X has density function f_X, the function g is differentiable, and the derivative g' is zero only at finitely many points. Then

$$f_Y(y) = \sum_{\substack{x: g(x)=y \\ g'(x) \neq 0}} f_X(x) \frac{1}{|g'(x)|}. \tag{5.14}$$

The sum ranges over all values of x such that $g(x) = y$ and $g'(x) \neq 0$. If there is no such x then the value of the sum is taken to be zero.

In cases where we can solve the inequality $g(x) \le y$ easily, formula (5.14) can be derived from the c.d.f. For example if $g(x) = x^2$ (which is not one-to-one) we can be sure that $Y = X^2 \ge 0$ so f_Y will be zero for negative values. If $y > 0$ then

$$F_Y(y) = P(X^2 \le y) = P(-\sqrt{y} \le X \le \sqrt{y}) = F_X(\sqrt{y}) - F(-\sqrt{y}).$$

Differentiating this gives

$$f_Y(y) = \frac{1}{2\sqrt{y}} f_X(\sqrt{y}) + \frac{1}{2\sqrt{y}} f_X(-\sqrt{y}) \qquad \text{for } y > 0.$$

This is exactly the same formula that (5.14) gives.

Here is another example where the density function of $g(X)$ has a case by case formula.

Example 5.28. Let $X \sim \text{Exp}(1)$ and $Y = (X-1)^2$. Find the density function of Y.

The density function and cumulative distribution function of X are

$$f_X(x) = \begin{cases} e^{-x}, & x \geq 0 \\ 0, & x < 0 \end{cases} \quad \text{and} \quad F_X(x) = \begin{cases} 1 - e^{-x}, & x \geq 0 \\ 0, & x < 0. \end{cases}$$

We solve this problem with the c.d.f. method and then with Fact 5.27.

Solution using the c.d.f. method. The cumulative distribution function $F_Y(y) = 0$ for $y \leq 0$ because $P(Y \leq 0) = P(Y = 0) = P(X = 1) = 0$.

For $y > 0$ we have

$$\begin{aligned} F_Y(y) &= P(Y \leq y) = P((X-1)^2 \leq y) = P(1 - \sqrt{y} \leq X \leq 1 + \sqrt{y}) \\ &= F_X(1 + \sqrt{y}) - F_X(1 - \sqrt{y}). \end{aligned}$$

If $0 < y < 1$ then both $1 + \sqrt{y}$ and $1 - \sqrt{y}$ are positive, and so

$$\begin{aligned} F_Y(y) &= (1 - e^{-(1+\sqrt{y})}) - (1 - e^{-(1-\sqrt{y})}) \\ &= e^{-1+\sqrt{y}} - e^{-1-\sqrt{y}} \qquad \text{for } 0 < y < 1. \end{aligned} \tag{5.15}$$

If $y > 1$ then $F_X(1 - \sqrt{y}) = 0$ because $1 - \sqrt{y} < 0$. Still, $1 + \sqrt{y} > 0$, and so

$$F_Y(y) = 1 - e^{-1-\sqrt{y}} \qquad \text{for } y > 1. \tag{5.16}$$

Differentiating (5.15) and (5.16) gives the density function

$$f_Y(y) = \begin{cases} 0, & \text{if } y < 0 \\ e^{-(1-\sqrt{y})}\frac{1}{2\sqrt{y}} + e^{-(1+\sqrt{y})}\frac{1}{2\sqrt{y}}, & \text{if } 0 < y < 1 \\ e^{-(1+\sqrt{y})}\frac{1}{2\sqrt{y}}, & \text{if } y > 1. \end{cases} \tag{5.17}$$

We can set the values $f_Y(y)$ at $y = 0$ and $y = 1$ to be whatever we like.

Solution using Fact 5.27. We need to worry only about $y > 0$ because $Y \geq 0$ and the single point $y = 0$ makes no difference to a density function. The function $g(x) = (x-1)^2$ is two-to-one: if $y > 0$ then $g(x) = y$ for $x = 1+\sqrt{y}$ and $x = 1-\sqrt{y}$. We also need $g'(x) = 2(x-1)$. Substituting everything into (5.14) gives

$$f_Y(y) = f_X(1 - \sqrt{y})\frac{1}{|2\sqrt{y}|} + f_X(1 + \sqrt{y})\frac{1}{|2\sqrt{y}|} \qquad \text{for } y > 0.$$

If $0 < y < 1$ then both $1 + \sqrt{y}$ and $1 - \sqrt{y}$ are positive, hence both $f_X(1 + \sqrt{y})$ and $f_X(1 - \sqrt{y})$ give nonzero terms. For $y > 1$ we have $1 - \sqrt{y} < 0$ and thus the $f_X(1 - \sqrt{y})$ term vanishes. The formula that results in the end is the same as (5.17). Again, we can set the values at $y = 0$ and $y = 1$ to be whatever we like. ▲

Generating random variables from a uniform distribution ♦

Many software packages have efficient methods for generating uniform random variables. Other distributions can then be produced by taking a suitable function of a uniform random variable.

We first discuss how to generate a discrete random variable with a given probability mass function using a Unif[0, 1] distributed random variable.

Example 5.29. Let $U \sim \text{Unif}[0, 1]$ and let X be a discrete random variable taking values in $\{0, 1, 2, \dots, n\}$ with probability mass function $P(X = j) = p_j$. Find a function $g : [0, 1] \to \{0, 1, 2, \dots, n\}$ such that $g(U)$ has the same distribution as X.

Define $g : [0, 1] \to \{0, 1, 2, \dots, n\}$ to be the step function

$$g(u) = \begin{cases} 0, & \text{if } 0 \le u \le p_0 \\ 1, & \text{if } p_0 < u \le p_0 + p_1 \\ 2, & \text{if } p_0 + p_1 < u \le p_0 + p_1 + p_2 \\ \vdots & \\ n, & \text{if } p_0 + p_1 + \cdots + p_{n-1} < u \le p_0 + p_1 + \cdots + p_n = 1. \end{cases}$$

For $0 \le k \le n$ we have

$$P(g(U) = k) = P(p_0 + \cdots + p_{k-1} < U \le p_0 + \cdots + p_k) = p_k = P(X = k).$$

Thus, $g(U)$ has the same distribution as X.

You can visualize the solution as follows. The probabilities $p_0, p_1, \dots, p_n$ add up to 1, so we can break up the unit interval [0, 1] into smaller pieces so that the length of the jth interval is p_j. Define the function g on [0, 1] so that it is equal to j on the jth interval. Then

$$P(g(U) = j) = P(U \text{ lies in the } j\text{th subinterval}) = p_j,$$

because the probability that the uniformly chosen point U lies in a given subinterval of [0, 1] equals the length of the subinterval. Figure 5.1 gives an example with $n = 6$.

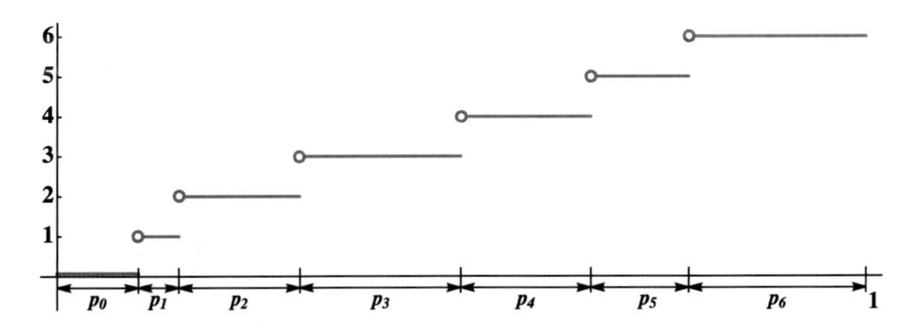

Figure 5.1. The green graph is the function g corresponding to probabilities $p_0 + \cdots + p_6 = 1$.

This construction works also for discrete random variables with arbitrary values, just by choosing the values $g(u)$ appropriately. It extends to discrete random variables with infinitely many possible values, by breaking $[0, 1]$ into infinitely many subintervals. ▲

The following example provides an algorithm that converts a uniform random variable into a random variable with a given continuous distribution.

Example 5.30. Let $U \sim \text{Unif}(0, 1)$ and let X be a continuous random variable with density function f. Assume that for some $-\infty \le a < b \le \infty$, $f(t) > 0$ on the interval (a, b) and $f(t) = 0$ for $t \notin (a, b)$. Find a function $g : (0, 1) \to (a, b)$ such that $g(U)$ has the same distribution as X.

Let F be the cumulative distribution function of X. Since $F(t) = \int_{-\infty}^{t} f(x)\,dx$, the assumptions on f tell us that over the interval (a, b) F is one-to-one and increases from 0 to 1. Hence, as a function from (a, b) onto $(0, 1)$, F has an inverse function F^{-1} from $(0, 1)$ onto (a, b). Let $g(u) = F^{-1}(u)$ for $0 < u < 1$.

Let us verify that $g(U)$ has cumulative distribution function F. Since by construction $g(U) \in (a, b)$, we must have $P(g(U) \le t) = 0$ for $t \le a$ and $P(g(U) \le t) = 1$ for $t \ge b$. For $t \in (a, b)$,

$$P(g(U) \le t) = P(F^{-1}(U) \le t) = P(U \le F(t)) = F(t).$$

Example 5.19 was a special case of this result. There $X \sim \text{Exp}(\lambda)$ has cumulative distribution function $F(t) = 1 - e^{-\lambda t}$ for $t > 0$, and zero otherwise. Therefore $F^{-1}(u) = -\lambda^{-1} \ln(1 - u)$ for $0 < u < 1$. ▲

5.3. Finer points ♣

Equality in distribution

As we have seen in Section 1.6, $P(X \in B)$ is in general defined only for Borel subsets B of $\mathbb{R}$. Hence, in the precise version of Definition 5.12 the set B must be a Borel set.

Transforms of random variables

In addition to the moment generating function, there are other important transforms of random variables. One of them is the *characteristic function* defined by $\varphi_X(t) = E[e^{itX}]$, where $i = \sqrt{-1}$ is the imaginary unit. (Basically the characteristic function is the same as the Fourier transform.) The expected value of a complex-valued random variable Z can be defined through the real and imaginary parts as $E[Z] = E[\text{Re}\, Z] + iE[\text{Im}\, Z]$. Formally, we have $\varphi_X(t) = M_X(it)$.

The characteristic function shares many of the useful properties of the moment generating function, but it has the added advantage of being well defined for all values of t for any random variable X. This is because the random variable $e^{itX} = \cos(tX) + i \sin(tX)$ is always bounded, so there is no problem with the existence of the expectation. If the nth moment of X exists then the nth derivative of $\varphi_X(t)$ at $t = 0$ satisfies

$$\varphi_X^{(n)}(0) = i^n E[X^n].$$

The analogue of Fact 5.14 holds for characteristic functions in the following form: if $\varphi_X(t) = \varphi_Y(t)$ for $t \in \mathbb{R}$ then $X \stackrel{d}{=} Y$. (See [Dur10] for the proof.) This also gives a proof of Fact 5.14. If $E(e^{sX}) = E(e^{sY}) < \infty$ for $s \in (-\delta, \delta)$ then methods of complex analysis extend this equality to $E(e^{(s+it)X}) = E(e^{(s+it)Y})$ for $s \in (-\delta, \delta)$ and all $t \in \mathbb{R}$. Taking $s = 0$ gives $\varphi_X(t) = \varphi_Y(t)$ for all t and so $X \stackrel{d}{=} Y$.

Another important transform is the *probability generating function* which is defined for a nonnegative integer valued random variable X by

$$g_X(s) = E[s^X] = \sum_{k=0}^{\infty} P(X = k)s^k. \tag{5.18}$$

This is a power series whose coefficients $P(X = k)$ are at most one in absolute value. Therefore, the radius of convergence is at least 1, and so $g_X(s)$ is well defined for $|s| \leq 1$. A convergent power series on an interval around the origin uniquely determines its coefficients. Consequently the function $g_X(s)$ determines the probability mass function of X.

Exercises

We start with some warm-up exercises arranged by section.

Section 5.1

Exercise 5.1. Let X be a discrete random variable with probability mass function

$$P(X = -6) = \tfrac{4}{9}, \quad P(X = -2) = \tfrac{1}{9}, \quad P(X = 0) = \tfrac{2}{9}, \quad \text{and} \quad P(X = 3) = \tfrac{2}{9}.$$

Find the moment generating function of X.

Exercise 5.2. Suppose that X has moment generating function

$$M_X(t) = \tfrac{1}{2} + \tfrac{1}{3}e^{-4t} + \tfrac{1}{6}e^{5t}.$$

(a) Find the mean and variance of X by differentiating the moment generating function to find moments.

(b) Find the probability mass function of X. Use the probability mass function to check your answer for part (a).

Exercise 5.3. Let $X \sim \text{Unif}[0, 1]$. Find the moment generating function $M(t)$ of X. Note that the calculation of $M(t)$ for $t \neq 0$ puts a t in the denominator, hence the value $M(0)$ has to be calculated separately.

Exercise 5.4. In parts (a)–(d) below, either use the information given to determine the distribution of the random variable, or show that the information given is not sufficient by describing at least two different random variables that satisfy the given condition.

(a) X is a random variable such that $M_X(t) = e^{6t^2}$ when $|t| < 2$.
(b) Y is a random variable such that $M_Y(t) = \frac{2}{2-t}$ for $t < 0.5$.
(c) Z is a random variable such that $M_Z(t) = \infty$ for $t \geq 5$.
(d) W is a random variable such that $M_W(2) = 2$.

Exercise 5.5. The moment generating function of the random variable X is $M_X(t) = e^{3(e^t-1)}$. Find $P(X = 4)$.

Hint. Look at Example 5.16.

Section 5.2

Exercise 5.6. Suppose that X is a discrete random variable with probability mass function

$$P(X = -1) = 1/7, \quad P(X = 0) = 1/14, \quad P(X = 2) = 3/14, \quad P(X = 4) = 4/7.$$

Compute the probability mass function of $(X - 1)^2$.

Exercise 5.7. Suppose $X \sim \text{Exp}(\lambda)$ and $Y = \ln X$. Find the probability density function of Y.

Exercise 5.8. Let $X \sim \text{Unif}[-1, 2]$. Find the probability density function of the random variable $Y = X^2$.

Further exercises

Exercise 5.9. Let $X \sim \text{Bin}(n, p)$.

(a) Find the moment generating function $M_X(t)$.
Hint. The binomial theorem from (D.10) could be useful.

(b) Use part (a) to find $E[X]$, $E[X^2]$ and $\text{Var}(X)$. Compare your answers with Examples 3.23 and 3.49.

Exercise 5.10. Suppose that X has moment generating function

$$M(t) = \left(\frac{1}{5} + \frac{4}{5}e^t\right)^{30}.$$

What is the distribution of X?

Hint. Do Exercise 5.9 first.

Exercise 5.11. The random variable X has the following probability density function:

$$f_X(x) = \begin{cases} xe^{-x}, & \text{if } x > 0 \\ 0, & \text{otherwise.} \end{cases}$$

(a) Find the moment generating function of X.
(b) Using the moment generating function of X, find $E[X^n]$ for all positive integers n. Your final answer should be an expression that depends only on n.

Exercise 5.12. Suppose the random variable X has density function

$$f(x) = \begin{cases} \frac{1}{2}x^2 e^{-x}, & \text{if } x \geq 0 \\ 0, & \text{otherwise.} \end{cases}$$

Find the moment generating function $M(t)$ of X. Be careful about the values of t for which $M(t) < \infty$.

Exercise 5.13. Suppose that Y has moment generating function

$$M_Y(t) = \frac{1}{2} + \frac{1}{16}e^{-34t} + \frac{1}{8}e^{-5t} + \frac{1}{100}e^{3t} + \frac{121}{400}e^{100t}.$$

Compute the mean of Y in two different ways. First, generate the first moment from the moment generating function. Next, provide the distribution of Y and calculate the mean directly.

Exercise 5.14. Suppose that $X \sim \text{Bin}(4, 1/2)$. Find the probability mass function of $Y = (X - 2)^2$.

Exercise 5.15. Suppose that X is a discrete random variable with probability mass function

$$P(X = -2) = 1/10, \quad P(X = -1) = 1/5, \quad P(X = 0) = 3/10, \quad P(X = 1) = 2/5.$$

(a) Compute the moment generating function of X.
(b) Find the probability mass function of $Y = |X + 1|$.

Exercise 5.16. Let $X \sim \text{Unif}[0, 1]$. Parts (a) and (b) below ask you to find the nth moment of X in two different ways.

(a) Use the formula $E(X^n) = \int_0^1 x^n \, dx$.
(b) Compute the moment generating function $M_X(t)$ (or use the result of Exercise 5.3). Find the Taylor series expansion of $M_X(t)$ and identify the coefficients.

Hint. $M_X(t)$ will be written in terms of e^t. Substitute the Taylor series for e^t into the formula and simplify the expression.

Exercise 5.17. Let X be a continuous random variable with probability density function

$$f(x) = \begin{cases} x/2, & \text{if } 0 < x < 2 \\ 0, & \text{otherwise.} \end{cases}$$

(a) Find the moment generating function $M_X(t)$ of X.
Hint. After setting up the integral, integrate by parts for $t \neq 0$

(b) Find the general formula for the moments of X from $M_X(t)$. You can either differentiate $M_X(t)$ repeatedly at $t = 0$, or use the exponential series to find the series expansion of $M_X(t)$.

(c) Verify the formula you found in part (b) by computing the expectation $E(X^k)$ for a general nonnegative integer k.

Exercise 5.18. Let $X \sim \text{Geom}(p)$.

(a) Compute the moment generating function $M_X(t)$ of X. Be careful about the possibility that $M_X(t)$ might be infinite.

(b) Use the moment generating function to compute the mean and the variance of X.

Exercise 5.19. Suppose that the random variable X has probability mass function given by

$$P(X = 0) = \frac{2}{5}, \qquad P(X = k) = \left(\frac{3}{4}\right)^k \frac{1}{5}, \qquad k = 1, 2, \ldots$$

(a) Compute the moment generating function $M_X(t)$ of X. Be careful about the possibility that $M_X(t)$ might be infinite.

(b) Use the moment generating function to compute the mean and the variance of X.

Exercise 5.20. Suppose that random variable X has density function $f(x) = \frac{1}{2}e^{-|x|}$.

(a) Compute the moment generating function $M_X(t)$ of X. Be careful about the possibility that $M_X(t)$ might be infinite.

(b) Use the moment generating function to compute the nth moment of X.

Exercise 5.21. Let $Y = aX + b$ where a, b are real numbers. Express the moment generating function $M_Y(t)$ of the random variable Y in terms of $M_X(t)$.

Hint. Use the definition of M_Y and the properties of expectation.

Exercise 5.22. Let $X \sim \text{Exp}(\lambda)$. Find the moment generating function of $Y = 3X - 2$.

Hint. Do not try to compute the probability density function of Y, instead use $M_X(t)$.

Exercise 5.23. The random variable Y has moment generating function $M_Y(t) = e^{2(e^{2t}-1)}$. Find $P(Y = 4)$.

Hint. Exercise 5.21 could be helpful.

Exercise 5.24. Let $X \sim \mathcal{N}(0, 1)$ and $Y = e^X$. Y is called a *log-normal* random variable.

(a) Find the probability density function of Y.

(b) Find the nth moment $E(Y^n)$ of Y.

Hint. Do not compute the moment generating function of Y. Instead relate the nth moment of Y to an expectation of X that you know.

Exercise 5.25. Suppose that X is uniform on $[-2,3]$ and let $Y = |X-1|$. Find the density function of Y.

Exercise 5.26. Let X be a random variable with density function

$$f_X(x) = \begin{cases} \frac{2}{9}x, & \text{for } 0 \le x \le 3 \\ 0, & \text{otherwise.} \end{cases}$$

Find the density function of $Y = X(X-3)$.

Exercise 5.27. Let $X \sim \text{Exp}(\lambda)$. Find the probability density function of the random variable $Y = \exp(X)$.

Exercise 5.28. Let $X \sim \text{Unif}[-1,2]$. Find the probability density function of the random variable $Y = X^4$.

Exercise 5.29. Let $Z \sim \mathcal{N}(0,1)$. Find the probability density function of $|Z|$.

Exercise 5.30. Suppose that X is uniform on $[-\pi, 2\pi]$. Find the probability density function of $Y = \sin(X)$.

Exercise 5.31. Suppose that $U \sim \text{Unif}[0,1]$. Let $Y = e^{\frac{U}{1-U}}$. Find the probability density function of Y.

Exercise 5.32. Suppose that X is a uniform random variable on the interval $(0,1)$ and let $Y = 1/X$. Find the probability density function of Y.

Exercise 5.33. Suppose that X is a discrete random variable with possible values $\{1,4,9\}$ and probability mass function

$$P\{X=1\} = \frac{1}{7}, \quad P\{X=4\} = \frac{2}{7}, \quad P\{X=9\} = \frac{4}{7}.$$

Let $U \sim \text{Uniform}[0,1]$. Find a function $g : [0,1] \to \mathbb{R}$ so that $g(U)$ has the same distribution as X.

Exercise 5.34. Let $U \sim \text{Unif}(0,1)$. Let X be a random variable with values in the interval $(1,3)$ and that satisfies

$$\begin{aligned} & P(X=2) = \tfrac{1}{3}, & \\ & P(1 < X < a) = \tfrac{1}{3}(a-1) & \text{for } 1 < a < 2, \\ \text{and} \quad & P(b < X < 3) = \tfrac{1}{3}(3-b) & \text{for } 2 < b < 3. \end{aligned}$$

Find a function g such that $g(U)$ has the same distribution as X.

Hint. Sketch F_X. Combine the recipes found in Examples 5.29 and 5.30.

Exercise 5.35. The floor function $\lfloor x \rfloor = \max\{n \in \mathbb{Z} : n \leq x\}$ gives the largest integer less than or equal to x. Let $X \sim \text{Exp}(\lambda)$. Find the probability mass function of $\lfloor X \rfloor$.

Exercise 5.36. Let $X \sim \text{Exp}(\frac{1}{100})$. Find the probability mass function of $g(X) = \lfloor \sqrt{X} \rfloor$, where $\lfloor x \rfloor$ is the floor function (defined in Exercise 5.35). Note: $g(X)$ is a discrete random variable. Be clear about its possible values.

Exercise 5.37. The fractional part of x is defined as $\{x\} = x - \lfloor x \rfloor$ with $\lfloor x \rfloor$ defined in Exercise 5.35. (For example $\{1.3\} = 0.3$, $\{-2.1\} = 0.9$.) Let $X \sim \text{Exp}(\lambda)$ and set $Y = \{X\}$. Find the probability density function of Y.

Exercise 5.38. Suppose that x has a probability density function

$$f(x) = \begin{cases} x^{-2}, & x > 1 \\ 0, & x \leq 1. \end{cases}$$

Find a function $g(x)$ so that $Y = g(X)$ is uniformly distributed on $[0, 1]$.

Challenging problems

Exercise 5.39. Determine whether the following functions are moment generating functions. If they are, give the corresponding probability distribution:

(a) $\cosh(t)$,
(b) $\sinh(t)$.

Exercise 5.40. Show that there is no random variable X with $M_X(1) = 3$ and $M_X(2) = 4$.

Exercise 5.41. Suppose that X is a continuous random variable with density function $f_X(x)$ and g is a differentiable, strictly increasing function for which $E[g(X)]$ is finite. Prove that $E[g(X)] = \int_{-\infty}^{\infty} g(x) f_X(x) dx$.
We have been using this formula without proof, but using Fact 5.24 we can actually derive it rigorously under the assumptions above.

Exercise 5.42. Let X be the random variable with density function

$$f(x) = \begin{cases} \frac{3}{2} x^{-4}, & x \leq -1 \text{ or } x \geq 1 \\ 0, & -1 < x < 1. \end{cases}$$

Compute the moment generating function $M_X(t)$ of X.

Exercise 5.43. Let X have cumulative distribution function

$$F(x) = \begin{cases} 0, & x < 0 \\ x, & 0 \leq x < \frac{1}{2} \\ 1, & x \geq \frac{1}{2}. \end{cases}$$

Find the moment generating function $M(t) = E[e^{tX}]$. Note that X is neither discrete nor continuous.

Exercise 5.44. Let X be a continuous random variable with density function $f_X(x)$. Let g be a differentiable, strictly increasing function and set $Y = g(X)$. Use the infinitesimal method introduced after Fact 3.7 to derive formula (5.13) for the density function of Y.

6

Joint distribution of random variables

We have come to a major juncture in the course where our point of view shifts from a single random variable to larger collections of random variables. We have seen some instances of this, for example in our discussions of sequences of trials. The remainder of this course (and really probability theory in general) is concerned with studying collections of random variables in various situations.

If $X_1, X_2, \dots, X_n$ are random variables defined on a sample space Ω, we can regard them as coordinates of the *random vector* $(X_1, X_2, \dots, X_n)$. This vector is again a "random variable" in the sense of being a function on Ω, but instead of being real valued it is $\mathbb{R}^n$ valued. The concepts related to a single random variable can be adapted to a random vector. The probability distribution of $(X_1, X_2, \dots, X_n)$ is now an assignment of probabilities $P((X_1, X_2, \dots, X_n) \in B)$ to subsets B of $\mathbb{R}^n$. The probability distribution of a random vector is called a *joint distribution*, and the probability distributions of the individual coordinates X_j are called *marginal distributions*. A joint distribution can again be described by a probability mass function in the discrete case, by a probability density function in the jointly continuous case, and in general one can define a multivariate cumulative distribution function.

6.1. Joint distribution of discrete random variables

The following definition is the multivariate extension of Definition 1.35 of the probability mass function of a single discrete random variable.

Definition 6.1. Let $X_1, X_2, \dots, X_n$ be discrete random variables, all defined on the same sample space. Their **joint probability mass function** is defined by

$$p(k_1, k_2, \dots, k_n) = P(X_1 = k_1, X_2 = k_2, \dots, X_n = k_n)$$

for all possible values $k_1, k_2, \dots, k_n$ of $X_1, X_2, \dots, X_n$.

As with our other notations, we can write $p_{X_1, X_2, \dots, X_n}(k_1, k_2, \dots, k_n)$ if we wish the notation to include the random variables. The joint probability mass function serves precisely the same purpose as the probability mass function of a single random variable. Of course, it must have nonnegative values and the values must sum up to one:

$$\sum_{k_1,k_2,\dots,k_n} p_{X_1,X_2,\dots,X_n}(k_1,k_2,\dots,k_n) = 1.$$

Once we have the joint probability mass function of the random variables $X_1,\dots,X_n$ we can (in principle) compute the probability of any event related to them. This is because any such event can be decomposed into a disjoint union of events of the form $\{X_1 = k_1, X_2 = k_2, \dots, X_n = k_n\}$ whose probabilities are given by the joint probability mass function.

If we have a description of the joint distribution of random variables $X_1,\dots,X_n$ via the joint probability mass function we can also compute expectations of random variables of the form $g(X_1,\dots,X_n)$. The formulas are similar to the single variable case, but now one needs to work with multiple sums. The following fact can be proved similarly as the single variable version Fact 3.33.

Fact 6.2. Let $g : \mathbb{R}^n \to \mathbb{R}$ be a real-valued function of an n-vector. If $X_1,\dots,X_n$ are discrete random variables with joint probability mass function p then

$$E[g(X_1,\dots,X_n)] = \sum_{k_1,\dots,k_n} g(k_1,\dots,k_n)\, p(k_1,\dots,k_n) \tag{6.1}$$

provided the sum is well defined.

We turn to our first example.

Example 6.3. Flip a fair coin three times. Let X be the number of tails in the first flip and Y the total number of tails observed. The possible values of X are $\{0, 1\}$ and the possible values of Y are $\{0, 1, 2, 3\}$. Let us first record which outcomes of the experiment lead to particular (X, Y)-values, with H for heads and T for tails.

X \ Y	0	1	2	3
0	HHH	HTH, HHT	HTT	
1		THH	THT, TTH	TTT

The way to read the table is that, for example, outcomes HTH and HHT give $(X, Y) = (0, 1)$ and no outcome results in $(X, Y) = (1, 0)$.

To obtain the joint probability mass function of (X, Y) record the probabilities of the events from the first table.

			Y		
		0	1	2	3
X	0	$\frac{1}{8}$	$\frac{2}{8}$	$\frac{1}{8}$	0
	1	0	$\frac{1}{8}$	$\frac{2}{8}$	$\frac{1}{8}$

From the table we can read the values of the joint probability mass function: $p_{X,Y}(0,0) = \frac{1}{8}$, $p_{X,Y}(0,1) = \frac{2}{8}$, ..., $p_{X,Y}(0,3) = 0$, ..., $p_{X,Y}(1,3) = \frac{1}{8}$.

Suppose each tails earns you 3 dollars, and if the first flip is tails, each reward is doubled. This reward is encoded by the function $g(x,y) = 3(1+x)y$. The expected reward is

$$\begin{aligned} E[g(X,Y)] &= \sum_{k=0}^{1}\sum_{\ell=0}^{3} g(k,\ell)\, p_{X,Y}(k,\ell) = \sum_{k=0}^{1}\sum_{\ell=0}^{3} 3(1+k)\ell\, p_{X,Y}(k,\ell) \\ &= \sum_{\ell=0}^{3} 3\ell\, p_{X,Y}(0,\ell) + \sum_{\ell=0}^{3} 6\ell\, p_{X,Y}(1,\ell) \\ &= \left(3\cdot\tfrac{2}{8} + 6\cdot\tfrac{1}{8}\right) + \left(6\cdot\tfrac{1}{8} + 12\cdot\tfrac{2}{8} + 18\cdot\tfrac{1}{8}\right) = 7\tfrac{1}{2}. \end{aligned}$$

Probabilities of X are obtained by summing up rows of the table. For example

$$P(X=0) = \sum_{\ell=0}^{3} p_{X,Y}(0,\ell) = \tfrac{1}{8} + \tfrac{2}{8} + \tfrac{1}{8} + 0 = \tfrac{1}{2}. \tag{6.2}$$

Probabilities of Y are obtained by summing up columns of the table. For example

$$P(Y=2) = P(X=0, Y=2) + P(X=1, Y=2) = \tfrac{1}{8} + \tfrac{2}{8} = \tfrac{3}{8}. \tag{6.3}$$

▲

As the example above illustrates, even if k is a possible value for X and ℓ is a possible value for Y, (k,ℓ) may be unattainable for (X,Y). For such values the joint probability mass function gives $p(k,\ell) = 0$.

The calculations in (6.2) and (6.3) show how the probability mass function of an individual random variable is derived from the joint probability mass function.

Fact 6.4. Let $p(k_1, \dots, k_n)$ be the joint probability mass function of $(X_1, \dots, X_n)$. Let $1 \le j \le n$. Then the probability mass function of X_j is given by

$$p_{X_j}(k) = \sum_{\ell_1,\dots,\ell_{j-1},\ell_{j+1},\dots,\ell_n} p(\ell_1, \dots, \ell_{j-1}, k, \ell_{j+1}, \dots, \ell_n), \tag{6.4}$$

where the sum is over the possible values of the other random variables. The function p_{X_j} is called the *marginal probability mass function* of X_j.

For example, for two random variables X and Y we get the formulas

$$p_X(x) = \sum_y p_{X,Y}(x, y) \quad \text{and} \quad p_Y(y) = \sum_x p_{X,Y}(x, y). \tag{6.5}$$

If the joint probability mass function of X and Y is presented as a table, then these are the row and column sums.

Proof of Fact 6.4. The proof falls back on the familiar idea of decomposing an event according to a partition of Ω. Now the partition is given by all the values of $X_1, X_2, \dots, X_{j-1}, X_{j+1}, \dots, X_n$.

$$\begin{aligned} p_{X_j}(k) &= P(X_j = k) \\ &= \sum_{\ell_1,\dots,\ell_{j-1},\ell_{j+1},\dots,\ell_n} P(X_1 = \ell_1, \dots, X_{j-1} = \ell_{j-1}, X_j = k, X_{j+1} = \ell_{j+1}, \dots, X_n = \ell_n) \\ &= \sum_{\ell_1,\dots,\ell_{j-1},\ell_{j+1},\dots,\ell_n} p(\ell_1, \dots, \ell_{j-1}, k, \ell_{j+1}, \dots, \ell_n). \end{aligned}$$

▲

We can also obtain the joint probability mass function of a subset of the random variables. Here is a particular case. The fact below is derived in a similar way as Fact 6.4 was proved.

Fact 6.5. Let $1 \le m < n$. The joint probability mass function of $(X_1, \dots, X_m)$ is obtained from

$$p_{X_1,\dots,X_m}(k_1, \dots, k_m) = \sum_{\ell_{m+1},\dots,\ell_n} p(k_1, \dots, k_m, \ell_{m+1}, \dots, \ell_n)$$

where the sum ranges over all possible values $\ell_{m+1}, \dots, \ell_n$ of the random variables $X_{m+1}, \dots, X_n$.

The next example illustrates, among other things, the derivation of the joint distribution of (Y_1, Y_2) that are functions of (X_1, X_2). It is a multivariate version of Example 5.17.

Example 6.6. A tetrahedron shaped die gives one of the numbers 1, 2, 3, 4 with equal probabilities. We roll two of these dice and denote the two outcomes by X_1 and X_2. Let $Y_1 = \min(X_1, X_2)$ and $Y_2 = |X_1 - X_2|$. Find (i) the joint probability mass function of X_1, X_2, (ii) the joint probability mass function of Y_1, Y_2, (iii) the marginal probability mass functions of Y_1 and Y_2, (iv) the probability $P(Y_1 Y_2 \le 2)$, and (v) $E[Y_1 Y_2]$.

(i) The joint probability mass function of X_1, X_2 is easy: the possible values for this pair are $\{(k_1, k_2) : 1 \le k_1, k_2 \le 4\}$ and all outcomes are equally likely. Thus $p_{X_1,X_2}(k_1, k_2) = (\frac{1}{4})^2 = \frac{1}{16}$ if $1 \le k_1, k_2 \le 4$.

(ii) In the case of (Y_1, Y_2) we start by identifying the possible values for the pair. Y_1 is between 1 and 4 and Y_2 is between 0 and 3 so we can describe the joint probability mass function by setting up a 4×4 table (see (6.7) below). For a given $1 \le k \le 4$ we have

$$p_{Y_1,Y_2}(k,0) = P(Y_1 = k, Y_2 = 0) = P(\text{both dice show } k) = \tfrac{1}{4^2} = \tfrac{1}{16}.$$

For $1 \le k_1 \le 4$ and $0 < k_2 \le 3$ we have

$$\begin{aligned} p_{Y_1,Y_2}(k_1,k_2) &= P(Y_1 = k_1, Y_2 = k_2) \\ &= P(\text{one die shows } k_1, \text{ the other shows } k_1 + k_2). \end{aligned}$$

If $k_1 + k_2 > 4$ then this probability equals zero since each die can only show numbers from 1 to 4. If $k_1 + k_2 \le 4$ then

$$\begin{aligned} p_{Y_1,Y_2}(k_1,k_2) &= P(\text{one die shows } k_1, \text{ the other shows } k_1 + k_2) \\ &= p_{X_1,X_2}(k_1, k_1+k_2) + p_{X_1,X_2}(k_1+k_2, k_1) = 2 \cdot (\tfrac{1}{4})^2 = \tfrac{1}{8}. \end{aligned}$$

To summarize:

$$p_{Y_1,Y_2}(k_1,k_2) = \begin{cases} \frac{1}{16}, & \text{if } 1 \le k_1 \le 4 \text{ and } k_2 = 0 \\ \frac{1}{8}, & \text{if } 1 \le k_1 \le 4,\ 0 < k_2 \le 3 \text{ and } k_1 + k_2 \le 4 \\ 0, & \text{if } 1 \le k_1 \le 4,\ 0 < k_2 \le 3 \text{ and } k_1 + k_2 > 4. \end{cases} \tag{6.6}$$

We can arrange the values in the following table:

(6.7)

Y_1 \ Y_2	0	1	2	3
1	$\frac{1}{16}$	$\frac{1}{8}$	$\frac{1}{8}$	$\frac{1}{8}$
2	$\frac{1}{16}$	$\frac{1}{8}$	$\frac{1}{8}$	0
3	$\frac{1}{16}$	$\frac{1}{8}$	0	0
4	$\frac{1}{16}$	0	0	0

Equation (6.6) and table (6.7) are both sufficient to describe the joint probability mass function.

(iii) The row sums of table (6.7) give the marginal probability mass function of Y_1 and the column sums give the marginal probability mass function of Y_2. For example,

$$p_{Y_1}(1) = P(Y_1 = 1) = \sum_{\ell=0}^{3} p_{Y_1,Y_2}(1,\ell) = \tfrac{1}{16} + \tfrac{1}{8} + \tfrac{1}{8} + \tfrac{1}{8} = \tfrac{7}{16}.$$

Check that your calculations agree with these remaining values:

$$\begin{aligned} &P(Y_1 = 2) = \tfrac{5}{16}, \quad P(Y_1 = 3) = \tfrac{3}{16}, \quad P(Y_1 = 4) = \tfrac{1}{16}, \\ &P(Y_2 = 0) = \tfrac{1}{4}, \quad P(Y_2 = 1) = \tfrac{3}{8}, \quad P(Y_2 = 2) = \tfrac{1}{4}, \quad P(Y_2 = 3) = \tfrac{1}{8}. \end{aligned}$$

(iv) To compute $P(Y_1 Y_2 \leq 2)$ we find the pairs (k_1, k_2) that satisfy $k_1 k_2 \leq 2$ with $1 \leq k_1 \leq 4$ and $0 \leq k_2 \leq 3$. These are $(1, 0), (1, 1), (1, 2), (2, 0), (2, 1), (3, 0), (4, 0)$. We then add up the probabilities of these pairs:

$$\begin{aligned} P(Y_1 Y_2 \leq 2) &= p_{Y_1,Y_2}(1,0) + p_{Y_1,Y_2}(1,1) + p_{Y_1,Y_2}(1,2) + p_{Y_1,Y_2}(2,0) \\ &\quad + p_{Y_1,Y_2}(2,1) + p_{Y_1,Y_2}(3,0) + p_{Y_1,Y_2}(4,0) \\ &= \tfrac{1}{16} + \tfrac{1}{8} + \tfrac{1}{8} + \tfrac{1}{16} + \tfrac{1}{8} + \tfrac{1}{16} + \tfrac{1}{16} = \tfrac{5}{8}. \end{aligned}$$

(v) To compute the desired expectation we sum over the possible values:

$$\begin{aligned} E[Y_1 Y_2] &= \sum_{k_1=1}^{4} \sum_{k_2=0}^{3} k_1 k_2 \, p_{Y_1,Y_2}(k_1, k_2) \\ &= 1 \cdot \tfrac{1}{8} + 2 \cdot \tfrac{1}{8} + 3 \cdot \tfrac{1}{8} + 2 \cdot \tfrac{1}{8} + 4 \cdot \tfrac{1}{8} + 3 \cdot \tfrac{1}{8} = \tfrac{15}{8}, \end{aligned}$$

where in the second equality we only kept the nonzero terms. ▲

Section 1.2 on sampling skipped sampling with replacement with order of the sample ignored. This is a special case of the multinomial experiment.

Example 6.7. (Multinomial distribution) Consider a trial with r possible outcomes labeled $1, \ldots, r$. Outcome j appears with probability p_j, and then of course $p_1 + p_2 + \cdots + p_r = 1$. Perform n independent repetitions of this trial. For $j = 1, \ldots, r$ let X_j denote the number of times we see outcome j among the n trials. Let us derive the joint distribution of $(X_1, \ldots, X_r)$.

Consider nonnegative integers $k_1, \ldots, k_r$ such that $k_1 + \cdots + k_r = n$. Any particular sequence of outcomes from the n trials that yields outcome j exactly k_j times has probability $p_1^{k_1} \cdots p_r^{k_r}$. By Fact C.16 from Appendix C, the number of such arrangements is given by the multinomial coefficient

$$\binom{n}{k_1, k_2, \ldots, k_r} = \frac{n!}{k_1! \, k_2! \, \cdots \, k_r!}.$$

Summing over the equally likely arrangements gives the formula

$$P(X_1 = k_1, X_2 = k_2, \ldots, X_r = k_r) = \binom{n}{k_1, k_2, \ldots, k_r} p_1^{k_1} p_2^{k_2} \cdots p_r^{k_r}.$$

These probabilities sum to 1 by the multinomial theorem (Fact D.3). ▲

Let us formalize this distribution with a definition.

Definition 6.8. Let n and r be positive integers and let $p_1, p_2, \ldots, p_r$ be positve reals such that $p_1 + p_2 + \cdots + p_r = 1$. Then the random vector $(X_1, \ldots, X_r)$ has the **multinomial distribution** with parameters n, r and $p_1, \ldots, p_r$ if the possible values are integer vectors $(k_1, \ldots, k_r)$ such that $k_j \geq 0$ and $k_1 + \cdots + k_r = n$, and the joint probability mass function is given by

$$P(X_1 = k_1, X_2 = k_2, \ldots, X_r = k_r) = \binom{n}{k_1, k_2, \ldots, k_r} p_1^{k_1} p_2^{k_2} \cdots p_r^{k_r}. \tag{6.8}$$

Abbreviate this by $(X_1, \ldots, X_r) \sim \text{Mult}(n, r, p_1, \ldots, p_r)$.

The special case with $r = 2$ is the binomial distribution. In that case $X_2 = n - X_1$ so only one of the variables is really needed. The special case of sampling n times with replacement from a set of r items has $p_1 = p_2 = \cdots = p_r = \frac{1}{r}$ and variable X_j counts how many times item j was sampled.

Example 6.9. Suppose an urn contains 1 green, 2 red and 3 yellow balls. We sample a ball with replacement 10 times. Find the probability that green appeared 3 times, red twice, and yellow 5 times.

Let X_G, X_R and X_Y denote the number of green, red and yellow balls in the sample. Then $(X_G, X_R, X_Y) \sim \text{Mult}(10, 3, \frac{1}{6}, \frac{2}{6}, \frac{3}{6})$ and

$$\begin{aligned} &P(\text{green appeared 3 times, red twice, and yellow 5 times}) \\ &\quad = P(X_G = 3, X_R = 2, X_Y = 5) = \frac{10!}{3!\,2!\,5!} \left(\frac{1}{6}\right)^3 \left(\frac{2}{6}\right)^2 \left(\frac{3}{6}\right)^5 \approx 0.0405. \end{aligned}$$

▲

The next two examples look at marginal distributions of a multinomial.

Example 6.10. Let $(X_1, \ldots, X_r) \sim \text{Mult}(n, r, p_1, \ldots, p_r)$. Find the distribution of X_1.

The random variable X_1 counts the number of times that outcome 1 appears among the n trials. Considering outcome 1 as a success and any other outcome as a failure, we see that X_1 is a binomial random variable with parameters n and p_1. ▲

Example 6.11. We roll a die 100 times. Find the probability that among the 100 rolls we observe exactly 22 ones and 17 fives.

Denote by X_i the number of times i appears among the 100 rolls. Then $(X_1, \ldots, X_6)$ has multinomial distribution with $n = 100$, $r = 6$ and $p_1 = \cdots = p_6 = \frac{1}{6}$. We need to compute $P(X_1 = 22, X_5 = 17)$. This can be computed from the joint probability mass function of $(X_1, \ldots, X_6)$ by summing over the possible values of X_2, X_3, X_4 and X_6. But there is a simpler way.

Since we are only interested in the rolls that are 1 or 5, we can combine the other outcomes into a new outcome: let $Y = X_2 + X_3 + X_4 + X_6$ denote the number of times we rolled something other than 1 or 5. The probability that a particular roll is 2, 3, 4, or 6 is $\frac{2}{3}$, and thus

$$(X_1, X_5, Y) \sim \text{Mult}(100, 3, \tfrac{1}{6}, \tfrac{1}{6}, \tfrac{2}{3}).$$

Now we can express the required probability using the multinomial joint probability mass function:

$$\begin{aligned} P(X_1 = 22, X_5 = 17) &= P(X_1 = 22, X_5 = 17, Y = 61) \\ &= \frac{100!}{22!\,17!\,61!} \left(\frac{1}{6}\right)^{22} \left(\frac{1}{6}\right)^{17} \left(\frac{2}{3}\right)^{61} \approx 0.0037. \end{aligned}$$

▲

6.2. Jointly continuous random variables

The continuous counterpart of a joint probability mass function is the joint density function.

Definition 6.12. Random variables $X_1, \dots, X_n$ are **jointly continuous** if there exists a **joint density function** f on $\mathbb{R}^n$ such that for subsets $B \subseteq \mathbb{R}^n$, ♣

$$P((X_1, \dots, X_n) \in B) = \int \cdots \int_B f(x_1, \dots, x_n)\, dx_1 \cdots dx_n. \tag{6.9}$$

We may write $f_{X_1,\dots,X_n}(x_1, \dots, x_n)$ when we wish to emphasize the random variables being considered. A joint density function must be nonnegative and its total integral must be 1:

$$f(x_1, \dots, x_n) \geq 0 \quad \text{and} \quad \int_{-\infty}^{\infty} \cdots \int_{-\infty}^{\infty} f(x_1, \dots, x_n)\, dx_1 \cdots dx_n = 1. \tag{6.10}$$

As in the discrete case, an expectation of a function of several random variables can be computed similarly to the one-dimensional setting.

Fact 6.13. Let $g : \mathbb{R}^n \to \mathbb{R}$ be a real-valued function of an n-vector. If $X_1, \dots, X_n$ are random variables with joint density function f then

$$E[g(X_1, \dots, X_n)] = \int_{-\infty}^{\infty} \cdots \int_{-\infty}^{\infty} g(x_1, \dots, x_n) f(x_1, \dots, x_n)\, dx_1 \cdots dx_n \tag{6.11}$$

provided the integral is well defined.

Example 6.14. To illustrate calculations with a joint density function, suppose X, Y have joint density function

$$f(x, y) = \begin{cases} \frac{3}{2}(xy^2 + y), & 0 \leq x \leq 1 \text{ and } 0 \leq y \leq 1 \\ 0, & \text{otherwise.} \end{cases}$$

The first task is to verify that f is a genuine joint density function:

$$\int_{-\infty}^{\infty}\int_{-\infty}^{\infty} f(x,y)\,dx\,dy = \tfrac{3}{2}\int_0^1 \left(\int_0^1 (xy^2+y)\,dx\right) dy = \tfrac{3}{2}\int_0^1 (\tfrac{1}{2}y^2+y)\,dy$$
$$= \tfrac{3}{2}\left(\tfrac{1}{6}+\tfrac{1}{2}\right) = 1.$$

Since $f \geq 0$ by its definition and integrates to 1, it passes the test. Next, find the probability $P(X < Y)$. To do this, apply (6.9) to the set $B = \{(x,y) : x < y\}$:

$$P(X < Y) = \iint_{x<y} f(x,y)\,dx\,dy = \tfrac{3}{2}\int_0^1 \left(\int_0^y (xy^2+y)\,dx\right) dy = \tfrac{3}{2}\int_0^1 (\tfrac{1}{2}y^4+y^2)\,dy$$
$$= \tfrac{3}{2}\left(\tfrac{1}{10}+\tfrac{1}{3}\right) = \tfrac{13}{20}.$$

Finally, calculate the expectation $E[X^2Y]$:

$$E[X^2Y] = \int_{-\infty}^{\infty}\int_{-\infty}^{\infty} x^2y\,f(x,y)\,dx\,dy = \int_0^1\int_0^1 x^2y\,\tfrac{3}{2}(xy^2+y)\,dx\,dy = \tfrac{25}{96}. \qquad \blacktriangle$$

Next we establish how to recover marginal density functions of individual random variables from the joint density function. The following fact is analogous to Fact 6.4 from the discrete case.

Fact 6.15. Let f be the joint density function of $X_1, \dots, X_n$. Then each random variable X_j has a density function f_{X_j} that can be obtained by integrating away the other variables from f:

$$f_{X_j}(x) = \underbrace{\int_{-\infty}^{\infty}\cdots\int_{-\infty}^{\infty}}_{n-1 \text{ integrals}} f(x_1,\dots,x_{j-1},x,x_{j+1},\dots,x_n)\,dx_1\dots dx_{j-1}\,dx_{j+1}\dots dx_n. \tag{6.12}$$

The formula says that to compute $f_{X_j}(x)$, place x in the jth coordinate inside f, and then integrate away the other $n-1$ variables.

For two random variables X and Y the formula is

$$f_X(x) = \int_{-\infty}^{\infty} f_{X,Y}(x,y)\,dy. \tag{6.13}$$

Proof of Fact 6.15. We demonstrate the proof for two random variables (X,Y) with joint density function $f_{X,Y}$. That is, we show (6.13).

By Definition 3.1, we need to show that for any real b,

$$P(X \leq b) = \int_{-\infty}^{b}\left(\int_{-\infty}^{\infty} f_{X,Y}(x,y)\,dy\right) dx.$$

This is straightforward since

$$P(X \le b) = P(-\infty < X \le b,\ -\infty < Y < \infty) = \int_{-\infty}^{b} \int_{-\infty}^{\infty} f_{X,Y}(x,y)\,dy\,dx,$$

where the final equality is identity (6.9) for the set $B = \{(x,y) : -\infty < x \le b,\ -\infty < y < \infty\}$. ▲

Example 6.16. To illustrate Fact 6.15, let us find the marginal density function of X in Example 6.14.

For $0 < x < 1$,

$$f_X(x) = \int_{-\infty}^{\infty} f(x,y)dy = \int_0^1 \tfrac{3}{2}(xy^2 + y)\,dy = \tfrac{3}{2}\left(\tfrac{1}{3}x + \tfrac{1}{2}\right) = \tfrac{1}{2}x + \tfrac{3}{4}.$$

For $x < 0$ and $x > 1$ the density is zero: $f_X(x) = \int_{-\infty}^{\infty} f(x,y)\,dy = 0$. ▲

In the next computational example (X, Y) is not restricted to a bounded region.

Example 6.17. Let the joint density function of the random variables X and Y be (see Figure 6.1)

$$f(x,y) = \begin{cases} 2xe^{x^2-y}, & \text{if } 0 < x < 1 \text{ and } y > x^2 \\ 0, & \text{else.} \end{cases} \tag{6.14}$$

Find the marginal density functions f_X of X and f_Y of Y, and compute the probability $P(Y < 3X^2)$.

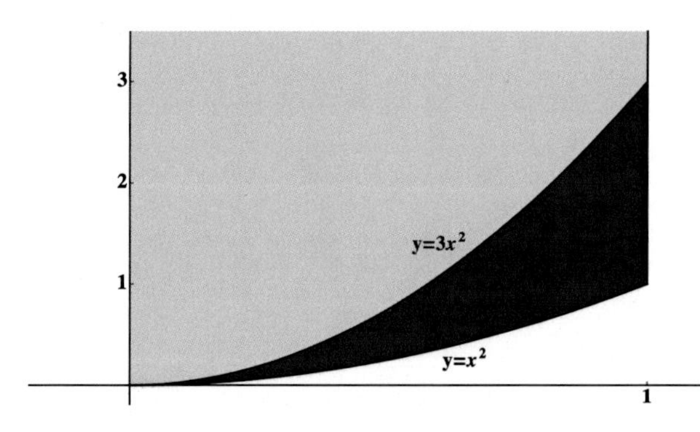

Figure 6.1. The relevant regions in Example 6.17. The dark blue region is the set of points (x, y) satisfying $0 \le x \le 1$ and $x^2 \le y \le 3x^2$.

Since the joint density function vanishes when x is outside $(0, 1)$, we can immediately conclude that $f_X(x) = 0$ for $x \le 0$ and $x \ge 1$. For $0 < x < 1$, from the general identity (6.12) or the simpler (6.13),

$$f_X(x) = \int_{-\infty}^{\infty} f(x,y)\,dy = 2xe^{x^2} \int_{x^2}^{\infty} e^{-y}\,dy = 2xe^{x^2} \cdot e^{-x^2} = 2x.$$

Next we find density function f_Y. Since $Y > 0$, we know right away that $f_Y(y) = 0$ for $y \le 0$. For $y > 0$ we have two cases to consider separately.

(i) If $0 < y < 1$, then $f(x, y) > 0$ for $0 < x < \sqrt{y}$. In this case

$$f_Y(y) = \int_{-\infty}^{\infty} f(x, y)\, dx = e^{-y} \int_0^{\sqrt{y}} 2xe^{x^2}\, dx = e^{-y} \cdot \left(e^{x^2}\right)\Big|_{x=0}^{x=\sqrt{y}} = 1 - e^{-y}.$$

(ii) If $y \geq 1$, then $f(x, y) > 0$ for $0 < x < 1$. In this case

$$f_Y(y) = e^{-y} \int_0^1 2xe^{x^2}\, dx = e^{-y} \cdot \left(e^{x^2}\right)\Big|_{x=0}^{x=1} = e^{1-y} - e^{-y}.$$

We summarize the answers:

$$f_X(x) = \begin{cases} 2x, & 0 < x < 1 \\ 0, & \text{else} \end{cases} \quad \text{and} \quad f_Y(y) = \begin{cases} 1 - e^{-y}, & 0 < y < 1 \\ e^{1-y} - e^{-y}, & y \geq 1 \\ 0, & \text{else.} \end{cases}$$

Lastly we compute the probability. To set up the correct integral, we combine the restriction $y < 3x^2$ of the event we desire with the definition (6.14):

$$P(Y < 3X^2) = \int_0^1 \left(\int_{x^2}^{3x^2} 2xe^{x^2 - y}\, dy \right) dx = \int_0^1 2xe^{x^2} \left(\int_{x^2}^{3x^2} e^{-y}\, dy \right) dx$$

$$= \int_0^1 2xe^{x^2} (e^{-x^2} - e^{-3x^2})\, dx = \left(x^2 + \tfrac{1}{2}e^{-2x^2}\right)\Big|_{x=0}^{x=1} = \tfrac{1}{2}(1 + e^{-2}).$$

▲

Uniform distribution in higher dimensions

To describe two- and three-dimensional generalizations of the uniform distribution, let area(D) denote the area of a set D in two dimensions and vol(A) the volume of a set A in three dimensions. In one dimension the density function of a uniform distribution is constant on an interval and zero elsewhere (recall Definition 3.5). It is then natural to define the uniform distribution on a higher dimensional set D by setting the density function equal to a constant c on D and zero elsewhere. The integral condition of (6.10) determines the value of the constant c. For example, to find the constant value c of the density function of the uniform distribution on a set $D \subset \mathbb{R}^2$, reason as follows:

$$1 = \iint_D c\, dx\, dy = c \iint_D dx\, dy = c \cdot \text{area}(D), \quad \text{which forces} \quad c = \frac{1}{\text{area}(D)}.$$

With these preparations we can state the definition of the uniform distribution in dimensions two and three.

Definition 6.18. Let D be a subset of the Euclidean plane $\mathbb{R}^2$ with finite area. Then the random point (X, Y) is **uniformly distributed on** D if its joint density function is

$$f(x,y) = \begin{cases} \dfrac{1}{\text{area}(D)}, & \text{if } (x,y) \in D \\ 0, & \text{if } (x,y) \notin D. \end{cases} \tag{6.15}$$

Let B be a subset of three-dimensional Euclidean space $\mathbb{R}^3$ with finite volume. Then the random point (X, Y, Z) is **uniformly distributed on** B if its joint density function is

$$f(x,y,z) = \begin{cases} \dfrac{1}{\text{vol}(B)}, & \text{if } (x,y,z) \in B \\ 0, & \text{if } (x,y,z) \notin B. \end{cases} \tag{6.16}$$

For the one-dimensional uniform distribution on an interval $[a, b]$, the connection between length and probability goes as follows: for any subinterval $[c, d]$ of $[a, b]$,

$$P(c \le X \le d) = \frac{d-c}{b-a} = \frac{\text{length of } [c,d]}{\text{length of } [a,b]}.$$

An analogous relation holds in higher dimensions in terms of area and volume. If (X, Y) is uniformly distributed on the set $D \subset \mathbb{R}^2$, then for any set $G \subset D$,

$$P((X,Y) \in G) = \frac{\text{area}(G)}{\text{area}(D)}. \tag{6.17}$$

Similarly, if (X, Y, Z) is uniformly distributed on $B \subset \mathbb{R}^3$, then for any set $H \subset B$,

$$P((X,Y,Z) \in H) = \frac{\text{vol}(H)}{\text{vol}(B)}.$$

The above relations follow from the definitions (6.15) and (6.16) of the density functions. For example, in two dimensions, for a set $G \subset D$,

$$\begin{aligned} P((X,Y) \in G) &= \iint_G f(x,y)\,dx\,dy = \iint_G \frac{1}{\text{area}(D)}\,dx\,dy \\ &= \frac{1}{\text{area}(D)} \iint_G 1\,dx\,dy = \frac{\text{area}(G)}{\text{area}(D)}. \end{aligned}$$

If desired, we can extend Definition 6.18 to any dimension because it is perfectly meaningful to talk mathematically about n-dimensional volume, even if we have a hard time visualizing it.

Let us do a couple of examples in two dimensions.

Example 6.19. Let (X, Y) be a uniform random point on a disk D centered at $(0, 0)$ with radius r_0. (This example continues the theme of Example 3.19.) Compute the marginal densities of X and Y.

The joint density function is

$$f(x,y) = \begin{cases} \dfrac{1}{\pi r_0^2}, & \text{if } (x,y) \in D \\ 0, & \text{if } (x,y) \notin D. \end{cases} \tag{6.18}$$

The marginal density function of X is

$$f_X(x) = \int_{-\infty}^{\infty} f(x,y)\,dy = \frac{1}{\pi r_0^2} \int_{-\sqrt{r_0^2-x^2}}^{\sqrt{r_0^2-x^2}} dy = \frac{2}{\pi r_0^2}\sqrt{r_0^2 - x^2}$$

for $-r_0 \le x \le r_0$ (see also Figure 6.2). Since the situation for Y is the same as for X, we also have

$$f_Y(y) = \frac{2}{\pi r_0^2}\sqrt{r_0^2 - y^2} \qquad \text{for } -r_0 \le y \le r_0.$$

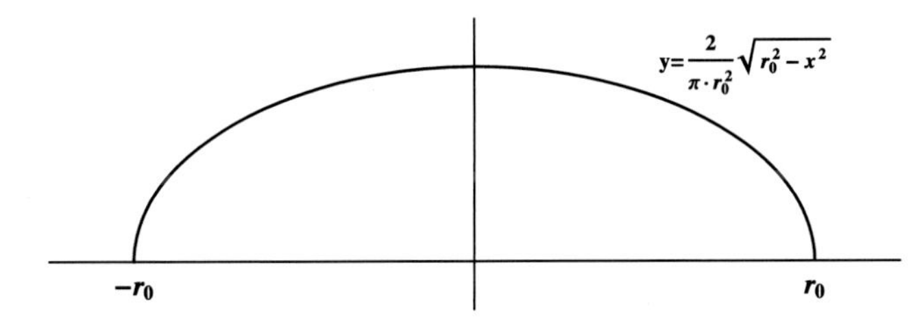

Figure 6.2. The marginal density function of X in Example 6.19.

▲

Example 6.20. Let (X,Y) be uniformly distributed on the triangle D with vertices $(1,0)$, $(2,0)$ and $(0,1)$. (See Figure 6.3.) Find the the joint density function of (X,Y) and the marginal density functions of both X and Y. Next, let A be the (random) area of the rectangle with corners $(0,0)$, $(X,0)$, $(0,Y)$, and (X,Y). Find $E[A]$.

Since the area of the triangle is $1/2$, the joint density function is

$$f(x,y) = \begin{cases} 2, & \text{if } (x,y) \in D \\ 0, & \text{if } (x,y) \notin D. \end{cases} \tag{6.19}$$

By Fact 6.15, the marginal density functions of X and Y are given by

$$f_X(x) = \int_{-\infty}^{\infty} f(x,y)\,dy \quad \text{and} \quad f_Y(y) = \int_{-\infty}^{\infty} f(x,y)\,dx.$$

Finding marginal densities requires care with limits of integration. From the description of D we see that nonzero marginal densities exist in the ranges $0 \le x \le 2$ and $0 \le y \le 1$. In particular,

$$f_Y(y) = \int_{1-y}^{2-2y} 2\,dx = 2(2-2y) - 2(1-y) = 2 - 2y \qquad \text{for } 0 \le y \le 1.$$

As illustrated by Figure 6.3, the integral for $f_X(x)$ needs to be evaluated separately in two cases. If $0 \le x < 1$ then

$$f_X(x) = \int_{1-x}^{1-x/2} 2\,dy = 2(1-x/2) - 2(1-x) = x$$

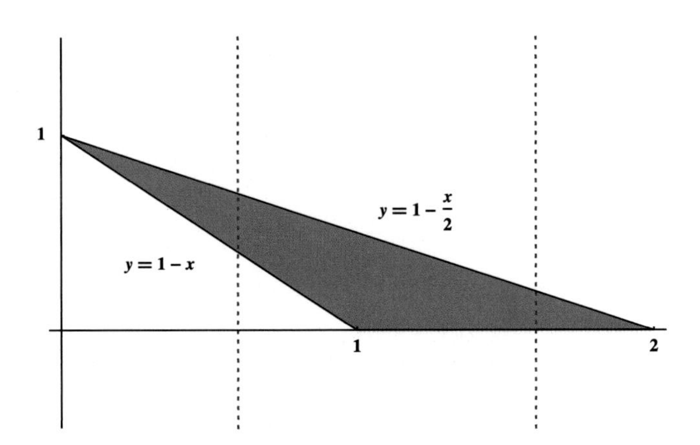

Figure 6.3. Integration over y along a vertical blue line at a given x coordinate gives the marginal density function $f_X(x)$ in Example 6.20. The two examples in the figure show the two cases with different lower limits of integration.

while for $1 \le x \le 2$

$$f_X(x) = \int_0^{1-x/2} 2\,dy = 2(1 - x/2) = 2 - x.$$

To summarize, we have (as also illustrated in Figure 6.4)

$$f_X(x) = \begin{cases} x, & 0 \le x < 1 \\ 2 - x, & 1 \le x \le 2 \\ 0, & x < 0 \text{ or } x > 2 \end{cases} \quad \text{and} \quad f_Y(y) = \begin{cases} 2 - 2y, & 0 \le y \le 1 \\ 0, & y < 0 \text{ or } y > 1. \end{cases}$$

Turning to the expected area $E[A]$, note that A is a function of (X, Y), namely $A = g(X, Y) = XY$. Then we evaluate $E[A]$ using (6.11), with joint density function f from (6.19):

$$E[A] = \iint_{\mathbb{R}^2} xyf(x, y)\,dx\,dy = \iint_D 2xy\,dx\,dy = \int_0^1 y\left(\int_{1-y}^{2-2y} 2x\,dx\right)dy$$
$$= \int_0^1 3y(1 - y)^2\,dy = \tfrac{1}{4}.$$

▲

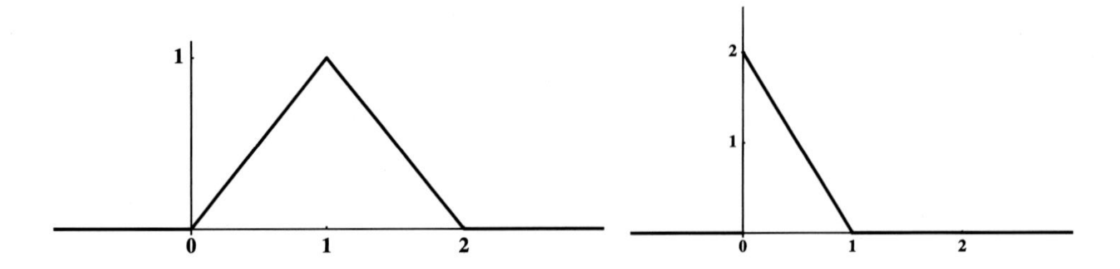

Figure 6.4. The marginal density functions of X (left graph) and Y (right graph) in Example 6.20

Nonexistence of joint density function ♦

The reader may have noticed a logical difference in the introduction of the joint distribution in the discrete and continuous cases. In Definition 6.1 the assumption for the definition of a joint probability mass function was simply that the individual random variables $X_1, \dots, X_n$ are discrete. By contrast, in Definition 6.12 we did not begin by assuming that the individual random variables $X_1, \dots, X_n$ are continuous. Instead, we began with the stronger assumption that a joint density function exists. This was done for a good reason: namely, a joint density may fail to exist even if the individual random variables have densities. Here is an example.

Example 6.21. Let X be a random variable with density function f_X. Let Y be a random variable defined simply by $Y = X$. In other words, Y is an exact copy of X, defined on the same sample space as X. Then Y has the same density function as X. However, the pair (X, Y) does *not* have a joint density function $f(x, y)$. Here is the proof.

By the definition of X and Y we have $P(X = Y) = 1$. Suppose that a joint density function f exists for (X, Y). Then we can calculate $P(X = Y)$ by formula (6.9). To do this, recognize first that $P(X = Y) = P\{(X, Y) \in D\}$ where $D = \{(x, y) : x = y\}$ is the diagonal line on the plane. Then we compute

$$\begin{aligned} P(X = Y) = P\{(X, Y) \in D\} &= \iint_D f(x, y)\, dx\, dy \\ &= \int_{-\infty}^{\infty} \left(\int_x^x f(x, y)\, dy \right) dx = \int_{-\infty}^{\infty} 0\, dx = 0. \end{aligned} \tag{6.20}$$

We have a contradiction and we must conclude that f cannot exist. Geometrically what is going on here is that D has zero area, and since the values of the pair (X, Y) live entirely on a set of zero area, (X, Y) cannot have a joint density function.

For future reference, note that calculation (6.20) shows that $P(X = Y) = 0$ whenever X, Y have a joint density function. ▲

6.3. Joint distributions and independence

This section provides conditions for independence in terms of joint probability mass functions and joint density functions. The condition for jointly continuous random variables is new, whereas the condition for discrete random variables is a repackaging of Fact 2.28. After the conditions and a few simple examples, we turn to a subsection of more involved examples of both the discrete and continuous kind.

Here is the condition for independence in the discrete case.

Fact 6.22. Let $p(k_1, \dots, k_n)$ be the joint probability mass function of the discrete random variables $X_1, \dots, X_n$. Let $p_{X_j}(k) = P(X_j = k)$ be the marginal probability mass function of the random variable X_j. Then $X_1, \dots, X_n$ are independent if and only if

$$p(k_1, \dots, k_n) = p_{X_1}(k_1) \cdots p_{X_n}(k_n) \tag{6.21}$$

for all possible values $k_1, \dots, k_n$.

Here are two examples.

Example 6.23. Check whether the random variables Y_1, Y_2 from Example 6.6 are independent or not.

We check whether $p_{Y_1,Y_2}(k_1, k_2) = p_{Y_1}(k_1)p_{Y_2}(k_2)$ for all k_1 and k_2. Since $p_{Y_1,Y_2}(4, 3) = 0$ while $p_{Y_1}(4)p_{Y_2}(3) > 0$, these random variables are not independent. ▲

Example 6.24. Roll two fair dice. Let X_1 and X_2 be the outcomes. We checked in Example 2.30 that X_1 and X_2 are independent. Let $S = X_1 + X_2$. Determine whether X_1 and S are independent.

For X_1 the possible values are $\{1, \dots, 6\}$, and for S they are $\{2, \dots, 12\}$. Independence of X_1 and S requires that $p_{X_1,S}(a, b) = p_{X_1}(a)p_S(b)$ for *all* possible values a and b. For $a = 1$ and $b = 12$ we derive that

$$p_{X_1,S}(1, 12) = P(X_1 = 1, S = 12) = 0$$

because $X_1 = 1, S = 12$ together imply $X_2 = 11$ which is impossible. On the other hand,

$$p_{X_1}(1)p_S(12) = P(X_1 = 1)P(S = 12) = \tfrac{1}{6} \cdot \tfrac{1}{36} \neq 0.$$

We conclude that X_1 and S are not independent.

Note that it is still possible that particular cases of events $\{X_1 = a\}$ and $\{S = b\}$ are independent. For example, we have

$$P(X_1 = 1, S = 7) = P(X_1 = 1, X_2 = 6) = \tfrac{1}{36} = \tfrac{1}{6} \cdot \tfrac{1}{6} = P(X_1 = 1)P(S = 7).$$ ▲

Next the independence criterion for continuous random variables.

Fact 6.25. Let $X_1, \dots, X_n$ be random variables on the same sample space. Assume that for each $j = 1, 2, \dots, n$, X_j has density function f_{X_j}.

(a) If $X_1, \dots, X_n$ have joint density function

$$f(x_1, x_2, \dots, x_n) = f_{X_1}(x_1)f_{X_2}(x_2) \cdots f_{X_n}(x_n) \tag{6.22}$$

then $X_1, \dots, X_n$ are independent.

(b) Conversely, if $X_1, \dots, X_n$ are independent, then they are jointly continuous with joint density function

$$f(x_1, x_2, \dots, x_n) = f_{X_1}(x_1) f_{X_2}(x_2) \cdots f_{X_n}(x_n).$$

Note the precise logic of the statements above.

(a) To use Fact 6.25 to check whether given continuous random variables $X_1, \dots, X_n$ are independent, we need to know that these random variables are jointly continuous and that their joint density function is given by the product in (6.22).

(b) On the other hand, if we know that $X_1, \dots, X_n$ are independent and have individual density functions, then both the existence and the product form of the joint density function follow. In particular, the product formula (6.22) can be used to *define* the joint density function of independent continuous random variables.

Proof of Fact 6.25 for the case $n = 2$. Suppose X has density function f_X and Y has density function f_Y. Let $A \subset \mathbb{R}$ and $B \subset \mathbb{R}$. Suppose the product form joint density function $f_{X,Y}(x, y) = f_X(x) f_Y(y)$ is given. Then

$$\begin{aligned} P(X \in A, Y \in B) &= \int_A \int_B f_{X,Y}(x, y)\, dy\, dx = \int_A \int_B f_X(x) f_Y(y)\, dy\, dx \\ &= \int_A f_X(x)\, dx \int_B f_Y(y)\, dy = P(X \in A) P(Y \in B), \end{aligned}$$

which shows that X and Y are independent. Conversely, if we assume that X and Y are independent, then

$$\begin{aligned} P(X \in A, Y \in B) &= P(X \in A) P(X \in B) = \int_A f_X(x)\, dx \int_B f_Y(y)\, dy \\ &= \int_A \int_B f_X(x) f_Y(y)\, dy\, dx. \end{aligned}$$

This tells us that $f_{X,Y}(x, y) = f_X(x) f_Y(y)$. ♣ ▲

Example 6.26. Suppose that X, Y have joint density function

$$f(x, y) = \begin{cases} \frac{7}{\sqrt{2\pi}} e^{-x^2/2 - 7y}, & -\infty < x < \infty \text{ and } y > 0 \\ 0, & \text{else.} \end{cases}$$

Are X and Y independent? Find the probability $P(X > 2, Y < 1)$.

The key is to recognize that $f(x, y)$ is the product of two familiar density functions, namely $\frac{1}{\sqrt{2\pi}} e^{-x^2/2}$ and the function that is $7e^{-7y}$ for $y > 0$ and zero otherwise. When this happens, these density functions must be the marginal density functions of X and Y because separately they integrate to 1. For example, here is the case for X. For $-\infty < x < \infty$,

$$f_X(x) = \int_{-\infty}^{\infty} f(x,y)\,dy = \int_0^{\infty} \frac{7}{\sqrt{2\pi}} e^{-x^2/2-7y}\,dy = \frac{1}{\sqrt{2\pi}} e^{-x^2/2} \int_0^{\infty} 7e^{-7y}\,dy$$
$$= \frac{1}{\sqrt{2\pi}} e^{-x^2/2}.$$

By a similar calculation, $f_Y(y) = 7e^{-7y}$ for $y > 0$ and $f_Y(y) = 0$ for $y \le 0$. Hence, we have confirmed that $f(x,y) = f_X(x)f_Y(y)$. This confirms that X and Y are independent and have marginal distributions $X \sim \mathcal{N}(0,1)$ and $Y \sim \text{Exp}(7)$.

Lastly,

$$P(X > 2, Y < 1) = P(X > 2)P(Y < 1) = (1 - \Phi(2)) \cdot (1 - e^{-7})$$
$$\approx (1 - 0.9772) \cdot (1 - e^{-7}) \approx 0.0228.$$ ▲

The next two examples explore the relationship of the geometric shape and the independence of the coordinates of a uniform random point. Exercise 6.13 continues this theme.

Example 6.27. (Continuing Example 6.20) Recall the setup of Example 6.20 where we choose a random point uniformly from a triangle with vertices $(1,0)$, $(2,0)$ and $(0,1)$. Check whether the random variables X and Y are independent or not.

The random variables X and Y are not independent. This should be evident from Figure 6.3 because the values X can take depend strongly on Y. We can show this either with the original definition of independence or with Fact 6.25.

In Example 6.20, we computed the marginal densities of X and Y. These were nonzero functions on the intervals $[0,2]$ and $[0,1]$, respectively. Thus their product is a nonzero function on the rectangle $[0,2] \times [0,1]$. However, the joint density of X and Y is nonzero only on the triangle D. See Figure 6.3. This shows that X and Y are not independent by Fact 6.25.

We can also use the original definition of independence, and show that the product property (2.14) fails by picking convenient sets. For example,

$$P(X < \tfrac{1}{4}, Y < \tfrac{1}{4}) = 0 \quad \text{but} \quad P(X < \tfrac{1}{4})P(Y < \tfrac{1}{4}) = \tfrac{1}{32} \cdot \tfrac{7}{16} > 0.$$ ▲

Example 6.28. Let (X,Y) be a uniform random point on the rectangle

$$D = [a,b] \times [c,d] = \{(x,y) : a \le x \le b, c \le y \le d\}.$$

Then X and Y are independent. We can see this from the density functions, which are

$$f(x,y) = \frac{1}{(b-a)(d-c)} \qquad \text{for } a \le x \le b,\ c \le y \le d,$$
$$f_X(x) = \frac{1}{b-a} \quad \text{for } a \le x \le b, \quad \text{and} \quad f_Y(y) = \frac{1}{d-c} \quad \text{for } c \le y \le d,$$

and zero elsewhere. Thus, $f(x,y) = f_X(x)f_Y(y)$ for all $x, y \in \mathbb{R}$. ▲

We revisit the notion that independent sources of randomness create new independent random variables.

Fact 6.29. Suppose that $X_1, \dots, X_{m+n}$ are independent random variables. Let $f : \mathbb{R}^m \to \mathbb{R}$ and let $g : \mathbb{R}^n \to \mathbb{R}$ be real-valued functions of multiple variables. Define random variables $Y = f(X_1, \dots, X_m)$ and $Z = g(X_{m+1}, \dots, X_{m+n})$. Then Y and Z are independent random variables.

Note carefully that Y and Z are functions of *distinct* independent random variables. Fact 6.29 is an extension of the idea introduced on page 65 in Section 2.5 where a similar statement was discussed for independent events.

Example 6.30. Consider a trial with success probability p that is repeated $m + n$ times. Let S be the number of successes among the first m trials, T the number of successes among the last n trials, and $Z = S + T$ the total number of successes. Check whether S and T are independent and check whether S and Z are independent.

Let X_j be the outcome of the jth trial. These random variables are independent. The independence of S and T follows from Fact 6.29 because $S = X_1 + \cdots + X_m$ and $T = X_{m+1} + \cdots + X_{m+n}$ depend on disjoint collections of X_js.

On the other hand, S and Z are not independent. For example

$$P(S = 1, Z = 0) = 0 \neq mp(1-p)^{m-1} \cdot (1-p)^{m+n} = P(S = 1)P(Z = 0). \qquad \blacktriangle$$

Further examples: the discrete case

The next example is an important result about geometric random variables. A continuous counterpart for exponential random variables is explored in Example 6.34 and Exercise 6.48.

Example 6.31. Suppose that $X_1, \dots, X_n$ are independent random variables with $X_i \sim \text{Geom}(p_i)$. Find the probability mass function of $Y = \min(X_1, \dots, X_n)$, the minimum of $X_1, \dots, X_n$.

The key observation to make is that the event $\{Y > k\}$ is the same as $\{X_1 > k, X_2 > k, \dots, X_n > k\}$. The independence of the random variables $X_1, \dots, X_n$ implies the independence of the events $\{X_1 > k\}, \dots, \{X_n > k\}$ (see Definition 2.27). Thus

$$P(Y > k) = P(X_1 > k, X_2 > k, \dots, X_n > k) = \prod_{i=1}^{n} P(X_i > k).$$

To find $P(X_i > k)$ think of $X_i \sim \text{Geom}(p_i)$ as the time of the first success in a sequence of trials. Then $P(X_i > k) = (1 - p_i)^k$ because $X_i > k$ means that the first k trials failed. Hence,

$$P(Y > k) = \prod_{i=1}^{n}(1-p_i)^k = \left(\prod_{i=1}^{n}(1-p_i)\right)^k = \left(1 - \left(1 - \prod_{i=1}^{n}(1-p_i)\right)\right)^k.$$

This identifies Y as a geometric random variable with success probability $r = 1 - \prod_{i=1}^{n}(1-p_i)$:

$$\begin{aligned} P(Y = k) &= P(Y > k-1) - P(Y > k) \\ &= (1-r)^{k-1} - (1-r)^k = r(1-r)^{k-1}. \end{aligned}$$

Note that we could have deduced this result another way. Imagine that we have n parallel independent sequences of trials where the success probability in the ith sequence is p_i. Let X_i be the time of the first success in the ith sequence. Then $Y = \min(X_1, \dots, X_n)$ is the time of the first success in the entire system. The probability that there is at least one success among the n trials at a given time is $1 - \prod_{i=1}^{n}(1-p_i)$. This explains the result that we got above.

The table below illustrates the case $n = 3$. Each row is a sequence of trials with ★ marking a success. Thus $X_1 = X_3 = 5$ and $X_2 = 7$. $Y = 5$ marks the first column with at least one success.

	1	2	3	4	5	6	7	8	...
1					★				...
2							★	★	...
3					★			★	...

▲

Sometimes independence shows up in places where we might not immediately expect it.

Example 6.32. Roll repeatedly a pair of fair dice until we see either a sum of two or a sum of six. Let N be the random number of rolls needed. Let X be the sum of the two dice of the Nth roll, that is, $X = 2$ or $X = 6$. Question: are N and X independent? It might appear the answer is no. After all, rolling a sum of six is much likelier than rolling a sum of two. Perhaps if $X = 6$ then N is likely to be smaller than when $X = 2$. In fact, X and N are independent. Let us do the math!

Let Y_i be the sum of the two dice on the ith roll. Then $P(Y_i = 2) = 1/36$ and $P(Y_i = 6) = 5/36$, and so $P(Y_i \in \{2,6\}) = 1/6$. Thus we know that $N \sim \text{Geom}(1/6)$, and so $P(N = n) = \left(\frac{5}{6}\right)^{n-1} \cdot \frac{1}{6}$. Let us compute the probability of the intersection $\{N = n, X = 6\}$.

$$\begin{aligned} P(N = n, X = 6) &= P\left(Y_1 \notin \{2,6\}, Y_2 \notin \{2,6\}, \dots, Y_{n-1} \notin \{2,6\}, Y_n = 6\right) \\ &= \left(\tfrac{5}{6}\right)^{n-1} \cdot \tfrac{5}{36}, \end{aligned}$$

where we used the independence of the random variables Y_i.

From the line above we can deduce the probability of the event $\{X = 6\}$ by decomposing this event according to the value of N.

$$P(X=6) = \sum_{n=1}^{\infty} P(N=n, X=6) = \sum_{n=1}^{\infty} \left(\tfrac{5}{6}\right)^{n-1} \cdot \tfrac{5}{36} = \frac{\frac{5}{36}}{1-\frac{5}{6}} = \tfrac{5}{6}.$$

Now combine all the facts from above to deduce this:

$$P(N=n, X=6) = \left(\tfrac{5}{6}\right)^{n-1} \cdot \tfrac{5}{36} = \left(\tfrac{5}{6}\right)^{n-1} \tfrac{1}{6} \cdot \tfrac{5}{6} = P(N=n)P(X=6).$$

This says that the events $\{N = n\}$ and $\{X = 6\}$ are independent. The same argument shows the events $\{N = n\}$ and $\{X = 2\}$ are independent. We could also show this via complements:

$$\begin{aligned} P(N=n, X=2) &= P(N=n) - P(N=n, X=6) \\ &= P(N=n) - P(N=n)P(X=6) = P(N=n)P(X=2). \end{aligned}$$

The conclusion is that the random variables N and X are independent. ▲

Further examples: the jointly continuous case

The next three examples illustrate elegant calculations that can be performed with exponential distributions. These calculations play a critical role in the theory and simulation of continuous time Markov chains, which are a type of stochastic process commonly used to model real-world phenomena.

Example 6.33. You get phone calls from your mother and your grandmother independently. Let X be the time until the next call from your mother, and Y the time until the next call from your grandmother. Let us assume the distributions $X \sim \text{Exp}(\lambda)$ and $Y \sim \text{Exp}(\mu)$. Find the joint density $f_{X,Y}(x, y)$ of the pair (X, Y). What is the probability that your mother calls before your grandmother?

The assumptions tell us that X and Y are independent with marginal densities $f_X(t) = \lambda e^{-\lambda t}$ and $f_Y(t) = \mu e^{-\mu t}$ for $t \geq 0$. Independence then implies that

$$f_{X,Y}(s,t) = f_X(s)f_Y(t) = \begin{cases} \lambda\mu e^{-\lambda s - \mu t}, & s,t \geq 0 \\ 0, & s < 0 \text{ or } t < 0. \end{cases}$$

Note the logical difference between this example and earlier ones. In Examples 6.26 and 6.28 the joint density function was given, and we *discovered* that X and Y were independent. In the present example we decided ahead of time that X and Y should be independent and this compelled us to *define* their joint density as the product of the marginal densities.

Next the probability:

$$\begin{aligned} P(X < Y) &= \iint_{s<t} f_{X,Y}(s,t)\,ds\,dt = \int_0^{\infty} \lambda e^{-\lambda s} \left(\int_s^{\infty} \mu e^{-\mu t}\,dt \right) ds \\ &= \int_0^{\infty} \lambda e^{-\lambda s} \cdot e^{-\mu s}\,ds = \frac{\lambda}{\lambda+\mu} \int_0^{\infty} (\lambda+\mu) e^{-(\lambda+\mu)s}\,ds \\ &= \frac{\lambda}{\lambda+\mu}. \end{aligned} \tag{6.23}$$

The last integral can be evaluated by noting that the integrand is the density function of the $\text{Exp}(\lambda + \mu)$ distribution or by direct calculation. ▲

Example 6.34. Continuing the previous example, let $T = \min(X, Y)$. T is the time until the next phone call comes from either your mother or your grandmother. Let us find the probability distribution of T.

$$\begin{aligned}P(T > t) &= P(\min(X, Y) > t) = P(X > t,\ Y > t) = P(X > t)P(Y > t)\\ &= e^{-\lambda t} \cdot e^{-\mu t} = e^{-(\lambda+\mu)t}.\end{aligned}$$

Because T was defined in terms of a minimum, it was convenient to compute its tail probability $P(T > t)$ and then use independence. The outcome tells us that $T \sim \text{Exp}(\lambda + \mu)$. This property is important for the applications of the exponential distribution. ▲

Example 6.35. Continuing the previous two examples, we discover a continuous analogue of the independence in Example 6.32. Define the indicator random variable

$$I = \begin{cases} 1, & X < Y \quad \text{(your mother calls first)} \\ 0, & X \geq Y \quad \text{(your grandmother calls first).} \end{cases}$$

The probability $P(X = Y) = 0$ so the case $X = Y$ does not actually appear. See the comments at the end of Example 6.21. The case $X = Y$ is included above simply to be logically complete.

In calculation (6.23) above we deduced the probability mass function of I:

$$P(I = 1) = \frac{\lambda}{\lambda + \mu}, \qquad P(I = 0) = 1 - P(I = 1) = \frac{\mu}{\lambda + \mu}.$$

It turns out that T and I are independent. In other words, knowing that your mother called first does not give any information about the time the call arrived. Here is the proof.

Since I is discrete while T is continuous, there is no joint probability mass function or joint density function for (I, T). In order to verify independence, we look at the probability of an intersection of events in terms of I and T. For I we need to consider only the two possible values 1 and 0. Let $0 \leq a < b$.

$$\begin{aligned}P(I = 1,\ a \leq T \leq b) &= P(a \leq X \leq b, Y > X) = \int_a^b \lambda e^{-\lambda s} \left(\int_s^\infty \mu e^{-\mu t}\, dt \right) ds\\ &= \int_a^b \lambda e^{-\lambda s} \cdot e^{-\mu s}\, ds = \frac{\lambda}{\lambda + \mu} \int_a^b (\lambda + \mu) e^{-(\lambda+\mu)s}\, ds\\ &= P(I = 1) P(a \leq T \leq b).\end{aligned}$$

A similar calculation works for the event $\{I = 0\}$, or we can take complements and use the result above:

$$\begin{aligned}P(I = 0,\ a \leq T \leq b) &= P(a \leq T \leq b) - P(I = 1,\ a \leq T \leq b)\\ &= P(a \leq T \leq b) - P(I = 1)P(a \leq T \leq b)\end{aligned}$$

$$= (1 - P(I = 1))\, P(a \le T \le b)$$
$$= P(I = 0)P(a \le T \le b).$$

The independence of T and I is again a very special property of the exponential distribution. ▲

6.4. Further multivariate topics ♦

We begin this section with the joint cumulative distribution function and then describe the standard bivariate normal distribution. Next, the infinitesimal method from Fact 3.7 is generalized to multiple dimensions. Lastly, we use multivariable calculus to find the joint density function of a transformation of a pair of random variables.

Joint cumulative distribution function

We have discussed two ways of describing the joint distribution of multiple random variables $X_1, X_2, \ldots, X_n$. Discrete random variables have a joint probability mass function and jointly continuous random variables have a joint probability density function. These are natural extensions of their one-dimensional counterparts. The cumulative distribution function (Section 3.2) also extends to several variables.

Definition 6.36. The *joint cumulative distribution function* of random variables $X_1, X_2, \ldots, X_n$ is a function of n variables defined as

$$F(s_1, \ldots, s_n) = P(X_1 \le s_1, \ldots, X_n \le s_n), \qquad \text{for all } s_1, \ldots, s_n \in \mathbb{R}. \tag{6.24}$$

The joint cumulative distribution function completely identifies the joint distribution. It is a bit cumbersome for computations, although it can help find the joint density function, when this exists. We discuss this connection for two jointly continuous random variables. The general case is similar.

If X and Y have joint density function f then their joint cumulative distribution function is

$$F(x, y) = P(X \le x, Y \le y) = \int_{-\infty}^{x} \int_{-\infty}^{y} f(s, t)\, dt\, ds.$$

In the opposite direction, if we assume that the joint density function is continuous at a point (x, y) then the density can be obtained from the joint cumulative distribution function by taking the mixed partial derivative and applying the fundamental theorem of calculus:

$$\frac{\partial^2}{\partial x \partial y} F(x, y) = f(x, y). \tag{6.25}$$

In Example 6.19 we derived the marginal distributions of a uniform random point (X, Y) on the disk D with radius r_0 and center $(0, 0)$. The coordinates X and Y are not independent (Exercise 6.13 asks you to show this). However, the next example shows that the polar coordinates (R, Θ) of the point (X, Y) are in fact independent. The joint cumulative distribution function is used as a tool for deriving the joint density function of (R, Θ).

Example 6.37. (Continuation of Example 6.19) Let (X, Y) be a uniform random point on a disk D centered at $(0, 0)$ with radius r_0. Let (R, Θ) denote the polar coordinates of the point (X, Y). Find the joint and marginal density functions of (R, Θ) and show that R and Θ are independent. We take for granted the existence of the joint density function.

We first compute the joint cumulative distribution function of (R, Θ). Since (X, Y) is on the disk $D = \{(x, y) : x^2 + y^2 \le r_0^2\}$, we have $0 \le R \le r_0$. We have $\Theta \in [0, 2\pi)$ by definition. Thus we just need to compute $F_{R,\Theta}(u, v) = P(R \le u, \Theta \le v)$ for $u \in [0, r_0]$ and $v \in [0, 2\pi)$. Let $(r(x, y), \theta(x, y))$ denote the functions that give the polar coordinates (r, θ) of a point (x, y). Then the set

$$A_{u,v} = \{(x, y) : r(x, y) \le u, \theta(x, y) \le v\}$$

is a circular sector of radius u bounded by the angles 0 and v. (See Figure 6.5.)

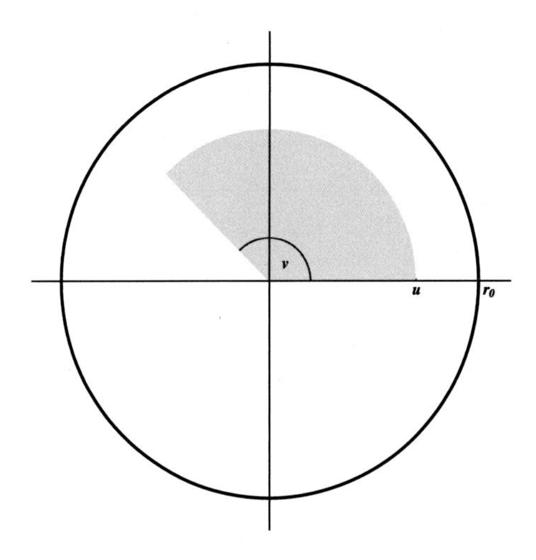

Figure 6.5. The shaded area is the set of points in the disk D whose polar coordinates (r, θ) satisfy $r \le u$ and $\theta \le v$.

The area of the circular sector with radius u and angle v is $\frac{1}{2}u^2 v$. Thus the probability that the random point (X, Y) is in this set is

$$F_{R,\Theta}(u, v) = P(R \le u, \Theta \le v) = P((X, Y) \in A_{u,v}) = \frac{\frac{1}{2}u^2 v}{r_0^2 \pi} = \frac{u^2 v}{2r_0^2 \pi}.$$

Taking the mixed partial derivative of $F_{R,\Theta}$ yields the joint density function

$$f_{R,\Theta}(r, \theta) = \frac{\partial^2}{\partial r \partial \theta}\left(\frac{r^2 \theta}{2r_0^2 \pi}\right) = \frac{r}{r_0^2 \pi} \qquad (6.26)$$

for $0 < r < r_0$ and $0 < \theta < 2\pi$. Integration gives the marginal density functions:

$$f_R(r) = \int_0^{2\pi} f_{R,\Theta}(r,\theta)\,d\theta = \int_0^{2\pi} \frac{r}{r_0^2\pi}\,d\theta = \frac{2r}{r_0^2}$$

(a formula we already discovered in Example 3.19) and

$$f_\Theta(\theta) = \int_0^{r_0} f_{R,\Theta}(r,\theta)\,dr = \int_0^{r_0} \frac{r}{r_0^2\pi}\,dr = \frac{1}{2\pi}.$$

Density function $f_\Theta(\theta) = \frac{1}{2\pi}$ tells us that the angle Θ is uniform on $[0, 2\pi)$. Moreover, since $f_{R,\Theta}(r,\theta) = f_R(r)f_\Theta(\theta)$, we get the independence of R and Θ.

The independence of R and Θ may at first surprise the reader. But on second thought it is not so entirely odd, for in terms of polar coordinates the disk has a rectangular representation $D = \{(r,\theta) : 0 \le r \le r_0, 0 \le \theta < 2\pi\}$. (Recall Example 6.28.) ▲

The joint cumulative distribution function can also be used to identify independence of random variables. The condition is the factorization of the joint cumulative distribution function as the product of the marginal cumulative distribution functions.

Fact 6.38. Random variables $X_1, X_2, \dots, X_n$ are independent if and only if

$$F(x_1, x_2, \dots, x_n) = \prod_{k=1}^{n} F_{X_k}(x_k) \tag{6.27}$$

for all real numbers $x_1, x_2, \dots, x_n$.

Standard bivariate normal distribution

This section describes a joint probability density function for two dependent normal random variables. Let $-1 < \rho < 1$ and suppose that the joint density function of the random variables X and Y is given by

$$f(x,y) = \frac{1}{2\pi\sqrt{1-\rho^2}} e^{-\frac{x^2+y^2-2\rho xy}{2(1-\rho^2)}}. \tag{6.28}$$

The multivariate distribution defined by the joint density function (6.28) is called the **standard bivariate normal with parameter** ρ. We check that this is indeed a probability density function, and that both X and Y have the standard normal distribution.

Because $|\rho| < 1$, the function f is real valued and positive for all x, y. To show that it is a joint density function we need to show that it integrates to 1. In order to show that X and Y are standard normals we need to check that their marginal densities are equal to the standard normal density function:

$$f_X(x) = \int_{-\infty}^{\infty} f(x,y)\,dy = \frac{1}{\sqrt{2\pi}} e^{-\frac{x^2}{2}}$$
$$\text{and} \quad f_Y(y) = \int_{-\infty}^{\infty} f(x,y)\,dx = \frac{1}{\sqrt{2\pi}} e^{-\frac{y^2}{2}}. \tag{6.29}$$

Since f is symmetric in x and y, it is enough to check the first equality in (6.29). Moreover, the first equality in (6.29) already implies that f integrates to 1, because then

$$\int_{-\infty}^{\infty}\int_{-\infty}^{\infty} f(x,y)\,dy\,dx = \int_{-\infty}^{\infty} \frac{1}{\sqrt{2\pi}} e^{-\frac{x^2}{2}}\,dx = 1.$$

Thus we focus on evaluating the integral $\int_{-\infty}^{\infty} f(x,y)dy$. As in Example 5.5 we complete the square in the exponential. Note that x is treated as a constant in this integral.

$$\begin{aligned}\int_{-\infty}^{\infty} \frac{1}{2\pi\sqrt{1-\rho^2}} e^{-\frac{x^2+y^2-2\rho xy}{2(1-\rho^2)}}\,dy &= \frac{1}{2\pi\sqrt{1-\rho^2}} e^{-\frac{x^2}{2(1-\rho^2)}} \int_{-\infty}^{\infty} e^{-\frac{y^2-2\rho xy}{2(1-\rho^2)}}\,dy \\ &= \frac{1}{2\pi\sqrt{1-\rho^2}} e^{-\frac{x^2}{2(1-\rho^2)}} \int_{-\infty}^{\infty} e^{-\frac{(y-\rho x)^2-\rho^2x^2}{2(1-\rho^2)}}\,dy \\ &= \frac{1}{\sqrt{2\pi}} e^{-\frac{x^2}{2(1-\rho^2)}-\frac{\rho^2x^2}{2(1-\rho^2)}} \int_{-\infty}^{\infty} \frac{1}{\sqrt{2\pi(1-\rho^2)}} e^{-\frac{(y-\rho x)^2}{2(1-\rho^2)}}\,dy = \frac{1}{\sqrt{2\pi}} e^{-\frac{x^2}{2}}.\end{aligned} \tag{6.30}$$

In the last step we simplified the exponent and recognized the integral of the $\mathcal{N}(\rho x, 1-\rho^2)$ density function that evaluates to 1.

If $\rho = 0$ the joint density function simplifies to $f(x,y) = \frac{1}{\sqrt{2\pi}} e^{-\frac{x^2}{2}} \cdot \frac{1}{\sqrt{2\pi}} e^{-\frac{y^2}{2}}$. Thus in this case (and only in this case) X and Y are independent standard normal random variables. The probabilistic meaning of the parameter ρ will be revealed in Example 8.38 in Section 8.4.

Infinitesimal method

As with single variable density functions, it should be kept in mind that the value $f(x_1,\dots,x_n)$ of a joint density function is *not* the probability of an event. But we can use the value $f(x_1,\dots,x_n)$ to approximate the probability that $(X_1,\dots,X_n)$ lies in a small n-dimensional cube around the point $(x_1,\dots,x_n)$. This is the multi-dimensional counterpart of Fact 3.7. The volume of an n-dimensional cube with side length ε is ε^n.

Fact 6.39. Suppose that $X_1,\dots,X_n$ have joint density function f and that f is continuous at the point $(a_1,\dots,a_n)$. Then for small $\varepsilon > 0$ we have

$$P(X_1 \in (a_1, a_1+\varepsilon), \dots, X_n \in (a_n, a_n+\varepsilon)) \approx f(a_1,\dots,a_n)\cdot\varepsilon^n.$$

As in the one-dimensional case, the precise meaning of the statement is the limit

$$\lim_{\varepsilon\to 0} \varepsilon^{-n} P(X_1 \in (a_1, a_1+\varepsilon), \dots, X_n \in (a_n, a_n+\varepsilon)) = f(a_1, \dots, a_n).$$

The next example illustrates the use of Fact 6.39 to derive a joint density function of a minimum and a maximum. We take the existence of this joint density function for granted.

Example 6.40. Let $X \sim \text{Exp}(\lambda)$ and $Y \sim \text{Exp}(\mu)$ be independent. Let $T = \min(X, Y)$ and $V = \max(X, Y)$. In the words of Example 6.33, T and V are the times of the next phone calls from your mother and grandmother, ordered in time. The task is to find the joint density function $f_{T,V}(t, v)$ of the pair (T, V).

We demonstrate an approach that relies on Fact 6.39. We reason on a general level and specialize to the given X, Y in the end. The minimum is at most as large as the maximum, so $T \le V$. By (6.20) $P(X = Y) = 0$, so the case $T = V$ has probability zero. We conclude that (T, V) is a random point from the set $L = \{(t, v) : 0 \le t < v\}$. The joint density function $f_{T,V}$ must equal zero outside of L.

To derive $f_{T,V}(t, v)$ on L, let $0 \le t < v$. Choose $\varepsilon > 0$ small enough so that $t + \varepsilon < v$. Then

$$\begin{aligned}
&P(T \in (t, t+\varepsilon), V \in (v, v+\varepsilon)) \\
&\quad = P(t < \min(X, Y) < t+\varepsilon,\ v < \max(X, Y) < v+\varepsilon) \\
&\quad = P(t < X < t+\varepsilon,\ v < Y < v+\varepsilon) + P(v < X < v+\varepsilon,\ t < Y < t+\varepsilon) \\
&\quad \approx f_{X,Y}(t, v)\varepsilon^2 + f_{X,Y}(v, t)\varepsilon^2.
\end{aligned}$$

In the last line we used Fact 6.39 for X and Y. So for $0 \le t < v$ and small enough $\varepsilon > 0$ we have

$$P(T \in (t, t+\varepsilon), V \in (v, v+\varepsilon)) \approx \big(f_{X,Y}(t, v) + f_{X,Y}(v, t)\big)\,\varepsilon^2.$$

By Fact 6.39 the probability above on the left is approximately $f_{T,V}(t, v)\varepsilon^2$. These approximations become accurate as ε is taken to zero and consequently we can claim that

$$f_{T,V}(t, v) = f_{X,Y}(t, v) + f_{X,Y}(v, t) \qquad \text{for } t < v. \tag{6.31}$$

The development above was general. Now return to the example at hand where the joint density function of (X, Y) is $f_{X,Y}(x, y) = f_X(x)f_Y(y) = \lambda\mu e^{-\lambda x - \mu y}$ for $x, y \ge 0$ and zero otherwise. Substituting $f_{X,Y}$ into (6.31) gives the answer

$$f_{T,V}(t, v) = \begin{cases} \lambda\mu e^{-\lambda t - \mu v} + \lambda\mu e^{-\mu t - \lambda v}, & \text{for } 0 < t < v < \infty \\ 0, & \text{else.} \end{cases}$$

▲

Transformation of a joint density function

Section 5.2 derived the density function of $g(X)$ from the density function of X. Here we solve the same problem for a function of several jointly continuous random variables.

Consider two jointly continuous random variables (X, Y) with joint density function $f_{X,Y}$. Let another pair of random variables (U, V) be defined as functions of (X, Y): $U = g(X, Y)$, $V = h(X, Y)$. The goal is to find the joint density function $f_{U,V}$ of (U, V) through a multivariate change of variable. We need to be slightly technical about the assumptions.

Let K be a region of the xy plane so that $f_{X,Y}(x, y) = 0$ outside K. This implies $P\{(X, Y) \in K\} = 1$. Let $G(x, y) = (g(x, y), h(x, y))$ be a one-to-one function that maps K onto a region L on the plane. Denote the inverse of G with $G^{-1}(u, v) = (q(u, v), r(u, v))$. This means that q and r are functions from the region L to K that satisfy

$$u = g(q(u, v), r(u, v)) \quad \text{and} \quad v = h(q(u, v), r(u, v)).$$

Assume that the functions q and r satisfy the following conditions.

(i) Functions q and r have continuous partial derivatives $\frac{\partial q}{\partial u}$, $\frac{\partial q}{\partial v}$, $\frac{\partial r}{\partial u}$ and $\frac{\partial r}{\partial v}$ on L.
(ii) The determinant of partial derivatives, called the *Jacobian*, defined by

$$J(u, v) = \det \begin{bmatrix} \frac{\partial q}{\partial u}(u, v) & \frac{\partial q}{\partial v}(u, v) \\ \frac{\partial r}{\partial u}(u, v) & \frac{\partial r}{\partial v}(u, v) \end{bmatrix} = \frac{\partial q}{\partial u}(u, v) \cdot \frac{\partial r}{\partial v}(u, v) - \frac{\partial q}{\partial v}(u, v) \cdot \frac{\partial r}{\partial u}(u, v),$$

does *not* vanish anywhere on L.

Fact 6.41. Under the assumptions explained above, the joint density function of (U, V) is given by

$$f_{U,V}(u, v) = f_{X,Y}(q(u, v), r(u, v))\, |J(u, v)| \tag{6.32}$$

for $(u, v) \in L$, and $f_{U,V}(u, v) = 0$ outside L.

To justify the statement above, we show that the formula given in (6.32) computes any expectation in accordance with (6.11). Let w be a real-valued function on the plane. Then we can compute the expectation of $w(U, V)$ by expressing (U, V) in terms of (X, Y):

$$\begin{aligned} E[w(U, V)] &= E[w(g(X, Y), h(X, Y))] \\ &= \int_{-\infty}^{\infty} \int_{\infty}^{\infty} w(g(x, y), h(x, y)) f_{X,Y}(x, y)\, dx\, dy \\ &= \iint_K w(g(x, y), h(x, y)) f_{X,Y}(x, y)\, dx\, dy \\ &= \iint_L w(u, v) f_{X,Y}(q(u, v), r(u, v))\, |J(u, v)|\, du\, dv. \end{aligned} \tag{6.33}$$

Above we first restricted the xy double integral to the region K because $f_{X,Y}(x,y)$ vanishes outside K. The last equality above is the change of variable formula from multivariable calculus. Comparison of line (6.33) with the general formula

$$E[w(U,V)] = \int_{-\infty}^{\infty}\int_{\infty}^{\infty} w(u,v) f_{U,V}(u,v)\,du\,dv \tag{6.34}$$

shows that the density function $f_{U,V}$ is the one specified in Fact 6.41.

Fact 6.41 generalizes to higher dimensions in a straightforward manner. In the case of n variables $(U_1,\dots,U_n) = G(X_1,\dots,X_n)$ the Jacobian is the determinant of the $n \times n$ matrix of partial derivatives of the inverse of G.

In the next example, we find the density function of the polar coordinates of a pair of independent standard normal random variables. We then show how this calculation leads to the well-known Box–Muller algorithm for the numerical generation of normal random variables.

Example 6.42. (Polar coordinates of independent standard normals) Suppose that X and Y are independent standard normals. Find the joint density function $f_{R,\Theta}$ of the polar coordinates (R,Θ) of the random point (X,Y).

Let G be the map $(x,y) \to (r,\theta)$ of changing from Cartesian to polar coordinates. To have one-to-one onto functions we can take as our regions $K = \{(x,y) : x^2 + y^2 > 0\}$ and $L = \{(r,\theta) : r > 0,\ 0 \le \theta < 2\pi\}$. We remove the origin from the Cartesian plane because it corresponds to all points $(0,\theta)$ in L and so keeping it would violate the one-to-one assumption. Excluding portions of regions with zero area makes no difference in these calculations.

In terms of the polar coordinates (r,θ) the Cartesian coordinates (x,y) are expressed as

$$x = r\cos\theta \qquad \text{and} \qquad y = r\sin\theta.$$

These equations give the coordinate functions of the inverse function $G^{-1}(r,\theta)$. The Jacobian is

$$J(r,\theta) = \det\begin{bmatrix} \frac{\partial x}{\partial r} & \frac{\partial x}{\partial \theta} \\ \frac{\partial y}{\partial r} & \frac{\partial y}{\partial \theta} \end{bmatrix} = \det\begin{bmatrix} \cos\theta & -r\sin\theta \\ \sin\theta & r\cos\theta \end{bmatrix} = r\cos^2\theta + r\sin^2\theta = r.$$

The joint density function of X,Y is $f_{X,Y}(x,y) = \varphi(x)\varphi(y) = \frac{1}{2\pi}e^{-\frac{x^2+y^2}{2}}$, so formula (6.32) gives

$$f_{R,\Theta}(r,\theta) = f_{X,Y}(r\cos\theta, r\sin\theta)\,|J(r,\theta)| = \tfrac{1}{2\pi} r e^{-\frac{r^2}{2}} \quad \text{for} \quad (r,\theta) \in L.$$

We recognize $f_{R,\Theta}$ as the product of two density functions: $f_{R,\Theta}(r,\theta) = f_R(r) f_\Theta(\theta)$ where

$$f_R(r) = \begin{cases} re^{-\frac{r^2}{2}}, & \text{for } r > 0 \\ 0, & \text{otherwise,} \end{cases} \qquad \text{and} \qquad f_\Theta(\theta) = \begin{cases} \frac{1}{2\pi}, & \text{for } 0 \le \theta < 2\pi \\ 0, & \text{otherwise.} \end{cases} \tag{6.35}$$

This tells us that R and Θ are independent, $\Theta \sim \text{Unif}(0, 2\pi)$, and R has the density function given in (6.35). ▲

Example 6.43. (Box–Muller algorithm) Example 5.30 showed how to generate a random variable with a given density function f as a function of a uniform random variable. The method required the inverse of the cumulative distribution function F. To apply the method to generate a standard normal random variable, one needs to invert the function Φ, which can only be done numerically.

Example 6.42 offers an alternative way to generate normal random variables. From two independent uniform random variables we generate the polar coordinate representation (R, Θ) of two independent standard normals. Then we transform (R, Θ) into Cartesian coordinates (X, Y) and get two independent standard normals X and Y.

Here are the details. Let U_1, U_2 be independent Unif(0, 1) random variables. By (6.35), $\Theta \sim \text{Unif}(0, 2\pi)$, so $V_2 = 2\pi U_2$ defines a random variable with the distribution of Θ. We use the result of Example 5.30 to generate a random variable with the density function f_R in (6.35). The cumulative distribution function F_R is given by

$$F_R(r) = \int_{-\infty}^{r} f_R(t)dt = \begin{cases} 1 - e^{-\frac{r^2}{2}}, & \text{for } r > 0 \\ 0, & \text{otherwise.} \end{cases} \tag{6.36}$$

The inverse function on $(0, 1)$ is $F_R^{-1}(u) = \sqrt{-2\ln(1-u)}$. By Example 5.30, the random variable $\sqrt{-2\ln(1-U_1)}$ has the same distribution as R. Since $U_1 \sim \text{Unif}(0, 1)$, the same is true for $1 - U_1$ as well, so $V_1 = \sqrt{-2\ln U_1}$ also has the same distribution as R. Since U_1, U_2 are independent, so are V_1, V_2. By comparison with the outcome of Example 6.42, we conclude that

$$V_1 = \sqrt{-2\ln(U_1)}, \qquad V_2 = 2\pi U_2$$

are random variables with the same joint distribution as R, Θ. Consequently, transforming (V_1, V_2) from polar to Cartesian coordinates gives two independent standard normals

$$X = \sqrt{-2\ln(U_1)}\cos(2\pi U_2) \quad \text{and} \quad Y = \sqrt{-2\ln(U_1)}\sin(2\pi U_2). \tag{6.37}$$

Transformation (6.37) is called the *Box–Muller transformation.* ▲

Example 6.44. Let X, Y be independent Exp(λ) random variables. Find the joint density function of $(U, V) = (\frac{X}{X+Y}, X + Y)$.

The joint density function of (X, Y) is $f_{X,Y}(x, y) = \lambda^2 e^{-\lambda(x+y)}$ for $x, y > 0$ and zero otherwise. The relevant mapping G and regions are now $G(x, y) = (\frac{x}{x+y}, x+y)$, $K = \{(x, y) : x > 0, y > 0\}$ and $L = \{(u, v) : 0 < u < 1, v > 0\}$ (the choice of L because $u = \frac{x}{x+y} \in (0, 1)$). The function G is inverted as follows:

$$u = \frac{x}{x+y}, \qquad v = x + y \quad \iff \quad x = uv, \quad y = (1-u)v.$$

This gives $G^{-1}(u,v) = (uv, (1-u)v)$. Its Jacobian is

$$J(u,v) = \det\begin{bmatrix} \frac{\partial x}{\partial u} & \frac{\partial x}{\partial v} \\ \frac{\partial y}{\partial u} & \frac{\partial y}{\partial v} \end{bmatrix} = \det\begin{bmatrix} v & u \\ -v & 1-u \end{bmatrix} = v(1-u) + uv = v.$$

Formula (6.32) now gives

$$f_{U,V}(u,v) = f_{X,Y}(uv, (1-u)v)\,|J(u,v)| = \lambda^2 e^{-\lambda v} v \quad \text{for} \quad 0 < u < 1 \text{ and } v > 0,$$

and $f_{U,V}(u,v) = 0$ for (u,v) outside L. We recognize $f_{U,V}$ as the product of two density functions: $f_{U,V}(u,v) = f_U(u) f_V(v)$ where

$$f_U(u) = \begin{cases} 1, & \text{for } 0 < u < 1 \\ 0, & \text{otherwise,} \end{cases} \qquad \text{and} \qquad f_V(v) = \begin{cases} \lambda^2 e^{-\lambda v}, & \text{for } v > 0 \\ 0, & \text{otherwise.} \end{cases}$$

This tells us that $U = \frac{X}{X+Y}$ and $V = X + Y$ are independent, $U \sim \text{Unif}(0,1)$, and $V \sim \text{Gamma}(2, \lambda)$ (recall Definition 4.26 for the gamma distribution).

Exercise 6.17 asks you to treat the inverse of this example and Exercise 6.50 generalizes these facts to beta and gamma variables. ▲

6.5. Finer points ♣

Zero probability events for jointly continuous random variables

Recall from Fact 3.2 that for a continuous random variable X, $P(X = c) = 0$ for any real c. The point was that the integral of the density function over a single point c is zero, or equivalently, the length of the interval $[c, c]$ is zero.

This generalizes to jointly continuous random variables $X_1, \dots, X_n$. If the set $A \subset \mathbb{R}^n$ has zero n-dimensional volume then $P((X_1, \dots, X_n) \in A) = 0$. A natural case of this is where A lives in a lower dimensional subspace of $\mathbb{R}^n$. The example in (6.20) is the one-dimensional diagonal D of the plane for which $P((X,Y) \in D) = 0$ whenever (X, Y) are jointly continuous.

This fact has implications for the uniqueness of the joint density function. Because integrals on zero volume sets are zero, changing the joint density function on such a set does not change the probability distribution. For example, if (X, Y) is a uniform random point on the unit square $[0,1]^2$, its joint density function can be either one of these two functions:

$$f(x,y) = \begin{cases} 1, & \text{for } x, y \in (0,1) \\ 0, & \text{otherwise,} \end{cases} \qquad \text{or} \qquad \tilde{f}(x,y) = \begin{cases} 1, & \text{for } x, y \in [0,1] \\ 0, & \text{otherwise.} \end{cases}$$

Measurable subsets of $\mathbb{R}^n$

The measurability issue introduced in Section 1.6 is present in the multivariate setting also. As explained there for a real-valued random variable, for jointly defined random variables $X_1, X_2, \dots, X_n$ a probability $P((X_1, \dots, X_n) \in B)$ makes sense only if B is a Borel subset of $\mathbb{R}^n$. This means that B is a member of the

σ-algebra generated by open rectangles $(a_1, b_1) \times (a_2, b_2) \times \cdots \times (a_n, b_n)$ in $\mathbb{R}^n$. Furthermore, all the functions that appear in integrals such as f and g in (6.11) have to be Borel functions on $\mathbb{R}^n$. One way to phrase this condition is that the set $\{\mathbf{x} \in \mathbb{R}^n : f(\mathbf{x}) \le t\}$ is a Borel subset of $\mathbb{R}^n$ for each real t.

The sketch of the proof given for Fact 6.25 relies on the following fact about Borel subsets in multiple dimensions: if

$$P(X \in A, Y \in B) = \int_A \int_B f_X(x) f_Y(y)\, dy\, dx$$

for all Borel sets $A, B \subset \mathbb{R}$, then

$$P((X, Y) \in C) = \iint_C f_X(x) f_Y(y)\, dx\, dy$$

for all Borel sets $C \subset \mathbb{R}^2$. The last statement implies that $f_X(x) f_Y(y)$ is the joint density function of (X, Y).

Joint moment generating function

The moment generating function can be extended to the multivariate setting. Let $X_1, \dots, X_n$ be random variables on the same probability space. Their joint moment generating function is the following function of n variables:

$$M_{X_1,\dots,X_n}(t_1, \dots, t_n) = E\left[e^{t_1 X_1 + \cdots + t_n X_n}\right]. \tag{6.38}$$

The key properties of the single variable moment generating function extend to the multivariate case. In particular, if it is finite in a neighborhood of the origin then it characterizes the joint distribution of $X_1, \dots, X_n$. Moreover, the joint moment generating function can also be used to compute expectations of monomials, see Exercise 6.59.

Exercises

We start with some warm-up exercises arranged by section.

Section 6.1

Exercise 6.1. Let X and Y be discrete random variables with joint probability mass function given by the table below.

$X \backslash Y$	0	1	2	3
1	0.10	0.15	0	0.05
2	0.20	0.05	0.05	0.20
3	0.05	0	0.10	0.05

To illustrate how to read the table, for example $P(X = 2, Y = 1) = 0.05$.

(a) Find the marginal probability mass function of X.
(b) Find the probability mass function of the random variable $Z = XY$.
(c) Calculate the expectation $E[Xe^Y]$.

Exercise 6.2. The joint probability mass function of the random variables (X, Y) is given by the following table:

X \ Y	0	1	2	3
1	$\frac{1}{15}$	$\frac{1}{15}$	$\frac{2}{15}$	$\frac{1}{15}$
2	$\frac{1}{10}$	$\frac{1}{10}$	$\frac{1}{5}$	$\frac{1}{10}$
3	$\frac{1}{30}$	$\frac{1}{30}$	0	$\frac{1}{10}$

(a) Find the marginal probability mass functions of X and Y.
(b) Calculate the probability $P(X + Y^2 \leq 2)$.

Exercise 6.3. For each lecture the professor chooses between white, yellow, and purple chalk, independently of previous choices. Each day she chooses white chalk with probability 0.5, yellow chalk with probability 0.4, and purple chalk with probability 0.1.

(a) What is the probability that over the next 10 days she will choose white chalk 5 times, yellow chalk 4 times, and purple chalk 1 time?
(b) What is the probability that over the next 10 days she will choose white chalk exactly 9 times?

Exercise 6.4. A sandwich shop offers 8 different sandwiches. Jamey likes them all equally. Each day for lunch he picks one randomly, independently of previous choices. Salami, falafel and veggie are three of the eight types of sandwiches. During a given week of 5 days, let X be the number of times he chooses salami, Y the number of times he chooses falafel, Z the number of times he chooses veggie, and W the number of times he chooses something else. Find the joint probability mass function of (X, Y, Z, W). Identify the distribution of (X, Y, Z, W) by name.

Section 6.2

Exercise 6.5. Suppose X, Y have joint density function

$$f(x, y) = \begin{cases} \frac{12}{7}(xy + y^2), & 0 \leq x \leq 1 \text{ and } 0 \leq y \leq 1 \\ 0, & \text{otherwise.} \end{cases}$$

(a) Check that f is a genuine joint density function.
(b) Find the marginal density functions of X and Y.
(c) Calculate the probability $P(X < Y)$.
(d) Calculate the expectation $E[X^2 Y]$.

Exercise 6.6. Suppose that X, Y are jointly continuous with joint probability density function

$$f(x,y) = \begin{cases} xe^{-x(1+y)}, & \text{if } x > 0 \text{ and } y > 0 \\ 0, & \text{otherwise.} \end{cases}$$

(a) Find the marginal density functions of X and Y.
(b) Calculate the expectation $E[XY]$.
(c) Calculate the expectation $E[\frac{X}{1+Y}]$.

Hint. For the last two parts, recalling moments of the exponential distribution can be helpful.

Exercise 6.7. Consider the triangle with vertices $(0, 0)$, $(1, 0)$, $(0, 1)$. Suppose that (X, Y) is a uniformly chosen random point from this triangle.

(a) Find the marginal density functions of X and Y.
(b) Calculate the expectations $E[X]$ and $E[Y]$.
(c) Calculate the expectation $E[XY]$.

Section 6.3

Exercise 6.8. Determine whether the following pairs of random variables X and Y are independent.

(a) Random variables X and Y of Exercise 6.2.
(b) Random variables X and Y of Exercise 6.5.
(c) Random variables X and Y of Exercise 6.6.
(d) Random variables X and Y of Exercise 6.7.

Exercise 6.9. Let X be the number of tails in three flips of a fair coin. Let Y be the outcome of a roll of a fair six-sided die. Assume X and Y are independent. Give the joint probability mass function of X, Y. Be precise about the values for which the joint probability mass function is defined.

Exercise 6.10. Let X and Y be independent uniform random variables on $(0, 1)$. Find their joint density function $f(x, y)$. Use the joint density function to calculate the probability $P(X < Y)$.

Exercise 6.11. Let X have density function

$$f_X(x) = \begin{cases} 2x, & \text{for } 0 < x < 1 \\ 0, & \text{else} \end{cases}$$

and let Y be uniformly distributed on the interval $(1, 2)$. Assume X and Y are independent. Give the joint density function of (X, Y). Calculate the probability $P(Y - X \geq \frac{3}{2})$.

Exercise 6.12. Suppose that X and Y are jointly continuous with joint probability density function

$$f(x, y) = \begin{cases} 2e^{-(x+2y)}, & \text{for } x > 0,\ y > 0 \\ 0, & \text{else.} \end{cases}$$

Show that X and Y are independent random variables and provide their marginal distributions.

Exercise 6.13. Recall Example 6.19 where we choose a uniform random point (X, Y) from a disk D with radius r_0 centered at $(0, 0)$. Determine whether random variables X and Y are independent or not.

Section 6.4

Exercise 6.14. Fix $a, b > 0$. Let (X, Y) be a uniform random point on the rectangle with corners $(0, 0)$, $(a, 0)$, $(0, b)$, and (a, b).

(a) Find the joint cumulative distribution function F of (X, Y). Express F as a single formula (no cases) by using the minimum and maximum functions. (These functions are $\min(x, y) =$ the smaller of x and y, and $\max(x, y) =$ the larger of x and y.)
(b) From F and (6.25) find the joint density function of (X, Y) recorded in Example 6.28.

Exercise 6.15. Let Z, W be independent standard normal random variables and $-1 < \rho < 1$. Check that if $X = Z$ and $Y = \rho Z + \sqrt{1 - \rho^2}\, W$ then the pair (X, Y) has standard bivariate normal distribution with parameter ρ.

Hint. You can use Fact 6.41 or arrange the calculation so that a change of variable in the inner integral of a double integral leads to the right density function.

Exercise 6.16. Example 6.37 used the joint cumulative distribution function to find the joint density function $f_{R,\Theta}$ of the polar coordinates (R, Θ) of a uniform random point (X, Y) on the disk $D = \{(x, y) : x^2 + y^2 \le r_0^2\}$. Give an alternative derivation of $f_{R,\Theta}$ using Fact 6.41.

Exercise 6.17. Let U and V be independent, $U \sim \text{Unif}(0, 1)$, and $V \sim \text{Gamma}(2, \lambda)$ which means that V has density function $f_V(v) = \lambda^2 v e^{-\lambda v}$ for $v > 0$ and zero elsewhere. Find the joint density function of $(X, Y) = (UV, (1 - U)V)$. Identify the joint distribution of (X, Y) in terms of named distributions. This exercise and Example 6.44 are special cases of the more general Exercise 6.50.

Further exercises

Exercise 6.18. Suppose that X and Y are integer-valued random variables with joint probability mass function given by

$$p_{X,Y}(a,b) = \begin{cases} \frac{1}{4a}, & \text{for} \quad 1 \le b \le a \le 4 \\ 0, & \text{otherwise.} \end{cases}$$

(a) Show that this is indeed a joint probability mass function.
(b) Find the marginal probability mass functions of X and Y.
(c) Find $P(X = Y + 1)$.

Exercise 6.19. The joint probability mass function of the random variables (X, Y) is given by the following table:

X \ Y	0	1	2
0	$\frac{1}{18}$	$\frac{1}{18}$	$\frac{4}{18}$
1	$\frac{2}{18}$	$\frac{5}{18}$	$\frac{5}{18}$

(a) Find the marginal probability mass functions of X and Y.
(b) Suppose Z and W are independent random variables and that (i) Z is equal to X in distribution, and (ii) W is equal to Y in distribution. Give the joint probability mass function $p_{Z,W}$ of (Z, W).

Exercise 6.20. Let $(X_1, X_2, X_3, X_4) \sim \text{Mult}(n, 4, \frac{1}{6}, \frac{1}{3}, \frac{1}{8}, \frac{3}{8})$. Derive the joint probability mass function of (X_3, X_4).

Hint. You can do this without computation.

Exercise 6.21. Let $(X_1, \dots, X_r) \sim \text{Mult}(n, r, p_1, \dots, p_r)$. Determine whether X_1 and X_2 are independent.

Exercise 6.22. Let $(X_1, X_2, X_3, X_4) \sim \text{Mult}(n, 4, p_1, p_2, p_3, p_4)$. What is the distribution of $X_1 + X_2$?

Hint. This can be done without any computation. Take a look at the argument at the end of Example 6.10.

Exercise 6.23. An urn contains 1 green ball, 1 red ball, and 1 yellow ball. I draw 4 balls with replacement. What is the probability that all three colors appear in the sample?

Exercise 6.24. An urn contains 1 green ball, 1 red ball, 1 yellow ball and 1 white ball. I draw 3 balls with replacement. What is the probability that exactly two balls are of the same color?

Exercise 6.25. We flip a fair coin three times. Let X be the number of heads among the first two coin flips, Y the number of heads in the last two coin flips.

(a) Find the joint probability mass function of (X, Y). (You can try to find a general formula, or display the function in a table.)
(b) Find the probability mass function of XY.

Exercise 6.26. Consider two urns, urn A and urn B. In urn A, there are 3 balls numbered $0, 1, 2$. In urn B there are 6 balls numbered $1, 2, 3, 4, 5, 6$. A ball is drawn from urn A, then a ball is drawn from urn B. Define X_A as the number of the ball from urn A and X_B the number of the ball drawn from urn B.

(a) What is the joint distribution of (X_A, X_B)?

(b) Let $Y_1 = X_A X_B$ and $Y_2 = \max\{X_A, X_B\}$. Find the joint distribution of (Y_1, Y_2).

Hint. Make a table.

(c) Find the marginal distribution of Y_1 and Y_2. Are Y_1 and Y_2 independent? Justify your answer using part (b).

Exercise 6.27. Suppose that X_1 and X_2 are independent random variables with $P(X_1 = 1) = P(X_1 = -1) = \frac{1}{2}$ and $P(X_2 = 1) = 1 - P(X_2 = -1) = p$ with $0 < p < 1$. Let $Y = X_1 X_2$. Show that X_2 and Y are independent.

Exercise 6.28. Let X and Y be independent Geom(p) random variables. Let $V = \min(X, Y)$ and

$$W = \begin{cases} 0, & \text{if } X < Y \\ 1, & \text{if } X = Y \\ 2, & \text{if } X > Y. \end{cases}$$

Find the joint probability mass function of V and W and show that V and W are independent.

Hint. Use the joint probability mass function of X and Y to compute the joint probability mass function of V and W.

Exercise 6.29. Suppose that $X \sim$ Geom(p) and $Y \sim$ Geom(r) are independent. Find the probability $P(X < Y)$.

Exercise 6.30. Suppose that $X \sim$ Geom(p) and $Y \sim$ Poisson(λ) are independent. Find the probability $P(X = Y + 1)$.

Exercise 6.31. We have an urn with 10 red, 15 green and 20 yellow balls. We take a sample of 8 balls *without replacement* and denote the number of red, yellow and green balls in the sample by X_1, X_2 and X_3. Find the joint probability mass function of (X_1, X_2, X_3).

Hint. The joint distribution is not a multinomial.

Exercise 6.32. Suppose I have an urn with 9 balls: 4 green, 3 yellow and 2 white ones. I draw a ball from the urn repeatedly with replacement, until I see the first green or yellow ball, and then I stop. Let N be the number draws I needed. Let Y equal 1 if the last draw is green and 2 if the last draw is yellow. Find the joint and marginal probability mass functions of N and Y and determine whether N and Y are independent. Is there an intuitive explanation for the probability mass function of Y that you discovered?

Exercise 6.33. Let the random variables X, Y have joint density function

$$f(x,y) = \begin{cases} 3(2-x)y, & \text{if } 0 < y < 1 \text{ and } y < x < 2-y \\ 0, & \text{otherwise.} \end{cases}$$

(a) Find the marginal density functions f_X and f_Y.
(b) Calculate the probability that $X + Y \le 1$.

Exercise 6.34. Let (X, Y) be a uniformly distributed random point on the quadrilateral D with vertices $(0, 0)$, $(2, 0)$, $(1, 1)$ and $(0, 1)$.

(a) Find the joint density function of (X, Y) and the marginal density functions of X and Y.
(b) Find $E[X]$ and $E[Y]$.
(c) Are X and Y independent?

Exercise 6.35. Suppose that X and Y are random variables with joint density function

$$f_{X,Y}(x,y) = \begin{cases} \frac{1}{4}(x+y), & 0 \le x \le y \le 2 \\ 0, & \text{otherwise.} \end{cases}$$

(a) Check that $f_{X,Y}$ is a joint density function.
(b) Calculate the probability $P\{Y < 2X\}$.
(c) Find the marginal density function $f_Y(y)$ of Y.

Exercise 6.36. Suppose that X, Y are jointly continuous with joint probability density function

$$f(x,y) = ce^{-\frac{x^2}{2} - \frac{(x-y)^2}{2}}, \quad x, y \in (-\infty, \infty),$$

for some constant c.

(a) Find the value of the constant c.
(b) Find the marginal density functions of X and Y.
(c) Determine whether X and Y are independent.

Exercise 6.37. Let h be a continuous function on $[a, b]$ such that $h(a) = h(b) = 0$ and $h(x) > 0$ for $a < x < b$. Let D be the region between the x-axis and the graph of h:

$$D = \{(x,y) : a < x < b,\ 0 < y < h(x)\}.$$

Let (X, Y) be a uniformly distributed random point on D. Find the marginal density function f_X of X.

Exercise 6.38. Suppose that X, Y are jointly continuous with joint probability density function given in Exercise 6.6. Find $E[Y]$.

Exercise 6.39. Suppose that X and Y are jointly continuous random variables with joint cumulative distribution function $F(x, y)$. Express the probability $P(a < X < b, c < Y < d)$ using the function F.

Exercise 6.40. Find the density function of the random variable $S = X + Y$ in the setting of Exercise 6.33.

Hint. Start by calculating $P(X + Y \leq s)$ for all $s \in \mathbb{R}$.

Exercise 6.41. Let $h > r > 0$ and let D be the vertical cylinder in three dimensions with radius r and height h, precisely described as the set

$$D = \{(x, y, z) : x^2 + y^2 < r^2, 0 < z < h\}.$$

Let (X, Y, Z) be a uniformly distributed random point in D. Find the probability that (X, Y, Z) lies inside the ball of radius r centered at the origin.

Exercise 6.42. Let (X, Y) be a uniform random point on the rectangle

$$D = [0, 2] \times [0, 3] = \{(x, y) : 0 \leq x \leq 2, 0 \leq y \leq 3\}.$$

Let $Z = X + Y$. Give the joint cumulative distribution function of (X, Z).

Hint. Recall the definition of the cumulative distribution function and draw a picture.

Exercise 6.43. Let the random variables X, Y have joint density function

$$f_{X,Y}(x, y) = \begin{cases} 2x^2y + \sqrt{y}, & \text{if } 0 < x < 1 \text{ and } 0 < y < 1 \\ 0, & \text{else.} \end{cases}$$

Let $T = \min(X, Y)$ and $V = \max(X, Y)$. Assuming that T, V are jointly continuous, find their joint density function.

Hint. Look at Example 6.40.

Exercise 6.44. You spend the night in a teepee shaped as a right circular cone whose base is a disk of radius r centered at the origin and the height at the apex is h. A fly is buzzing around the teepee at night. At some time point the fly dies in mid-flight and falls directly on the floor of the teepee at a random location (X, Y). Assume that the position of the fly at the moment of its death was uniformly random in the volume of the teepee. Derive the joint density function $f_{XY}(x, y)$ of the point (X, Y) where you find the dead fly in the morning. Let Z be the height from which the dead fly fell to the floor. Find the density function $f_Z(z)$ of Z.

Exercise 6.45. Suppose that X and Y are independent continuous random variables with density functions f_X and f_Y. Let $T = \min(X, Y)$ and $V = \max(X, Y)$. Find the marginal density functions f_T of T and f_V of V using the following outline.

(a) Find the cumulative distribution functions of T and V.
Hint. Example 6.34 is helpful for T. For V express the inequality $\max(X, Y) \leq v$ in terms of separate statements for X and Y.

(b) Differentiate the cumulative distribution functions to get the probability density functions.

Exercise 6.46. Let X, Y be independent random variables with density functions f_X and f_Y. Let $T = \min(X, Y)$ and $V = \max(X, Y)$. Use the joint density function $f_{T,V}$ in equation (6.31) to compute the marginal density functions f_T of T and f_V of V.

Exercise 6.47. Let $X_1, X_2, \ldots, X_n$ be independent random variables with the same cumulative distribution function F, and let us assume that F is continuous. Denote the minimum and the maximum by

$$Z = \min(X_1, X_2, \ldots, X_n) \quad \text{and} \quad W = \max(X_1, X_2, \ldots, X_n).$$

(a) Find the cumulative distribution functions F_Z of Z and F_W of W.
(b) Assume additionally that the random variables X_i are continuous, so that, consequently, they all have the same density function f. Now find the density functions f_Z of Z and f_W of W.

Hint. Solving Exercise 6.45 could be helpful.

Exercise 6.48. Let $X_1, X_2, \ldots, X_n$ be independent exponential random variables, with parameter λ_i for X_i. Let Y be the minimum of these random variables. Show that $Y \sim \text{Exp}(\lambda_1 + \cdots + \lambda_n)$.

Exercise 6.49. Let X, Y have joint density function $f_{X,Y}$. Let $T = \min(X, Y)$ and $V = \max(X, Y)$. Find the joint density function $f_{T,V}$ by using Fact 6.41.

Hint. Consider the two-to-one mapping $G(x, y) = (\min(x, y), \max(x, y))$ separately from regions $K_1 = \{(x, y) : x < y\}$ and $K_2 = \{(x, y) : x > y\}$ onto $L = \{(t, v) : t < v\}$.

Exercise 6.50. This exercise establishes the most basic facts of the *beta-gamma algebra.*

Let $r, s > 0$. A random variable X has the *beta distribution* with parameters (r, s) if $0 < X < 1$ and X has density function

$$f(x) = \frac{\Gamma(r+s)}{\Gamma(r)\Gamma(s)} x^{r-1}(1-x)^{s-1}, \qquad 0 < x < 1.$$

This is abbreviated $X \sim \text{Beta}(r, s)$.

The gamma distribution was introduced in Definition 4.37.

(a) Let X, Y be independent with $X \sim \text{Gamma}(r, \lambda)$ and $Y \sim \text{Gamma}(s, \lambda)$. Find the joint distribution of $(B, G) = (\frac{X}{X+Y}, X + Y)$.

(b) Let B, G be independent with $B \sim \text{Beta}(r, s)$ and $G \sim \text{Gamma}(r+s, \lambda)$. Find the joint distribution of $(X, Y) = (BG, (1-B)G)$.

Exercise 6.51. Let X and Y be random variables defined on the same probability space and g a function on the range of X. The expectation $E[g(X)]$ can now be computed either as in Fact 3.33 or in terms of the joint distribution of (X, Y) as in Facts 6.2 or 6.13. Here we ask you to check that these are consistent.

(a) Assume that X and Y are discrete. Show that formula (6.1) implies that $E[g(X)] = \sum_k g(k)P(X = k)$.
(b) Assume that X and Y are jointly continuous. Show that formula (6.11) implies that $E[g(X)] = \int_{-\infty}^{\infty} g(x) f_X(x)\,dx$ where f_X is the marginal density function of X.

Exercise 6.52. Let $(X_1, \dots, X_r) \sim \text{Mult}(n, r, p_1, \dots, p_r)$, and suppose that $t_1, \dots, t_r$ are real numbers. Compute the expectation $E\left[e^{t_1 X_1 + \cdots + t_r X_r}\right]$.

The resulting expectation, as a function of $(t_1, \dots, t_r)$, is the joint moment generating function of $X_1, \dots, X_r$ defined in Section 6.5.

Exercise 6.53. Prove Fact 6.5.

Exercise 6.54. Let $X_1, \dots, X_n$ be jointly continuous random variables with joint density function f. State and prove a theorem that derives the joint density function $f_{X_1,\dots,X_m}$ of random variables $X_1, \dots, X_m$ from f, for any $1 \le m \le n$.

Exercise 6.55. On any given day, Adam receives a call from his brother Ben with probability 0.3, and a call from his sister Daisy with probability 0.8. Introduce the indicator random variables

$$X_B = \begin{cases} 1, & \text{if Ben calls tomorrow} \\ 0, & \text{otherwise,} \end{cases} \quad \text{and} \quad X_D = \begin{cases} 1, & \text{if Daisy calls tomorrow} \\ 0, & \text{otherwise.} \end{cases}$$

(a) Is there a joint probability mass function for (X_B, X_D) under which Adam is certain to receive a call from at least one of his siblings tomorrow?
(b) Does the answer to part (a) change when Daisy reduces her calling probability to 0.7? To 0.6?

Exercise 6.56. Let $p(x, y)$ be the joint probability mass function of (X, Y). Assume that there are two functions a and b such that $p(x, y) = a(x)b(y)$ for all possible values x of X and y of Y. Show that X and Y are independent. Do not assume that a and b are probability mass functions.

Exercise 6.57. Let $X_1, \dots, X_n$ be independent random variables, each one distributed uniformly on $[0, 1]$. Let Z be the minimum and W the maximum of these numbers. Find the joint density function of Z and W.

Hint. Try to find the joint cumulative distribution function first.

Challenging problems

Exercise 6.58. Let $X_1, \dots, X_{99}$ be independent random variables, each one distributed uniformly on $[0, 1]$. Let Y denote the 50th largest among the 99 numbers. Find the probability density function of Y.

Exercise 6.59. Suppose that X and Y are discrete random variables defined on the same probability space with finitely many values. Let $M_{X,Y}(s, t) = E[e^{sX+tY}]$ be the joint moment generating function of X, Y.

(a) Show that $E[XY] = \frac{\partial^2}{\partial s \partial t} M_{X,Y}(0, 0)$.
(b) Generalize the previous statement to $E[X^a Y^b]$ for positive integers a, b.

Exercise 6.60. Let $X_1, X_2, \dots, X_n$ be independent exponential random variables, with parameter λ_i for X_i. Let Y be the minimum of these random variables, and let I be the index i for which $Y = X_i$. Find $P(I = i)$ for $1 \le i \le n$.

Hint. This exercise generalizes Example 6.35.

Exercise 6.61. Suppose that the joint distribution of X, Y is standard bivariate normal with parameter ρ (see (6.28)). Compute the probability $P(X > 0, Y > 0)$.

Exercise 6.62. Suppose that the joint density function of the random variables X, Y is given by the function

$$f(x, y) = \begin{cases} \frac{1}{\sqrt{2\pi}} e^{-\frac{(x-y)^2}{2} - y}, & x \in \mathbb{R}, y > 0 \\ 0, & \text{otherwise.} \end{cases}$$

Find the marginal density function of X.

7

Sums and symmetry

In this chapter we derive distributions of sums of independent random variables and demonstrate how symmetry simplifies complicated calculations. Our new tools bring additional insight on the Poisson process.

7.1. Sums of independent random variables

If we know the joint distribution of two random variables X and Y, then we have all the information about them, so in theory we can describe any random variable of the form $g(X, Y)$. In this section we take a closer look at the distribution of $X + Y$. We describe the distribution of $X + Y$ in both the discrete and the jointly continuous cases. We also specialize to the case of independent X and Y. Finally, we look at some important examples.

Suppose first that X and Y are discrete with joint probability mass function $p_{X,Y}$. Then $X+Y$ is also discrete and its probability mass function can be computed by breaking up the event $\{X + Y = n\}$ into the disjoint union of the events $\{X = k, Y = n - k\}$ over all possible values of k:

$$p_{X+Y}(n) = P(X + Y = n) = \sum_k P(X = k, Y = n - k) = \sum_k p_{X,Y}(k, n - k).$$

If X and Y are independent then we can write the joint probability mass function as the product of the marginal probability mass functions and get

$$p_{X+Y}(n) = \sum_k p_X(k)\, p_Y(n - k).$$

The same argument also gives $p_{X+Y}(n) = \sum_\ell p_X(n - \ell) p_Y(\ell)$.

The operation above that produces the new probability mass function p_{X+Y} from p_X and p_Y is a kind of product of probability mass functions. This operation is called the *convolution* and it is denoted by $p_X * p_Y$:

$$p_X * p_Y(n) = \sum_k p_X(k)\, p_Y(n - k) = \sum_\ell p_X(n - \ell) p_Y(\ell).$$

Next we look at jointly continuous random variables X and Y with joint density function $f_{X,Y}$. First we identify the cumulative distribution function of $X + Y$:

$$
\begin{aligned}
F_{X+Y}(z) &= P(X + Y \leq z) = \iint_{x+y\leq z} f_{X,Y}(x, y)\, dx\, dy \\
&= \int_{-\infty}^{\infty} \left(\int_{-\infty}^{z-x} f_{X,Y}(x, y)\, dy \right) dx = \int_{-\infty}^{\infty} \left(\int_{-\infty}^{z} f_{X,Y}(x, w - x)\, dw \right) dx \\
&= \int_{-\infty}^{z} \left(\int_{-\infty}^{\infty} f_{X,Y}(x, w - x)\, dx \right) dw. \qquad (7.1)
\end{aligned}
$$

The first line of the calculation above comes from (6.9) with the set $B = \{(x, y) : x + y \leq z\}$. Then we write the integral as an iterated integral. We change variables in the inner integral from y to $w = y + x$. The last step switches around the integrals which is possible because their limits no longer depend on each other. By definition, the cumulative distribution function F_{X+Y} and the density function f_{X+Y} are related by

$$F_{X+Y}(z) = \int_{-\infty}^{z} f_{X+Y}(w)\, dw.$$

Comparing the right-hand side above with (7.1) shows that the probability density function of $X + Y$ is

$$f_{X+Y}(z) = \int_{-\infty}^{\infty} f_{X,Y}(x, z - x)\, dx.$$

If X and Y are independent we can write the joint density function as the product of the marginal densities, so in that case

$$f_{X+Y}(z) = \int_{-\infty}^{\infty} f_X(x) f_Y(z - x)\, dx.$$

The right hand side above is again called the convolution of f_X and f_Y and denoted by $f_X * f_Y$. A similar argument yields $f_{X+Y}(z) = \int_{-\infty}^{\infty} f_X(z - x) f_Y(x) dx$.

We summarize our findings in the following statement.

Fact 7.1. (Convolution of distributions) If X and Y are independent discrete random variables with probability mass functions p_X and p_Y, then the probability mass function of $X + Y$ is

$$p_{X+Y}(n) = p_X * p_Y(n) = \sum_k p_X(k)\, p_Y(n - k) = \sum_\ell p_X(n - \ell)\, p_Y(\ell). \qquad (7.2)$$

If X and Y are independent continuous random variables with density functions f_X and f_Y then the density function of $X + Y$ is

$$f_{X+Y}(z) = f_X * f_Y(z) = \int_{-\infty}^{\infty} f_X(x) f_Y(z - x)\, dx = \int_{-\infty}^{\infty} f_X(z - x) f_Y(x)\, dx. \qquad (7.3)$$

The remainder of this section treats examples, beginning with discrete ones.

Example 7.2. (Convolution of Poisson random variables) Suppose that $X \sim \text{Poisson}(\lambda)$, $Y \sim \text{Poisson}(\mu)$ and these are independent. Find the distribution of $X + Y$.

Since $X + Y \geq 0$, we only need to consider $P(X + Y = n)$ for $n \geq 0$. Using Fact 7.1,

$$P(X+Y=n) = \sum_{k=-\infty}^{\infty} P(X=k)P(Y=n-k) = \sum_{k=0}^{n} P(X=k)P(Y=n-k)$$

$$= \sum_{k=0}^{n} e^{-\lambda}\frac{\lambda^k}{k!}e^{-\mu}\frac{\mu^{n-k}}{(n-k)!} = \frac{e^{-(\lambda+\mu)}}{n!}\sum_{k=0}^{n}\frac{n!}{k!(n-k)!}\lambda^k\mu^{n-k}.$$

The second equality holds because $P(X = j) = P(Y = j) = 0$ for any $j < 0$. The last sum equals $(\lambda + \mu)^n$ by the binomial theorem, so we get

$$P(X+Y=n) = e^{-(\lambda+\mu)}\frac{(\lambda+\mu)^n}{n!}.$$

We found that $X+Y \sim \text{Poisson}(\lambda+\mu)$. In other words, *the sum of two independent Poisson random variables is also Poisson, and the parameter of the sum is the sum of the parameters.* ▲

Example 7.3. (Continuation of Example 4.21) Suppose a factory experiences on average λ accidents each month and we assume that the number of accidents per month is Poisson distributed. What is the probability that during a particular 2-month period there are exactly 3 accidents?

Let X denote the number of accidents in this 2-month period. One could imagine two approaches. We can use the Poisson model again, but with double the accident rate, so with mean 2λ. Then the answer is

$$P(X=3) = e^{-2\lambda}\frac{(2\lambda)^3}{6}.$$

On the other hand, we could write $X = X_1 + X_2$ where X_1 and X_2 are the Poisson(λ) numbers of accidents in individual months and assume X_1 and X_2 are independent with probability mass functions p_{X_1} and p_{X_2}. Then by the previous example $X \sim \text{Poisson}(2\lambda)$ which gives the same answer.

It is reassuring that the two approaches lead to the same answer. It tells us that the model is consistent with itself. Of course, if we assume accidents occur according to a Poisson process, then this result follows from the discussion about Poisson processes; see Definition 4.34. ▲

Example 7.4. (Convolution of binomials with the same success probability) Let $X \sim \text{Bin}(m_1, p)$ and $Y \sim \text{Bin}(m_2, p)$ be independent. Find the distribution of $X + Y$.

We can bypass the convolution by thinking about independent trials. Imagine $m_1 + m_2$ independent trials with success probability p. Let X be the number of successes among the first m_1 trials and Y the number of successes among the last

m_2 trials. Then by Example 6.30 X and Y are independent with distributions $X \sim \text{Bin}(m_1, p)$ and $Y \sim \text{Bin}(m_2, p)$. The random variable $X+Y$ counts the total number of successes and so its distribution is $\text{Bin}(m_1 + m_2, p)$.

Note that it was important that X and Y have the same success probability. If the success probabilities are not equal then the distribution of $X + Y$ is more complicated. ▲

Example 7.5. (Convolution of geometric random variables) Let X and Y be independent geometric random variables with the same success parameter $p < 1$. Find the distribution of $X + Y$.

The common probability mass function of X and Y is given by

$$P(X = k) = P(Y = k) = p(1-p)^{k-1} \qquad \text{for } k \geq 1.$$

The possible values for $X + Y$ are $n \geq 2$. Using Fact 7.1 we get

$$P(X + Y = n) = \sum_{k=-\infty}^{\infty} P(X = k)P(Y = n - k) = \sum_{k=1}^{n-1} P(X = k)P(Y = n - k),$$

where the second equality follows because $P(X = k) = 0$ unless $k \geq 1$ and $P(Y = n - k) = 0$ unless $n - k \geq 1$. Taken together, the condition on k is $1 \leq k \leq n - 1$. Continuing,

$$\begin{aligned} P(X + Y = n) &= \sum_{k=1}^{n-1} P(X = k)P(Y = n - k) = \sum_{k=1}^{n-1} p(1-p)^{k-1} p(1-p)^{n-k-1} \\ &= \sum_{k=1}^{n-1} p^2(1-p)^{n-2} = (n-1)p^2(1-p)^{n-2}. \end{aligned}$$

Can we interpret the final result? Recall that for repeated independent trials with success probability p, the number of trials needed until the first success is a geometric random variable X with parameter p. Let Y denote the number of trials performed *after* the first success until the *second* success. Y does not care about the earlier trials, hence $Y \sim \text{Geom}(p)$ and X and Y are independent. $X + Y$ is exactly the number of trials needed until we get two successes. The probability mass function of this random variable can be computed directly: $X+Y = n$ means that among the first $n - 1$ trials we have exactly one success and $n - 2$ failures, and the nth trial is a success. The probability of this event is

$$\binom{n-1}{1} p(1-p)^{n-2} \cdot p = (n-1)p^2(1-p)^{n-2},$$

exactly as given by the convolution formula above.

The second approach generalizes easily. Let $X_1, X_2, \ldots, X_k$ be independent $\text{Geom}(p)$ random variables with $p < 1$. Then $X_1 + X_2 + \cdots + X_k$ has the same distribution as the number of trials needed to get exactly k successes. The event that the kth success arrives on the nth trial is the same as having $k-1$ successes among the first $n-1$ trials and then a success at the nth . Consequently, for $n \geq k$,

$$P(X_1 + \cdots + X_k = n) = \binom{n-1}{k-1} p^{k-1}(1-p)^{n-k} \cdot p = \binom{n-1}{k-1} p^k (1-p)^{n-k}.$$

▲

The distribution of the number of trials needed to get exactly k successes is called the *negative binomial* distribution.

Definition 7.6. Let k be a positive integer and $0 < p < 1$. A random variable X has the **negative binomial distribution** with parameters (k, p) if the set of possible values of X is the set of integers $\{k, k+1, k+2, \dots\}$ and the probability mass function is

$$P(X = n) = \binom{n-1}{k-1} p^k (1-p)^{n-k} \qquad \text{for } n \geq k.$$

Abbreviate this by $X \sim \text{Negbin}(k, p)$.

Note that the Negbin$(1, p)$ distribution is the same as the Geom(p) distribution.

Example 7.7. The Red Sox and Brewers are playing in the World Series. This is a best-of-seven series, which means that the teams play each other until one team accumulates 4 wins. The probability that the Red Sox win a given game is 0.55. Assuming outcomes are independent, what is the probability that the Brewers win the series in 6 games? What is the probability that the Brewers win the series?

For modeling purposes, imagine that the Red Sox and Brewers play repeatedly until the Brewers get to 4 wins, even if this takes more than seven games. Let X be the number of games played. Then $X \sim \text{Negbin}(4, 0.45)$. The Brewers win the World Series in 6 games exactly if $X = 6$. The probability of this can be computed as

$$P(\text{Brewers win in 6}) = P(X = 6) = \binom{5}{3} 0.45^4 (0.55)^{6-4} \approx 0.12.$$

The Brewers win the series if X is equal to 4, 5, 6 or 7. These are the cases when the Brewers get to 4 wins before the Red Sox. We already have $P(X = 6)$, and we can compute the other probabilities as

$$\begin{aligned} P(X = 4) &= \binom{3}{3} 0.45^4 (0.55)^0 \approx 0.04, \\ P(X = 5) &= \binom{4}{3} 0.45^4 (0.55)^1 \approx 0.09, \\ \text{and} \quad P(X = 7) &= \binom{6}{3} 0.45^4 (0.55)^3 \approx 0.14. \end{aligned}$$

Therefore, the probability that the Brewers win the series is

$$P(\text{Brewers Win}) = P(X=4) + P(X=5) + P(X=6) + P(X=7)$$
$$\approx 0.04 + 0.09 + 0.12 + 0.14 = 0.39. \qquad \blacktriangle$$

We turn to continuous examples.

Example 7.8. (Convolution of normal random variables) Suppose that $X \sim \mathcal{N}(\mu_1, \sigma_1^2)$ and $Y \sim \mathcal{N}(\mu_2, \sigma_2^2)$, and X and Y are independent. Find the distribution of $X + Y$.

To simplify the computation we first consider the case

$$\mu_1 = 0, \quad \sigma_1 = 1, \quad \mu_2 = 0, \quad \text{and} \quad \sigma_2 = \sigma.$$

Then $f_X(x) = \varphi(x) = \frac{1}{\sqrt{2\pi}} e^{-\frac{x^2}{2}}$ and $f_Y(y) = \frac{1}{\sqrt{2\pi\sigma^2}} e^{-\frac{y^2}{2\sigma^2}}$. By the convolution formula (7.3) we need to compute

$$f_{X+Y}(z) = \int_{-\infty}^{\infty} \frac{1}{\sqrt{2\pi}} e^{-\frac{x^2}{2}} \frac{1}{\sqrt{2\pi\sigma^2}} e^{-\frac{(z-x)^2}{2\sigma^2}}\,dx = \int_{-\infty}^{\infty} \frac{1}{2\pi\sigma} e^{-\left(\frac{x^2}{2} + \frac{(z-x)^2}{2\sigma^2}\right)}\,dx.$$

This integral looks complicated, but it can be handled after completing the square in the exponent:

$$\begin{aligned}
\tfrac{x^2}{2} + \tfrac{(z-x)^2}{2\sigma^2} &= \tfrac{1}{2\sigma^2}\left((\sigma^2+1)x^2 - 2xz + z^2\right) \\
&= \tfrac{1+\sigma^2}{2\sigma^2}\left(x - \tfrac{z}{1+\sigma^2}\right)^2 + \tfrac{z^2}{2(1+\sigma^2)}.
\end{aligned}$$

This means that

$$\begin{aligned}
\int_{-\infty}^{\infty} \frac{1}{2\pi\sigma} e^{-\left(\frac{x^2}{2} + \frac{(z-x)^2}{2\sigma^2}\right)}\,dx &= \int_{-\infty}^{\infty} \frac{1}{2\pi\sigma} e^{-\frac{1+\sigma^2}{2\sigma^2}\left(x - \frac{z}{1+\sigma^2}\right)^2 - \frac{z^2}{2(1+\sigma^2)}}\,dx \\
&= \frac{1}{\sqrt{2\pi(1+\sigma^2)}} e^{-\frac{z^2}{2(1+\sigma^2)}} \int_{-\infty}^{\infty} \frac{1}{\sqrt{2\pi\frac{\sigma^2}{1+\sigma^2}}} e^{-\frac{1+\sigma^2}{2\sigma^2}\left(x - \frac{z}{1+\sigma^2}\right)^2}\,dx.
\end{aligned}$$

The first term on the last line is the density function of the $\mathcal{N}(0, 1+\sigma^2)$ distribution. The integral may still look complicated at first glance, but it is actually the total integral of the density of the $\mathcal{N}(\frac{z}{1+\sigma^2}, \frac{\sigma^2}{1+\sigma^2})$ distribution, and hence equals 1. This means that

$$f_{X+Y}(z) = \frac{1}{\sqrt{2\pi(1+\sigma^2)}} e^{-\frac{z^2}{2(1+\sigma^2)}},$$

and thereby $X + Y \sim \mathcal{N}(0, 1+\sigma^2)$.

To treat the general case of independent $X \sim \mathcal{N}(\mu_1, \sigma_1^2)$ and $Y \sim \mathcal{N}(\mu_2, \sigma_2^2)$ we use the special case we just proved together with Fact 3.61. First, observe that $Z_1 = \frac{X-\mu_1}{\sigma_1}$ and $Z_2 = \frac{Y-\mu_2}{\sigma_2}$ are independent standard normals. From $X = \sigma_1 Z_1 + \mu_1$ and $Y = \sigma_2 Z_2 + \mu_2$ you can deduce

$$X + Y = \sigma_1 Z_1 + \mu_1 + \sigma_2 Z_2 + \mu_2 = \sigma_1\left(Z_1 + \tfrac{\sigma_2}{\sigma_1} Z_2\right) + (\mu_1 + \mu_2).$$

Since $\frac{\sigma_2}{\sigma_1} Z_2 \sim \mathcal{N}(0, \frac{\sigma_2^2}{\sigma_1^2})$ and it is independent of Z_1, by the special case that we proved above

$$Z_1 + \tfrac{\sigma_2}{\sigma_1} Z_2 \sim \mathcal{N}(0, 1 + \tfrac{\sigma_2^2}{\sigma_1^2}).$$

Since the linear transform of a normal is also normal, we have

$$X + Y = \sigma_1 \left(Z_1 + \tfrac{\sigma_2}{\sigma_1} Z_2 \right) + (\mu_1 + \mu_2) \sim \mathcal{N} \left(\mu_1 + \mu_2, \left(1 + \tfrac{\sigma_2^2}{\sigma_1^2} \right) \sigma_1^2 \right).$$

Simplifying the parameters we get that $X + Y \sim \mathcal{N}(\mu_1 + \mu_2, \sigma_1^2 + \sigma_2^2)$. Thus *the sum of independent normals is also normal*, and the new mean and variance are the sums of the means and variances. ▲

The example above generalizes to a sum of any number of independent normal random variables. In the case of three random variables we would have independent $X_i \sim \mathcal{N}(\mu_i, \sigma_i^2)$ for $i = 1, 2, 3$. The example above implies first $X_1 + X_2 \sim \mathcal{N}(\mu_1 + \mu_2, \sigma_1^2 + \sigma_2^2)$. Next we apply the same point again to $X_1 + X_2$ and X_3 to conclude that $X_1 + X_2 + X_3 \sim \mathcal{N}(\mu_1 + \mu_2 + \mu_3, \sigma_1^2 + \sigma_2^2 + \sigma_3^2)$. We can combine this with the linear transformation of normals (Fact 3.61) to reach the following conclusion.

Fact 7.9. Assume $X_1, X_2, \dots, X_n$ are independent random variables with $X_i \sim \mathcal{N}(\mu_i, \sigma_i^2)$, $a_i \neq 0$, and $b \in \mathbb{R}$. Let $X = a_1 X_1 + \cdots + a_n X_n + b$. Then $X \sim \mathcal{N}(\mu, \sigma^2)$ where

$$\mu = a_1 \mu_1 + \cdots + a_n \mu_n + b \quad \text{and} \quad \sigma^2 = a_1^2 \sigma_1^2 + \cdots + a_n^2 \sigma_n^2.$$

Example 7.10. Let $X \sim \mathcal{N}(1, 3)$ and $Y \sim \mathcal{N}(0, 4)$ be independent and let $W = \frac{1}{2} X - Y + 6$. Identify the distribution of W.

According to Fact 7.9, W has normal distribution with mean

$$E[W] = \tfrac{1}{2} E[X] - E[Y] + 6 = \tfrac{1}{2} - 0 + 6 = \tfrac{13}{2}$$

and variance

$$\mathrm{Var}(W) = \tfrac{1}{4} \mathrm{Var}(X) + \mathrm{Var}(Y) = \tfrac{3}{4} + 4 = \tfrac{19}{4}.$$

Note that all terms in the variance calculation come with a plus sign, even though one random variable comes with a minus sign in the expression for W. ▲

Example 7.11. Adult border collies have mean weight 40 pounds with standard deviation 4 pounds, while adult pugs have mean weight 15 pounds with standard deviation 2 pounds. (Border collies and pugs are dog breeds.) What is the probability that a border collie weighs at least 12 pounds more than two pugs? Assume that the weights of dogs within a breed are well modeled by a normal distribution, and that the weights are independent of each other.

The question boils down to finding the probability $P(X_1 \geq X_2 + X_3 + 12)$ where X_1, X_2, X_3 are independent with marginal distributions $X_1 \sim \mathcal{N}(40, 16)$ and $X_2, X_3 \sim \mathcal{N}(15, 4)$.

Let $Y = X_1 - X_2 - X_3$. By Fact 7.9, $Y \sim \mathcal{N}(10, 24)$. (Be sure to realize that means may be added or subtracted but variances are only added.) Consequently

$$P(X_1 \geq X_2 + X_3 + 12) = P(Y \geq 12) = P\left(\frac{Y-10}{\sqrt{24}} \geq \frac{12-10}{\sqrt{24}}\right)$$
$$= P(Z > 0.41) = 1 - \Phi(0.41) = 1 - 0.6591 \approx 0.34.$$

In the computation above we used the fact that $Z = \frac{1}{\sqrt{24}}(Y - 10)$ is a standard normal. ▲

As the last examples of the section we look at convolution of independent exponential random variables and convolution of independent uniform random variables.

Example 7.12. (Convolution of exponential random variables) Suppose that X and Y are independent Exp(λ) random variables. Find the density of $X + Y$.

Since X and Y are nonnegative, $X + Y$ is also nonnegative. This means that $f_{X+Y}(z) = 0$ if $z < 0$. If $z \geq 0$ then

$$f_{X+Y}(z) = \int_{-\infty}^{\infty} f_X(x) f_Y(z-x)dx = \int_0^z f_X(x) f_Y(z-x)dx$$
$$= \int_0^z \lambda e^{-\lambda x} \lambda e^{-\lambda(z-x)} dx = \lambda^2 \int_0^z e^{-\lambda z} dx = \lambda^2 z e^{-\lambda z}.$$

In the second equality we restricted the integral to $[0, z]$ because $f_X(x) f_Y(z-x) = 0$ if $x < 0$ or $x > z$.

Recall Definition 4.37 of the gamma distribution. The result above shows that $X + Y$ has the Gamma($2, \lambda$) distribution. ▲

Example 7.13. (Convolution of uniform random variables) Suppose that X and Y are independent and distributed as Unif[0, 1]. Find the distribution of $X + Y$.

The density functions for X and Y are

$$f_X(x) = f_Y(x) = \begin{cases} 1, & 0 \leq x \leq 1 \\ 0, & x < 0 \text{ or } x > 1. \end{cases}$$

$X + Y$ is always between 0 and 2, so the density function $f_{X+Y}(z)$ is zero for $z < 0$ and $z > 2$. It remains to consider $0 \leq z \leq 2$. By the convolution formula,

$$f_{X+Y}(z) = \int_{-\infty}^{\infty} f_X(x) f_Y(z-x)\, dx.$$

We must take care with the integration limits. The product $f_X(x) f_Y(z-x)$ is nonzero and equal to 1 if and only if $0 \leq x \leq 1$ and $0 \leq z - x \leq 1$. The second inequality is equivalent to $z - 1 \leq x \leq z$. To simultaneously satisfy $0 \leq x \leq 1$ and $z - 1 \leq x \leq z$, we take the larger of the lower bounds and the smaller of the upper bounds on x. Thus the integrand $f_X(x)f_Y(z-x)$ is 1 if and only if

$$\max(0, z-1) \le x \le \min(1, z).$$

Now for $0 \le z \le 2$

$$f_{X+Y}(z) = \int_{-\infty}^{\infty} f_X(x)f_Y(z-x)dx = \int_{\max(0,z-1)}^{\min(1,z)} dx = \min(1,z) - \max(0, z-1).$$

We can simplify the result by considering the cases $0 \le z \le 1$ and $1 < z \le 2$ separately:

$$f(z) = \begin{cases} z - 0 = z, & \text{if} \quad 0 \le z \le 1 \\ 1-(z-1) = 2-z, & \text{if} \quad 1 < z \le 2 \\ 0, & \text{otherwise.} \end{cases}$$

The graph of the density function looks like a triangle on the interval $[0, 2]$ (see Figure 7.1). ▲

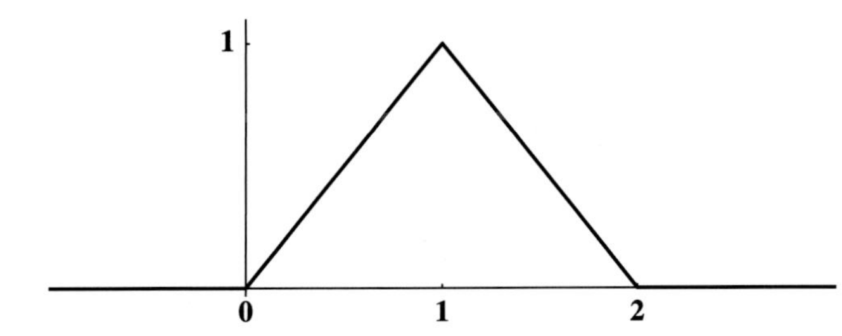

Figure 7.1. The p.d.f of the convolution of two Unif[0, 1] distributions.

7.2. Exchangeable random variables

Consider this example for motivation.

Example 7.14. We flip over cards from a shuffled deck one by one. What is the probability that the 23rd card is a spade?

We could write down the joint probability mass function of the first 23 cards and then take the marginal of the 23rd card to find the answer. On the other hand, it seems clear that the probability of a spade as the 23rd card should be the same as the probability of a spade as the first card, namely $\frac{1}{4}$. (For why should spades be any likelier to come in one spot rather than in another?) The concept of *exchangeability* will allow us to make this argument rigorous. ▲

Exchangeability is a type of symmetry among random variables. When it is present, it is extremely useful for calculations. We begin this section with a general discussion and then focus on two particular examples of exchangeable sequences: (i) independent, identically distributed (i.i.d.) sequences and (ii) sampling without replacement.

In order to define an exchangeable sequence of random variables, we must recall two notions: equality in distribution and permutations.

First, recall Definition 5.12 for equality in distribution of two real-valued random variables. Analogously, we say that two random vectors $(X_1, \dots, X_n)$ and $(Y_1, \dots, Y_n)$ are equal in distribution if

$$P((X_1, \dots, X_n) \in B) = P((Y_1, \dots, Y_n) \in B) \tag{7.4}$$

for all subsets B of $\mathbb{R}^n$. We abbreviate this as

$$(X_1, \dots, X_n) \stackrel{d}{=} (Y_1, \dots, Y_n).$$

As in the single variable case, equality in distribution implies that corresponding probabilities and expectations are the same for both $(X_1, \dots, X_n)$ and $(Y_1, \dots, Y_n)$. That is, for $g : \mathbb{R}^n \to \mathbb{R}$,

$$E[g(X_1, \dots, X_n)] = E[g(Y_1, \dots, Y_n)],$$

so long as the expectation exists.

Next, recall that a *permutation* is a rearrangement of a sequence. For example, $(2, 5, 1, 3, 4)$ is a permutation of $(1, 2, 3, 4, 5)$.

Definition 7.15. A sequence of random variables $X_1, X_2, \dots, X_n$ is **exchangeable** if the following condition holds: for every permutation $(k_1, k_2, \dots, k_n)$ of $(1, 2, \dots, n)$, we have

$$(X_1, X_2, \dots, X_n) \stackrel{d}{=} (X_{k_1}, X_{k_2}, \dots, X_{k_n}).$$

Example 7.16. We illustrate Definition 7.15 for $n = 3$. The 6 permutations of $(1, 2, 3)$ are

$$(1, 2, 3), \quad (1, 3, 2), \quad (2, 1, 3), \quad (2, 3, 1), \quad (3, 1, 2), \quad (3, 2, 1).$$

Thus the random variables X_1, X_2, X_3 are exchangeable if and only if the following six vectors all have the same distribution:

$$\begin{array}{lll} (X_1, X_2, X_3), & (X_1, X_3, X_2), & (X_2, X_1, X_3), \\ (X_2, X_3, X_1), & (X_3, X_1, X_2), & (X_3, X_2, X_1). \end{array} \tag{7.5}$$

▲

Having stated this somewhat abstract definition of exchangeability, we address two questions.

1. How do we check for exchangeability?
2. What does exchangeability allow us to do in calculations?

The following fact answers the first question for discrete and jointly continuous random variables. We first recall that the real-valued function $f : \mathbb{R}^n \to \mathbb{R}$ is *symmetric* if for any permutation $(k_1, k_2, \dots, k_n)$ of $(1, 2, \dots, n)$, we have

$$f(x_{k_1}, x_{k_2}, \dots, x_{k_n}) = f(x_1, x_2, \dots, x_n) \tag{7.6}$$

for all values $(x_1, x_2, \dots, x_n)$. For example, $f_1(x_1, x_2, x_3) = x_1 x_2 x_3$ is symmetric, whereas $f_2(x_1, x_2, x_3) = x_1 x_2 + x_3$ is not.

Fact 7.17. (How to check exchangeability)

Suppose $X_1, X_2, \dots, X_n$ are discrete random variables with joint probability mass function p. Then these random variables are exchangeable if and only if p is a symmetric function.

Suppose $X_1, X_2, \dots, X_n$ are jointly continuous random variables with joint density function f. Then these random variables are exchangeable if and only if f is a symmetric function.

Proof. We prove here the case of three discrete random variables. The main idea is already present in this simple case.

Let X_1, X_2, X_3 have joint probability mass function p_{X_1,X_2,X_3}. The joint probability mass function of the permuted vector (X_3, X_1, X_2) is given by

$$\begin{aligned} p_{X_3,X_1,X_2}(a, b, c) &= P(X_3 = a, X_1 = b, X_2 = c) \\ &= P(X_1 = b, X_2 = c, X_3 = a) = p_{X_1,X_2,X_3}(b, c, a), \end{aligned}$$

where a, b, c are arbitrary possible values of the random variables.

The calculation above shows that the function p_{X_3,X_1,X_2} is obtained by permuting the arguments of p_{X_1,X_2,X_3}. Thus the functions p_{X_1,X_2,X_3} and p_{X_3,X_1,X_2} agree if and only if this permutation leaves p_{X_1,X_2,X_3} unchanged. Agreement of p_{X_1,X_2,X_3} and p_{X_3,X_1,X_2} is the same as the equality in distribution of (X_1, X_2, X_3) and (X_3, X_1, X_2). Hence the symmetry of the function p_{X_1,X_2,X_3} is equivalent to equality in distribution of all the permutations of (X_1, X_2, X_3). These permutations were listed in (7.5). ▲

The practical meaning of exchangeability is that we can rearrange the random variables $X_1, X_2, \dots, X_n$ in *any probability or expectation* without changing its value. For example, for any sets $B_1, \dots, B_n$,

$$\begin{aligned} &P(X_1 \in B_1, X_2 \in B_2, \dots, X_n \in B_n) \\ &\qquad = P(X_{k_1} \in B_1, X_{k_2} \in B_2, \dots, X_{k_n} \in B_n). \end{aligned} \tag{7.7}$$

If we take $B_1 = B$ and $B_2 = \cdots = B_n = \mathbb{R}$ in (7.7) we get

$$P(X_1 \in B) = P(X_{k_1} \in B)$$

for any index k_1. In other words, individual random variables in an exchangeable sequence all have the same probability distribution. The following fact summarizes the key consequences of exchangeability.

Fact 7.18. Suppose that $X_1, X_2, \ldots, X_n$ are exchangeable. Then the following statements hold.

(i) The random variables X_k are identically distributed, that is, they have the same marginal distributions. More concretely, if the random variables are discrete then each X_k has the same marginal probability mass function, while if they are continuous then each X_k has the same marginal density function.

(ii) For any $1 \le k \le n$ the random variables $(X_1, \ldots, X_k)$ have the same joint distribution as any other k-tuple $(X_{i_1}, X_{i_2}, \ldots, X_{i_k})$. Consequently, for any function g of k variables, the expectation does not change under permutations:

$$E\left[g(X_1, X_2, \ldots, X_k)\right] = E\left[g(X_{i_1}, X_{i_2}, \ldots, X_{i_k})\right]. \tag{7.8}$$

We turn to study two particular cases of exchangeable random variables that are common in applications.

Independent identically distributed random variables

We begin with the basic definition.

Definition 7.19. Random variables $X_1, X_2, X_3, \ldots$ are **independent and identically distributed** (abbreviated **i.i.d.**) if they are independent and each X_k has the same probability distribution.

We have seen examples of i.i.d. sequences. Outcomes X_j of independent repeated trials in Section 2.4 are i.i.d. (provided the success probability p is the same in each trial).

Theorem 7.20. *An i.i.d. sequence of random variables is exchangeable.*

Proof. Our tools allow us to prove the theorem for discrete or continuous random variables. By Fact 7.17 we need to show that either the joint probability mass function (discrete case) or joint density function (continuous case) is symmetric.

We demonstrate the proof for discrete random variables with $n = 3$. Let X_1, X_2, X_3 be i.i.d. discrete random variables with common marginal probability mass function p. We check what happens to the joint probability mass function of X_1, X_2, X_3 when we replace (x_1, x_2, x_3) with the permuted variables (x_2, x_3, x_1):

$$\begin{aligned}p_{X_1,X_2,X_3}(x_2,x_3,x_1) &= p_{X_1}(x_2)p_{X_2}(x_3)p_{X_3}(x_1) = p(x_2)p(x_3)p(x_1)\\ &= p(x_1)p(x_2)p(x_3) = p_{X_1}(x_1)p_{X_2}(x_2)p_{X_3}(x_3) = p_{X_1,X_2,X_3}(x_1,x_2,x_3).\end{aligned}$$

The computation shows that the joint probability mass function of X_1, X_2, X_3 does not change if we substitute in the permuted variables (x_2, x_3, x_1). The same argument works for any other permutation. This verifies that p_{X_1,X_2,X_3} is symmetric, and so X_1, X_2, X_3 are indeed exchangeable.

Similar proofs work if we have more than 3 random variables, or if the distributions are continuous. ▲

Example 7.21. Suppose that X_1, X_2, X_3 are independent Unif[0, 1] random variables. Find the probability that X_1 is the largest.

We could compute this probability by integrating the density function

$$f(x_1, x_2, x_3) = \begin{cases} 1, & \text{if each } x_i \text{ lies in } [0, 1] \\ 0, & \text{else} \end{cases}$$

on the set $\{(x_1, x_2, x_3) : x_1 > x_2, x_1 > x_3\}$. This integral would be

$$\iiint_{\substack{1 \geq x_1 > \max(x_2, x_3),\\ x_2 \geq 0,\, x_3 \geq 0}} dx_1\, dx_2\, dx_3.$$

This is not a difficult integral, but the answer comes quicker from symmetry. Since X_1, X_2, X_3 are jointly continuous, the probability that any two of them will be the same is 0. (See the comment at the end of Example 6.21.) Thus

$$1 = P(X_1 \text{ is the largest}) + P(X_2 \text{ is the largest}) + P(X_3 \text{ is the largest}).$$

But these three probabilities are the same by the exchangeability of X_1, X_2, X_3, so $P(X_1 \text{ is the largest}) = \frac{1}{3}$. ▲

Note that the solution of the previous problem did not use the fact that the random variables are uniform, only that they are i.i.d. and have a continuous distribution. Thus if $X_1, \dots, X_n$ are i.i.d. with a continuous distribution then the same argument shows that $P(X_1 \text{ is the largest}) = \frac{1}{n}$. In fact, we can extend the argument to the ordering of the entire sequence $X_1, \dots, X_n$. Any specific ordering of the random variables will have probability $\frac{1}{n!}$. For example, $P(X_1 < \cdots < X_n) = \frac{1}{n!}$.

Sampling without replacement

We return to the context of Example 2.30 where $X_1, \dots, X_k$ denote the outcomes of successive draws *without* replacement from the set $\{1, \dots, n\}$. In the example we checked that these random variables are identically distributed, but not independent. We now show that they are exchangeable.

Theorem 7.22. *Let $1 \leq k \leq n$. Let $X_1, \dots, X_k$ denote the outcomes of successive draws without replacement from the set $\{1, \dots, n\}$. Then $X_1, \dots, X_k$ are exchangeable.*

Proof. By Fact 7.17 we need to show that the joint probability mass function is symmetric. This is immediate because for any selection $x_1, \dots, x_k$ of k distinct elements from $\{1, \dots, n\}$

$$P(X_1 = x_1, X_2 = x_2, \dots, X_k = x_k) = \frac{1}{n(n-1)\cdots(n-k+1)},$$

which shows that every permutation of the values x_j will give the same probability (since the probability is constant). ▲

We often sample in a situation where not all items are distinct. (An example would be an urn where some balls have the same color.) Then we can imagine that at first all items are distinct, but then new labels, given by a function, are attached to the items. For example, if we want an urn with 5 red and 3 green balls, we start with balls labeled $\{1, 2, \dots, 8\}$ in the urn and then paint balls 1–5 red and balls 6–8 green by defining the function

$$g(x) = \begin{cases} \texttt{red}, & x \in \{1, 2, 3, 4, 5\} \\ \texttt{green}, & x \in \{6, 7, 8\}. \end{cases}$$

When a ball is drawn we record its color, in other words we record $g(x)$ and not x. The next result tells us that the resulting sequence of random variables is also exchangeable.

Theorem 7.23. *Suppose that the random variables $X_1, \dots, X_n$ are exchangeable. Then, for any function g, the sequence of random variables $g(X_1), \dots, g(X_n)$ is exchangeable.*

Exercise 7.37 asks you to prove this theorem for discrete random variables for $n = 3$.

Using the exchangeability of sampling without replacement enables us to easily calculate probabilities for which counting would be cumbersome.

Example 7.24. Suppose an urn contains 37 red balls, 61 green balls, and 50 yellow balls. We sample 20 balls without replacement. What is the probability that the 9th draw is yellow, the 12th draw is red, and the 20th draw is yellow?

If X_i denotes the color of the ith draw, then exchangeability gives us

$$\begin{aligned} &P(X_9 = \texttt{yellow}, X_{12} = \texttt{red}, X_{20} = \texttt{yellow}) \\ &= P(X_1 = \texttt{yellow}, X_2 = \texttt{yellow}, X_3 = \texttt{red}) = \frac{50 \cdot 49 \cdot 37}{148 \cdot 147 \cdot 146}. \end{aligned}$$

By exchangeability, we were allowed to replace the indices 9, 12, 20 with 1, 2, 3. This allows us to compute the final answer by only considering the outcomes of the first three draws. ▲

Exchangeability is also helpful when dealing with conditional probabilities.

Example 7.25. Suppose we deal 10 cards from a standard deck of 52. What is the probability that the sixth card is an ace, given that the fifth and tenth cards are both aces?

Let X_j be the value of the jth card dealt.

$$\begin{aligned}P(X_6 = \text{ace} \mid X_5 = \text{ace}, X_{10} = \text{ace}) &= \frac{P(X_6 = \text{ace}, X_5 = \text{ace}, X_{10} = \text{ace})}{P(X_5 = \text{ace}, X_{10} = \text{ace})} \\ &= \frac{P(X_1 = \text{ace}, X_2 = \text{ace}, X_3 = \text{ace})}{P(X_1 = \text{ace}, X_2 = \text{ace})} \\ &= P(X_3 = \text{ace} \mid X_1 = \text{ace}, X_2 = \text{ace}) = \tfrac{2}{50} = \tfrac{1}{25}.\end{aligned}$$

Again we can permute the indices so that the calculation is as convenient as possible. Once 2 aces have been removed for the first two picks, there are 2 aces left among the remaining 50 cards. ▲

7.3. Poisson process revisited ♦

This section returns to the Poisson process introduced in Section 4.6. Example 7.12 showed that the sum of two independent Exp(λ) random variables has the Gamma($2, \lambda$) distribution. We first encountered the gamma distribution in Example 4.36. There we saw that the nth arrival time T_n of a Poisson process with intensity λ has the Gamma(n, λ) distribution.

Thus, the first point T_1 in a Poisson process with intensity λ has the Exp(λ) distribution, while the second point T_2 has the distribution of the sum of two independent Exp(λ) random variables. This suggests that T_1 and $T_2 - T_1$ are independent Exp(λ) random variables. This is indeed correct and generalizes further, as made precise in the next statement.

If the points of the Poisson process are thought of as arrival times, then we can call $T_n - T_{n-1}$ the *waiting time* between the $(n-1)$st and the nth arrival.

> **Fact 7.26.** (Waiting times of a Poisson process) Let $0 < T_1 < T_2 < T_3 < \cdots$ denote the positions of the random points of a Poisson process on $[0, \infty)$ with intensity λ. Then the random variables $T_1, T_2 - T_1, T_3 - T_2, \dots, T_n - T_{n-1}, \dots$ are i.i.d. with distribution Exp(λ).

Fact 7.26 gives a practical construction of a Poisson process. Take an i.i.d. sequence $Y_1, Y_2, Y_3, \dots$ with distribution Exp(λ). Then the values Y_1, $Y_1 + Y_2$,

$Y_1 + Y_2 + Y_3, \ldots, \sum_{i=1}^{n} Y_i, \ldots$ give the locations of the points of a Poisson process with intensity λ, in increasing order.

Sketch of the proof of Fact 7.26. Let $W_1 = T_1$ and $W_n = T_n - T_{n-1}$ for $n > 1$. We assume that the joint density function f of $(W_1, \ldots, W_n)$ exists and derive a formula for it by the infinitesimal approach based on Fact 6.39. Once we observe that this joint density function is a product of Exp(λ) densities, Fact 7.26 is proved.

With this in mind, let $0 \le t_1 < t_2 < \cdots < t_n$ and $w_1 = t_1$, $w_2 = t_2 - t_1$, $\ldots$, $w_n = t_n - t_{n-1}$. For small $\varepsilon > 0$, by Fact 6.39 we have,

$$P(W_1 \in (w_1, w_1+\varepsilon), W_2 \in (w_2, w_2 + \varepsilon), \ldots, W_n \in (w_n, w_n + \varepsilon))$$
$$\approx f(w_1, w_2, \ldots, w_n)\varepsilon^n.$$

Note that if $W_1, \ldots, W_n$ are close to $w_1, \ldots, w_n$, respectively, then $T_1, \ldots, T_n$ are close to $t_1, \ldots, t_n$, and vice versa. Hence converting from W_i to T_i we have

$$\begin{aligned} &P(W_1 \in (w_1, w_1 + \varepsilon), W_2 \in (w_2, w_2 + \varepsilon), \ldots, W_n \in (w_n, w_n + \varepsilon)) \\ &\quad\approx P(T_1 \in (t_1, t_1 + \varepsilon), T_2 \in (t_2, t_2 + \varepsilon), \ldots, T_n \in (t_n, t_n + \varepsilon)). \end{aligned}$$

One can make this approximation precise by using the tools of Section 6.4. Thus, we may conclude that

$$\begin{aligned} &f(w_1, w_2, \ldots, w_n)\varepsilon^n \\ &\quad\approx P\left(T_1 \in (t_1, t_1 + \varepsilon), T_2 \in (t_2, t_2 + \varepsilon), \ldots, T_n \in (t_n, t_n + \varepsilon)\right), \end{aligned} \tag{7.9}$$

where the approximation becomes precise as ε goes to zero.

The event on the right-hand side of (7.9) can be described as follows: there are no points in the intervals

$$[0, t_1],\ [t_1 + \varepsilon, t_2],\ [t_2 + \varepsilon, t_3], \ldots, [t_k + \varepsilon, t_{k+1}], \ldots, [t_{n-1} + \varepsilon, t_n]$$

and exactly one point in each of the intervals $(t_k, t_k + \varepsilon)$. See Figure 7.2.

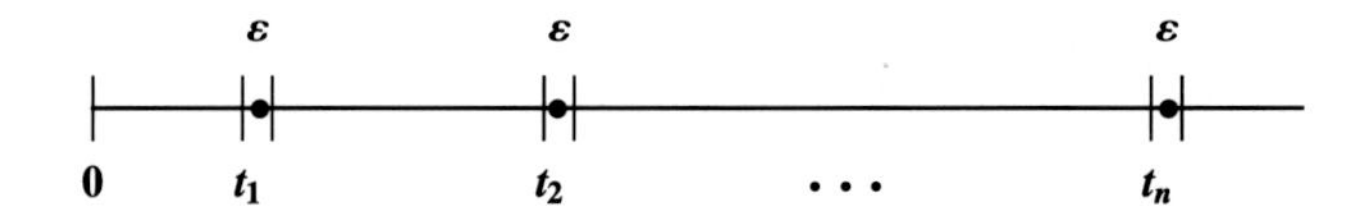

Figure 7.2. The first n points of the Poisson process are confined to the intervals $(t_k, t_k + \varepsilon)$ for $k = 1, \ldots, n$.

These are statements about disjoint intervals, so the probability of the event can be written down from basic facts about the Poisson process. Recall that (i) the number of points in an interval is a Poisson random variable with parameter λ times the length of the interval, and (ii) for disjoint intervals these random variables are independent.

Now continue from (7.9) by multiplying probabilities of independent events:

$$\begin{aligned}
f(w_1, w_2, \dots, w_n)\varepsilon^n &\approx P(T_1 \in (t_1, t_1+\varepsilon), T_2 \in (t_2, t_2+\varepsilon), \dots, T_n \in (t_n, t_n+\varepsilon)) \\
&= e^{-\lambda t_1} \cdot \prod_{k=2}^{n} e^{-\lambda(t_k - t_{k-1} - \varepsilon)} \cdot \prod_{k=1}^{n} \lambda \varepsilon e^{-\lambda\varepsilon} \\
&= \varepsilon^n \, e^{-\lambda\varepsilon} \, \lambda e^{-\lambda t_1} \prod_{k=2}^{n} \lambda e^{-\lambda(t_k - t_{k-1})} \\
&\approx \varepsilon^n \, \lambda e^{-\lambda t_1} \prod_{k=2}^{n} \lambda e^{-\lambda(t_k - t_{k-1})} = \varepsilon^n \prod_{k=1}^{n} \lambda e^{\lambda w_k}.
\end{aligned}$$

The last approximation step above is $e^{-\lambda\varepsilon} \approx 1$. Everything above becomes accurate as $\varepsilon \to 0$, so we get

$$f(w_1, \dots, w_n) = \prod_{k=1}^{n} \lambda e^{\lambda w_k}.$$

This says that $W_1, \dots, W_n$ are independent Exp(λ) random variables. ▲

The description of the Poisson process using the independent exponential waiting times is useful for calculations.

Example 7.27. Customers arrive at a store according to a Poisson process with rate 5/hr. Find the probability that the fourth customer arrives within 10 minutes of the third.

The waiting time, $W_4 = T_4 - T_3$, between the third and fourth arrival is exponential with parameter 5 (measured in hours). We need to find $P(W_4 \le 1/6)$. This is given by

$$P(W_4 \le 1/6) = 1 - e^{-\frac{5}{6}}.$$ ▲

A consequence of Fact 7.26 is that the nth arrival time T_n of a Poisson process with rate λ has the same distribution as the sum of n independent Exp(λ) random variables. Since we already identified the distribution of T_n as Gamma(n, λ), we get the following conclusion. This generalizes Example 7.12 that demonstrated the case $n = 2$.

Fact 7.28. The sum of n independent Exp(λ) variables has the Gamma(n, λ) distribution.

As a consequence, if we have independent random variables $X \sim$ Gamma(m, λ) and $Y \sim$ Gamma(n, λ) with positive integers m and n, then $X + Y \sim$ Gamma($m + n, \lambda$). This conclusion also holds without the integer assumption on m and n, as shown by the next example.

Example 7.29. (Convolution of gamma random variables) Let a, b, λ be positive reals. Let $X \sim \text{Gamma}(a, \lambda)$ and $Y \sim \text{Gamma}(b, \lambda)$ be independent. Then $X + Y \sim \text{Gamma}(a + b, \lambda)$. Note that X and Y have the same parameter λ.

We have $f_X(x) = \frac{\lambda^a x^{a-1}}{\Gamma(a)} e^{-\lambda x}$ for $x \geq 0$, and $f_Y(y) = \frac{\lambda^b y^{b-1}}{\Gamma(b)} e^{-\lambda y}$, for $y \geq 0$. Using the convolution formula from Fact 7.1 we get for $z \geq 0$:

$$\begin{aligned} f_{X+Y}(z) &= \int_0^z \frac{\lambda^a x^{a-1}}{\Gamma(a)} e^{-\lambda x} \frac{\lambda^b (z-x)^{b-1}}{\Gamma(b)} e^{-\lambda(z-x)} dx \\ &= \frac{\lambda^{a+b}}{\Gamma(a)\Gamma(b)} e^{-\lambda z} \int_0^z x^{a-1}(z-x)^{b-1} dx. \end{aligned}$$

Using the change of variables $x = zt$ in the last integral we get

$$\int_0^z x^{a-1}(z-x)^{b-1} dx = z^{a+b-1} \int_0^1 t^{a-1}(1-t)^{b-1} dt,$$

which yields

$$f_{X+Y}(z) = \frac{\lambda^{a+b} z^{a+b-1}}{\Gamma(a)\Gamma(b)} e^{-\lambda z} \int_0^1 t^{a-1}(1-t)^{b-1} dt.$$

Note that the integral now depends only on a and b. Let us denote its value by $B(a, b)$. Then $f_{X+Y}(z) = \frac{\lambda^{a+b} z^{a+b-1} B(a,b)}{\Gamma(a)\Gamma(b)} e^{-\lambda z}$. This is almost the same as the density of a $\text{Gamma}(a + b, \lambda)$ random variable, which would be $g(z) = \frac{\lambda^{a+b} z^{a+b-1}}{\Gamma(a+b)} e^{-\lambda z}$, for $z \geq 0$. In fact, the ratio of the two densities is constant:

$$\frac{f_{X+Y}(z)}{g(z)} = \frac{B(a, b)\Gamma(a + b)}{\Gamma(a)\Gamma(b)}.$$

But since both f_{X+Y} and g are probability density functions, they both integrate to 1. Thus this constant can only be equal to 1 and $f_{X+Y}(z) = g(z)$. This means that $X + Y \sim \text{Gamma}(a + b, \lambda)$.

Another consequence of the last step is the following identity:

$$B(a, b) = \int_0^1 t^{a-1}(1-t)^{b-1} dt = \frac{\Gamma(a)\Gamma(b)}{\Gamma(a+b)} \tag{7.10}$$

for any $a, b > 0$. The function $B(a, b)$ is called the *beta integral.* ▲

Remark 7.30. By repeating the statement of Example 7.29, we see that for mutually independent $X_i \sim \text{Gamma}(\alpha_i, \lambda)$, where $i = 1, \ldots, n$, we have $X_1 + \cdots + X_n \sim \text{Gamma}(\alpha_1 + \cdots + \alpha_n, \lambda)$. ▲

Example 7.31. (The χ_n^2 distribution) Suppose that $X_1, \ldots, X_n$ are i.i.d. $\mathcal{N}(0, 1)$ random variables. What is the distribution of $Y = X_1^2 + \cdots + X_n^2$?

We have seen in Example 5.20 and Remark 5.21 that $X_i^2 \sim \text{Gamma}(1/2, 1/2)$. By Remark 7.30, we can conclude that $Y \sim \text{Gamma}(n/2, 1/2)$. This distribution is also called the *chi-squared distribution with parameter n*, denoted by χ_n^2. Note that χ_n^2 is just another name for $\text{Gamma}(n/2, 1/2)$. In fact, the χ_n^2 notation is used for general n and not just for integer values. Thus the density of a χ_p^2 random variable is given by

$$f_p(x) = \frac{2^{-p/2} x^{p/2-1}}{\Gamma(p/2)} e^{-x/2}, \qquad \text{if } x \geq 0$$

and $f_p(x) = 0$ for $x < 0$.

The square root of such a random variable is said to have *chi distribution with parameter p* or χ_p distribution for short. For integer p it gives the distribution of the length of a p-dimensional random vector with i.i.d. $\mathcal{N}(0, 1)$ entries.

The probability density function of the χ_p distribution can easily be computed using Fact 5.24:

$$g_p(x) = f_p(x^2) 2x = \frac{2^{1-p/2} x^{p-1}}{\Gamma(p/2)} e^{-x^2/2}, \qquad \text{if } x \geq 0$$

and $g_p(x) = 0$ for $x < 0$. ▲

Exercises

We start with some warm-up exercises arranged by section.

Section 7.1

Exercise 7.1. Let X be a Poisson random variable with parameter $\lambda = 2$, and let Y be a geometric random variable with parameter $p = \frac{2}{3}$. Suppose that X and Y are independent, and let $Z = X + Y$. Find $P(Z = 3)$.

Exercise 7.2. Let X and Y be independent Bernoulli random variables with parameters p and r, respectively. Find the distribution of $X + Y$.

Exercise 7.3. The dollar value of a barrel of crude oil changes during a given day according to the following distribution.

Value change	-2	-1	0	1	2	3
probability	$\frac{1}{8}$	$\frac{1}{4}$	$\frac{1}{4}$	$\frac{1}{4}$	$\frac{1}{16}$	$\frac{1}{16}$

Supposing that price movements from day to day are independent, find the probability that the price will go up by a total of \$2 over the next two days.

Exercise 7.4. Suppose that X and Y are independent exponential random variables with parameters $\lambda \neq \mu$. Find the density function of $X + Y$.

Exercise 7.5. Let X, Y and Z be independent normal random variables with distributions $X \sim \mathcal{N}(1, 2)$, $Y \sim \mathcal{N}(2, 1)$, and $Z \sim \mathcal{N}(0, 7)$. Let $W = X - 4Y + Z$.

(a) Identify the distribution of W.
(b) Find the probability $P(W > -2)$.

Section 7.2

Exercise 7.6. We deal 5 cards from a standard deck of 52, one by one. What is the probability that the third is a king, and the fifth is the ace of spades?

Exercise 7.7. Let X_1, X_2, X_3, and X_4 be independent standard normal random variables. What is the probability that X_3 is the second largest?

Exercise 7.8. An urn contains 7 red balls, 6 green balls, and 9 yellow balls. Ten balls are chosen without replacement. What is the probability that the 3rd ball chosen was green, given that the 5th ball chosen was yellow?

Section 7.3

Exercise 7.9. Complaints arrive at a call center according to a Poisson process with rate 6/hr.

(a) Find the probability that the 5th call comes within 10 minutes of the 4th.
(b) Find the probability that the 9th call comes within 15 minutes of the 7th.

Exercise 7.10. Carter's desk lamp uses a lightbulb that has an exponential lifetime with a mean of 6 months. When the lightbulb goes out, it is immediately replaced. It is now New Year's Eve. What is the probability that exactly three bulbs will be replaced before the end of March?

Further exercises

Exercise 7.11. Let $0 < p < 1$ and $0 < r < 1$ with $p \neq r$. You repeat a trial with success probability p until you see the first success. I repeat a trial with success probability r until I see the first success. All the trials are independent of each other.

(a) What is the probability that you and I performed the same number of trials?
(b) Let Z be the total number of trials you and I performed altogether. Find the possible values and the probability mass function of Z.

Exercise 7.12. Let Z be a Bernoulli random variable with parameter p and W an independent binomial random variable with parameters n and p. What is the distribution of $Z + W$?

Exercise 7.13. Suppose that $X \sim \text{Negbin}(k, p)$ and $Y \sim \text{Negbin}(m, p)$. Assume that X and Y are independent, find the probability mass function of $X + Y$.

Hint. You do not need complicated calculations for this one, just recall how we introduced the negative binomial distribution.

Exercise 7.14. Repeat the analysis of Example 7.7 under the assumption that the probability that the Brewers win a given game is $p = 0.40$. Solve the problem again under the assumption that $p = 0.35$ and then again under the assumption that $p = 0.30$.

Exercise 7.15. Let X be an integer chosen uniformly at random from the set $\{1, 2, \dots, n\}$ and Y be an independent integer chosen uniformly at random from the set $\{1, 2, \dots, m\}$. Find the probability mass function of $X + Y$.

Exercise 7.16. Let X be a Poisson random variable with parameter λ and Y an independent Bernoulli random variable with parameter p. Find the probability mass function of $X + Y$.

Exercise 7.17. Independent trials with success probability p are repeated, $0 < p < 1$. Fix positive integers k and ℓ. Find the probability that the number of successes reaches k before the number of failures reaches ℓ. Express the probability as a sum.

Exercise 7.18. Suppose that X and Y are independent random variables with density functions

$$f_X(x) = \begin{cases} 2e^{-2x}, & x \geq 0 \\ 0, & x < 0, \end{cases} \quad \text{and} \quad f_Y(x) = \begin{cases} 4xe^{-2x}, & x \geq 0 \\ 0, & x < 0. \end{cases}$$

Find the density function of $X + Y$.

Exercise 7.19. Let X and Y be independent exponential random variables with parameter 1.

(a) Calculate the probability $P(Y \geq X \geq 2)$.
(b) Find the density function of the random variable $X - Y$.

Exercise 7.20. Let X have density $f_X(x) = 2x$ for $0 < x < 1$ and 0 outside of $(0, 1)$, and let Y be uniform on the interval $(1, 2)$. Assume X and Y independent.

(a) Give the joint density function of (X, Y). Calculate $P(Y - X \geq \frac{3}{2})$.
(b) Find the density function of $X + Y$.

Exercise 7.21. Let X and Y be independent normal random variables with distributions $X \sim \mathcal{N}(6, 3)$ and $Y \sim \mathcal{N}(-2, 2)$. Let $W = 3X + 4Y$.

(a) Identify the distribution of W.
(b) Find the probability $P(W > 15)$.

Exercise 7.22. Let X and Y be two independent normals with the *same* $\mathcal{N}(\mu, \sigma^2)$ distribution, with μ real and $\sigma^2 > 0$. Are there values μ and σ^2 for which $2X$ and $X + Y$ have the same distribution?

Exercise 7.23. Suppose that X and Y are independent standard normals. Find $P(X > Y + 2)$.

Exercise 7.24. Suppose that X, Y, and Z are independent mean zero normal random variables. Find $P(X + 2Y - 3Z > 0)$.

Exercise 7.25. Let X be a uniform random variable on the interval $[0, 1]$ and Y a uniform random variable on the interval $[8, 10]$. Suppose that X and Y are independent. Find the density function f_{X+Y} of $X + Y$ and sketch its graph. Check that your answer is a legitimate probability density function.

Exercise 7.26. Let X be a uniform random variable on the interval $[1, 3]$ and Y a uniform random variable on the interval $[9, 10]$. Suppose that X and Y are independent. Find the density function f_{X+Y} of $X + Y$ and sketch its graph. Check that your answer is a legitimate probability density function.

Exercise 7.27. Suppose X has probability density function f, and Y is uniformly distributed on $[0, 1]$. Supposing X and Y are independent, find the probability density function of $X + Y$.

Exercise 7.28. Let X_1, X_2, X_3 be independent Exp(λ) distributed random variables. Find the probability that $P(X_1 < X_2 < X_3)$.

Exercise 7.29. Let $X_1, \dots, X_{100}$ be independent standard normal random variables. Find the probability that X_{20} is the 50th largest number among these 100 numbers.

Exercise 7.30. We deal five cards, one by one, from a standard deck of 52. (Dealing cards from a deck means sampling without replacement.)

(a) Find the probability that the second card is an ace and the fourth card is a king.
(b) Find the probability that the first and the fifth cards are both spades.
(a) Find the conditional probability that the second card is a king given that the last two cards are both aces.

Exercise 7.31. We have an urn with 20 red, 10 black and 15 green balls. We take a sample of 30, without replacement, with order. Find the probability that the 3rd, 10th and 23rd picks are of different colors.

Exercise 7.32. 23 chips numbered 1 through 23 are placed into an urn. The chips are then selected one by one, without replacement, until all 23 have been selected. What is the probability that the numerical values of the 9th, 14th, and 21st chips are all less than or equal to five?

Exercise 7.33. An urn contains 7 red chips, 4 black chips, and 13 green chips. You sample 20 chips without replacement. What is the probability the 5th chip is black, given that the 3rd and 10th chips are red?

Exercise 7.34. We pick five cards out of a deck of 52 randomly without replacement. There are four aces in the deck: $\heartsuit, \clubsuit, \diamondsuit, \spadesuit$, let us number them from 1 to 4. For each $1 \le i \le 4$ let X_i be 1 if the appropriate ace is among the chosen

5 cards, and zero otherwise. (For example, $X_3 = 1$ if the ace of diamonds was chosen.) Show that the random variables X_1, X_2, X_3, X_4 are exchangeable.

Exercise 7.35. 20 chips numbered 1 through 20 are placed into an urn. The chips are then selected one by one, without replacement, until all 20 have been selected. What is the probability that the numerical values of the 3rd, 15th, and 19th chips are increasing?

Exercise 7.36. A porch light is left on all the time. The lightbulbs used by the homeowner have an exponential lifetime with a mean of 2 years. When one lightbulb goes out, it is immediately replaced.

(a) What is the probability that we will have to replace a lightbulb during the next calendar year?
(b) Given that we change lightbulbs exactly once during the next calendar year, what is the probability that we will have to change lightbulbs exactly two times during the following calendar year?

Exercise 7.37. Suppose that the discrete random variables X_1, X_2, X_3 are exchangeable. Show that for any function g, the random variables $g(X_1), g(X_2), g(X_3)$ are also exchangeable.

Hint. Suppose that (k_1, k_2, k_3) is a permutation of $(1, 2, 3)$. Show that the joint probability mass function of $g(X_{k_1}), g(X_{k_2}), g(X_{k_3})$ is the same as the joint p.m.f of X_1, X_2, X_3.

Challenging problems

Exercise 7.38. Annie and Bert play the following game. They first shuffle a deck of 52 cards. Annie flips the cards one by one until she finds a red ace. After that Bert takes the remaining deck and flips the cards one by one until he finds the other red ace. Annie wins if she flipped fewer cards than Bert, Bert wins if he flipped fewer cards than Annie. It is a tie if they needed the same number of cards.

(a) Denote by X_1 the number of cards Annie flipped, by X_2 the number of cards Bert flipped and set $X_3 = 53 - X_1 - X_2$. Show that (X_1, X_2, X_3) is exchangeable.
(b) Who has the higher probability of winning in this game?

Exercise 7.39. Suppose that $X_1, X_2, \ldots$ are i.i.d. Unif(0, 1) random variables.

(a) For a given $n \geq 2$ find the probability that $X_1 > X_2 > \cdots > X_{n-1} < X_n$. This is the probability that the first $n - 1$ terms are in decreasing order, but the first n terms are not.
(b) Let $N \geq 2$ denote the smallest positive integer for which $X_{N-1} < X_N$. (This is the first time the sequence is increasing.) Find $E[N]$.

Exercise 7.40. Suppose that X and Y are independent positive random variables with probability density functions f_X and f_Y. Show that $Z = X/Y$ is a continuous random variable and find its probability density function.

Exercise 7.41. This exercise is a probabilistic derivation of the formula

$$\sum_{k=1}^{\infty} \frac{1}{k^2} = \frac{\pi^2}{6}. \tag{7.11}$$

It is based on article [Pac11].

(a) Let X and Y be independent Cauchy random variables. This means that their density function is $f(x) = \frac{1}{\pi} \cdot \frac{1}{1+x^2}$ on the real line. Let $Z = |X| / |Y|$. Derive for Z the density function

$$f_Z(z) = \frac{4}{\pi^2} \cdot \frac{\ln z}{z^2 - 1} \qquad \text{for } z > 0.$$

(b) Find the probability $P(Z < 1)$ by symmetry and by integrating the density in the form $f_Z(z) = -\frac{4}{\pi^2} \sum_{k=0}^{\infty} z^{2k} \ln z$ for $0 < z < 1$. This gives a formula for the series $\sum_{k=0}^{\infty} \frac{1}{(2k+1)^2}$.

(c) Complete the derivation of (7.11).

Hint. $\sum_{k=1}^{\infty} \frac{1}{(2k)^2} = \frac{1}{4} \sum_{k=1}^{\infty} \frac{1}{k^2}$.

8

Expectation and variance in the multivariate setting

This chapter collects several topics related to means and variances in multivariate situations, both fundamental properties and techniques for computation. We begin by repeating the definition of the expectation of a function of multiple random variables.

Let $X_1, X_2, \dots, X_n$ be random variables defined on the same sample space and $g : \mathbb{R}^n \to \mathbb{R}$ a real-valued n-variable function. Recall from Fact 6.2 that if the random variables are discrete with joint probability mass function p then

$$E[g(X_1, \dots, X_n)] = \sum_{k_1, k_2, \dots, k_n} g(k_1, k_2, \dots, k_n) p(k_1, k_2, \dots, k_n) \tag{8.1}$$

while from Fact 6.13 recall that if they are jointly continuous with joint density function f then

$$E[g(X_1, \dots, X_n)] = \int_{-\infty}^{\infty} \cdots \int_{-\infty}^{\infty} g(x_1, \dots, x_n) f(x_1, \dots, x_n) \, dx_1 \cdots dx_n, \tag{8.2}$$

provided the expected values are well defined. In this chapter we take a closer look at two special cases:

(i) $g(x_1, \dots, x_n) = g_1(x_1) + g_2(x_2) + \cdots + g_n(x_n)$,
(ii) $g(x_1, \dots, x_n) = g_1(x_1) \cdot g_2(x_2) \cdots g_n(x_n)$.

In the first case we can always decompose the expectation, which leads to useful applications. In the second case we can decompose the expectation if the random variables are independent, and this also generates valuable corollaries.

8.1. Linearity of expectation

The next fact is one of the most useful properties of expectations.

Fact 8.1. (Linearity of expectation) Let $g_1, g_2, \dots, g_n$ be single variable functions and $X_1, X_2, \dots, X_n$ random variables defined on the same sample space. Then

$$E[g_1(X_1) + g_2(X_2) + \cdots + g_n(X_n)] = E[g_1(X_1)] + E[g_2(X_2)] + \cdots + E[g_n(X_n)], \tag{8.3}$$

provided that all the expectations are finite. In particular,

$$E[X_1 + X_2 + \cdots + X_n] = E[X_1] + E[X_2] + \cdots + E[X_n], \tag{8.4}$$

provided all the expectations are finite.

In short, the expectation of a sum is *always* the sum of expectations. It is important that the identities in Fact 8.1 hold with no assumptions on the random variables, beyond that the expectations are finite. In particular, *independence need not be assumed.*

We show the proof of Fact 8.1 for discrete random variables. Exercise 8.34 asks you to prove the jointly continuous case.

Proof of Fact 8.1 in the discrete case. Start with two random variables X_1, X_2 with joint probability mass function p. Beginning with (8.1) we manipulate the expectation as follows:

$$\begin{aligned}
E[g_1(X_1) + g_2(X_2)] &= \sum_{k,\ell} (g_1(k) + g_2(\ell))\, p(k,\ell) \\
&= \sum_{k,\ell} g_1(k)p(k,\ell) + \sum_{k,\ell} g_2(\ell)p(k,\ell) \\
&= \sum_{k} g_1(k) \sum_{\ell} p(k,\ell) + \sum_{\ell} g_2(\ell) \sum_{k} p(k,\ell) \\
&= \sum_{k} g_1(k)\, p_{X_1}(k) + \sum_{\ell} g_2(\ell)\, p_{X_2}(\ell) = E[g_1(X_1)] + E[g_2(X_2)].
\end{aligned} \tag{8.5}$$

We relied on equation (6.5) that showed how to find a marginal probability mass function from the joint probability mass function. This type of issue appeared also in Exercise 6.51.

The general case comes by induction. For example, to extend to three random variables, first regard $g_2(X_2) + g_3(X_3)$ as a single random variable and apply the just proved case of two variables, and then apply this same again to X_2 and X_3:

$$\begin{aligned}
&E[g_1(X_1) + g_2(X_2) + g_3(X_3)] = E[g_1(X_1) + (g_2(X_2) + g_3(X_3))] \\
&\quad = E[g_1(X_1)] + E[g_2(X_2) + g_3(X_3)] = E[g_1(X_1)] + E[g_2(X_2)] + E[g_3(X_3)].
\end{aligned}$$

In this manner we extend the statement to any number of random variables. ▲

Certain calculations that were difficult in the past are now easy.

Example 8.2. Adam must pass both a written test and a road test to get his driver's license. Each time he takes the written test he passes with probability $\frac{4}{10}$, independently of other tests. Each time he takes the road test he passes with probability $\frac{7}{10}$, also independently of other tests. What is the total expected number of tests Adam must take before earning his license?

Let X_1 and X_2 be the number of written and road tests Adam takes, respectively. Then $X_1 \sim \text{Geom}(\frac{4}{10})$ and $X_2 \sim \text{Geom}(\frac{7}{10})$ and $X = X_1 + X_2$ is the total number of tests Adam takes. Using the techniques from Chapter 7, we could compute the probability mass function of X and then compute the expectation. However, that would be cumbersome. Fact 8.1 provides a quicker way.

From Example 3.25, the expected value of a geometric random variable with parameter p is $1/p$. Hence, $E[X_1] = \frac{1}{4/10} = \frac{5}{2}$ and $E[X_2] = \frac{1}{7/10} = \frac{10}{7}$. Thus, by Fact 8.1,

$$E[X] = E[X_1] + E[X_2] = \tfrac{5}{2} + \tfrac{10}{7} = \tfrac{55}{14} \approx 3.93. \qquad \blacktriangle$$

Example 8.3. (Expected value of the binomial, easy calculation) Let $X \sim \text{Bin}(n, p)$. The linearity of the expectation yields a short proof that $E[X] = np$.

The key observation is that X counts the number of successes among n independent trials with success probability p. Let $X_1, X_2, \ldots, X_n$ be the outcomes of the trials ($X_i = 1$ if the ith trial is a success, $X_i = 0$ if it is a failure). Then $X = X_1 + X_2 + \cdots + X_n$ and by linearity of expectation

$$E[X] = E[X_1] + E[X_2] + \cdots + E[X_n].$$

But $X_1, X_2, \ldots, X_n$ are all $\text{Ber}(p)$ random variables so their expectations are equal to p. Hence $E[X] = p + p + \cdots + p = np$. ▲

Note that the independence of the trials is not relevant for the example above because linearity of expectation does not require independence.

Remark 8.4. The technique used in the previous example is the *indicator method.* If a nonnegative discrete random variable X can be written as a sum of indicator random variables (random variables with values 0 and 1) then $E[X]$ is the sum of the expectations of the indicators. Since the expected value of an indicator is the probability that the indicator equals one, this sum can be convenient to compute, especially if there is some symmetry in the problem. We cannot give a general rule for deciding when the indicator method works. The method is often useful when a random variable counts random objects. ▲

We present next a number of examples where the indicator method allows a painless computation of seemingly complicated expectations.

Example 8.5. There are 15 workers in the office. Management allows a birthday party on the last day of a month if there were birthdays in the office during that month. How many parties are there on average during a year? Assume for simplicity that the birth months of the individuals are independent and each month is equally likely.

Let Y denote the number of parties during a year. The probability mass function of Y would be hard to compute, but the expected value is easy with the indicator method. Let J_i be the indicator of the event that there is a birthday in the ith month. Then

$$Y = J_1 + J_2 + \cdots + J_{12}, \quad \text{and} \quad E[Y] = E[J_1] + E[J_2] + \cdots + E[J_{12}].$$

Since $J_1, \ldots, J_{12}$ all have the same distribution (by the assumed symmetry), we have $E[Y] = 12E[J_1]$. Since J_1 is an indicator random variable, we have

$$\begin{aligned} E[J_1] &= P(\text{there is a birthday in January}) = 1 - P(\text{no birthdays in January}) \\ &= 1 - \left(\tfrac{11}{12}\right)^{15}. \end{aligned}$$

Above we used the assumption that the birth months of different people are independent. Altogether we get $E[Y] = 12\left(1 - \left(\frac{11}{12}\right)^{15}\right) \approx 8.75$. ▲

There can be multiple ways to represent a random variable as a sum of indicators. This is illustrated by the next example.

Example 8.6. We deal five cards from a deck of 52 without replacement. Let X denote the number of aces among the chosen cards. Find the expected value of X.

We present two solutions.

(i) The distribution of X does not depend on whether we choose cards in order or without order, so we can assume there is a first card, a second card and so on. Let I_i, $i = 1, \ldots, 5$, be the indicator of the event that the ith card is an ace, that is,

$$I_i = \begin{cases} 1, & \text{if the } i\text{th card is an ace} \\ 0, & \text{otherwise.} \end{cases}$$

Then

$$X = I_1 + I_2 + \cdots + I_5$$

and by the linearity of expectation

$$E[X] = E[I_1] + E[I_2] + \cdots + E[I_5].$$

The random variables $I_1, \ldots, I_5$ are exchangeable because we are sampling without replacement (Theorem 7.22). Then $E[I_i] = E[I_1]$ for each i which implies

$$E[X] = 5E[I_1].$$

But $E[I_1] = P(\text{the first card is ace}) = \frac{4}{52} = \frac{1}{13}$. Thus $E[X] = \frac{5}{13}$.

(ii) The random variable X can be rewritten as a sum of indicators in a different way which leads to an alternative method to arrive at the same answer. Let us label the four aces in the deck with the numbers $1, 2, 3, 4$ and denote by J_i, $i = 1, 2, 3, 4$, the indicator of the event that the ith ace is among the five chosen cards. Then $X = J_1 + J_2 + J_3 + J_4$ and $E[X] = E[J_1] + E[J_2] + E[J_3] + E[J_4]$. Again, by symmetry,

$$E[X] = 4E[J_1] = 4P(\text{the ace of spades is among the five chosen cards}).$$

We compute the last probability by imagining an unordered sample of five cards:

$$P(\text{the ace of spades is among the five}) = \frac{\binom{1}{1}\binom{51}{4}}{\binom{52}{5}} = \frac{\frac{51\cdot 50\cdot 49\cdot 48}{4!}}{\frac{52\cdot 51\cdot 50\cdot 49\cdot 48}{5!}} = \frac{5}{52}.$$

We obtain again $E[X] = 4 \cdot \frac{5}{52} = \frac{5}{13}$. ▲

The random variable X from Example 8.6 has a hypergeometric distribution since we are sampling without replacement. (This distribution was introduced in Definition 2.42 on page 65.) The indicator method works in general for the expected value of a hypergeometric random variable.

Example 8.7. (Expected value of a hypergeometric random variable) Let X be a random variable with Hypergeom(N, N_A, n) distribution. By the discussion on page 65 we can represent X as the number of type A items in a sample of n taken without replacement from a set of N items that contains N_A type A items. Assume that we take an ordered sample so that there is a first pick, a second pick, all the way to an nth pick. Denote by I_j the indicator of the event that the jth pick is of type A. Then $X = I_1+I_2+\cdots+I_n$ and $E[X] = E[I_1]+\cdots+E[I_n]$. The exchangeability of sampling without replacement implies that the random variables I_j all have the same distribution. Hence $E[X] = nE[I_1]$. For I_1 we have

$$E[I_1] = P(\text{the first pick is of type } A) = \frac{N_A}{N}$$

which gives $E[X] = \frac{nN_A}{N}$. We could derive the same formula with an argument like that in solution (ii) of Example 8.6. ▲

We close this section with two more examples demonstrating the indicator method.

Example 8.8. The sequence of coin flips $\underline{\text{HH}}\text{T}\underline{\text{HHH}}\text{TTT}\underline{\text{H}}\text{T}\underline{\text{HHH}}\text{T}$ contains two runs of heads of length three, one run of length two and one run of length one. The head runs are underlined. If a fair coin is flipped 100 times, what is the expected number of runs of heads of length 3?

Let T be the number of runs of heads of length 3. Let I_i be one if a run of length 3 begins at flip $i \in \{1, \ldots, 98\}$, and zero otherwise. Then, $T = \sum_{i=1}^{98} I_i$ and

$$E[T] = \sum_{i=1}^{98} E[I_i] = \sum_{i=1}^{98} P(I_i = 1).$$

There are three cases to consider. A run of length 3 begins at flip $i = 1$ only if there are heads on flips $1, 2$, and 3, and tails on flip 4. A run of length 3 begins at flip $i = 98$ only if there is a tails on flip 97, and then heads on flips 98, 99, and

100. Finally, a run of length 3 begins at flip $i \in \{2, \dots, 97\}$ only if there are heads on flips i, $i+1$, and $i+2$ and tails on flips $i-1$ and $i+3$. Thus,

$$E[I_1] = 2^{-4}, \quad E[I_{98}] = 2^{-4}, \text{ and } E[I_i] = 2^{-5}, \text{ for } i \in \{2, \dots, 97\}.$$

Therefore, $E[T] = \sum_{i=1}^{98} E[I_i] = 2 \cdot 2^{-4} + 96 \cdot 2^{-5} = \frac{25}{8} = 3.125.$ ▲

Example 8.9. There are n guests at a party. Assume that each pair of guests know each other with probability $\frac{1}{2}$, independently of the other guests. Let X denote the number of groups of size three where everybody knows everybody else. Find $E[X]$.

For three distinct numbers $1 \le a < b < c \le n$ let $I_{a,b,c}$ denote the indicator of the event that guests a, b and c all know each other. Then $X = \sum_{a,b,c} I_{a,b,c}$ and $E[X] = \sum_{a,b,c} E[I_{a,b,c}]$. The random variables $I_{a,b,c}$ have the same distribution and there are $\binom{n}{3}$ of them. So $E[X] = \binom{n}{3} E[I_{1,2,3}]$. We have

$$E[I_{1,2,3}] = P(\text{the first three guests all know each other}) = \tfrac{1}{2} \cdot \tfrac{1}{2} \cdot \tfrac{1}{2} = \tfrac{1}{8}$$

which gives $E[X] = \frac{1}{8}\binom{n}{3}$.

Here is the real math story behind this example. Start with n nodes labeled $1, 2, \dots, n$. Build a *random graph* or *network* as follows: for each pair of distinct nodes i, j flip a fair coin to determine whether to connect i and j with an edge or not. A *triangle* in a graph is any collection of three nodes i, j, k with all three edges present (see Figure 8.1). We computed above the expected number of triangles in this graph.

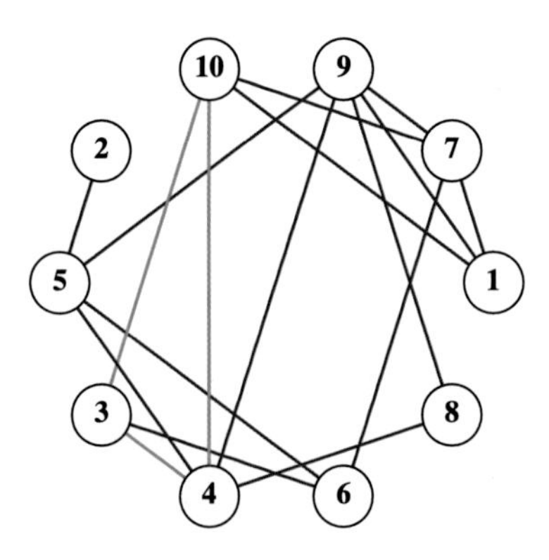

Figure 8.1. An example of a network (or graph) with ten nodes numbered $1, 2, \dots, 10$. The three green edges between the nodes $3, 4$ and 10 form a triangle.

This example and Exercise 8.33 come from the subject of *random graph theory*. Random graph theory is an important part of probability theory with many applications in computer science, biology and engineering. ▲

8.2. Expectation and independence

An essential feature of independence is that it breaks expectations of products into products of expectations.

Fact 8.10. Let $X_1, \ldots, X_n$ be independent random variables. Then for all functions $g_1, \ldots, g_n$ for which the expectations below are well defined,

$$E\left[\prod_{k=1}^{n} g_k(X_k)\right] = \prod_{k=1}^{n} E\left[g_k(X_k)\right]. \tag{8.6}$$

Note the difference in the conditions of Fact 8.1 and Fact 8.10. In Fact 8.1, independence was not required. However, in Fact 8.10 independence is crucial.

Proof. Our tools allow us to prove the fact for n continuous or n discrete random variables. We show the proof for two continuous random variables.

Suppose X and Y are independent with density functions f_X and f_Y. Then by Fact 6.25 their joint density function is $f_{X,Y}(x, y) = f_X(x)f_Y(y)$. We compute

$$\begin{aligned} E[g(X)h(Y)] &= \int_{-\infty}^{\infty}\int_{-\infty}^{\infty} g(x)h(y)f_{X,Y}(x, y)\,dx\,dy \\ &= \int_{-\infty}^{\infty}\int_{-\infty}^{\infty} g(x)h(y)f_X(x)f_Y(y)\,dx\,dy \\ &= \int_{-\infty}^{\infty} g(x)f_X(x)\,dx \int_{-\infty}^{\infty} h(y)f_Y(y)\,dy = E[g(X)] \cdot E[h(Y)]. \end{aligned}$$

In the last step we were able to separate the double integral into a product of two single variable integrals which are exactly the expected values of $g(X)$ and $h(Y)$. ▲

A consequence of Fact 8.10 is that the variance of a sum of independent random variables is the same as the sum of the variances. Previously we knew this for Bernoulli random variables (Example 3.49) and normal random variables (Fact 7.9).

Fact 8.11. (Variance of a sum of independent random variables) Assume the random variables $X_1, \ldots, X_n$ are independent and have finite variances. Then

$$\operatorname{Var}(X_1 + X_2 + \cdots + X_n) = \operatorname{Var}(X_1) + \operatorname{Var}(X_2) + \cdots + \operatorname{Var}(X_n). \tag{8.7}$$

Proof. We first demonstrate the proof for two random variables. Suppose that X_1 and X_2 are independent and their expected values μ_1, μ_2 and variances σ_1^2, σ_2^2 are finite. Then by the linearity of expectation we have

$$E[X_1 + X_2] = E[X_1] + E[X_2] = \mu_1 + \mu_2.$$

By the definition of the variance of the random variable $X_1 + X_2$ we have

$$\begin{aligned}\mathrm{Var}(X_1 + X_2) &= E\left[(X_1 + X_2 - E(X_1 + X_2))^2\right] = E\left[(X_1 + X_2 - (\mu_1 + \mu_2))^2\right] \\ &= E\left[((X_1 - \mu_1) + (X_2 - \mu_2))^2\right] \\ &= E\left[(X_1 - \mu_1)^2\right] + E\left[(X_2 - \mu_2)^2\right] + 2E\left[(X_1 - \mu_1)(X_2 - \mu_2)\right].\end{aligned}$$

The first two terms in the last expression are exactly $\mathrm{Var}(X_1)$ and $\mathrm{Var}(X_2)$. The third term is the expectation of a product of independent random variables (as X_1 and X_2 are independent). Therefore

$$2E[(X_1 - \mu_1)(X_2 - \mu_2)] = 2E[(X_1 - \mu_1)]\, E[(X_2 - \mu_2)] = 0,$$

since both $E(X_1 - \mu_1) = E[X_1] - \mu_1$ and $E(X_2 - \mu_2) = E[X_2] - \mu_2$ are zero. This gives $\mathrm{Var}(X_1 + X_2) = \mathrm{Var}(X_1) + \mathrm{Var}(X_2)$ which is exactly what we wanted to prove.

For the general case (with n terms) we can repeat the same argument using

$$\begin{aligned}\mathrm{Var}\left(\sum_{i=1}^n X_i\right) &= E\left[\left(\sum_{i=1}^n (X_i - \mu_i)\right)^2\right] \\ &= \sum_{i=1}^n E\left[(X_i - \mu_i)^2\right] + 2 \sum_{1 \le i < j \le n} E\left[(X_i - \mu_i)(X_j - \mu_j)\right] \\ &= \sum_{i=1}^n \mathrm{Var}(X_i).\end{aligned}$$

Alternatively, we could appeal to induction as we did in the proof of Fact 8.1. ▲

In Example 8.3 the indicator method gave a short derivation of the mean of a binomial random variable. Now we can provide a short computation of the variance.

Example 8.12. (Variance of the binomial, easy calculation) The random variable $X \sim \mathrm{Bin}(n, p)$ counts the number of successes among n independent trials with success probability p. Write $X = X_1 + \cdots + X_n$ where X_i is the outcome of the ith trial. Since $X_i \sim \mathrm{Ber}(p)$ we have $\mathrm{Var}(X_i) = p(1 - p)$ by Example 3.47. The random variables X_i are independent and so by Fact 8.11

$$\mathrm{Var}(X) = \mathrm{Var}(X_1) + \cdots + \mathrm{Var}(X_n) = np(1 - p).$$ ▲

Example 8.13. (Expectation and variance of the negative binomial) In Definition 7.6 we introduced the negative binomial distribution with parameters (k, p). Here we compute its expectation and variance. This can be done with the probability mass function, but it is easier to use the techniques introduced in this chapter.

Recall that if $X_1, X_2, \dots, X_k$ are independent $\mathrm{Geom}(p)$ random variables then $Y = X_1 + \cdots + X_k$ has distribution $\mathrm{Negbin}(k, p)$. Recall from Examples 3.25 and

3.51 that $E[X_1] = \frac{1}{p}$ and $\mathrm{Var}(X_1) = \frac{1-p}{p^2}$. Using the linearity of expectation and the fact that the random variables are identically distributed, we get

$$E[Y] = E[X_1 + \cdots + X_k] = kE[X_1] = \frac{k}{p}.$$

Using independence we also get

$$\mathrm{Var}(Y) = \mathrm{Var}(X_1 + \cdots + X_k) = k\,\mathrm{Var}(X_1) = \frac{k(1-p)}{p^2}.$$

▲

Sample mean and sample variance

We encountered the average $\frac{1}{n}(X_1 + \cdots + X_n)$ of i.i.d. Bernoulli random variables in Section 4.1 as an estimate of an unknown success probability. Here we compute the mean and variance of $\frac{1}{n}(X_1 + \cdots + X_n)$ for general i.i.d. samples to verify that it provides a sensible estimate of the true mean. The average $\frac{1}{n}(X_1 + \cdots + X_n)$ is called the *sample mean* of $X_1, \dots, X_n$. We will introduce a similar object called the *sample variance* below in Example 8.16. The sample mean and sample variance are examples of *statistics*. In this usage the term statistic means a number calculated as a function of the random variables $X_1, \dots, X_n$. Hence a statistic is itself a random variable.

Fact 8.14. (Expectation and variance of the sample mean) Consider a sequence of independent and identically distributed random variables $X_1, X_2, X_3, \dots$. Assume these random variables have a finite mean μ and variance σ^2. Denote the average of the first n random variables by

$$\bar{X}_n = \frac{X_1 + \cdots + X_n}{n}.$$

This is called the *sample mean* of the first n observations. Then

$$E[\bar{X}_n] = \mu \qquad \text{and} \qquad \mathrm{Var}(\bar{X}_n) = \frac{\sigma^2}{n}. \tag{8.8}$$

The fact that $E[\bar{X}_n] = \mu$ is expressed by saying that $\bar{X}_n$ is an *unbiased estimator* of the true mean μ.

Proof. Using linearity and the fact that the random variables X_i are identically distributed:

$$\begin{aligned} E[\bar{X}_n] &= E\left[\tfrac{1}{n}(X_1 + \cdots + X_n)\right] = \tfrac{1}{n}E[X_1 + \cdots + X_n] \\ &= \tfrac{1}{n} \cdot nE[X_1] = E[X_1] = \mu. \end{aligned}$$

For the variance we first use (3.37) and then Fact 8.11:

$$\begin{aligned} \mathrm{Var}(\bar{X}_n) &= \mathrm{Var}\left(\tfrac{1}{n}(X_1 + \cdots + X_n)\right) = \tfrac{1}{n^2}\,\mathrm{Var}(X_1 + \cdots + X_n) \\ &= \tfrac{1}{n^2} \cdot n\,\mathrm{Var}(X_1) = \frac{\sigma^2}{n}. \end{aligned}$$

▲

Remark 8.15. (The effect of averaging) There is something very important in the formula $\text{Var}(\bar{X}_n) = \sigma^2/n$. Namely, as more and more random variables are averaged, *the variance of the sample mean $\bar{X}_n$ goes down.* Why? Because random fluctuations above and below the mean tend to average out. This is the phenomenon behind the law of large numbers, which we first saw in Theorem 4.8, and will see in more generality in Section 9.2. As $n \to \infty$, $\text{Var}(\bar{X}_n)$ vanishes, and so $\bar{X}_n$ converges to its mean μ. ▲

Example 8.16. (Sample variance) Let $X_1, X_2, X_3, \dots$ be a sequence of independent and identically distributed (i.i.d.) random variables with unknown mean μ and unknown variance σ^2. Fact 8.14 established that the sample mean $\bar{X}_n = \frac{1}{n}(X_1 + \cdots + X_n)$ is an unbiased estimator of the true mean μ. This means that $E[\bar{X}_n] = \mu$.

An unbiased estimator of the variance σ^2 is the sample variance

$$s_n^2 = \frac{1}{n-1}\sum_{i=1}^{n}(X_i - \bar{X}_n)^2. \tag{8.9}$$

The normalization by $n-1$ may surprise the reader, but as we verify next, this is the correct way to obtain $E[s_n^2] = \sigma^2$. It is traditional to denote the sample variance by a lower case letter s_n^2 even though it is a random variable.

We start by computing the expectation of a single term on the right-hand side of (8.9). Recall from Fact 8.14 that $\text{Var}(\bar{X}_n) = \sigma^2/n$. Then

$$\begin{aligned}
E[(X_i - \bar{X}_n)^2] &= E[(X_i - \mu + \mu - \bar{X}_n)^2] \\
&= E[(X_i - \mu)^2] + 2E[(X_i - \mu)(\mu - \bar{X}_n)] + E[(\bar{X}_n - \mu)^2] \\
&= \sigma^2 - 2E[(\bar{X}_n - \mu)(X_i - \mu)] + \sigma^2/n \\
&= \tfrac{n+1}{n}\sigma^2 - 2E[(\bar{X}_n - \mu)(X_i - \mu)].
\end{aligned}$$

Now add these up over i. Notice below that $\sum_{i=1}^n (X_i - \mu) = n(\bar{X}_n - \mu)$:

$$\begin{aligned}
(n-1)E[s_n^2] &= \sum_{i=1}^{n} E[(X_i - \bar{X}_n)^2] = (n+1)\sigma^2 - 2E\left[(\bar{X}_n - \mu)\sum_{i=1}^{n}(X_i - \mu)\right] \\
&= (n+1)\sigma^2 - 2nE[(\bar{X}_n - \mu)^2] = (n+1)\sigma^2 - 2n \cdot \sigma^2/n \\
&= (n-1)\sigma^2.
\end{aligned}$$

This shows that $E[s_n^2] = \sigma^2$ and thereby that s_n^2 is an unbiased estimator of σ^2. ▲

Coupon collector's problem

We finish this section with the coupon collector's problem, a classical example that illustrates several techniques that we have encountered.

Example 8.17. (Coupon collector's problem) A cereal company is performing a promotion, and they have put a toy in each box of cereal they make. There are n different toys altogether and each toy is equally likely to show up in any given box, independently of the other boxes. Let T_n be the number of boxes I need to buy in order to collect the complete set of n toys. How large are $E[T_n]$ and $\mathrm{Var}(T_n)$, and how do these values grow as $n \to \infty$?

Finding the probability mass function of T_n is complicated. (Althought not impossible, see Exercise 8.66.) However, we can compute the expectation and the variance by decomposing T_n into a sum of simpler random variables. For $1 \le k \le n$ let T_k denote the number of boxes opened when I first reach k *different* toys. In other words: $T_1, T_2, \ldots, T_n$ are the "times" when I get a toy I had not seen previously, and time is now measured in terms of opened cereal boxes.

$T_1 = 1$ since I get my first toy from the first box. How long do I have to wait until T_2? Think of the boxes *after* the first one as a sequence of independent trials. The first successful trial is the one that gives me a toy that is different from the one I already received from the first box. Thus, the success probability is $\frac{n-1}{n}$ and so $T_2 - 1 = T_2 - T_1$ is a $\mathrm{Geom}(\frac{n-1}{n})$ random variable.

This argument generalizes readily. For notational convenience let $W_k = T_{k+1} - T_k$. The random variable W_k is the number of boxes I need to open to see a new toy *after* I have collected k distinct toys. At that point the chances of a new toy in a box are $\frac{n-k}{n}$. Hence $W_k \sim \mathrm{Geom}(\frac{n-k}{n})$ and thereby $E[W_k] = \frac{n}{n-k}$.

From

$$T_n = 1 + W_1 + W_2 + \cdots + W_{n-1} \tag{8.10}$$

and linearity of expectation we can compute the mean:

$$\begin{aligned} E[T_n] &= 1 + E[W_1] + E[W_2] + \cdots + E[W_{n-1}] \\ &= 1 + \sum_{k=1}^{n-1} \frac{n}{n-k} = n \cdot \frac{1}{n} + n \sum_{k=1}^{n-1} \frac{1}{n-k} = n \sum_{k=1}^{n} \frac{1}{j}. \end{aligned} \tag{8.11}$$

We turn to the variance of T_n. Example 3.51 calculated the variance of a geometric random variable. Applying this to the individual terms in the sum (8.10) gives

$$\mathrm{Var}(W_k) = \frac{\frac{k}{n}}{\frac{(n-k)^2}{n^2}} = \frac{kn}{(n-k)^2} \qquad \text{for } 1 \le k \le n-1. \tag{8.12}$$

According to Fact 8.11 the variance of a sum of independent random variables is the sum of the individual variances. The independence of the random variables W_k should appear intuitively natural. For, once I have k different toys, a new sequence of trials with success probability $\frac{n-k}{n}$ starts and the previous waiting times $W_1, W_2, \ldots, W_{k-1}$ have no influence on the number of boxes I need to open for the next new toy. Proof of independence of $W_1, W_2, \ldots, W_{n-1}$ is left as Exercise 8.60.

Fact 8.11 and the independence of $W_1, W_2, \ldots, W_{n-1}$ allow us to write $\mathrm{Var}(T_n) = \sum_{k=1}^{n-1} \mathrm{Var}(W_k)$. Then by (8.12) and some algebra we obtain

$$\operatorname{Var}(T_n) = \sum_{k=1}^{n-1} \operatorname{Var}(W_k) = \sum_{k=1}^{n-1} \frac{kn}{(n-k)^2} = \sum_{j=1}^{n-1} \frac{(n-j)n}{j^2}$$
$$= n^2 \sum_{j=1}^{n-1} \frac{1}{j^2} - n \sum_{j=1}^{n-1} \frac{1}{j}. \tag{8.13}$$

Above the summation index changed to $j = n - k$ and then the sum was split in two.

Now we wish to understand the growth rates of $E[T_n]$ and $\operatorname{Var}(T_n)$ as $n \to \infty$ (in other words, as the number of toys increases). From (8.11) and (8.13) we see that we need to understand how the sums $\sum_{j=1}^{n} \frac{1}{j^2}$ and $\sum_{j=1}^{n} \frac{1}{j}$ behave as $n \to \infty$. Here are the relevant asymptotics:

$$\sum_{k=1}^{\infty} \frac{1}{j^2} = \frac{\pi^2}{6} \quad \text{and} \quad \lim_{n\to\infty} \frac{1}{\ln n} \sum_{j=1}^{n} \frac{1}{j} = 1. \tag{8.14}$$

The integral test shows that $\sum_{j=1}^{\infty} \frac{1}{j^2}$ converges and $\sum_{j=1}^{\infty} \frac{1}{j}$ diverges (Example D.5 in Appendix D). The exact value $\sum_{j=1}^{\infty} \frac{1}{j^2} = \pi^2/6$ was first proved by Euler in the eighteenth century. Exercise 7.41 outlines a probabilistic derivation. The second limit in (8.14) comes by carefully approximating $\sum_{k=1}^{n} \frac{1}{k}$ with the integral $\int_1^n \frac{1}{x} dx = \ln n$.

Applying limits (8.14) and recalling also that $(\ln n)/n \to 0$ as $n \to \infty$ gives these asymptotics:

$$\lim_{n\to\infty} \frac{E[T_n]}{n \log n} = 1 \quad \text{and} \quad \lim_{n\to\infty} \frac{\operatorname{Var}(T_n)}{n^2} = \frac{\pi^2}{6}. \tag{8.15}$$

To summarize: the expected value of T_n grows at rate $n \ln n$ while its variance grows at rate n^2. Exercise 8.61 sketches how to fill in the details of these limits. ▲

8.3. Sums and moment generating functions

Recall from our discussion in Section 5.1 that the moment generating function can be used to identify distributions. This gives us alternative derivations of some of the facts presented in Section 7.1 about the sum of independent random variables. Here is our main tool.

Fact 8.18. Suppose that X and Y are independent random variables with moment generating functions $M_X(t)$ and $M_Y(t)$. Then for all real numbers t,

$$M_{X+Y}(t) = M_X(t) M_Y(t). \tag{8.16}$$

Proof. The proof follows from the independence of X and Y and Fact 8.10:

$$M_{X+Y}(t) = E[e^{t(X+Y)}] = E[e^{tX} e^{tY}] = E[e^{tX}]E[e^{tY}] = M_X(t) M_Y(t).$$ ▲

If we know the moment generating functions of independent random variables X and Y, then using Fact 8.18 we can easily compute the moment generating function of $X+Y$. If we can identify the resulting function as the moment generating function of another random variable Z, then by the uniqueness of the moment generating function (Fact 5.14) we can be sure that $X+Y$ has the same distribution as Z (provided that $M_{X+Y}(t)$ is finite in a neighborhood of zero). In certain cases this provides a quick way to determine the distribution of $X+Y$.

We first revisit Example 7.2.

Example 8.19. (Convolution of Poisson random variables revisited) Suppose that $X \sim \text{Poisson}(\lambda)$, $Y \sim \text{Poisson}(\mu)$ and these are independent.

By Example 5.4 we have $M_X(t) = e^{\lambda(e^t-1)}$ and $M_Y(t) = e^{\mu(e^t-1)}$. Then $M_{X+Y}(t) = M_X(t)M_Y(t) = e^{\lambda(e^t-1)}e^{\mu(e^t-1)} = e^{(\lambda+\mu)(e^t-1)}$. But this is the same as the moment generating function of a Poisson$(\lambda+\mu)$ random variable, hence $X+Y \sim$ Poisson$(\lambda+\mu)$. ▲

The same method also works for sums of independent normal random variables which we first considered in Example 7.8.

Example 8.20. (Convolution of normal random variables revisited) Suppose that $X \sim \mathcal{N}(\mu_1, \sigma_1^2)$ and $Y \sim \mathcal{N}(\mu_2, \sigma_2^2)$, and these are independent.

By Example 5.5 we have $M_X(t) = e^{\mu_1 t + \frac{1}{2}\sigma_1^2 t^2}$ and $M_Y(t) = e^{\mu_2 t + \frac{1}{2}\sigma_2^2 t^2}$. Thus,

$$M_{X+Y}(t) = M_X(t)M_Y(t) = e^{\mu_1 t + \frac{1}{2}\sigma_1^2 t^2} e^{\mu_2 t + \frac{1}{2}\sigma_2^2 t^2} = e^{(\mu_1+\mu_2)t + \frac{1}{2}(\sigma_1^2+\sigma_2^2)t^2}.$$

This shows that $X+Y$ has the moment generating function of a $\mathcal{N}(\mu_1+\mu_2, \sigma_1^2+\sigma_2^2)$ random variable. Consequently $X+Y \sim \mathcal{N}(\mu_1+\mu_2, \sigma_1^2+\sigma_2^2)$. ▲

Note that although this technique is very quick, it only applies in situations where we can recognize $M_{X+Y}(t)$ as the moment generating function of a known random variable. This can happen for some of the discussed named distributions, but also if we have a discrete distribution with not-too-many possible values.

Example 8.21. Let X and Y be independent with probability mass functions

$$p_X(1) = \tfrac{1}{3}, \quad p_X(2) = \tfrac{1}{4}, \quad p_X(3) = \tfrac{1}{6}, \quad p_X(4) = \tfrac{1}{4}$$

and

$$p_Y(1) = \tfrac{1}{2}, \quad p_Y(2) = \tfrac{1}{3}, \quad p_Y(3) = \tfrac{1}{6}.$$

Find the probability mass function of $X+Y$.

We start by computing the moment generating functions of X and Y:

$$M_X(t) = \tfrac{1}{3}e^t + \tfrac{1}{4}e^{2t} + \tfrac{1}{6}e^{3t} + \tfrac{1}{4}e^{4t} \quad \text{and} \quad M_Y(t) = \tfrac{1}{2}e^t + \tfrac{1}{3}e^{2t} + \tfrac{1}{6}e^{3t}.$$

Then $M_{X+Y}(t) = M_X(t)M_Y(t)$ gives

$$\begin{aligned} M_{X+Y}(t) &= (\tfrac{1}{3}e^t + \tfrac{1}{4}e^{2t} + \tfrac{1}{6}e^{3t} + \tfrac{1}{4}e^{4t})(\tfrac{1}{2}e^t + \tfrac{1}{3}e^{2t} + \tfrac{1}{6}e^{3t}) \\ &= \tfrac{1}{6}e^{2t} + \tfrac{17}{72}e^{3t} + \tfrac{2}{9}e^{4t} + \tfrac{2}{9}e^{5t} + \tfrac{1}{9}e^{6t} + \tfrac{1}{24}e^{7t} \end{aligned}$$

where we just expanded the product. By reading off the coefficients of e^{kt} for $k = 2, 3, \dots, 7$ we get the probability mass function of $X + Y$:

$$\begin{aligned} &p_{X+Y}(2) = \tfrac{1}{6}, \quad p_{X+Y}(3) = \tfrac{17}{72}, \quad p_{X+Y}(4) = \tfrac{2}{9}, \\ &p_{X+Y}(5) = \tfrac{2}{9}, \quad p_{X+Y}(6) = \tfrac{1}{9}, \quad p_{X+Y}(7) = \tfrac{1}{24}. \end{aligned}$$

▲

Fact 8.18 extends readily to sums of arbitrarily many random variables: if $X_1, X_2, \dots, X_n$ are independent with sum $S = X_1 + X_2 + \cdots + X_n$, then

$$M_S(t) = M_{X_1}(t)M_{X_2}(t) \cdots M_{X_n}(t). \tag{8.17}$$

Example 8.22. Let $X_1, X_2, \dots, X_{20}$ be independent random variables with probability mass function $P(X_i = 2) = \frac{1}{3}$ and $P(X_i = 5) = \frac{2}{3}$. Let $S = X_1 + \cdots + X_{20}$. Find the moment generating function of S.

For each X_i we have $M_{X_i}(t) = \frac{1}{3}e^{2t} + \frac{2}{3}e^{5t}$. By (8.17), $M_S(t) = (\frac{1}{3}e^{2t} + \frac{2}{3}e^{5t})^{20}$. ▲

8.4. Covariance and correlation

Section 3.3 introduced variance as quantifying the strength of the random fluctuations of a single random variable. Covariance and correlation quantify the strength and type of dependence between two random variables.

Definition 8.23. Let X and Y be random variables defined on the same sample space with expectations μ_X and μ_Y. The **covariance** of X and Y is defined by

$$\mathrm{Cov}(X, Y) = E\left[(X - \mu_X)(Y - \mu_Y)\right] \tag{8.18}$$

if the expectation on the right is finite.

When $X = Y$, the covariance is the variance:

$$\mathrm{Cov}(X, X) = E\left[(X - \mu_X)^2\right] = \mathrm{Var}(X). \tag{8.19}$$

Analogously with variance, covariance has an alternative formula obtained by expansion and linearity.

Fact 8.24. (Alternative formula for the covariance)

$$\mathrm{Cov}(X, Y) = E[XY] - \mu_X\mu_Y. \tag{8.20}$$

Showing Fact 8.24 is straightforward:

$$\begin{aligned}\operatorname{Cov}(X,Y) &= E[(X-\mu_X)(Y-\mu_Y)] = E[XY - \mu_X Y - \mu_Y X + \mu_X\mu_Y] \\ &= E[XY] - \mu_X E[Y] - \mu_Y E[X] + \mu_X\mu_Y \\ &= E[XY] - \mu_X\mu_Y.\end{aligned}$$

Formula (8.18) reveals that whether covariance is positive or negative carries meaning. Inside the expectation we have

$$(X-\mu_X)(Y-\mu_Y) > 0 \quad \text{if } X-\mu_X \text{ and } Y-\mu_Y \text{ have the same sign}$$

$$\text{while} \quad (X-\mu_X)(Y-\mu_Y) < 0 \quad \text{if } X-\mu_X \text{ and } Y-\mu_Y \text{ have the opposite sign.}$$

Thus $\operatorname{Cov}(X,Y) > 0$ if on average X and Y tend to deviate together above or below their means, while $\operatorname{Cov}(X,Y) < 0$ if on average X and Y tend to deviate in opposite directions. Here is the commonly used terminology. X and Y are

- *positively correlated* if $\operatorname{Cov}(X,Y) > 0$,
- *negatively correlated* if $\operatorname{Cov}(X,Y) < 0$, and
- *uncorrelated* if $\operatorname{Cov}(X,Y) = 0$.

We illustrate these three possibilities with indicator random variables.

Example 8.25. (Covariance of indicator random variables) Let A and B be two events and consider their indicator random variables I_A and I_B. Find $\operatorname{Cov}(I_A, I_B)$.

First note that $I_A I_B = I_{A\cap B}$ since $I_A I_B = 1$ exactly when both indicator random variables are equal to one, which happens precisely on $A \cap B$. Next remember that the expectation of an indicator is the probability of the corresponding event, and compute:

$$\begin{aligned}\operatorname{Cov}(I_A, I_B) &= E[I_A I_B] - E[I_A]E[I_B] = E[I_{A\cap B}] - E[I_A]E[I_B] \\ &= P(A\cap B) - P(A)P(B) = P(B)[P(A|B) - P(A)].\end{aligned}$$

For the last equality assume that $P(B) > 0$ so that $P(A|B)$ makes sense.

From above we see that events A and B are positively correlated if $P(A|B) > P(A)$, in other words, if the occurrence of B increases the chances of A, and negatively correlated if the occurrence of B decreases the chances of A. Furthermore, $\operatorname{Cov}(I_A, I_B) = 0$ if and only if $P(A\cap B) = P(A)P(B)$, which is exactly the definition of independence of A and B. In other words, in the case of indicator random variables independence is equivalent to being uncorrelated. ▲

We illustrate the computation of covariance with one more example before continuing with the general properties.

Example 8.26. Let (X, Y) be a random point uniformly distributed on the triangle D with vertices $(0,0)$, $(1,0)$ and $(0,1)$. Calculate $\operatorname{Cov}(X,Y)$. Before calculation, the reader should determine from the geometry whether one should expect X and Y to be positively or negatively correlated.

The joint density function is $f(x, y) = 2$ on the triangle and zero outside. We compute the needed expectations, leaving calculus details to the reader:

$$E[X] = \iint_D x f(x, y)\, dx\, dy = \int_0^1 \int_0^{1-y} 2x\, dx\, dy = \tfrac{1}{3},$$

$E[Y] = \frac{1}{3}$ by symmetry (by the symmetry of the triangle across the diagonal, X and Y have the same distribution), and

$$E[XY] = \iint_D xy f(x, y)\, dx\, dy = \int_0^1 \int_0^{1-y} 2xy\, dx\, dy = \tfrac{1}{12}.$$

The covariance is

$$\mathrm{Cov}(X, Y) = E[XY] - E[X]E[Y] = \tfrac{1}{12} - (\tfrac{1}{3})^2 = -\tfrac{1}{36}.$$ ▲

Variance of a sum

With the help of covariance, we can express the variance of a sum of random variables succinctly, without any assumptions of independence.

Fact 8.27. (Variance of a sum) Let $X_1, \dots, X_n$ be random variables with finite variances and covariances. Then

$$\mathrm{Var}\left(\sum_{i=1}^{n} X_i\right) = \sum_{i=1}^{n} \mathrm{Var}(X_i) + 2 \sum_{1 \le i < j \le n} \mathrm{Cov}(X_i, X_j). \tag{8.21}$$

In the case of two random variables formula (8.21) is particularly clean and simple:

$$\mathrm{Var}(X + Y) = \mathrm{Var}(X) + \mathrm{Var}(Y) + 2\,\mathrm{Cov}(X, Y). \tag{8.22}$$

In the case of three random variables the formula is:

$$\begin{aligned}\mathrm{Var}(X_1 + X_2 + X_3) &= \mathrm{Var}(X_1) + \mathrm{Var}(X_2) + \mathrm{Var}(X_3) \\ &\quad + 2\,\mathrm{Cov}(X_1, X_2) + 2\,\mathrm{Cov}(X_1, X_3) + 2\,\mathrm{Cov}(X_2, X_3).\end{aligned}$$

Proof of Fact 8.27. The proof uses the definition of variance. By the linearity of the expectation we have $E[\sum_{i=1}^n X_i] = \sum_{i=1}^n \mu_{X_i}$. Hence, the definition of the variance combined with expansion of the square gives

$$\begin{aligned}\mathrm{Var}\left(\sum_{i=1}^{n} X_i\right) &= E\left[\left(\sum_{i=1}^{n} X_i - \sum_{i=1}^{n} \mu_{X_i}\right)^2\right] = E\left[\left(\sum_{i=1}^{n} (X_i - \mu_{X_i})\right)^2\right] \\ &= \sum_{i=1}^{n} E[(X_i - \mu_{X_i})^2] + \sum_{i=1}^{n} \sum_{j \ne i} E[(X_i - \mu_{X_i})(X_j - \mu_{X_j})] \\ &= \sum_{i=1}^{n} \mathrm{Var}(X_i) + 2 \sum_{1 \le i < j \le n} \mathrm{Cov}(X_i, X_j).\end{aligned}$$

The last sum is over all pairs i, j with $i < j$. ▲

As a corollary of Fact 8.27, we can generalize the earlier Fact 8.11.

Fact 8.28. (Variance of a sum of uncorrelated random variables) Let $X_1, \dots, X_n$ be uncorrelated random variables with finite variances. Then

$$\text{Var}(X_1 + X_2 + \cdots + X_n) = \text{Var}(X_1) + \text{Var}(X_2) + \cdots + \text{Var}(X_n).$$

We turn to examples.

Example 8.29. An urn has 5 balls, 3 red ones and 2 green ones. Draw 2 balls. Let X denote the number of red balls in the sample. Compute Var(X) when the sampling is done (a) with replacement and (b) without replacement.

Start by introducing indicator random variables for the outcomes of the draws:

$$Y_1 = \begin{cases} 1, & \text{first draw is red} \\ 0, & \text{first draw is green} \end{cases} \quad \text{and} \quad Y_2 = \begin{cases} 1, & \text{second draw is red} \\ 0, & \text{second draw is green,} \end{cases}$$

so that $X = Y_1 + Y_2$. Y_1 is a Bernoulli random variable with success probability $\frac{3}{5}$. By exchangeability, so is Y_2. Thus $\text{Var}(Y_1) = \text{Var}(Y_2) = \frac{3}{5} \cdot \frac{2}{5} = \frac{6}{25}$.

(a) In sampling with replacement, Y_1 and Y_2 are independent. Consequently $\text{Var}(X) = \text{Var}(Y_1) + \text{Var}(Y_2) = 2 \cdot \frac{6}{25} = \frac{12}{25}$.

(b) In sampling without replacement, Y_1 and Y_2 are not independent. Calculate their covariance:

$$\begin{aligned} \text{Cov}(Y_1, Y_2) &= E[Y_1 Y_2] - E[Y_1]E[Y_2] \\ &= P(\text{both draws are red}) - \tfrac{3}{5} \cdot \tfrac{3}{5} = \tfrac{3 \cdot 2}{5 \cdot 4} - \tfrac{9}{25} = -\tfrac{6}{100}. \end{aligned}$$

The covariance is negative because drawing a red ball reduces the chances of further red draws. We apply (8.22) to calculate Var(X):

$$\begin{aligned} \text{Var}(X) &= \text{Var}(Y_1 + Y_2) = \text{Var}(Y_1) + \text{Var}(Y_2) + 2\,\text{Cov}(Y_1, Y_2) \\ &= 2 \cdot \tfrac{6}{25} - 2 \cdot \tfrac{6}{100} = \tfrac{9}{25}. \end{aligned}$$

The qualitative conclusion is that there is less variability in the case of sampling without replacement. ▲

We can generalize the previous example, which calculated the variance of a particular hypergeometric distribution.

Example 8.30. (Variance of the hypergeometric distribution) Recall the setting of a hypergeometric random variable. An urn contains N_A type A items, N_B type B items, and $N = N_A + N_B$ items altogether. We sample n items without replacement and let X denote the number of type A items in our sample. Compute Var(X).

In Example 8.7 we computed the mean $E[X] = nN_A/N$. As in that calculation, imagine the sampling done in order, and let

$$I_j = \begin{cases} 1, & \text{if } j\text{th sample in of type } A \\ 0, & \text{if } j\text{th sample in of type } B \end{cases}$$

so that $X = I_1 + \cdots + I_n$.

It is convenient for the calculation and also for comparison with the binomial to introduce the notations $p = N_A/N$ and $q = 1 - p = N_B/N$. By exchangeability the indicators have common distribution

$$P(I_j = 1) = \frac{N_A}{N} = p, \qquad P(I_j = 0) = \frac{N_B}{N} = q = 1 - p.$$

This situation is very similar to that of Example 8.12 where we computed the variance of a binomial random variable using indicator random variables. The big difference is that here the indicators are not independent.

From (8.21) and exchangeability,

$$\begin{aligned} \text{Var}(X) &= \sum_{j=1}^{n} \text{Var}(I_j) + \sum_{i=1}^{n} \sum_{j \neq i} \text{Cov}(I_i, I_j) \\ &= n \,\text{Var}(I_1) + n(n-1) \,\text{Cov}(I_1, I_2). \end{aligned} \tag{8.23}$$

By the variance of an indicator (Example 3.47), $\text{Var}(I_1) = pq$. Note that the product $I_1 I_2 = 1$ precisely when the first two samples are both of type A. Hence

$$E[I_1 I_2] = P(I_1 = 1, I_2 = 1) = \frac{N_A(N_A - 1)}{N(N-1)},$$

and

$$\begin{aligned} \text{Cov}(I_1, I_2) &= E[I_1 I_2] - E[I_1]E[I_2] = \frac{N_A(N_A - 1)}{N(N-1)} - \frac{N_A^2}{N^2} \\ &= p\left(\frac{N_A - 1}{N - 1} - \frac{N_A}{N}\right) = p\frac{-N + N_A}{N(N-1)} = -\frac{pq}{N-1}. \end{aligned}$$

Substituting this back into (8.23):

$$\text{Var}(X) = npq - n(n-1)\frac{pq}{N-1} = \frac{N-n}{N-1} npq. \tag{8.24}$$

Now recall that the variance of $\text{Bin}(n, p)$ is npq. The hypergeometric variance above is smaller. Why? Because the indicators are negatively correlated. In sampling without replacement, every draw reduces the remaining variability. Imagine the extreme case where $n = N$ (every item is sampled): then there is *no* randomness in the outcome, and the variance formula above gives zero.

At the other extreme we see another manifestation of the hypergeometric-to-binomial limit of Remark 4.16. If we let $N \to \infty$ in (8.24) while keeping n and p constant we arrive at npq, the variance of the binomial. ▲

Uncorrelated versus independent random variables

Example 8.25 observed that indicator random variables are uncorrelated precisely when the corresponding events are independent. There is a general relationship between being independent and uncorrelated, but only in one direction, as stated in the next fact.

Fact 8.31. Independent random variables are uncorrelated. The converse *does not hold in general.* That is, there are uncorrelated random variables that are not independent.

To verify the first claim above, use Fact 8.10 to observe that the covariance of independent random variables X and Y vanishes:

$$\text{Cov}(X, Y) = E[XY] - \mu_X\mu_Y = E[X]E[Y] - \mu_X\mu_Y = 0.$$

The next example illustrates random variables that are uncorrelated but not independent.

Example 8.32. Let X be uniform on the set $\{-1, 0, 1\}$ and $Y = X^2$. The random variables X and Y are certainly not independent, because for example

$$P(X = 1, Y = 0) = 0 \neq \tfrac{1}{3} \cdot \tfrac{1}{3} = P(X = 1)P(Y = 0).$$

Since $E[X] = 0$ and $X^3 = X$, the covariance is

$$\begin{aligned}\text{Cov}(X, Y) &= E[XY] - E[X] \cdot E[Y] = E[X^3] - E[X] \cdot E[X^2] \\ &= E[X] - E[X] \cdot E[X^2] = 0.\end{aligned}$$

▲

Properties of the covariance

The next fact lists properties of the covariance.

Fact 8.33. The following statements hold when the covariances are well defined.

(i) $\text{Cov}(X, Y) = \text{Cov}(Y, X)$.

(ii) $\text{Cov}(aX + b, Y) = a\,\text{Cov}(X, Y)$ for any real numbers a, b.

(iii) For random variables X_i and Y_j and real numbers a_i and b_j,

$$\text{Cov}\left(\sum_{i=1}^{m} a_i X_i, \sum_{j=1}^{n} b_j Y_j\right) = \sum_{i=1}^{m}\sum_{j=1}^{n} a_i b_j \,\text{Cov}(X_i, Y_j). \tag{8.25}$$

Property (iii) is called the *bilinearity* of covariance.

Proof. The symmetry in (i) is immediate from definition (8.18). For (ii), we can use the linearity of expectation to get

$$\begin{aligned}\mathrm{Cov}(aX+b,Y) &= E[(aX+b)Y]-E[aX+b]E[Y]\\ &= aE[XY]+bE[Y]-aE[X]E[Y]-bE[Y]=a(E[XY]-E[X]E[Y])\end{aligned}$$

which gives the statement we wanted to show.

For (iii) let us introduce the notations $\mu_{X_i}=E[X_i]$, $\mu_{Y_j}=E[Y_j]$ and note that

$$E\left[\sum_{i=1}^m a_iX_i\right]=\sum_{i=1}^m a_i\mu_{X_i} \quad\text{and}\quad E\left[\sum_{j=1}^n b_jY_j\right]=\sum_{j=1}^n b_j\mu_{Y_j}.$$

Then we do some algebra inside the expectation, and use linearity of expectation:

$$\begin{aligned}\mathrm{Cov}\left(\sum_{i=1}^m a_iX_i,\sum_{j=1}^n b_jY_j\right) &= E\left[\left(\sum_{i=1}^m a_iX_i-\sum_{i=1}^m a_i\mu_{X_i}\right)\left(\sum_{j=1}^n b_jY_j-\sum_{j=1}^n b_j\mu_{Y_j}\right)\right]\\ &= E\left[\left(\sum_{i=1}^m a_i(X_i-\mu_{X_i})\right)\left(\sum_{j=1}^n b_j(Y_j-\mu_{Y_j})\right)\right]\\ &= E\left[\sum_{i=1}^m\sum_{j=1}^n a_ib_j(X_i-\mu_{X_i})(Y_j-\mu_{Y_j})\right]\\ &= \sum_{i=1}^m\sum_{j=1}^n a_ib_jE[(X_i-\mu_{X_i})(Y_j-\mu_{Y_j})]\\ &= \sum_{i=1}^m\sum_{j=1}^n a_ib_j\,\mathrm{Cov}(X_i,Y_j).\end{aligned}$$

▲

Example 8.34. (Multinomial random variables) Recall Example 6.7 where we considered the multinomial experiment. We have a trial with r possible outcomes, labeled $1,\dots,r$. Outcome i appears with probability p_i, and $p_1+\cdots+p_r=1$. We perform n independent repetitions of this trial and denote by X_i the number of appearances of the ith outcome. Find $\mathrm{Cov}(X_i,X_j)$.

From Example 6.10 we know that the marginal distribution of X_i is $\mathrm{Bin}(n,p_i)$, and so

$$\mathrm{Cov}(X_i,X_i)=\mathrm{Var}(X_i)=np_i(1-p_i).$$

Consider now $i\neq j$. The bilinearity of covariance suggests that decomposition into sums of indicators might be useful. Define the indicator random variables

$$I_{k,i}=\begin{cases}1, & \text{if trial } k \text{ gives outcome } i\\ 0, & \text{if trial } k \text{ gives an outcome other than } i.\end{cases}$$

Then $X_i = \sum_{k=1}^n I_{k,i}$ and $X_j = \sum_{k=1}^n I_{k,j}$. Bilinearity of covariance gives

$$\operatorname{Cov}(X_i, X_j) = \operatorname{Cov}\left(\sum_{k=1}^n I_{k,i}, \sum_{\ell=1}^n I_{\ell,j}\right) = \sum_{k=1}^n \sum_{\ell=1}^n \operatorname{Cov}(I_{k,i}, I_{\ell,j}). \tag{8.26}$$

Let us determine $\operatorname{Cov}(I_{k,i}, I_{\ell,j})$. If $k \neq \ell$, $I_{k,i}$ and $I_{\ell,j}$ are independent because they depend on distinct trials, and trials are independent by design. Hence in the last sum on line (8.26), only terms with $k = \ell$ are nonzero and

$$\operatorname{Cov}(X_i, X_j) = \sum_{k=1}^n \operatorname{Cov}(I_{k,i}, I_{k,j}).$$

Since $i \neq j$, we have $I_{k,i} I_{k,j} = 0$ because trial k cannot simultaneously yield both outcome i and outcome j. We deduce

$$\operatorname{Cov}(I_{k,i}, I_{k,j}) = E[I_{k,i} I_{k,j}] - E[I_{k,i}]E[I_{k,j}] = 0 - p_i p_j = -p_i p_j.$$

Return back to line (8.26) to conclude:

$$\operatorname{Cov}(X_i, X_j) = \sum_{k=1}^n \operatorname{Cov}(I_{k,i}, I_{k,j}) = -np_i p_j \qquad \text{for } i \neq j.$$

In particular we see that if $i \neq j$, then X_i and X_j are negatively correlated. This is natural since the more often outcome i appears, the fewer opportunities there are for outcome j.

Here is an alternative solution, which is quicker. Note that $X_i + X_j$ is binomial with parameter $p_i + p_j$ (success means either outcome i or j). By (8.21),

$$\operatorname{Var}(X_i + X_j) = \operatorname{Var}(X_i) + \operatorname{Var}(X_j) + 2\operatorname{Cov}(X_i, X_j)$$

from which

$$n(p_i + p_j)(1 - p_i - p_j) = np_i(1 - p_i) + np_j(1 - p_j) + 2\operatorname{Cov}(X_i, X_j).$$

Now solve for $\operatorname{Cov}(X_i, X_j) = -np_i p_j$. ▲

Correlation

As a measure of strength of dependence, covariance has a drawback. Bilinearity gives $\operatorname{Cov}(10X, 7Y) = 70\operatorname{Cov}(X, Y)$. In other words, a huge covariance can be simply the result of the size of the random variables, and not signify an especially strong dependence. We can eliminate this problem by normalizing the covariance in the right way.

Definition 8.35. The **correlation** (or **correlation coefficient**) of two random variables X and Y with positive finite variances is defined by

$$\operatorname{Corr}(X, Y) = \frac{\operatorname{Cov}(X, Y)}{\sqrt{\operatorname{Var}(X)}\sqrt{\operatorname{Var}(Y)}}. \tag{8.27}$$

The correlation coefficient is sometimes denoted by $\rho(X, Y)$ or simply by ρ.

Definition 8.35 requires both X and Y to have positive variance. Recall from Fact 3.54 that $\operatorname{Var}(X) = 0$ if and only if $P(X = E[X]) = 1$. That is, $\operatorname{Var}(X) = 0$ only when X is a degenerate random variable (not random at all).

The next fact reveals that $\operatorname{Corr}(X, Y)$ lies in $[-1, 1]$, and furthermore, there is a precise description for the extreme values ± 1: these happen if and only if X and Y are linearly related.

Theorem 8.36. *Let X and Y be two random variables with positive finite variances. Then we have these properties.*

(a) Let a, b be real numbers with $a \neq 0$. Then $\operatorname{Corr}(aX + b, Y) = \frac{a}{|a|} \operatorname{Corr}(X, Y)$.
(b) $-1 \leq \operatorname{Corr}(X, Y) \leq 1$.
(c) $\operatorname{Corr}(X, Y) = 1$ if and only if there exist $a > 0$ and $b \in \mathbb{R}$ such that $Y = aX + b$.
(d) $\operatorname{Corr}(X, Y) = -1$ if and only if there exist $a < 0$ and $b \in \mathbb{R}$ such that $Y = aX + b$.

Proof. The proof of (a) is immediate from the properties of the variance and the covariance:

$$\operatorname{Corr}(aX + b, Y) = \frac{\operatorname{Cov}(aX + b, Y)}{\sqrt{\operatorname{Var}(aX + b)\operatorname{Var}(Y)}} = \frac{a\operatorname{Cov}(X, Y)}{\sqrt{a^2 \operatorname{Var}(X)\operatorname{Var}(Y)}} = \frac{a}{|a|}\operatorname{Corr}(X, Y).$$

Note that $\frac{a}{|a|}$ is either 1 or -1 depending on the sign of a.

The proof of (b) gives us practice in manipulating expectations. Abbreviate $\sigma_X^2 = \operatorname{Var}(X)$ and $\sigma_Y^2 = \operatorname{Var}(Y)$. The trick is to work with the standardized random variables

$$\widetilde{X} = \frac{X - \mu_X}{\sigma_X} \quad \text{and} \quad \widetilde{Y} = \frac{Y - \mu_Y}{\sigma_Y}.$$

Observe these properties:

$$E[\widetilde{X}^2] = E\left[\frac{(X - \mu_X)^2}{\sigma_X^2}\right] = 1, \qquad E[\widetilde{Y}^2] = E\left[\frac{(Y - \mu_Y)^2}{\sigma_Y^2}\right] = 1,$$

and

$$E[\widetilde{X}\widetilde{Y}] = E\left[\frac{X - \mu_X}{\sigma_X} \cdot \frac{Y - \mu_Y}{\sigma_Y}\right] = \frac{E[(X - \mu_X)(Y - \mu_Y)]}{\sigma_X \sigma_Y} = \operatorname{Corr}(X, Y).$$

Now we can derive the bounds in part (b). We will derive the upper bound. We have

$$0 \leq E[(\widetilde{X} - \widetilde{Y})^2] = E[\widetilde{X}^2] + E[\widetilde{Y}^2] - 2E[\widetilde{X}\widetilde{Y}] = 2\,[1 - \operatorname{Corr}(X, Y)]. \tag{8.28}$$

From this $0 \leq 1 - \operatorname{Corr}(X, Y)$, and $\operatorname{Corr}(X, Y) \leq 1$. The lower bound can be derived similarly by first considering $E[(\widetilde{X} + \widetilde{Y})^2]$.

Now turn to part (c). Suppose first that $Y = aX + b$ for $a > 0$. Then $\sigma_Y^2 = a^2\sigma_X^2$ and $\sigma_Y = |a|\,\sigma_X = a\sigma_X$, and

$$\mathrm{Cov}(X, Y) = \mathrm{Cov}(X, aX + b) = a\,\mathrm{Cov}(X, X) + \mathrm{Cov}(X, b) = a\sigma_X^2 + 0 = a\sigma_X^2.$$

From these,

$$\mathrm{Corr}(X, Y) = \frac{\mathrm{Cov}(X, Y)}{\sigma_X\sigma_Y} = \frac{a\sigma_X^2}{\sigma_X \cdot a\sigma_X} = 1.$$

Conversely, suppose $\mathrm{Corr}(X, Y) = 1$. Then, following the calculation in (8.28) in reverse order,

$$0 = 2\,[1 - \mathrm{Corr}(X, Y)] = E[(\widetilde{X} - \widetilde{Y})^2].$$

Hence, by Fact 3.54, we have $\widetilde{X} - \widetilde{Y} = 0$ with probability 1 (note that $E[\widetilde{X} - \widetilde{Y}] = 0$). The conclusion now follows from a rearrangement:

$$\widetilde{X} = \widetilde{Y} \iff \frac{X - \mu_X}{\sigma_X} = \frac{Y - \mu_Y}{\sigma_Y} \iff Y = \frac{\sigma_Y}{\sigma_X}X - \frac{\sigma_Y}{\sigma_X}\mu_X + \mu_Y.$$

The relation above gives $Y = aX + b$ with $a = \sigma_Y/\sigma_X > 0$.

The proof of part (d) follows similarly. ▲

Example 8.37. (Continuing Example 8.34 of multinomial variables) In Example 8.34 we deduced

$$\mathrm{Cov}(X_i, X_j) = \begin{cases} np_i(1 - p_i), & i = j \\ -np_ip_j, & i \neq j. \end{cases}$$

Thus, for $i \neq j$ we have

$$\mathrm{Corr}(X_i, X_j) = \frac{\mathrm{Cov}(X_i, X_j)}{\sqrt{\mathrm{Var}(X_i)}\sqrt{\mathrm{Var}(X_j)}} = \frac{-np_ip_j}{\sqrt{np_i(1 - p_i)}\sqrt{np_j(1 - p_j)}}$$
$$= -\sqrt{\frac{p_ip_j}{(1 - p_i)(1 - p_j)}}.$$

Notice that the correlation removes the factor n that measures the number of trials.

Consider the case $r = 2$. This is the binomial case, with $p_2 = 1 - p_1$, and we get

$$\mathrm{Corr}(X_1, X_2) = -\sqrt{\frac{p_1p_2}{(1 - p_1)(1 - p_2)}} = -1.$$

This reflects the linear relationship $X_2 = n - X_1$ of the binomial case. ▲

Example 8.38. (Correlation in the standard bivariate normal) Let $-1 < \rho < 1$. Let the joint distribution of the random variables X and Y be the standard bivariate normal with parameter ρ introduced in Section 6.4 on page 229, with joint density function

$$f(x, y) = \frac{1}{2\pi\sqrt{1 - \rho^2}}e^{-\frac{x^2+y^2-2\rho xy}{2(1-\rho^2)}}. \tag{8.29}$$

Find the covariance and correlation of X and Y.

By (6.29) X and Y are both standard normals. This means that $\mathrm{Var}(X) = \mathrm{Var}(Y) = 1$, and hence $\mathrm{Cov}(X, Y) = \mathrm{Corr}(X, Y)$. Moreover, since $E[X] = E[Y] = 0$, we also have $\mathrm{Cov}(X, Y) = E[XY]$. To compute $E[XY]$ we use the joint density function (8.29):

$$E[XY] = \int_{-\infty}^{\infty}\int_{-\infty}^{\infty} xy \frac{1}{2\pi\sqrt{1-\rho^2}} e^{-\frac{x^2+y^2-2\rho xy}{2(1-\rho^2)}}\, dy\, dx.$$

Completing the square in the exponential and grouping the terms the same way as in computation (6.30), the above equals

$$\int_{-\infty}^{\infty} x \frac{1}{\sqrt{2\pi}} e^{-\frac{x^2}{2}} \left(\int_{-\infty}^{\infty} y \frac{1}{\sqrt{2\pi(1-\rho^2)}} e^{-\frac{(y-\rho x)^2}{2(1-\rho^2)}}\, dy \right) dx.$$

The inner integral with respect to y above is the expected value of a $\mathcal{N}(\rho x, 1-\rho^2)$ random variable, and hence equal to ρx. This leads to

$$E[XY] = \int_{-\infty}^{\infty} \rho x^2 \frac{1}{\sqrt{2\pi}} e^{-\frac{x^2}{2}}\, dx = \rho E[X^2] = \rho,$$

where we again used that $E[X^2] = \mathrm{Var}(X) = 1$. Thus, in the standard bivariate normal distribution the parameter ρ is the correlation (and covariance) of the two random variables. ▲

8.5. The bivariate normal distribution ♦

Chapter 4 demonstrated the central role of the normal distribution. The *multivariate normal distribution* has a similarly important role. This section gives a short introduction to the bivariate case with two random variables. It generalizes the standard bivariate normal distribution of Section 6.4 and Example 8.38. Section 8.6 explains the generalization to more variables.

In the single variable case, the general normal distribution is determined by two parameters, the mean μ and variance σ^2. It was introduced as the linear transform $X = \sigma Z + \mu$ of a standard normal Z, from which we deduced its density function. We take the analogous route here.

The normal distribution of a pair (X, Y) is determined by five parameters, namely the means, the variances and the correlation:

$$E[X] = \mu_X,\ E[Y] = \mu_Y,\ \mathrm{Var}(X) = \sigma_X^2,\ \mathrm{Var}(Y) = \sigma_Y^2,\ \rho = \mathrm{Corr}(X, Y). \tag{8.30}$$

The distribution can be obtained by a linear transformation from independent standard normals. So assume we are given five parameters: real μ_X and μ_Y, positive σ_X and σ_Y, and $-1 < \rho < 1$. Let Z and W be independent standard normal random variables and define

$$\begin{aligned} X &= \sigma_X Z + \mu_X \\ Y &= \sigma_Y \rho Z + \sigma_Y \sqrt{1-\rho^2}\, W + \mu_Y. \end{aligned} \tag{8.31}$$

By Fact 7.9, the marginal distributions are normal: $X \sim \mathcal{N}(\mu_X, \sigma_X^2)$ and $Y \sim \mathcal{N}(\mu_Y, \sigma_Y^2)$. A quick check also verifies $\text{Corr}(X, Y) = \rho$.

Our task is to find the joint density function $f_{X,Y}$ of (X, Y). While the Jacobian technique of Section 6.4 could be used, here we choose an alternative route and instead first compute the joint cumulative distribution function $F_{X,Y}$. The joint density will then be found by differentiation.

The calculation of $F_{X,Y}$ is relatively straightforward, utilizing (i) the joint density function $f_{Z,W}(z, w) = e^{-(z^2+w^2)/2}/2\pi$ of (Z, W) and (ii) a change of variables:

$$
\begin{aligned}
&F_{X,Y}(s,t) = P(X \le s,\ Y \le t)\\
&= P\left(\sigma_X Z + \mu_X \le s,\ \sigma_Y \rho Z + \sigma_Y\sqrt{1-\rho^2}\, W + \mu_Y \le t\right)\\
&= P\left(Z \le \frac{s-\mu_X}{\sigma_X},\ W \le \frac{t-\mu_Y-\sigma_Y\rho Z}{\sigma_Y\sqrt{1-\rho^2}}\right)\\
&= \frac{1}{2\pi}\int_{-\infty}^{\frac{s-\mu_X}{\sigma_X}} e^{-z^2/2}\left(\int_{-\infty}^{\frac{t-\mu_Y-\sigma_Y\rho z}{\sigma_Y\sqrt{1-\rho^2}}} e^{-w^2/2}\,dw\right)dz\\
&= \frac{1}{2\pi\sigma_Y\sqrt{1-\rho^2}}\int_{-\infty}^{\frac{s-\mu_X}{\sigma_X}} e^{-z^2/2}\left(\int_{-\infty}^{t}\exp\left\{-\frac{(y-\mu_Y-\sigma_Y\rho z)^2}{2\sigma_Y^2(1-\rho^2)}\right\}dy\right)dz\\
&= \frac{1}{2\pi\sigma_X\sigma_Y\sqrt{1-\rho^2}}\int_{-\infty}^{s}\int_{-\infty}^{t}\exp\left\{-\frac{(x-\mu_X)^2}{2\sigma_X^2}-\frac{(y-\mu_Y-\sigma_Y\rho\frac{x-\mu_X}{\sigma_X})^2}{2\sigma_Y^2(1-\rho^2)}\right\}dy\,dx.
\end{aligned}
$$

First the w variable of the inner integral was changed to $y = \sigma_Y\rho z + \sigma_Y\sqrt{1-\rho^2}\,w + \mu_Y$, with z a fixed constant. Then the z variable of the outer integral was changed to $x = \sigma_X z + \mu_X$, and the integrand written as a single exponential.

Taking partial derivatives (recall (6.25)) and some rearranging in the exponential gives this expression for the joint density function:

$$
f_{X,Y}(x,y) = \frac{1}{2\pi\sigma_X\sigma_Y\sqrt{1-\rho^2}}\exp\left\{-\frac{1}{2(1-\rho^2)}\left(\left(\frac{x-\mu_X}{\sigma_X}\right)^2+\left(\frac{y-\mu_Y}{\sigma_Y}\right)^2-2\rho\frac{(x-\mu_X)(y-\mu_Y)}{\sigma_X\sigma_Y}\right)\right\}. \tag{8.32}
$$

The density function above defines the general *bivariate normal distribution*. The special case with $\mu_X = \mu_Y = 0$ and $\sigma_X^2 = \sigma_Y^2 = 1$ is the standard bivariate normal distribution with correlation ρ introduced in Section 6.4 and studied in Example 8.38.

Note from (8.32) that if $\rho = 0$ then $f_{X,Y}(x, y) = f_X(x)f_Y(y)$. In other words, a bivariate normal (X, Y) has the special feature that if the coordinates X and Y are uncorrelated, they are also independent.

Formula (8.32) is clumsy. The natural way to present the multivariate normal distribution involves matrices. This is explained briefly in Section 8.6 below.

8.6. Finer points ♣

Independence and expectations of products

Under independence, expectations of products become products of expectations, as stated in Fact 8.10. The converse also holds in the following sense. If $X_1, \dots, X_n$ are random variables for which $E[\prod_{k=1}^{n} g_k(X_k)] = \prod_{k=1}^{n} E[g_k(X_k)]$ for all bounded (measurable) functions $g_1, \dots, g_n$, then $X_1, \dots, X_n$ are mutually independent. This follows because if we take indicator functions

$$g_k(x) = I_{B_k}(x) = \begin{cases} 1, & x \in B_k \\ 0, & x \notin B_k \end{cases} \tag{8.33}$$

for arbitrary (Borel) subsets $B_1, \dots, B_n$ of $\mathbb{R}$, then the product property becomes $P(X_1 \in B_1, \dots, X_n \in B_n) = \prod_{k=1}^{n} P(X_k \in B_k)$. This property is the definition of independence (Definition 2.27).

Multivariate normal distribution

To lead to the general multivariate normal distribution, we first rewrite the joint density function $f_{X,Y}$ of (8.32) in terms of the means and the *covariance matrix* given by

$$S = \begin{bmatrix} \text{Cov}(X,X) & \text{Cov}(X,Y) \\ \text{Cov}(Y,X) & \text{Cov}(Y,Y) \end{bmatrix} = \begin{bmatrix} \sigma_X^2 & \sigma_X\sigma_Y\rho \\ \sigma_X\sigma_Y\rho & \sigma_Y^2 \end{bmatrix}.$$

Matrix S has determinant $\det S = \sigma_X^2\sigma_Y^2(1-\rho^2)$ and inverse matrix

$$S^{-1} = \frac{1}{1-\rho^2} \begin{bmatrix} \frac{1}{\sigma_X^2} & -\frac{\rho}{\sigma_X\sigma_Y} \\ -\frac{\rho}{\sigma_X\sigma_Y} & \frac{1}{\sigma_Y^2} \end{bmatrix}.$$

In terms of these ingredients $f_{X,Y}$ of (8.32) can be expressed as

$$f_{\mathbf{X}}(\mathbf{x}) = \frac{1}{(2\pi)^{n/2}\sqrt{\det S}} e^{-\frac{1}{2}(\mathbf{x}-\boldsymbol{\mu})^T S^{-1}(\mathbf{x}-\boldsymbol{\mu})}, \qquad \mathbf{x} \in \mathbb{R}^n, \tag{8.34}$$

where the dimension $n = 2$. We employed matrix notation: the random column vector is $\mathbf{X} = \begin{bmatrix} X \\ Y \end{bmatrix}$, its mean vector is $\boldsymbol{\mu} = \begin{bmatrix} \mu_X \\ \mu_Y \end{bmatrix}$, the $\mathbb{R}^2$-valued variable is $\mathbf{x} = \begin{bmatrix} x \\ y \end{bmatrix}$, and superscript T denotes transpose.

Formula (8.34) gives also the general multivariate normal density function of the $\mathbb{R}^n$-valued normal vector $\mathbf{X} = [X_1, \dots, X_n]^T$ with mean vector $\boldsymbol{\mu} = [\mu_1, \dots, \mu_n]^T$ and $n \times n$ covariance matrix $S = [\text{Cov}(X_i, X_j)]_{1 \le i,j \le n}$, as long as S is invertible. This is abbreviated by $\mathbf{X} \sim \mathcal{N}(\boldsymbol{\mu}, S)$.

The $\mathcal{N}(\boldsymbol{\mu}, S)$ distribution can be defined for all real n-vectors $\boldsymbol{\mu}$ and all symmetric, real $n \times n$ matrices S. If S is singular (that is, noninvertible), this distribution lives on a lower dimensional subspace of $\mathbb{R}^n$ and does not have a density function on $\mathbb{R}^n$. In that case it is defined in terms of a transform such as its moment generating or characteristic function. A $\mathcal{N}(\mu, S)$-distributed random vector $\mathbf{X}$ can be constructed from a vector $\mathbf{Z} = [Z_1, \ldots, Z_n]^T$ of independent standard normals. Diagonalization of S shows the existence of a matrix A such that $S = AA^T$. Then $\mathbf{X} = A\mathbf{Z} + \mu$ satisfies $\mathbf{X} \sim \mathcal{N}(\mu, S)$.

Details of expectations and measure theory

Just like expectations, covariances can blow up or be undefined. Throughout Section 8.4 we have tacitly assumed in all the results that the expectations, variances and covariances are well defined and finite, even if we have not always stated this explicitly.

The fact that some identities and inequalities in probability theory can only hold with probability one and not on every point of the sample space is a measure-theoretic subtlety. Occasionally it spills into view. For example, in the last step of the proof of Theorem 8.36 we concluded that $\widetilde{X} = \widetilde{Y}$ from $E[(\widetilde{X} - \widetilde{Y})^2] = 0$. However, since the conclusion comes from an expectation, it can only hold with probability one. Consider this example.

Example 8.39. Let $\Omega = \{a, b, c\}$ with $P(a) = P(b) = 1/2$ and $P(c) = 0$. Define $X(a) = X(b) = 2$ and $X(c) = 3$, while $Y(a) = Y(b) = Y(c) = 2$. Then

$$E[(X - Y)^2] = (2 - 2)^2 \cdot (P(a) + P(b)) + (3 - 2)^2 \cdot P(c) = 0.$$

Since $X(c) \neq Y(c)$ we cannot say that $X = Y$ on all of Ω, but $P(X = Y) = P\{a, b\} = 1$. ▲

To be precise, Theorem 8.36(c)–(d) should state that $Y = aX + b$ with probability one.

Exercises

We start with some warm-up exercises arranged by section.

Section 8.1

Exercise 8.1. Let n be a positive integer and p and r two real numbers in $(0, 1)$. Two random variables X and Y are defined on the same sample space. All we know about them is that $X \sim \text{Geom}(p)$ and $Y \sim \text{Bin}(n, r)$. For each expectation in parts (a)–(d) below, decide whether it can be calculated with this information, and if it can, give its value.

(a) $E[X + Y]$
(b) $E[XY]$
(c) $E[X^2 + Y^2]$
(d) $E[(X + Y)^2]$

Exercise 8.2. Caleb has a four-sided die, a six-sided die, and a 12-sided die. He rolls the three dice and adds up the numbers showing. What is the expected value of the sum of the rolls?

Exercise 8.3. Adam has two brothers, Ben and Chris, and one sister, Daisy. On any given day, Ben calls Adam with probability 0.3, Chris calls Adam with probability 0.4, and Daisy calls Adam with probability 0.7. Let X be the number of siblings that call Adam tomorrow. Find the expectation of X.

Hint. Express X as the sum of suitable indicator random variables.

Exercise 8.4. Jessica has a four-sided die, a six-sided die, and a 12-sided die. She rolls the three dice once. Let Z be the number of fours showing. Find $E[Z]$ using the indicator method.

Section 8.2

Exercise 8.5. Let X be a geometric random variable with parameter $p = \frac{1}{3}$ and let Y be a Poisson random variable with parameter $\lambda = 4$. Assume X and Y independent. A rectangle is drawn with side lengths X and $Y + 1$. Find the expected values of the perimeter and the area of the rectangle.

Exercise 8.6. Let X and Y be as in Exercise 8.1. Assume additionally that X and Y are independent. Evaluate all the expectations in parts (a)–(d) of Exercise 8.1.

Exercise 8.7. In the setting of Exercise 8.3, assume that all calls are made independently. Calculate the mean and variance of X, the number of siblings that call Adam tomorrow.

Exercise 8.8. Our faucet is broken, and a plumber has been called. The arrival time of the plumber is uniformly distributed between 1 PM and 7 PM. Independently of when the plumber arrives, the time it takes to fix the broken faucet is exponentially distributed with mean 30 minutes. Find the expectation and variance of the time at which the plumber completes the project.

Exercise 8.9. Suppose X and Y are independent random variables with $E[X] = 3$, $E[Y] = 5$, $\text{Var}(X) = 2$ and $\text{Var}(Y) = 3$. Compute the following quantities.

(a) $E[3X - 2Y + 7]$
(b) $\text{Var}(3X - 2Y + 7)$
(c) $\text{Var}(XY)$

Section 8.3

Exercise 8.10. Let X_1, X_2 be independent random variables, all with the same distribution

$$P(X_i = 0) = \tfrac{1}{2}, \quad P(X_i = 1) = \tfrac{1}{3}, \quad P(X_i = 2) = \tfrac{1}{6}.$$

Let $S = X_1+X_2$. Use the moment generating function of S to find the probability mass function of S.

Exercise 8.11. In the setting of Exercise 8.3, assume that all calls are made independently. Find the moment generating function of X, the number of siblings that call Adam tomorrow.

Exercise 8.12.

(a) Let Z be Gamma$(2, \lambda)$ distributed. That is, Z has the density function

$$f_Z(z) = \begin{cases} \lambda^2 z e^{-\lambda z}, & z \geq 0 \\ 0, & \text{otherwise.} \end{cases}$$

Use the definition of the moment generating function to calculate $M_Z(t) = E(e^{tZ})$.

(b) Let X and Y be two independent Exp(λ) random variables. Recall the moment generating function of X and Y from Example 5.6 in Section 5.1. Using the approach from Section 8.3 show that Z and $X + Y$ have the same distribution.

Exercise 8.13. Let Z have moment generating function

$$M_Z(t) = \left(\tfrac{1}{2}e^{-t} + \tfrac{2}{5} + \tfrac{1}{10}e^{t/2}\right)^{36}.$$

Describe Z as a sum of certain independent random variables.

Section 8.4

Exercise 8.14. Let the joint probability mass function of (X, Y) be given by the table below. Calculate Cov(X, Y) and Corr(X, Y).

X \ Y	0	1	2	3
1	$\frac{1}{15}$	$\frac{1}{15}$	$\frac{2}{15}$	$\frac{1}{15}$
2	$\frac{1}{10}$	$\frac{1}{10}$	$\frac{1}{5}$	$\frac{1}{10}$
3	$\frac{1}{30}$	$\frac{1}{30}$	0	$\frac{1}{10}$

Exercise 8.15. Let (X, Y) be a uniformly distributed random point on the quadrilateral D with vertices $(0, 0)$, $(2, 0)$, $(1, 1)$ and $(0, 1)$. Calculate the covariance of X and Y. Based on the description of the experiment, should it be negative or positive?

Exercise 8.16. Let $E[X] = 1$, $E[X^2] = 3$, $E[XY] = -4$ and $E[Y] = 2$. Find Cov$(X, 2X + Y - 3)$.

Exercise 8.17. Suppose that A and B are events on the same sample space with $P(A) = 0.5$, $P(B) = 0.2$ and $P(AB) = 0.1$. Let $X = I_A + I_B$ be the random variable that counts how many of the events A and B occur. Find $\text{Var}(X)$.

Section 8.5

Exercise 8.18. Suppose that X and Y are independent standard normals. Show that $U = \frac{1}{\sqrt{2}}(X - Y)$ and $V = \frac{1}{\sqrt{2}}(X + Y)$ are also independent standard normals. You can take for granted that (U, V) has bivariate normal distribution.

Exercise 8.19. Let $-1 < \rho < 1$, and suppose that Z_1, Z_2 are independent standard normals. Let

$$X = Z_1, \qquad Y = \rho Z_1 + \sqrt{1 - \rho^2} Z_2.$$

Show that the joint distribution of X, Y is standard bivariate normal with parameter ρ.

Further exercises

Exercise 8.20. I have an urn with 30 chips numbered from 1 to 30. The chips are then selected one by one, without replacement, until all 30 have been selected. Let X_i denote the value of the ith pick. Find $E(X_3 + X_{10} + X_{22})$.

Hint. Use the linearity of expectation and exchangeability.

Exercise 8.21. I have 5 dimes, 3 quarters and 2 pennies in my pocket. I pick randomly three coins. (This is a sample of three without replacement.) Find the expected total value of the three chosen coins.

Exercise 8.22. In a lottery 5 numbers are chosen from the set $\{1, \dots, 90\}$ without replacement. We order the five numbers in increasing order and we denote by X the number of times the difference between two neighboring numbers is 1. (For example, for $\{1, 2, 3, 4, 5\}$ we have $X = 4$, for $\{3, 6, 8, 9, 24\}$ we have $X = 1$ and for $\{12, 14, 43, 75, 88\}$ we have $X = 0$.) Find $E[X]$.

Exercise 8.23. We have an urn with 20 red and 30 green balls. We draw all the balls one by one and record the colors we see. (That is, we take a sample of 50 without replacement.)

(a) Find the probability that the 28th and 29th balls are of different color.
(b) Let X be the number of times two consecutive balls are of different color. Find $E[X]$.

Exercise 8.24. A bucket contains 30 red balls and 50 white balls. Sam and Jane take turns drawing balls until all the balls are drawn. Sam goes first. Let N be the number of times that Jane draws a ball that matches the color of the ball that Sam drew on the turn before. Calculate the expectation of N.

Exercise 8.25. We have a bucket with 100 balls which are numbered from 1 to 100. We draw 10 balls one by one, without replacement, and we look at their numbers.

(a) What is the probability that the number of the 5th pick is larger than that of the 4th pick?
(b) Let X denote the number of times that during the 10 picks the number on the sampled ball is larger than the number that was drawn just before it. (For example, if we pick the numbers 43, 72, 12, 10, 14, 4, 67, 92, 100, 31 in that order then $X = 5$.) Find $E[X]$.

Exercise 8.26. Suppose I have an urn with 9 balls: 4 green, 3 yellow and 2 white ones. I draw a ball from the urn repeatedly with replacement.

(a) Suppose I draw n times. Let X_n be the number of times I saw a green ball followed by a yellow ball. Calculate the expectation $E[X_n]$.
(b) Let Y be the number of times I drew a green ball before the first white draw. Calculate $E[Y]$. Can you give an intuitive explanation for your answer?

Exercise 8.27. Let n be a positive integer. We sample n numbers $a_1, \dots, a_n$ from the set $\{1, \dots, n\}$ uniformly at random, with replacement. We say that picks i and j with $i < j$ are a match if $a_i = a_j$. What is the expected total number of matches?

Exercise 8.28. We have an urn with 25 yellow, 30 green and 40 red marbles. We take a sample of 10 without replacement and denote the number of yellow and green marbles in the sample by X and Y, respectively. Find $E(X - Y)$.

Exercise 8.29. We shuffle a deck of 52 cards and then flip them one by one. Let X denote the number of times when we see three number cards in a row (the numbered cards are $2, 3, \dots, 10$). Find the expected value of X.

Exercise 8.30. A bucket contains 20 balls; two are numbered 1, two are numbered 2, two are numbered 3, ..., two are numbered 10. The balls are drawn one at a time without replacement. An *increase* occurs on the kth draw if the number on the kth draw is (strictly) larger than the number on the $(k-1)$st draw. Let N be the total number of draws on which an increase occurs. What is $E[N]$?

Exercise 8.31. We choose a number from the set $\{0, \dots, 9999\}$ randomly, and denote by X the sum of its digits. Find the expected value of X.

Exercise 8.32. This exercise shows how to prove the inclusion-exclusion formulas of Fact 1.23 with indicator variables. Parts (a)–(b) below show you the ropes for the case of two events and then you generalize.

(a) Prove the identity $I_{A\cup B} = I_A + I_B - I_{A\cap B}$. You can do this from de Morgan's laws and the identities $I_{A^c} = 1 - I_A$ and $I_{A\cap B} = I_A I_B$. Alternatively, you can evaluate all indicators at a given sample point ω by assuming that ω lies in AB, AB^c, A^cB or A^cB^c.
(b) To prove the inclusion-exclusion formula (1.16) for two events take expectations of both sides of $I_{A\cup B} = I_A + I_B - I_{A\cap B}$.
(c) Prove the inclusion-exclusion formula (1.17) for three events.

Exercise 8.33. 10 tennis players participate in a round robin tournament. This means that every player plays everybody else exactly once, hence each player plays 9 matches. Assume that the outcomes of the matches are random: in each match each of the two players wins with probability $\frac{1}{2}$ independently of the outcomes of the other matches.

(a) Find the expected number of players that win exactly 2 matches.
(b) We say that three players A, B and C form a 3-cycle if A beat B, B beat C and C beat A. Find the expected number of 3-cycles.
(c) We say that k distinct players $A_1, \dots, A_k$ form a k-path if A_i beat A_{i+1} for $i = 1, 2, \dots, k-1$. For each $2 \le k \le n$ find the expected number of k-paths.

Hint. Use the idea of Example 8.9. Set up a random *directed* network among the ten players with arrows between the nodes expressing who won.

Exercise 8.34. Prove Fact 8.1 for jointly continuous random variables $X_1, X_2, \dots, X_n$.

Hint. Imitate the proof given for discrete random variables.

Exercise 8.35. I have four different sweaters. Every day I choose one of the four sweaters at random to wear. Let X be the number of different sweaters I wore during a 5-day week. (For example, if my 5-day sweater sequence is $(3, 2, 4, 4, 2)$ then $X = 3$ because that week I wore sweaters 2, 3 and 4.)

(a) Find the mean of X.
(b) Find the variance of X.

Exercise 8.36. I roll a fair die four times. Let X be the number of different outcomes that I see. (For example, if the die rolls are 5,3,6,6 then $X = 3$ because the different outcomes are 3, 5 and 6.)
(a) Find the mean of X.
(b) Find the variance of X.

Exercise 8.37. A cereal company announces a series of 10 different toys for prizes. Every box of cereal contains two different randomly chosen toys from the series of 10. Let X be the number of different toys I have accumulated after buying 4 boxes of cereal.

(a) Find the mean of X.

(b) Find the variance of X.

Exercise 8.38. We roll a die repeatedly until we see all 6 numbers as outcomes. Find the mean and variance of the number of rolls needed.

Hint. Example 8.17 could help.

Exercise 8.39. Suppose that a professor chooses a random student in a class of 40 students (there are 23 girls and 17 boys in the class) to perform a calculation on the board. The professor repeats this procedure 15 times, choosing a new student each time (i.e. no student will go twice). Let X be the total number of boys chosen. Calculate the mean and variance of X.

Exercise 8.40. In a certain lottery 5 numbers are drawn without replacement out of the set $\{1, 2, \dots, 90\}$ randomly each week. Let X be the number of different numbers drawn during the first four weeks of the year. (For example, if the draws are $\{1, 2, 3, 4, 5\}, \{2, 3, 4, 5, 6\}, \{3, 4, 5, 6, 7\}$ and $\{4, 5, 6, 7, 8\}$ then $X = 8$.)

(a) Find $E[X]$.
(b) Find $\mathrm{Var}(X)$.

Exercise 8.41. The random variables $X_1, \dots, X_n$ are i.i.d. We also know that $E[X_1] = 0$, $E[X_1^2] = a$ and $E[X_1^3] = b$. Let $\bar{X}_n = \frac{X_1+\cdots+X_n}{n}$. Find the third moment of $\bar{X}_n$.

Exercise 8.42. The random variables $X_1, \dots, X_n$ are i.i.d. We also know that $E[X_1] = 0$, $E[X_1^2] = a$, $E[X_1^3] = b$ and $E[X_1^4] = c$. Let $\bar{X}_n = \frac{X_1+\cdots+X_n}{n}$. Find the fourth moment of $\bar{X}_n$.

Exercise 8.43. Let $Z_1, Z_2, \dots, Z_n$ be independent normal random variables with mean 0 and variance 1. Let

$$Y = Z_1^2 + \cdots + Z_n^2.$$

(a) Using that Y is the sum of independent random variables, compute both the mean and variance of Y.
(b) Find the moment generating function of Y and use it to compute the mean and variance of Y.

Exercise 8.44. Suppose that X has possible values $\{1, 2, 3\}$ and associated probabilities

$$p_X(1) = \frac{1}{4}, \quad p_X(2) = \frac{1}{4}, \quad p_X(3) = \frac{1}{2},$$

and that Y has possible values $\{2, 3, 4\}$ with associated probabilities

$$p_Y(2) = \frac{1}{7}, \quad p_Y(3) = \frac{2}{7}, \quad p_Y(4) = \frac{4}{7}.$$

Suppose that X and Y are independent.

(a) Give the moment generating functions $M_X(t)$ of X and $M_Y(t)$ of Y.

(b) Find $M_{X+Y}(t)$ and using this moment generating function identify the possible values and probability mass function of $X + Y$.

Exercise 8.45. The joint probability mass function of the random variables (X, Y) is given by the following table. Calculate $\text{Cov}(X, Y)$ and $\text{Corr}(X, Y)$.

$X \backslash Y$	0	1	2
1	0	$\frac{1}{9}$	$\frac{1}{9}$
2	$\frac{1}{3}$	$\frac{2}{9}$	0
3	0	$\frac{1}{9}$	$\frac{1}{9}$

Exercise 8.46. We have an urn with 6 red balls and 4 green balls. We draw balls from the urn one by one without replacement, noting the order of the colors, until the urn is empty. Let X be the number of red balls in the first five draws, and Y the number of red balls in the last five draws. Compute the covariance $\text{Cov}(X, Y)$.

Exercise 8.47. Recall the setting of Exercise 8.3. On any given day, Ben calls Adam with probability 0.3, Chris calls Adam with probability 0.4, and Daisy calls Adam with probability 0.7. Let X be the number of siblings that call Adam tomorrow. Make the following assumptions. (i) Daisy calls Adam independently of the brothers. (ii) If Ben calls Adam, then Chris calls Adam with probability 0.8. Under these assumptions, calculate the mean and variance of X.

Exercise 8.48. Suppose there are 100 cards numbered 1 through 100. Draw a card at random. Let X be the number of digits on the card (between 1 and 3) and let Y be the number of zeros. Find $\text{Cov}(X, Y)$.

Exercise 8.49. Let (X, Y) be a randomly chosen point from the triangle D with vertices $(0, 0)$, $(1, 1)$ and $(-1, 1)$. Find $\text{Cov}(X, Y)$. After you get the answer, try to give another solution without computation.

Exercise 8.50. Consider the points $A = (0, 0)$, $B = (1, 0)$, and $C = (0, 1)$. Now choose a random point (X, Y) uniformly from the union of the line segments AB and AC. Find $\text{Cov}(X, Y)$.

Hint. Look at XY first.

Exercise 8.51. Suppose that X, Y and Z are random variables for which

$$\begin{aligned} &E[X] = 1, \quad E[X^2] = 2, \quad E[XY] = 2, \\ &E[Y] = 3, \quad E[Y^2] = 12, \quad E[XZ] = 4, \\ &E[Z] = 3, \quad E[Z^2] = 12, \quad E[YZ] = 9. \end{aligned}$$

Find $\text{Var}(X + 2Y + Z)$.

Exercise 8.52. A fair coin is flipped 30 times. Let X denote the number of heads among the first 20 coin flips and Y denote the number of heads among the last 20 coin flips. Compute the correlation coefficient of X and Y.

Exercise 8.53. Suppose that X and Y are random variables with

$$E[X] = 1, \quad E[Y] = 2, \quad E[X^2] = 3, \quad E[Y^2] = 13, \quad E[XY] = -1.$$

(a) Compute $\text{Cov}(3X + 2, 2Y - 3)$.
(b) Compute $\text{Corr}(X, Y)$.

Exercise 8.54. Suppose that for the random variables X, Y we have $E[X] = 2, E[Y] = 1$, $E[X^2] = 5$, $E[Y^2] = 10$ and $E[XY] = 1$.

(a) Compute $\text{Corr}(X, Y)$.
(b) Find a number c so that X and $X + cY$ are uncorrelated.

Exercise 8.55. Let I_A be the indicator of the event A. Show that for any A, B we have

$$\text{Corr}(I_A, I_B) = \text{Corr}(I_{A^c}, I_{B^c}).$$

Exercise 8.56. Let X and Y be random variables with positive variances. Furthermore, let a, b, c, d be real numbers with $a, b \neq 0$. Show that

$$\text{Corr}(aX + c, bY + d) = \pm\text{Corr}(X, Y)$$

where the sign on the right is the same as the sign of ab.

Exercise 8.57. Is it possible for two random variables X and Y to satisfy $E[X] = 1$, $E[Y] = 2$, $E[X^2] = 3$, $E[Y^2] = 5$, and $E[XY] = -1$?

Hint. What would be the correlation of X and Y?

Exercise 8.58. Let (X, Y) have bivariate normal distribution with marginals $X \sim \mathcal{N}(\mu_X, \sigma_X^2)$ and $Y \sim \mathcal{N}(\mu_Y, \sigma_Y^2)$ and correlation $\text{Corr}(X, Y) = \rho$. Let $U = 2X + Y$, $V = X - Y$. Determine the joint distribution of (U, V).

Exercise 8.59. Let Z and W be independent standard normal random variables. Assume the following given five parameters: real μ_X and μ_Y, positive σ_X and σ_Y, and $-1 < \rho < 1$. Define random variables

$$\begin{aligned} X &= \sigma_X Z + \mu_X \\ Y &= \sigma_Y \rho Z + \sigma_Y \sqrt{1 - \rho^2}\, W + \mu_Y. \end{aligned}$$

Use the Jacobian technique of Section 6.4 to derive the joint density function of (X, Y).

Exercise 8.60. Show that the random variables $W_1, W_2, \dots, W_{n-1}$ defined in the coupon collector's problem Example 8.17 are mutually independent.

Hint. The goal is to show that for any positive integers $a_1, a_2, \dots, a_{n-1}$ we have

$$P(W_1 = a_1, W_2 = a_2, \dots, W_{n-1} = a_{n-1}) = \prod_{k=1}^{n-1} P(W_k = a_k).$$

Calculate the probability on the left by counting the number of ways in which toys can be chosen. The event requires that new toys appear at times $1, 1 + a_1, 1 + a_1 + a_2, \dots, 1 + a_1 + \dots + a_{n-1}$. So there are n choices for the first toy, 1 choice for each of the next $a_1 - 1$ toys, $n - 1$ choices for the second new toy, 2 choices for each of the next $a_2 - 1$ toys, and so on. On the other hand, the total number of boxes opened is $1 + a_1 + \dots + a_{n-1}$ and each box has one of n possible toys. Once the probability on the left has been found, it can be rearranged to the product on the right.

Exercise 8.61. This exercise fills in the details of the asymptotics for $E(T_n)$ and $\mathrm{Var}(T_n)$ in the coupon collector's problem Example 8.17.

(a) Use the Riemann sum approximation that appears in the proof of Fact D.4 in Appendix D to show that $0 \le \sum_{k=1}^{n} \frac{1}{k} - \ln n \le 1$ for all positive integers n. From this deduce that $\frac{1}{\ln n} \sum_{k=1}^{n} \frac{1}{k} \to 1$ as $n \to \infty$.

(b) Use part (a) and the identity $\sum_{k=1}^{\infty} \frac{1}{k^2} = \frac{\pi^2}{6}$ to prove that

$$\lim_{n\to\infty} \frac{E(T_n)}{n \ln n} = 1 \quad \text{and} \quad \lim_{n\to\infty} \frac{\mathrm{Var}(T_n)}{n^2} = \frac{\pi^2}{6}.$$

Challenging problems

Exercise 8.62. Is it possible to write positive integers on the faces of two six-sided dice (which are not the usual numbers), so that when we roll these dice their sum has the same distribution as the sum of two regular dice?

Hint. Moment generating functions could help.

Exercise 8.63. We shuffle a deck of 52 cards, and we flip the cards one by one until we find the ace of hearts. Find the expected number of cards that were flipped.

Exercise 8.64. We shuffle a deck of 52 cards, and we flip the cards one by one until we find a red ace. Find the expected number of cards that were flipped.

Exercise 8.65. A child who gets into an elite preschool will have a lifetime earning potential given by a Poisson random variable with mean \$3,310,000. If the child does not get into the preschool, her earning potential will be Poisson with mean \$2,700,000. Let $X = 1$ if the child gets into the elite preschool, and zero otherwise, and assume that $P(X = 1) = p$.

(a) Find the covariance of X and the child's lifetime earnings.
(b) Find the correlation of X and the child's lifetime earnings.

Exercise 8.66. Recall the coupon collector's problem from Example 8.17. In this exercise you will derive the probability mass function of the random variable T_n, the number of boxes we need to buy to collect all n toys. Label the toys $1, 2, \dots, n$. Let Z_k denote the number of boxes we need to buy until we find a box with the toy labelled with k. (Note that this is not the same as the random variable T_k from Example 8.17.) Let ℓ be a positive integer.

(a) Find the probability $P(Z_k > \ell)$ for a given $1 \le k \le n$.
(b) Let $1 \le k_1 < \cdots < k_j \le n$ be positive integers. Find the probability $P(Z_{k_i} > \ell \text{ for all } 1 \le i \le j)$.
(c) Use the inclusion-exclusion formula to calculate $P(T_n > \ell)$, and use this to find the probability mass function of T_n.
(d) Use the result of part (c) to find $E[T_n]$.

Exercise 8.67. Let $p_1, \dots, p_n$ be probabilities that sum up to one. Consider the following generalization of the coupon collector's problem from Example 8.17. Each box contains one of n toys, but the probability of containing the ith toy is p_i (independently of the other boxes). Let T_n denote the number of boxes needed to be bought to collect all n toys. Use the outline of Exercise 8.66 to derive the probability mass function of T_n and a formula for $E[T_n]$.

9

Tail bounds and limit theorems

In Chapter 4 we saw various ways to approximate a binomial random variable X. In particular, we saw how to estimate tail probabilities like $P(X > a)$. The methods we discussed there relied heavily on the fact that we knew the probability mass function of X. This chapter introduces tools that can be used when we know very little about the distribution of the random variable in question. In the first section we discuss the Markov and Chebyshev inequalities. Using only knowledge of the mean and the variance of X, these inequalities give estimates on probabilities of the form $P(X > a)$. In Sections 9.2 and 9.3 we return to the law of large numbers and the central limit theorem, and explain why they hold for a wide variety of distributions, not just for binomials.

9.1. Estimating tail probabilities

Situations arise where one would like to estimate the tail probability $P(X > a)$ of a random variable. If the probability distribution is available then, of course, we know the probability $P(X > a)$ precisely. However, it is also possible to get bounds on $P(X > a)$ simply from the expectation and variance of X. In this section we prove two famous inequalities of probability theory that give such estimates.

As preparation, we state a rather obvious fact.

Fact 9.1. Suppose that X and Y are random variables with finite means and that the inequality $X \geq Y$ holds with probability one. Then $E[X] \geq E[Y]$.

Fact 9.1 holds for all random variables. With our tools we can prove it in the cases where X and Y are discrete or jointly continuous. We provide the proof in the discrete case.

Proof of Fact 9.1 in the discrete case. The fact will follow from an even simpler statement: if a random variable Z is nonnegative with probability one, then $E[Z] \geq 0$. We prove this when Z is discrete. The expectation of Z is $\sum_k kP(Z = k)$, and since $P(Z = k) = 0$ for any $k < 0$, the sum includes only nonnegative terms. Thus, $E[Z] \geq 0$.

Returning to Fact 9.1, set $Z = X - Y$, which is nonnegative with probability one. Thus,

$$0 \le E[Z] = E[X - Y] = E[X] - E[Y],$$

and rearranging terms gives the statement. ▲

We are ready for the main result of this section.

Theorem 9.2. (Markov's inequality) *Let X be a nonnegative random variable. Then for any $c > 0$*

$$P(X \ge c) \le \frac{E[X]}{c}. \tag{9.1}$$

Proof. Recall that an indictor random variable $I(A)$ is 1 if the event A holds, and 0 otherwise. Fix $c > 0$ and note that

$$X \ge X \cdot I(X \ge c) \ge c \cdot I(X \ge c).$$

The first inequality above relies on X being nonnegative. Setting $Y = c \cdot I(X \ge c)$, we therefore have $X \ge Y$. Applying Fact 9.1 now gives

$$E[X] \ge E[Y] = cE[I(X \ge c)] = cP(X \ge c),$$

and dividing both sides by the positive constant c gives (9.1). ▲

The value of Markov's inequality lies in the fact that it requires knowledge only about the mean of the random variable. However, we also need to recognize its limitations: it only gives an upper bound on $P(X \ge c)$, and it can be far from the truth.

Example 9.3. Markov's inequality can provide a useful bound, or a useless bound. Suppose X is a Bernoulli random variable with $P(X = 1) = p$ and $P(X = 0) = 1-p$. Then Markov's inequality gives the bound

$$P(X \ge 1) \le \frac{E[X]}{1} = p$$

which agrees with the true value $P(X \ge 1) = p$. On the other hand, Markov's inequality also gives

$$P(X \ge 0.01) \le \frac{E[X]}{0.01} = 100p.$$

The upper bound $100p$ is way off the true value $P(X \ge 0.01) = p$. In fact, $100p$ could even be greater than 1, giving a useless estimate on a probability. ▲

Example 9.4. An ice cream parlor has, on average, 1000 customers per day. What can be said about the probability that it will have at least 1400 customers tomorrow.

Let X be the number of customers tomorrow. Then $E[X] = 1000$. By Markov's inequality

$$P(X \geq 1400) \leq \frac{E(X)}{1400} = \frac{1000}{1400} = \frac{5}{7}.$$

▲

If in addition to the mean we know something about the variance, we can often get a better estimate in the form of Chebyshev's inequality. Its proof is an application of Markov's inequality.

Theorem 9.5. (Chebyshev's inequality) *Let X be a random variable with a finite mean μ and a finite variance σ^2. Then for any $c > 0$ we have*

$$P(|X - \mu| \geq c) \leq \frac{\sigma^2}{c^2}. \tag{9.2}$$

Proof. Since $|X - \mu| \geq 0$ and $c > 0$, the inequality $|X - \mu| \geq c$ is equivalent to the inequality $(X - \mu)^2 \geq c^2$. Now we apply Markov's inequality to the random variable $(X - \mu)^2$:

$$P(|X - \mu| \geq c) = P((X - \mu)^2 \geq c^2) \leq \frac{E[(X - \mu)^2]}{c^2} = \frac{\sigma^2}{c^2},$$

which proves (9.2). ▲

The following inequalities are direct consequences of Chebyshev's inequality:

$$P(X \geq c + \mu) \leq \frac{\sigma^2}{c^2} \quad \text{and} \quad P(X \leq \mu - c) \leq \frac{\sigma^2}{c^2}. \tag{9.3}$$

These inequalities follow from the fact that the event $\{|X - \mu| \geq c\}$ contains both $\{X \geq \mu + c\}$ and $\{X \leq \mu - c\}$.

As in Markov's inequality, the tail bound in Chebyshev's inequality is also just an upper bound, and it can be far from the truth.

Example 9.6. Suppose a nonnegative random variable X has mean 50.

(a) Give an upper bound for the probability $P(X \geq 60)$.
(b) Suppose that we also know $\text{Var}(X) = 25$. How does your bound for $P(X \geq 60)$ change?
(c) Suppose that we also know that X is binomially distributed. Compare the upper bounds from (a) and (b) with the normal approximation and the precise value of $P(X \geq 60)$.

In part (a) we can use Markov's inequality to get

$$P(X \geq 60) \leq \frac{E[X]}{60} = \frac{50}{60} = \frac{5}{6}.$$

In part (b) we can use Chebyshev's inequality to get

$$P(X \geq 60) = P(X - E[X] \geq 10) \leq P(|X - E[X]| \geq 10) \leq \frac{\mathrm{Var}(X)}{10^2} = \frac{25}{100} = \frac{1}{4}.$$

Note that this upper bound is better than the one we got from Markov's inequality.

Finally, if X is binomially distributed, then $np = 50$, $np(1-p) = 25$ which gives $p = 1/2$ and $n = 100$ for the parameters of X. The normal approximation gives

$$P(X \geq 60) = P(\tfrac{X-50}{5} \geq 2) \approx 1 - \Phi(2) \approx 0.0228.$$

Note that the last estimate is an approximation (there is an error because we replaced a scaled binomial with a normal), while the first two were rigorous bounds. However, the last one is the closest to the truth. The precise probability is

$$P(X \geq 60) = \sum_{k=60}^{100} \binom{100}{k} 2^{-100} \approx 0.0284.$$

With the continuity correction the normal approximation would give an even better estimate $P(X \geq 60) \approx 1 - \Phi(1.9) \approx 0.0287$.

The estimates in parts (a) and (b) are far from the accurate result of part (c), but they were obtained with very little knowledge about the distribution of X. ▲

Example 9.7. Continuing Example 9.4, we now suppose that the variance of the number of customers to the ice cream parlor on a given day is 200. What can be said about the probability that there will be between 950 and 1050 customers tomorrow? What can we say about the probability that at least 1400 customers arrive?

We want

$$\begin{aligned} P(950 < X < 1050) &= P(-50 < X - 1000 < 50) \\ &= P(|X - 1000| < 50) \\ &= 1 - P(|X - 1000| \geq 50). \end{aligned}$$

By Chebyshev's inequality (Theorem 9.5), we have

$$P(|X - 1000| \geq 50) \leq \frac{200}{50^2} = \frac{2}{25} = 0.08.$$

Thus,

$$P(950 < X < 1050) \geq 1 - 0.08 = 0.92.$$

Now to bound the probability that at least 1400 customers arrive. We have by Chebyshev's inequality

$$P(X \geq 1400) = P(X - 1000 \geq 400) \leq \frac{200}{400^2} = \frac{1}{800}.$$

Note how much sharper this result is than the conclusion of Example 9.4. ▲

The next example shows that symmetry improves the bound from Chebyshev's inequality.

Example 9.8. Suppose the random variable X is discrete and symmetric, meaning that $P(X = a) = P(X = -a)$ for any a. Estimate $P(X \geq 5)$ if $\text{Var}(X) = 3$.

Since X is symmetric, its expectation is zero: in the sum $\sum_k kP(X = k)$ the positive and negative terms cancel out exactly. (To make this fully precise, we would need the finiteness of $\sum_k |k|P(X = k)$, which follows from the existence of the variance.) We can use Chebyshev's inequality to get

$$P(X \geq 5) = P(X - E[X] \geq 5) \leq P(|X - E[X]| \geq 5) \leq \frac{\text{Var}(X)}{5^2} = \frac{3}{25}.$$

This is certainly an upper bound, but we can actually do a little bit better. Because of the symmetry of X, we have $P(X \geq 5) = P(X \leq -5)$ and thus

$$P(X \geq 5) = \tfrac{1}{2}(P(X \geq 5) + P(X \leq -5)) = \tfrac{1}{2}P(|X| \geq 5).$$

Chebyshev's inequality gives $P(|X| \geq 5) \leq \frac{3}{25}$, and so we obtain the sharper bound $P(X \geq 5) \leq \frac{3}{50}$. ▲

9.2. Law of large numbers

In Section 4.2 we considered i.i.d. Bernoulli random variables $X_1, X_2, X_3, \ldots$ with success probability $0 < p < 1$. We let $S_n = X_1 + \cdots + X_n$ and showed that $\frac{S_n}{n}$ converges to p in the following sense:

$$\lim_{n\to\infty} P\left(\left|\tfrac{S_n}{n} - p\right| < \varepsilon\right) = 1 \qquad \text{for any fixed } \varepsilon > 0.$$

This was the weak law of large numbers for binomial random variables. It expresses the fact that the average of i.i.d. Bernoulli random variables converges to the mean, which is the success probability. In this section we show that this "law of averages" holds much more generally.

Theorem 9.9. (Law of large numbers with finite variance) *Suppose that we have i.i.d. random variables $X_1, X_2, X_3, \ldots$ with finite mean $E[X_1] = \mu$ and finite variance $\text{Var}(X_1) = \sigma^2$. Let $S_n = X_1 + \cdots + X_n$. Then for any fixed $\varepsilon > 0$ we have*

$$\lim_{n\to\infty} P\left(\left|\tfrac{S_n}{n} - \mu\right| < \varepsilon\right) = 1. \tag{9.4}$$

The proof of this theorem is elegant and short, and uses several tools introduced in recent chapters.

Proof. Fix the value of $\varepsilon > 0$. We use Chebyshev's inequality to show that the complementary probability $P\left(\left|\frac{S_n}{n} - \mu\right| \geq \varepsilon\right)$ converges to zero. Recall from Fact 8.14 that the expectation and variance of the sample mean $\frac{S_n}{n} = \frac{X_1+\cdots+X_n}{n}$ are

$$E\left(\tfrac{S_n}{n}\right) = \mu \qquad \text{and} \qquad \text{Var}\left(\tfrac{S_n}{n}\right) = \tfrac{\sigma^2}{n}.$$

Then Chebyshev's inequality (9.2) gives

$$P\left(\left|\tfrac{S_n}{n}-\mu\right|\ge\varepsilon\right)\le\frac{1}{\varepsilon^2}\operatorname{Var}\left(\frac{S_n}{n}\right)=\frac{\sigma^2}{n\varepsilon^2}. \tag{9.5}$$

Since $\varepsilon > 0$ is fixed, we have $\frac{\sigma^2}{n\varepsilon^2} \to 0$ as $n \to \infty$. The probability on the left of (9.5) is nonnegative and bounded above by a sequence converging to 0, so it also has to converge to 0. This proves (9.4). ▲

The theorem actually holds without the assumption of finite variance as long as the expectation $E(X_1) = \mu$ is finite. The law of large numbers that we have just proved is the *weak law of large numbers*. The *strong* version of the LLN states that $P(\lim_{n\to\infty} \frac{S_n}{n} = \mu) = 1$, meaning that the sequence of numbers $\frac{S_n}{n}$ converges to μ with probability one. We prove a version of the strong law of large numbers in Section 9.5 below.

We return to our study of the sample mean as described in Fact 8.14.

Example 9.10. Suppose we want to estimate the mean of a random variable X from a finite number of independent samples from a population, using the sample mean. For example, perhaps we are measuring average IQ score, average income, etc. Suppose also that we know an upper bound on the variance of X: $\operatorname{Var}(X) \le \hat{\sigma}^2$. (This happens for example when $|X|$ is bounded by $\hat{\sigma}$.) Show that for a large enough sample the sample mean is within 0.05 of the correct value with a probability larger than 0.99. How many samples do we need to take to be at least 99% sure that our estimated value is within 0.05 of the correct value?

The first statement is a simple consequence of the weak law of large numbers. Choosing $\varepsilon = 0.05$ we see that the probability of the sample mean being closer than ε to the actual mean converges to 1 as the sample size goes to infinity. Of course this implies that eventually this probability has to be larger than 0.99, which is what we wanted to show.

To answer the final question we introduce some notation. Let X_i denote the ith sample from the population and suppose that $E[X_i] = \mu$ and $\operatorname{Var}(X_i) = \sigma^2$. Note that both μ and σ^2 are unknown, though $\sigma^2 \le \hat{\sigma}^2$. Let

$$\bar{X}_n = \frac{X_1 + \cdots + X_n}{n}$$

be the sample mean. We know from Fact 8.14 that

$$E[\bar{X}_n] = \mu \qquad \text{and} \qquad \operatorname{Var}(\bar{X}_n) = \frac{\sigma^2}{n}.$$

By Chebyshev's inequality (and the proof of Theorem 9.9) we have

$$P(|\bar{X}_n - \mu| \ge \varepsilon) \le \frac{\sigma^2}{\varepsilon^2 n}.$$

We want to find an n for which

$$P(|\bar{X}_n - \mu| < 0.05) > 0.99$$

or, equivalently, for which $P(|\bar{X}_n - \mu| \geq 0.05) \leq 0.01$. Thus, it is sufficient to find an n so that

$$\frac{\sigma^2}{(0.05)^2 n} \leq 0.01 \quad \text{or} \quad n \geq \frac{\sigma^2}{(0.05)^2(0.01)} = 40{,}000\sigma^2.$$

Since $\sigma^2 \leq \hat{\sigma}^2$ by our assumption, we can choose $n = 40{,}000 \times \hat{\sigma}^2$. For example, if $\hat{\sigma}^2 = 2$, then n can be taken to be 80,000. ▲

9.3. Central limit theorem

Just as the law of large numbers holds for general i.i.d. random variables and not only for binomial random variables, so does the central limit theorem.

Theorem 9.11. (Central limit theorem) *Suppose that we have i.i.d. random variables* $X_1, X_2, X_3, \ldots$ *with finite mean* $E[X_1] = \mu$ *and finite variance* $\mathrm{Var}(X_1) = \sigma^2$. *Let* $S_n = X_1 + \cdots + X_n$. *Then for any fixed* $-\infty \leq a \leq b \leq \infty$ *we have*

$$\lim_{n\to\infty} P\left(a \leq \frac{S_n - n\mu}{\sigma\sqrt{n}} \leq b\right) = \Phi(b) - \Phi(a) = \int_a^b \frac{1}{\sqrt{2\pi}} e^{-\frac{y^2}{2}}\,dy. \tag{9.6}$$

The statement of the theorem is exactly the same as the central limit theorem learned in the binomial case, Theorem 4.1. Take a sum S_n of n i.i.d. random variables, subtract the mean $n\mu$, and divide by the standard deviation $\sqrt{n\sigma^2}$. Then for large n the standardized sum $\frac{S_n - n\mu}{\sigma\sqrt{n}}$ is approximately a standard normal random variable. As in the binomial case, we can use the theorem to give approximate probabilities.

Example 9.12. We roll 1000 dice and add up the values. Estimate the probability that the sum is at least 3600.

Let $X_1, X_2, \ldots, X_{1000}$ denote the outcomes of the rolls and $S = \sum_{i=1}^{1000} X_i$. The random variables X_i are i.i.d. with $E[X_1] = \frac{7}{2}$ and $\mathrm{Var}(X_1) = \frac{35}{12}$. Thus, $E[S] = 3500$ and $\mathrm{Var}(S) = \frac{35000}{12}$, and so

$$\begin{aligned} P(S \geq 3600) &= P\left(\frac{S - E[S]}{\sqrt{\mathrm{Var}\, S}} \geq \frac{3600 - 3500}{\sqrt{35000/12}}\right) = P\left(\frac{S - E[S]}{\sqrt{\mathrm{Var}\, S}} \geq 1.852\right) \\ &\approx 1 - \Phi(1.852) \approx 0.03. \end{aligned}$$

▲

Example 9.13. A new diner specializing in waffles opens on our street. It will be open 24 hours a day, seven days a week. It is assumed that the inter-arrival times between customers will be i.i.d. exponential random variables with mean 10 minutes. Approximate the probability that the 120th customer will arrive after the first 21 hours of operation.

We could solve this problem by recognizing that the arrival time of the 120th customer is a gamma random variable with parameters $n = 120$ and $\lambda = 6$ (per hour). However, we can also approximate the probability using the central limit theorem.

Recall that if $X \sim \text{Exp}(\lambda)$ then $E[X] = 1/\lambda$ and $\text{Var}(X) = 1/\lambda^2$. Hence if X_k is the wait (in hours) between the $(k-1)$st and kth customer, then $\mu = E[X_k] = 1/6$ and $\sigma^2 = \text{Var}(X_k) = 1/36$. The total wait until the 120th customer is $S_{120} = \sum_{k=1}^{120} X_k$. With $n = 120$, $n\mu = 20$, and $\sigma\sqrt{n} = \frac{\sqrt{120}}{6}$, the normal approximation goes as follows:

$$
\begin{aligned}
P(S_{120} > 21) &= P\left(\tfrac{S_{120}-20}{\sqrt{120}/6} > \tfrac{21-20}{\sqrt{120}/6}\right) \\
&\approx P(Z > 0.55) = 1 - \Phi(0.55) \approx 1 - 0.71 = 0.29.
\end{aligned}
$$

Letting $Z \sim \text{Gamma}(120, 6)$, the actual probability can be evaluated by a computer as

$$
P(Z > 21) = \int_{21}^{\infty} \frac{6^{120} \cdot x^{119}}{119!} e^{-6x} dx \approx 0.285,
$$

which demonstrates the quality of the approximation. ▲

Example 9.14. (Normal approximation of a Poisson) We know that the sum of independent Poisson random variables is Poisson, with a parameter equal to the sum of the parameters. Thus if $X_n \sim \text{Poisson}(n)$ for an integer n, then X_n has the same distribution as a sum of n i.i.d. Poisson(1) random variables. In other words, $X_n \overset{d}{=} Y_1 + \cdots + Y_n$, with $Y_i \sim \text{Poisson}(1)$. Since both the mean and the variance of Y_1 are equal to 1, by the central limit theorem we get

$$
P\left(\tfrac{X_n - n}{\sqrt{n}} \le a\right) \approx \Phi(a)
$$

if n is large. This means that a standardized Poisson random variable with a large parameter is close to a standard normal. ▲

Proof of the central limit theorem ♦

We give a partial proof of the CLT by showing that the moment generating function of the standardized sum $\frac{S_n - n\mu}{\sigma\sqrt{n}}$ converges to the moment generating function of the standard normal distribution. The convergence of moment generating functions implies the convergence of probabilities. This is the content of the next theorem, which we state without proof. The theorem extends powerfully Fact 5.14 of Section 5.1 where we learned that the moment generating function determines the probability distribution of a random variable. The theorem below states that the moment generating function can also be used to verify distributional limits of random variables.

Theorem 9.15. (Continuity theorem for moment generating functions) *Suppose the random variable X has a continuous cumulative distribution function, and that its moment generating function $M_X(t)$ is finite in an interval $(-\varepsilon, \varepsilon)$ for some $\varepsilon > 0$. Assume further that the moment generating functions of the random variables $Y_1, Y_2, Y_3, \dots$ satisfy*

$$\lim_{n\to\infty} M_{Y_n}(t) = M_X(t)$$

for all t in the interval $(-\varepsilon, \varepsilon)$. Then for any $a \in \mathbb{R}$

$$\lim_{n\to\infty} P(Y_n \le a) = P(X \le a).$$

Specializing Theorem 9.15 to a standard normal X gives the following corollary.

Theorem 9.16. *Assume that the moment generating functions of the random variables $Y_1, Y_2, Y_3, \dots$ satisfy*

$$\lim_{n\to\infty} M_{Y_n}(t) = e^{\frac{t^2}{2}} \tag{9.7}$$

for all t in an interval $(-\varepsilon, \varepsilon)$ for some $\varepsilon > 0$. Then for any $a \in \mathbb{R}$

$$\lim_{n\to\infty} P(Y_n \le a) = \Phi(a) = \int_{-\infty}^{a} \frac{1}{\sqrt{2\pi}} e^{-\frac{y^2}{2}}\, dy.$$

We give a rough sketch of the proof of the CLT for the case where our random variables have finite moment generating function. The reader should appreciate how easy this is, compared to potentially trying to show that the distribution function of $\frac{S_n - n\mu}{\sigma\sqrt{n}}$ converges to Φ.

Sketch of the proof of Theorem 9.11. We assume that $M_{X_i}(t)$ is finite and take $Y_n = \frac{S_n - n\mu}{\sigma\sqrt{n}}$. We prove limit (9.7) from which the conclusion (9.6) follows.

From the definition of the moment generating function and the fact that the random variables $X_1, \dots, X_n$ are independent,

$$\begin{aligned} M_{Y_n}(t) &= E\left[e^{tY_n}\right] = E\left[e^{t\frac{S_n - n\mu}{\sigma\sqrt{n}}}\right] = E\left[\exp\left(\frac{t}{\sigma\sqrt{n}} \sum_{k=1}^{n} (X_k - \mu)\right)\right] \\ &= E\left[\prod_{k=1}^{n} e^{\frac{t}{\sigma\sqrt{n}}(X_k - \mu)}\right] = \prod_{k=1}^{n} E\left[e^{\frac{t}{\sigma\sqrt{n}}(X_k - \mu)}\right]. \end{aligned} \tag{9.8}$$

Inside each expectation we insert the second order Taylor approximation $e^x \approx 1 + x + x^2/2$ which is valid near 0 and then use additivity of expectation:

$$E\left[e^{\frac{t}{\sigma\sqrt{n}}(X_k-\mu)}\right] \approx E\left[1 + \frac{t}{\sigma\sqrt{n}}(X_k - \mu) + \frac{t^2}{2\sigma^2 n}(X_k - \mu)^2\right]$$
$$= 1 + \frac{t}{\sigma\sqrt{n}}E[X_k - \mu] + \frac{t^2}{2\sigma^2 n}E\left[(X_k - \mu)^2\right]$$
$$= 1 + \frac{t}{\sigma\sqrt{n}} \cdot 0 + \frac{t^2}{2\sigma^2 n} \cdot \sigma^2$$
$$= 1 + \frac{t^2}{2n}.$$

Put this back above in (9.8) to get

$$M_{Y_n}(t) \approx \left(1 + \frac{t^2}{2n}\right)^n \longrightarrow e^{t^2/2} \qquad \text{as } n \to \infty.$$

We have verified (9.7).

Making this proof watertight requires error bounds for the $\approx$ steps. Note that as $n \to \infty$ the Taylor approximation gains accuracy because the quantity $\frac{t}{\sigma\sqrt{n}}(X_k-\mu)$ is converging to zero. ▲

9.4. Monte Carlo method ♦

The Monte Carlo method is a technique for numerical approximation of a quantity that is hard or impossible to compute. It involves the repeated sampling of a random variable. The basic idea is the following. Let μ be the quantity to be approximated. Assume that we can generate a random variable X so that the expected value of X is exactly equal to μ. If we now generate a sequence of i.i.d. random variables $X_1, X_2, X_3, \ldots$ with the same distribution as X then the law of large numbers tells us that the average

$$\bar{X}_n = \frac{X_1 + \cdots + X_n}{n} \tag{9.9}$$

approaches μ as $n \to \infty$.

In practice we cannot evaluate the limit, so we take $\bar{X}_n$ for large n as the estimate of μ. To control the error $\bar{X}_n - \mu$ we might want to follow Section 4.3 or 9.3 and use the central limit theorem to provide a confidence interval. However, the CLT requires knowledge of the variance $\sigma^2 = \mathrm{Var}(X)$, and if $\mu = E[X]$ is hard to compute, it is unlikely that σ^2 is available to us. In Section 4.3 we got around this problem by bounding the variance, but this technique was special to Bernoulli random variables. In the general case what we have available is the sample variance from Example 8.16:

$$s_n^2 = \frac{1}{n-1}\sum_{i=1}^{n}(X_i - \bar{X}_n)^2. \tag{9.10}$$

Exercise 9.33 will ask you to prove that s_n^2 converges to σ^2 with probability one. The theorem below shows that we can replace σ^2 with s_n^2 in the central limit theorem.

Theorem 9.17. *Let $X_1, X_2, X_3, \ldots$ be i.i.d. random variables with finite mean $E[X_1] = \mu$ and finite variance $\mathrm{Var}(X_1) = \sigma^2$. Let $\bar{X}_n$ and s_n^2 be the sample mean and variance defined in (9.9) and (9.10). Then for any fixed $-\infty \le a \le b \le \infty$ we have*

$$\lim_{n\to\infty} P\left(a \le \frac{(\bar{X}_n - \mu)\sqrt{n}}{s_n} \le b \right) = \Phi(b) - \Phi(a) = \int_a^b \frac{1}{\sqrt{2\pi}} e^{-\frac{y^2}{2}}\, dy. \tag{9.11}$$

Note that $(\bar{X}_n - \mu)\sqrt{n} = \frac{(X_1+\cdots+X_n)-n\mu}{\sqrt{n}}$. Thus Theorem 9.17 can be viewed as a version of Theorem 9.11, with σ replaced by s_n. The proof of Theorem 9.17 is beyond the scope of this text and relies on Slutsky's theorem.

Confidence intervals

We now use Theorem 9.17 to find a confidence interval for the approximation $\bar{X}_n \approx \mu$. For a given $\varepsilon > 0$ and large n we get

$$\begin{aligned} P(\bar{X}_n - \varepsilon \le \mu \le \bar{X}_n + \varepsilon) &= P(\mu - \varepsilon \le \bar{X}_n \le \mu + \varepsilon) \\ &= P\left(-\frac{\varepsilon\sqrt{n}}{s_n} \le \frac{(\bar{X}_n - \mu)\sqrt{n}}{s_n} \le \frac{\varepsilon\sqrt{n}}{s_n} \right) \approx 2\Phi\left(\frac{\varepsilon\sqrt{n}}{s_n}\right) - 1. \end{aligned}$$

To find the $1-\alpha$ confidence interval, let z_α be the number for which

$$2\Phi(z_\alpha) - 1 = 1 - \alpha$$

and set $\varepsilon = \frac{z_\alpha s_n}{\sqrt{n}}$. Then inside Φ we have $\frac{\varepsilon\sqrt{n}}{s_n} = z_\alpha$, and so our $1-\alpha$ confidence interval is $[\bar{X}_n - \varepsilon, \bar{X}_n + \varepsilon]$, sometimes denoted as $\bar{X}_n \pm \varepsilon$. A common choice of α is 0.05, which gives a 95% confidence interval and $z_\alpha = 1.96$.

Monte Carlo is presently a critical computational method for science and industry. The next two examples highlight one specific application, the computation of integrals.

Example 9.18. (Estimation of π with Monte Carlo) The integral $\int_0^1 \sqrt{1-x^2}\, dx$ gives 1/4 of the area of the unit circle, and hence

$$\pi = \int_0^1 4\sqrt{1-x^2}\, dx. \tag{9.12}$$

Moreover, $\int_0^1 4\sqrt{1-x^2}\, dx = E\left[4\sqrt{1-U^2}\right]$, where $U \sim \mathrm{Unif}[0,1]$. This suggests the following Monte Carlo estimator. Let $U_1, U_2, \ldots, U_n$ be independent Unif[0, 1] random variables, let $X_i = 4\sqrt{1-U_i^2}$, and set

$$\bar{X}_n = \frac{1}{n}\sum_{i=1}^n X_i.$$

Performing this experiment with $n = 10^6$ yields

$$\bar{X}_n = 3.14091 \quad \text{and} \quad \frac{1.96 \cdot s_n}{\sqrt{n}} = 0.00175.$$

Therefore our 95% confidence interval is 3.1409 ± 0.0018. ▲

The example above is easily generalized. Let $-\infty \le a < b \le \infty$ and suppose we wish to find $\int_a^b g(x)dx$. Let f be a nonnegative function for which (i) $\int_a^b f(x)dx = 1$ and (ii) $f(x) = 0$ implies $g(x) = 0$. Then,

$$\int_a^b g(x)\,dx = \int_a^b \frac{g(x)}{f(x)} f(x)\,dx = E\left[\frac{g(X)}{f(X)}\right],$$

where X is a random variable with density function f. Thus, if we can generate i.i.d. copies of X, the Monte Carlo method may be used to estimate the integral.

Example 9.19. (Monte Carlo integration in one dimension) Suppose that we wish to estimate the integral

$$I = \int_0^\infty e^{\sin(\ln(x)) - x^2}\,dx.$$

Letting $g(x) = e^{-x}$ we have that

$$I = \int_0^\infty e^{\sin(\ln(x)) - x^2} e^x e^{-x}\,dx = E\left[e^{\sin(\ln(X)) - X^2 + X}\right],$$

where $X \sim \text{Exp}(1)$. Then let X_i be independent Exp(1) random variables and perform Monte Carlo with the independent random variables $Y_i = e^{\sin(\ln(X_i)) - X_i^2 + X_i}$. Taking $n = 10^6$ samples leads to an estimate with

$$\bar{Y}_n = \frac{1}{n}\sum_{i=1}^n Y_i = 0.65464 \quad \text{and} \quad \frac{1.96 \cdot s_n}{\sqrt{n}} = 0.00078.$$

Thus, our 95% confidence interval is 0.65464 ± 0.00078. The actual value of I is approximately 0.65405, obtained with numerical integration by a computer. ▲

We have seen two examples of computing one-dimensional integrals with Monte Carlo. In low dimensions there are more efficient methods than Monte Carlo for integration. However, as the dimension increases, numerical integration methods become inefficient, and Monte Carlo becomes a preferred choice.

9.5. Finer points ♣

Generalizations of Markov's inequality

Let $f(x)$ be a strictly increasing positive function and X a random variable. Then the event $\{X \ge a\}$ is the same as $\{f(X) \ge f(a)\}$. If $E[f(X)]$ is finite, we can use Markov's inequality to obtain

$$P(X \ge a) = P(f(X) \ge f(a)) \le \frac{E[f(X)]}{f(a)}.$$

A commonly used special case is $f(x) = e^{\theta x}$ for $\theta > 0$. This yields a bound in terms of the moment generating function:

$$P(X \ge a) \le e^{-\theta a} E\left[e^{\theta X}\right] \qquad \text{for any } \theta > 0. \tag{9.13}$$

The strong law of large numbers

Theorem 9.9, the weak law of large numbers, guarantees that for large n the random variable $(X_1 + \cdots + X_n)/n$ is close to μ with high probability. The strong law of large numbers, which guarantees $S_n/n \to \mu$ with a probability of one, is stated now.

Theorem 9.20. (Strong law of large numbers) *Suppose that we have i.i.d. random variables $X_1, X_2, \ldots$, with finite mean $E[X_1] = \mu$. Let $S_n = X_1 + \cdots + X_n$. Then*

$$P\left(\lim_{n\to\infty} \frac{X_1 + \cdots + X_n}{n} = \mu\right) = 1.$$

Proof. We give a proof under the additional assumption that $E[X_1^4]$ is finite. Most examples in this book, and found in practice, satisfy this assumption. By the remark on page 125, the existence of the finite fourth moment implies that the second and third moments of X_1 are finite.

We first suppose that $\mu = 0$. The general case of $\mu \neq 0$ will follow easily at the end of the proof.

We show that the random variable $\sum_{n=1}^{\infty} \frac{S_n^4}{n^4}$ has a finite expectation. This is sufficient due to the following.

(i) If $E\left[\sum_{n=1}^{\infty} \frac{S_n^4}{n^4}\right]$ is finite, then the positive random variable $\sum_{n=1}^{\infty} \frac{S_n^4}{n^4}$ must be finite with probability one (otherwise the expectation would be infinite).

(ii) If a series $\sum_{n=1}^{\infty} a_n$ converges, then $a_n \to 0$. Thus, by (*i*) we have $S_n^4/n^4 \to 0$ with probability one. This implies that $S_n/n \to 0$ with probability one, which was our desired result.

We turn to bounding $E[S_n^4]$. Expanding S_n^4 gives terms of the form

$$X_i^4, \quad X_i^2 X_j^2, \quad X_i^3 X_j, \quad X_i^2 X_j X_k, \quad X_i X_k X_j X_\ell, \qquad \text{for distinct } i, j, k, \ell \in \{1, \ldots, n\}.$$

Because the random variables $X_1, \ldots, X_n$ are i.i.d. and have mean zero, only the terms X_i^4 and $X_i^2 X_j^2$ have nonzero expectation. Denote these by $E[X_i^4] = E[X_1^4] = c_1$ and $E[X_i^2 X_j^2] = (E[X_1^2])^2 = c_2$. Hence, after collecting terms

$$\begin{aligned} E[S_n^4] &= E[(X_1 + \cdots + X_n)^4] = nE[X_1^4] + 6\binom{n}{2}(E[X_1^2])^2 \\ &= nc_1 + 3n(n-1)c_2 \le nc_1 + 3n^2 c_2. \end{aligned}$$

Switching the expectation and the infinite sum (see the monotone convergence theorem in Rudin [Rud87] for justification) yields

$$E\left[\sum_{n=1}^{\infty} \frac{S_n^4}{n^4}\right] = \sum_{i=1}^{\infty} E\left[\frac{S_n^4}{n^4}\right] \le \sum_{n=1}^{\infty} \frac{c_1}{n^3} + \sum_{n=1}^{\infty} \frac{3c_2}{n^2} < \infty,$$

the desired result.

We had assumed that $\mu = 0$. We can drop this assumption now by applying the previous argument to the mean zero random variables $X_i - \mu$. This yields

$$\lim_{n\to\infty} \sum_{i=1}^{n} \frac{(X_i - \mu)}{n} = 0 \quad \text{or equivalently} \quad \lim_{n\to\infty} \sum_{i=1}^{n} \frac{X_i}{n} = \mu,$$

with probability one. ▲

Proof of the central limit theorem

The proof sketch for Theorem 9.11 presented in Section 9.3 assumed that the moment generating function is finite at least in an interval around the origin. In fact the theorem requires only that the mean and variance are finite. The proof in the general case is very similar to the one presented. The difference is that instead of the moment generating function one uses the characteristic function introduced in Section 5.3. As we discussed there, the characteristic function shares most of the useful properties of the moment generating function and has the additional advantage of being always finite.

Error bound in the central limit theorem

In Section 4.7 we added precision to the normal approximation of the binomial with the explicit error bound of Theorem 4.38. The following theorem is the analogous error estimate for the general case. An early version of it was proved by Berry in 1941 and Esseen in 1942.

Theorem 9.21. *Suppose that $X_1, X_2, X_3, \ldots$ are i.i.d. random variables with $E[X_1] = \mu$ and $\text{Var}(X_1) = \sigma^2$. Let $S_n = X_1 + \cdots + X_n$. Then for any $x \in \mathbb{R}$ and positive integer n we have*

$$\left| P\left(\frac{S_n - n\mu}{\sqrt{n\sigma^2}} \le x\right) - \Phi(x) \right| \le \frac{3E[|X_1 - \mu|^3]}{\sigma^3\sqrt{n}}.$$

Note that the statement is meaningful only if $E[|X_1 - \mu|^3] < \infty$. In that case it shows that the speed of convergence in the central limit theorem is at least $\frac{1}{\sqrt{n}}$. Exercise 9.32 asks you to derive Theorem 4.38 from Theorem 9.21.

Exercises

We start with some warm-up exercises arranged by section.

Section 9.1

Exercise 9.1. Let Y be a geometric random variable with parameter $p = 1/6$.

(a) Use Markov's inequality to find an upper bound for $P(Y \ge 16)$.

(b) Use Chebyshev's inequality to find an upper bound for $P(Y \geq 16)$.
(c) Explicitly compute the probability $P(Y \geq 16)$ and compare with the upper bounds you derived.

Exercise 9.2. Let X be an exponential random variable with parameter $\lambda = \frac{1}{2}$.

(a) Use Markov's inequality to find an upper bound for $P(X > 6)$.
(b) Use Chebyshev's inequality to find an upper bound for $P(X > 6)$.
(c) Explicitly compute the probability above and compare with the upper bounds you derived.

Section 9.2

Exercise 9.3. The unit price of a certain commodity evolves randomly from day to day with a general downward drift but with an occasional upward jump when some unforeseen event excites the markets. Long term records suggest that, independently of the past, the daily price increases by a dollar with probability 0.45, declines by 2 dollars with probability 0.5, but jumps up by 10 dollars with probability 0.05. Let C_0 be the price today and C_n the price n days into the future. How does the probability $P(C_n > C_0)$ behave as $n \to \infty$?

Exercise 9.4. The European style roulette wheel has the following probabilities: a red number appears with probability $\frac{18}{37}$, a black number appears with probability $\frac{18}{37}$, and a green number appears with probability $\frac{1}{37}$. Ben bets exactly \$1 on black each round. Explain why this is not a good long term strategy.

Section 9.3

Exercise 9.5. Suppose that X is a nonnegative random variable with $E[X] = 10$.

(a) Give an upper bound on the probability that X is larger than 15.
(b) Suppose that we also know that $\operatorname{Var}(X) = 3$. Give a better upper bound on $P(X > 15)$ than in part (a).
(c) Suppose that $Y_1, Y_2, \ldots, Y_{300}$ are i.i.d. random variables with the same distribution as X so that, in particular, $E(Y_i) = 10$ and $\operatorname{Var}(Y_i) = 3$. Estimate the probability that $\sum_{i=1}^{300} Y_i$ is larger than 3030.

Exercise 9.6. Nate is a competitive eater specializing in eating hot dogs. From past experience we know that it takes him on average 15 seconds to consume one hot dog, with a standard deviation of 4 seconds. In this year's hot dog eating contest he hopes to consume 64 hot dogs in just 15 minutes. Use the CLT to approximate the probability that he achieves this feat of skill.

Exercise 9.7. A car insurance company has 2500 policy holders. The expected claim paid to a policy holder during a year is \$1000 with a standard deviation of \$900. What premium should the company charge each policy holder to assure that with probability 0.999, the premium income will cover the cost of the claims? Compute the answer both with Chebyshev's inequality and with the CLT.

Section 9.4

Exercise 9.8. In Example 9.18 we estimated π by approximating the integral $\int_0^1 4\sqrt{1-x^2}\,dx$ via Monte Carlo. Here we estimate π with Monte Carlo via the following double integral

$$\pi = \int_0^1 \int_0^1 4 \cdot I(x^2 + y^2 \leq 1)\,dx\,dy.$$

For this exercise, you need access to a computer software package that can generate independent samples from the Unif[0, 1] distribution.

(a) Explain why the double integral is equal to π.
(b) Rewrite the double integral as an expectation of the form $E[g(U_1, U_2)]$, where U_1, U_2 are independent Unif[0, 1] random variables, and g is an appropriately chosen function.
(c) Using g from part (b) and $n = 10^6$ independent samples of the random variable $X = g(U_1, U_2)$, produce a 95% confidence interval for π.

Further exercises

Exercise 9.9. Let X_i be the amount of money earned by a food truck on State Street on day i. From past experience, the owner of the cart knows that $E[X_i] = \$5000$.

(a) Give the best possible upper bound for the probability that the cart will earn at least \$7000 tomorrow.
(b) Answer part (a) again with the extra knowledge that $\mathrm{Var}(X_i) = \$4500$.
(c) Continue to assume that for all i we have $E[X_i] = 5000$ and $\mathrm{Var}(X_i) = 4500$. Assuming that the amount earned on any given day is independent of the earning on other days, how many days does the cart have to be on State Street to ensure, with a probability at least 0.95, that the cart's average earnings would be between \$4950 and \$5050. (Do not use the central limit theorem for this problem.)

Exercise 9.10. Answer Exercise 9.9(c) again, but now under the assumption that there is a dependence between the earnings on consecutive days. Specifically, suppose that $\mathrm{Corr}(X_i, X_{i+1}) = 0.5$ and $\mathrm{Corr}(X_i, X_j) = 0$ for $|i-j| \geq 2$.

Hint. Start by computing $\mathrm{Var}(X_1 + \cdots + X_n)$.

Exercise 9.11. Suppose the random variable X is positive and has moment generating function $M_X(t) = (1-2t)^{-3/2}$ for $t < \frac{1}{2}$ and $M_X(t) = \infty$ for $t \geq \frac{1}{2}$.

(a) Use Markov's inequality to bound $P(X > 8)$.
(b) Use Chebyshev's inequality to bound $P(X > 8)$.

Exercise 9.12. Let X be a Poisson(2) random variable, and Y an independent Unif(0, 1) random variable.

(a) Use Markov's inequality to bound $P(X + Y > 10)$.
(b) Use Chebyshev's inequality to bound $P(X + Y > 10)$.

Exercise 9.13. Let $X \sim \text{Bin}(10, 1/3)$ and $Y \sim \text{Exp}(3)$. Assume that these are independent. Use Chebyshev's inequality to bound $P(X - Y < -1)$.

Exercise 9.14. Chebyshev's inequality does not always give a better estimate than Markov's inequality. Let X be a positive random variable with $E[X] = 2$ and $\text{Var}(X) = 9$. Find the values of $t > 2$ where Markov's inequality gives a better bound for $P(X > t)$ than Chebyshev's inequality.

Exercise 9.15. Omar and Cheryl each run a food truck on a university campus. Suppose that the number of customers that come to Omar's truck on a given day has a Poisson(220) distribution, whereas the number of customers that come to Cheryl's truck on a given day has a Poisson(210) distribution. What can be said about the probability that more customers will come to Omar's truck than Cheryl's truck in the long run? Provide random variables for the number of customers arriving to each food truck each day and give a clear mathematical statement. Be precise about the assumptions you are making.

Exercise 9.16. Every morning I take either bus number 5 or bus number 8 to work. Every morning the waiting time for the number 5 is exponential with mean 10 minutes, while the waiting time for the number 8 is exponential with mean 20 minutes. Assume all waiting times are independent of each other. Let S_n be the total amount of bus-waiting (in minutes) that I have done during n mornings, and let T_n be the number of times I have taken the number 5 bus during n mornings.

(a) Find the limit $\lim_{n\to\infty} P(S_n \le 7n)$.
(b) Find the limit $\lim_{n\to\infty} P(T_n \ge 0.6n)$.

Hint. Recall Examples 6.33 and 6.34.

Exercise 9.17. Let X be a Poisson random variable with a parameter of 100.

(a) Use Markov's inequality to find a bound on $P(X > 120)$.
(b) Use Chebyshev's inequality to find a bound on $P(X > 120)$.
(c) Using the fact that for $X_i \sim \text{Poisson}(1)$, independent random variables, $X_1 + X_2 + \cdots + X_n \sim \text{Poisson}(n)$, use the central limit theorem to approximate the value of $P(X > 120)$.

Exercise 9.18. Suppose that $X \sim \text{NegBin}(100, \frac{1}{3})$.

(a) Give an upper bound for $P(X > 500)$ using Markov's inequality.
Hint. Example 8.13 could be helpful.

(b) Give an upper bound for $P(X > 500)$ using Chebyshev's inequality.
(c) Recall that X can be written as the sum of 100 i.i.d. random variables with Geom($\frac{1}{3}$) distribution. Use this representation and the CLT to estimate $P(X > 500)$.
(d) Recall that X can be represented as the number of trials needed to get the 100th success in a sequence of i.i.d. trials of success probability 1/3. Estimate $P(X > 500)$ by looking at the distribution of the number of successes within the first 500 trials.

Exercise 9.19. A four year old is going to spin around with his arms stretched out 100 times. From past experience, his father knows it takes approximately 1/2 second to perform one full spin, with a standard deviation of 1/3 second. Consider the probability that it will take this child over 55 seconds to complete spinning. Give an upper bound with Chebyshev's inequality and an approximation with the CLT.

Exercise 9.20. Let $X_1, X_2, X_3, \dots$ be i.i.d. random variables with mean zero and finite variance σ^2. Let $S_n = X_1 + \cdots + X_n$. Determine the limits below, with precise justifications.

(a) $\lim_{n\to\infty} P(S_n \geq 0.01n)$
(b) $\lim_{n\to\infty} P(S_n \geq 0)$
(c) $\lim_{n\to\infty} P(S_n \geq -0.01n)$

Exercise 9.21. Let $X_1, \dots, X_{500}$ be i.i.d. random variables with expected value 2 and variance 3. The random variables $Y_1, \dots, Y_{500}$ are independent of the X_i variables, also i.i.d., but they have expected value 2 and variance 2. Use the CLT to estimate $P\left(\sum_{i=1}^{500} X_i > \sum_{i=1}^{500} Y_i + 50\right)$.

Hint. Use the CLT for the random variables $X_1 - Y_1, X_2 - Y_2, \dots$.

Exercise 9.22. In this exercise, you estimate the integral $\int_0^\infty x^6 e^{-x^{1.5}} dx$ via Monte Carlo. For this exercise, you need access to a computer software package that can generate independent samples from the Unif[0, 1] distribution.

Observe that

$$\int_0^\infty x^6 e^{-x^{1.5}} dx = \int_0^\infty x^6 e^{-x^{1.5}+x} e^{-x} dx = E\left[X^6 e^{-X^{1.5}+X}\right]$$

where $X \sim \text{Exp}(1)$. Set $Y = X^6 e^{-X^{1.5}+X}$ and sample $n = 10^5$ independent copies of Y in order to provide a 95% confidence interval for the integral.

Challenging problems

Exercise 9.23. By mimicking the proof of Theorem 9.5, prove the following variant of Chebyshev's inequality.

Theorem. *Let $c > 0$ and $n > 0$ and let X be a random variable with a finite mean μ and for which $E[|X - \mu|^n] < \infty$. Then we have*

$$P(X \geq \mu + c) \leq \frac{E[|X - \mu|^n]}{c^n}.$$

Exercise 9.24. Suppose that we have independent and identically distributed (i.i.d.) random variables $X_1, X_2, X_3, \ldots$ with finite mean $E[X_1] = \mu$ and variance $\text{Var}(X_1) = \sigma^2$. Let $S_n = X_1 + \cdots + X_n$. Prove that for any fixed $\varepsilon > 0$ and $1/2 < \alpha < 1$ we have

$$\lim_{n\to\infty} P\left(\left|\frac{S_n - n\mu}{n^\alpha}\right| < \varepsilon\right) = 1.$$

Exercise 9.25. By mimicking the proof of Theorem 9.9, prove the following variant of the weak law of large numbers, in which the independence assumption is weakened.

Theorem. *Suppose that we have random variables $X_1, X_2, \ldots$ each with finite mean $E[X_i] = \mu$ and variance $Var(X_i) = \sigma^2$. Suppose further that $Cov(X_i, X_j) = 0$ whenever $|i - j| \geq 2$ and that there is a constant $c > 0$ so that $|\text{Cov}(X_i, X_{i+1})| < c$ for all i. Let $S_n = X_1 + \cdots + X_n$. Then for any fixed $\varepsilon > 0$ we have*

$$\lim_{n\to\infty} P\left(\left|\frac{S_n}{n} - \mu\right| < \varepsilon\right) = 1.$$

Exercise 9.26. Consider the coupon collector's problem of Example 8.17. Assume that each box contains one of n possible toys and let T_n denote the number of boxes needed to collect all the toys.

(a) Use Markov's inequality to estimate $P(T_n > (1 + \varepsilon)n \ln(n))$.
(b) Use Chebyshev's inequality to estimate $P(|T_n - E[T_n]| > \varepsilon n \ln(n))$.
(c) Assume now that we have a sequence of random variables T_n where T_n is distributed as the number of boxes needed to collect all the toys in the coupon collector's problem with n toys. Show that for any $\varepsilon > 0$ we have

$$\lim_{n\to\infty} P\left(\left|\frac{T_n}{n \ln n} - 1\right| > \varepsilon\right) = 0.$$

This is a weak law of large numbers for the coupon collector's problem.

Exercise 9.27. Here is a limit theorem that one can prove without complicated tools. Suppose that $X_1, X_2, \ldots$ are i.i.d. random variables with distribution $\text{Exp}(1)$, and let $M_n = \max(X_1, \ldots, X_n)$. Show that for any $x \in \mathbb{R}$ we have

$$\lim_{n\to\infty} P(M_n - \ln n \leq x) = \exp(-e^{-x}).$$

$F(x) = \exp(-e^{-x})$ is the cumulative distribution function of the *Gumbel distribution.* Because it arises as the limit distribution of a maximum of a large number of random variables, it is an example of an *extreme value distribution.*

Hint. Use Exercise 6.47 to compute $P(M_n \leq \ln n + x)$ explicitly, and then evaluate the limit as $n \to \infty$.

Exercise 9.28. Evaluate the limit below for all positive real values of α:

$$\lim_{n\to\infty} \sum_{k:0\leq k\leq \alpha n} e^{-n}\frac{n^k}{k!}.$$

Hint. Find a useful probabilistic representation for the sum.

Exercise 9.29. The toll booth operator cannot take a break as long as there are cars in her queue. The times between arrivals of cars are independent with mean 5 minutes. It takes her on average 2 minutes to process each customer. These service times are independent for different customers and independent of the arrival times of the cars. Let p_n be the probability that the toll booth operator has had to serve n customers in a row without a break. Show that $p_n \to 0$ as $n \to \infty$.

Exercise 9.30. Let $g : [0, \infty)^{100} \to \mathbb{R}$ be

$$g(x_1, \ldots, x_{100}) = \frac{10^4}{1 + \ln(x_1 + \cdots + x_{100} + 1) + (x_1 + \cdots + x_{100})^2}.$$

Using Monte Carlo, approximate $I = \int_{[0,1]^{100}} g(x_1, \ldots, x_{100})\, dx_1 \ldots dx_{100}$, the integral of g over the 100-dimensional unit cube. Produce a 95% confidence interval $\bar{X}_n \pm \varepsilon$, with $\varepsilon = 0.005$.

Exercise 9.31. In the proof of Theorem 9.20 presented on page 321 we assumed that $E[X_1^4] < \infty$. Try to repeat the steps of that proof under the weaker assumption $E[X_1^2] < \infty$, together with $E[X_1] = 0$. Why does this argument now fail?

Exercise 9.32. This exercise demonstrates that Theorem 4.38 is a corollary of Theorem 9.21. Suppose that $X_1, X_2, \ldots$ are independent Ber(p) random variables so $X_1 + \cdots + X_n = S_n \sim \text{Bin}(n, p)$.

(a) Show that $E[|X_1 - p|^3] = p(1-p)(p^2 + (1-p)^2)$.
(b) Derive the bound in Theorem 4.38 by applying Theorem 9.21.

Exercise 9.33. Suppose that $X_1, X_2, \ldots,$ are i.i.d. random variables with finite mean μ and finite variance σ^2. Consider the sample variance s_n^2 defined in (9.10). Use the strong law of large numbers, Theorem 9.20, to prove that s_n^2 converges with probability one to σ^2.

10

Conditional distribution

This chapter combines two of our central themes, conditional probabilities and joint distributions of random variables, in order to define and study conditional distributions of random variables. We gain new concepts and tools for modeling and for answering questions about our models. Beyond the present text, conditional distributions are fundamental to the study of stochastic processes.

In the next two sections we first study conditional distributions of discrete random variables and then conditional distributions of continuous random variables. Section 10.3 develops a conditional expectation that unifies the treatment of discrete and continuous variables. Section 10.4 contains additional interesting examples, including some that mix discrete and continuous random variables.

10.1. Conditional distribution of a discrete random variable

This section develops conditional probability mass functions and conditional expectations of discrete random variables. First we condition on an event and then, building on this, on the outcome of another random variable.

Recall that the conditional probability of the event A, given the event B, is defined by

$$P(A \mid B) = \frac{P(AB)}{P(B)}, \quad \text{whenever } P(B) > 0. \tag{10.1}$$

One application of conditioning was to calculate probabilities through the averaging identity of Fact 2.10. Recall that $B_1, \dots, B_n$ is a partition of the sample space Ω if the sets B_i are pairwise disjoint and $\Omega = \bigcup_{i=1}^n B_i$. Then if each $P(B_i) > 0$,

$$P(A) = \sum_{i=1}^{n} P(A|B_i)P(B_i). \tag{10.2}$$

Conditioning on an event

Our first new definition comes by applying (10.1) to an event $A = \{X = k\}$ for a discrete random variable X.

Definition 10.1. Let X be a discrete random variable and B an event with $P(B) > 0$. Then the **conditional probability mass function of X, given B,** is the function $p_{X|B}$ defined as follows for all possible values k of X:

$$p_{X|B}(k) = P(X = k \mid B) = \frac{P(\{X = k\} \cap B)}{P(B)}.$$

The conditional probability mass function $p_{X|B}(k)$ behaves like a regular probability mass function. The values $p_{X|B}(k)$ are nonnegative and sum up to one:

$$\begin{aligned}\sum_k p_{X|B}(k) &= \sum_k P(X = k \mid B) = \sum_k \frac{P(\{X = k\} \cap B)}{P(B)} \\ &= \frac{1}{P(B)} P\left(\bigcup_k [\{X = k\} \cap B]\right) = \frac{P(B)}{P(B)} = 1. \end{aligned} \tag{10.3}$$

The key point above was that the events $\{X = k\} \cap B$ are disjoint for different values of k and their union over k is B.

We can use the conditional probability mass function to compute an expectation. This expectation is called a conditional expectation.

Definition 10.2. Let X be a discrete random variable and B an event with $P(B) > 0$. Then **the conditional expectation of X, given the event B,** is denoted by $E[X|B]$ and defined as

$$E[X \mid B] = \sum_k k\, p_{X|B}(k) = \sum_k k\, P(X = k \mid B) \tag{10.4}$$

where the sum ranges over all possible values k of X.

Applying the averaging principle (10.2) to the event $A = \{X = k\}$ gives the following identity for conditional probability mass functions.

Fact 10.3. Let Ω be a sample space, X a discrete random variable on Ω, and $B_1, \dots, B_n$ a partition of Ω such that each $P(B_i) > 0$. Then the (unconditional) probability mass function of X can be calculated by averaging the conditional probability mass functions:

$$p_X(k) = \sum_{i=1}^{n} p_{X|B_i}(k)\, P(B_i). \tag{10.5}$$

The averaging idea extends to expectations.

Fact 10.4. Let Ω be a sample space, X a discrete random variable on Ω, and $B_1, \dots, B_n$ a partition of Ω such that each $P(B_i) > 0$. Then

$$E[X] = \sum_{i=1}^{n} E[X|B_i]P(B_i). \tag{10.6}$$

To be fully precise, we need to assume that all expectations and conditional expectations are well defined in Fact 10.4.

Proof. The proof is a combination of the definition of the expectation, identity (10.5), and rearranging a double sum:

$$\begin{aligned} E(X) &= \sum_k k\, P(X = k) = \sum_k k \sum_{i=1}^{n} P(X = k|B_i)P(B_i) \\ &= \sum_{i=1}^{n} \sum_k k\, P(X = k|B_i)P(B_i) = \sum_{i=1}^{n} E[X|B_i]P(B_i). \end{aligned}$$

▲

The next example illustrates a situation where the conditional information is naturally available.

Example 10.5. Let X denote the number of customers that arrive in my store tomorrow. If the day is rainy X is Poisson(λ), and if the day is dry X is Poisson(μ). Suppose that the probability it rains tomorrow is 10%. Find the probability mass function and expectation of X.

Let B be the event that it rains tomorrow. We have $P(B) = 0.1$. The conditional probability mass functions and conditional expectations for tomorrow's arrivals are

$$\begin{aligned} p_{X|B}(k) &= e^{-\lambda}\frac{\lambda^k}{k!}, \qquad & p_{X|B^c}(k) &= e^{-\mu}\frac{\mu^k}{k!}, \\ E(X \mid B) &= \lambda, \qquad \text{and} & E(X \mid B^c) &= \mu. \end{aligned}$$

From (10.5) the unconditional probability mass function is

$$p_X(k) = P(B)\, p_{X|B}(k) + P(B^c)\, p_{X|B^c}(k) = \frac{1}{10} \cdot e^{-\lambda}\frac{\lambda^k}{k!} + \frac{9}{10} \cdot e^{-\mu}\frac{\mu^k}{k!}.$$

From (10.6) the expected number of customers is

$$E[X] = P(B)E(X \mid B) + P(B^c)E(X \mid B^c) = \frac{\lambda}{10} + \frac{9\mu}{10}.$$

▲

In the following example conditioning assists a calculation that originally does not ask for conditioning.

Example 10.6. (Expectation of a geometric random variable revisited) Let X_1, X_2, $X_3, \dots$ be i.i.d. Bernoulli trials with success probability $0 < p \le 1$. Denote by N the number of trials needed for the first success. Then $N \sim \text{Geom}(p)$. We compute $E(N)$.

Let B be the event that the first trial is a success. $N = 1$ on the event B, and so $E[N|B] = 1$.

Conditioning on B^c means that the first trial fails. Since the trials are independent we can imagine that the process starts all over again. Hence the distribution of N given B^c is the same as the distribution of $N + 1$ *without* conditioning (the plus one is for the first failed trial). To see this rigorously, observe that for $k \ge 2$,

$$P(N = k \mid B^c) = \frac{P(N = k, X_1 = 0)}{P(X_1 = 0)} = \frac{(1-p)^{k-1}p}{(1-p)} = (1-p)^{k-2}p = P(N + 1 = k).$$

Apply (10.6) to derive

$$\begin{aligned} E(N) &= E[N|B]P(B) + E[N|B^c]P(B^c) = 1 \cdot p + E(N+1) \cdot (1-p) \\ &= E(N)(1-p) + 1. \end{aligned}$$

This gives a linear equation that yields $E(N) = \frac{1}{p}$. The same approach applies also to $E(N^2)$ (Exercise 10.26). ▲

Conditioning on a random variable

The next step is to let the partition in Facts 10.3 and 10.4 come from another discrete random variable Y. Here are the key definitions.

Definition 10.7. Let X and Y be discrete random variables. Then the **conditional probability mass function of X given $Y = y$** is the following two-variable function:

$$p_{X|Y}(x \mid y) = P(X = x \mid Y = y) = \frac{P(X = x, Y = y)}{P(Y = y)} = \frac{p_{X,Y}(x,y)}{p_Y(y)}. \tag{10.7}$$

The **conditional expectation of X given $Y = y$** is

$$E[X|Y = y] = \sum_x x\, p_{X|Y}(x \mid y). \tag{10.8}$$

The definitions above are valid for y such that $P(Y = y) > 0$.

As previously noted in (10.3), the conditional probability mass function $p_{X|Y}(x \mid y)$ is an honest probability mass function in x for each fixed value of y, whenever $p_Y(y) > 0$. The conditional expectation (10.8) is the usual expectation computed with this conditional probability mass function. It satisfies familiar properties. For example, Exercise 10.47 asks you to verify that for any function g,

$$E[g(X) \mid Y = y] = \sum_x g(x)\, p_{X|Y}(x \mid y), \tag{10.9}$$

where the sum is over the possible values of X.

A small numerical example illustrates the mechanics of these definitions.

Example 10.8. Suppose X and Y are $\{0,1\}$-valued random variables with joint probability mass function given in the table below.

X \ Y	0	1
0	$\frac{3}{10}$	$\frac{2}{10}$
1	$\frac{1}{10}$	$\frac{4}{10}$

Find the conditional probability mass function and conditional expectation of X, given that $Y = y$.

The marginal probability mass function of Y comes from column sums of the table:

$$p_Y(0) = p_{X,Y}(0,0) + p_{X,Y}(1,0) = \tfrac{3}{10} + \tfrac{1}{10} = \tfrac{4}{10}$$
$$p_Y(1) = p_{X,Y}(0,1) + p_{X,Y}(1,1) = \tfrac{2}{10} + \tfrac{4}{10} = \tfrac{6}{10}.$$

The y-values for which conditioning on $Y = y$ is possible are 0 and 1.

The conditional probability mass function of X given that $Y = 0$ is

$$p_{X|Y}(0|0) = \frac{p_{X,Y}(0,0)}{p_Y(0)} = \frac{3/10}{4/10} = \frac{3}{4}, \qquad p_{X|Y}(1|0) = \frac{p_{X,Y}(1,0)}{p_Y(0)} = \frac{1/10}{4/10} = \frac{1}{4}.$$

Similarly, the conditional probability mass function of X given that $Y = 1$ is

$$p_{X|Y}(0|1) = \frac{p_{X,Y}(0,1)}{p_Y(1)} = \frac{1}{3}, \qquad p_{X|Y}(1|1) = \frac{p_{X,Y}(1,1)}{p_Y(1)} = \frac{2}{3}.$$

Notice that the conditional probability mass functions of X are obtained by normalizing the columns of the table to sum to 1.

The conditional expectations of X come by computing expectations with the conditional probability mass functions:

$$E[X|Y=0] = 0 \cdot p_{X|Y}(0|0) + 1 \cdot p_{X|Y}(1|0) = 0 \cdot \tfrac{3}{4} + 1 \cdot \tfrac{1}{4} = \tfrac{1}{4}$$
$$E[X|Y=1] = 0 \cdot p_{X|Y}(0|1) + 1 \cdot p_{X|Y}(1|1) = 0 \cdot \tfrac{1}{3} + 1 \cdot \tfrac{2}{3} = \tfrac{2}{3}. \qquad \blacktriangle$$

Example 10.9. (Multinomial) Consider n independent repetitions of a trial with three possible outcomes labeled 1, 2, and 3. In each trial these outcomes appear with probabilities p_1, p_2, and p_3, respectively, where $p_1+p_2+p_3 = 1$. Let X_i be the number of times outcome i appears among the n trials. Find the conditional probability mass function $p_{X_2|X_1}(\ell \mid m)$ and the conditional expectation $E[X_2 \mid X_1 = m]$. Before the calculation you should try to guess the answer from intuition.

Let $0 \le m \le n$. Since $X_1 + X_2 + X_3 = n$ and from Example 6.10 the marginal distribution of X_1 is $\text{Bin}(n, p_1)$, we deduce that for $0 \le \ell \le n - m$,

$$\begin{aligned}
p_{X_2|X_1}(\ell \mid m) = P(X_2 = \ell \mid X_1 = m) &= \frac{P(X_1 = m,\ X_2 = \ell)}{P(X_1 = m)} \\
&= \frac{P(X_1 = m,\ X_2 = \ell,\ X_3 = n - m - \ell)}{P(X_1 = m)} \\
&= \frac{\binom{n}{m,\ \ell,\ n-m-\ell} p_1^m\, p_2^\ell\, p_3^{n-m-\ell}}{\binom{n}{m} p_1^m\, (p_2 + p_3)^{n-m}} \\
&= \frac{(n-m)!}{\ell!(n-m-\ell)!} \left(\frac{p_2}{p_2 + p_3}\right)^{\ell} \left(\frac{p_3}{p_2 + p_3}\right)^{n-m-\ell}.
\end{aligned}$$

In the third line we used the joint probability mass function for the multinomial distribution. The formula tells us that, given $X_1 = m$, $X_2 \sim \text{Bin}(n - m, \frac{p_2}{p_2+p_3})$. Consequently, by the formula for the mean of a binomial, $E[X_2 \mid X_1 = m] = (n-m)\frac{p_2}{p_2+p_3}$. ▲

As y varies the events $\{Y = y\}$ form a partition of Ω. Hence, Facts 10.3 and 10.4 give the following statements.

Fact 10.10. Let X and Y be discrete random variables. Then

$$p_X(x) = \sum_y p_{X|Y}(x \mid y)\, p_Y(y) \tag{10.10}$$

and

$$E(X) = \sum_y E[X|Y = y]\, p_Y(y). \tag{10.11}$$

The sums extend over those values y such that $p_Y(y) > 0$.

Example 10.11. Suppose that X and Y are independent Poisson random variables with parameters λ and μ and let $Z = X + Y$. Find the conditional probability mass function of X and conditional expectation of X, given $Z = \ell$. Verify the averaging identity (10.10) for p_X.

The joint probability mass function of X, Z comes by taking advantage of the independence of X and Y. Since $0 \le X \le Z$, we only need to consider $0 \le k \le \ell$ in the probability below:

$$\begin{aligned}
P(X = k, Z = \ell) &= P(X = k, X + Y = \ell) = P(X = k, Y = \ell - k) \\
&= P(X = k)P(Y = \ell - k) = \tfrac{\lambda^k}{k!} e^{-\lambda} \tfrac{\mu^{\ell-k}}{(\ell-k)!} e^{-\mu}.
\end{aligned}$$

Recall from Example 7.2 or Example 8.19 that $Z = X + Y \sim \text{Poisson}(\lambda + \mu)$. Now we have the information needed for deducing the conditional probability mass function of X. For $0 \le k \le \ell$,

$$p_{X|Z}(k|\ell) = \frac{p_{X,Z}(k,\ell)}{p_Z(\ell)} = \frac{\frac{\lambda^k}{k!}e^{-\lambda}\frac{\mu^{\ell-k}}{(\ell-k)!}e^{-\mu}}{\frac{(\lambda+\mu)^\ell}{\ell!}e^{-(\lambda+\mu)}} = \frac{\lambda^k\mu^{\ell-k}}{(\lambda+\mu)^\ell}\frac{\ell!}{k!(\ell-k)!}$$
$$= \binom{\ell}{k}\left(\tfrac{\lambda}{\lambda+\mu}\right)^k\left(\tfrac{\mu}{\lambda+\mu}\right)^{\ell-k}.$$

This says that, given $Z = \ell$, the conditional distribution of X is $\text{Bin}(\ell, \frac{\lambda}{\lambda+\mu})$. Recalling that the mean of $\text{Bin}(n, p)$ equals np, we can write down the conditional expectations without further computation: $E[X|Z = \ell] = \ell\frac{\lambda}{\lambda+\mu}$.

We check that the averaging identity (10.10) gives back the probability mass function of X:

$$\sum_\ell p_{X|Z}(k|\ell)\,p_Z(\ell) = \sum_{\ell=k}^{\infty}\binom{\ell}{k}\left(\frac{\lambda}{\lambda+\mu}\right)^k\left(\frac{\mu}{\lambda+\mu}\right)^{\ell-k}\frac{(\lambda+\mu)^\ell}{\ell!}e^{-(\lambda+\mu)}$$
$$= \frac{\lambda^k e^{-(\lambda+\mu)}}{k!}\sum_{\ell=k}^{\infty}\frac{\mu^{\ell-k}}{(\ell-k)!} = \frac{\lambda^k e^{-(\lambda+\mu)}}{k!}\sum_{j=0}^{\infty}\frac{\mu^j}{j!} = \frac{\lambda^k e^{-(\lambda+\mu)}}{k!}\cdot e^{\mu}$$
$$= e^{-\lambda}\frac{\lambda^k}{k!}.$$

On the last line we recognize the Poisson(λ) probability mass function $p_X(k)$.

Here is a story to illuminate the example. Count cars and trucks passing an intersection during a fixed time interval. Let X be the number of cars, Y the number of trucks, and Z their total. Assume that X and Y are independent Poisson random variables. The calculation above can be interpreted as follows. Given that there were $Z = \ell$ total vehicles, the distribution of the random number X of cars is the same as that obtained by marking each vehicle independently as a car with probability $\frac{\lambda}{\lambda+\mu}$. This is a special feature of the Poisson distribution, investigated further in Example 10.13 below. ▲

Constructing joint probability mass functions

Rearranging (10.7) expresses the joint probability mass function in terms of the conditional and marginal probability mass functions:

$$p_{X,Y}(x,y) = p_{X|Y}(x|y)\,p_Y(y). \tag{10.12}$$

Even though (10.7) was valid only when $p_Y(y) > 0$, this is not an issue now because if $p_Y(y) = 0$ then $p_{X,Y}(x,y) = 0$ as well.

Identity (10.12) can be used to define a joint probability mass function when a marginal and a conditional probability mass function are given. The next example illustrates this idea and the averaging identity (10.11).

Example 10.12. Suppose n people apply for a job. Each applicant has to pass two tests. Each person independently passes the first test with probability p. Only those who pass the first test can take the second test. Each person independently passes the second test with probability r. Let M be the number of people who pass the first test and L the number of people who pass the second test after passing the first one. Find the joint probability mass function of M, L and the mean of L.

The information tells us that $M \sim \text{Bin}(n, p)$, and given that $M = m$, $L \sim \text{Bin}(m, r)$. In terms of probability mass functions, for $0 \le \ell \le m \le n$,

$$p_M(m) = \binom{n}{m} p^m (1-p)^{n-m} \quad \text{and} \quad p_{L|M}(\ell|m) = \binom{m}{\ell} r^\ell (1-r)^{m-\ell}.$$

Furthermore, from knowing the mean of a binomial, we deduce $E[L|M = m] = mr$.

By (10.12) the joint probability mass function is, for $0 \le \ell \le m \le n$,

$$p_{M,L}(m, \ell) = p_{L|M}(\ell|m)\, p_M(m) = \binom{m}{\ell} r^\ell (1-r)^{m-\ell} \cdot \binom{n}{m} p^m (1-p)^{n-m}.$$

By the averaging identity (10.11) and since the mean of a Bin(n, p) is np,

$$E(L) = \sum_m E[L|M = m] p_M(m) = r \sum_{m=0}^{n} m\, p_M(m) = rE(M) = npr.$$

Exercise 10.29 asks you to find the marginal probability mass function p_L of L. ▲

Marking Poisson arrivals

We close this section with a property of Poisson random variables that is important for applications. Example 10.14 below requires familiarity with Section 4.6 on the Poisson process.

Example 10.13. Suppose the number of customers that arrive in my store during the course of a day is a Poisson(λ) random variable X. At the door each customer receives a randomly chosen coupon for a discount. The coupons come in three types labeled 1, 2 and 3. A coupon is of type i with probability p_i for $i = 1, 2, 3$, and naturally $p_1 + p_2 + p_3 = 1$. Assume that the coupons given to customers are independent. Let X_i be the number of customers who received a type i coupon, so that $X_1 + X_2 + X_3 = X$. Find the joint distribution of (X_1, X_2, X_3) and check whether they are independent or not.

First we need to understand the nature of the experiment. Suppose $X = n$ is given. Then the types of coupons come from n repeated independent trials, each with possible outcomes 1, 2, or 3, which have probabilities p_1, p_2, p_3. The random variable X_i is the number of trials that resulted in a type i outcome. Thus, given $X = n$, we have $(X_1, X_2, X_3) \sim \text{Mult}(n, 3, p_1, p_2, p_3)$.

Now let $k_1, k_2, k_3 \in \{0, 1, 2, \dots\}$ and set $k = k_1 + k_2 + k_3$:

$$\begin{aligned}
P(X_1 = k_1, X_2 = k_2, X_3 = k_3)& \\
&= P(X_1 = k_1, X_2 = k_2, X_3 = k_3, X = k) \\
&= P(X = k)\, P(X_1 = k_1, X_2 = k_2, X_3 = k_3 \mid X = k) \\
&= \frac{e^{-\lambda}\lambda^k}{k!} \cdot \frac{k!}{k_1!k_2!k_3!} p_1^{k_1} p_2^{k_2} p_3^{k_3} \\
&= \frac{e^{-p_1\lambda}(p_1\lambda)^{k_1}}{k_1!} \cdot \frac{e^{-p_2\lambda}(p_2\lambda)^{k_2}}{k_2!} \cdot \frac{e^{-p_3\lambda}(p_3\lambda)^{k_3}}{k_3!}.
\end{aligned} \tag{10.13}$$

In the passage from line 3 to line 4 we used the *conditional joint probability mass function* of (X_1, X_2, X_3), given that $X = k$, namely

$$P(X_1 = k_1, X_2 = k_2, X_3 = k_3 \mid X = k) = \frac{k!}{k_1!k_2!k_3!} p_1^{k_1} p_2^{k_2} p_3^{k_3},$$

which came from the description of the problem. In the last equality of (10.13) we canceled $k!$ and then used both $k = k_1 + k_2 + k_3$ and $p_1 + p_2 + p_3 = 1$.

We recognize Poisson probabilities on the last line of (10.13). To verify that the marginal distribution of X_1 is Poisson, sum away the other variables:

$$\begin{aligned} P(X_1 = k) &= \sum_{k_2=0}^{\infty} \sum_{k_3=0}^{\infty} P(X_1 = k, X_2 = k_2, X_3 = k_3) \\ &= \frac{e^{-p_1\lambda}(p_1\lambda)^k}{k!} \cdot \sum_{k_2=0}^{\infty} \frac{e^{-p_2\lambda}(p_2\lambda)^{k_2}}{k_2!} \cdot \sum_{k_3=0}^{\infty} \frac{e^{-p_3\lambda}(p_3\lambda)^{k_3}}{k_3!} \\ &= \frac{e^{-p_1\lambda}(p_1\lambda)^k}{k!}. \end{aligned}$$

This last calculation works the same for X_2 and X_3. Hence (10.13) can be completed to read

$$P(X_1 = k_1, X_2 = k_2, X_3 = k_3) = P(X_1 = k_1)P(X_2 = k_2)P(X_3 = k_3).$$

The conclusion is that X_1, X_2, X_3 are independent with Poisson marginals $X_i \sim \text{Poisson}(p_i\lambda)$.

By any rights, the conclusion that the Poisson distribution was preserved by the operation of handing out the coupons should surprise the reader. However, it seems downright counterintuitive that the random variables X_i are independent. Since $X_2 + X_3 = X - X_1$, should it not be the case that if X_1 is way above its mean $p_1\lambda$, we would expect X_2 and X_3 to be below their means? Or, going in the other direction, if $\lambda = 100$ we would expect around 100 customers on a given day. However, if we are told that 1000 type 1 coupons were given, we may think that there was a rush of customers, implying an increased number of type 2 and type 3 coupons. Yet neither of these "intuitive" explanations accounts for what really happens.

This process of cataloging each customer as a particular *type* is called *marking* in the probability literature. There was nothing special about splitting the Poisson number of customers into three categories. It works the same way for any number of labels attached to the arrivals. If you are comfortable with indices, rework the example for a general number r of types (Exercise 10.28). ▲

Example 10.14. (Thinning a Poisson process) The marking property of Example 10.13 works also for the Poisson process. Suppose that customers arrive to a store according to a Poisson process with intensity 5/hour. Each customer rolls a die and receives a coupon if the roll is a six. What can we say about the arrival times of those customers that receive coupons? We argue that this is also a Poisson process, but with rate $\frac{1}{6} \cdot 5 = \frac{5}{6}$. To do so, we check that the new process satisfies the properties from Definition 4.34.

The number of customers who arrive in a time interval I is Poisson with parameter $\lambda = 5|I|$. By Example 10.13, the number of customers who receive a coupon in the time interval I is Poisson with parameter $\frac{1}{6}\lambda = \frac{5}{6}|I|$. The independence over disjoint intervals also works: since the numbers of arrived customers are independent and coupons are assigned independently to each customer, the same holds for the numbers of customers who win coupons. Thus the process of coupon-winning customers is also Poisson, with intensity $\frac{5}{6}$.

By the same argument those customers who do not win coupons obey a Poisson process with intensity $\frac{25}{6}$. Furthermore, these new Poisson processes are independent, by analogy with Example 10.13. ▲

10.2. Conditional distribution for jointly continuous random variables

In the previous section we defined a conditional probability mass function $p_{X|Y}(x|y)$ for X by conditioning on the event $\{Y = y\}$ for a discrete random variable Y. If Y is a continuous random variable we run into a problem: $P(Y = y) = 0$, so we cannot condition on this event. However, ratios of density functions turn out to be meaningful, and enable us to define conditional density functions.

Definition 10.15. Let X and Y be jointly continuous random variables with joint density function $f_{X,Y}(x, y)$. The **conditional density function of X, given $Y = y$,** is denoted by $f_{X|Y}(x \mid y)$ and defined as

$$f_{X|Y}(x \mid y) = \frac{f_{X,Y}(x, y)}{f_Y(y)} \quad \text{for those } y \text{ such that } f_Y(y) > 0. \tag{10.14}$$

We check that the conditional density function integrates to 1, as any legitimate density function should. Suppose that $f_Y(y) > 0$ and recall that $f_Y(y) = \int_{-\infty}^{\infty} f_{X,Y}(w, y)dw$. Then

$$\int_{-\infty}^{\infty} f_{X|Y}(x \mid y)\, dx = \int_{-\infty}^{\infty} \frac{f_{X,Y}(x, y)}{\int_{-\infty}^{\infty} f_{X,Y}(w, y)dw}\, dx = \frac{\int_{-\infty}^{\infty} f_{X,Y}(x, y)dx}{\int_{-\infty}^{\infty} f_{X,Y}(w, y)dw} = 1.$$

A heuristic justification of formula (10.14) can be found at the end of this section.

Using the conditional density function we establish the continuous analogues of the definitions and facts of Section 10.1. Just as an ordinary density function is used to calculate probabilities and expectations, a conditional density function is used to calculate conditional probabilities and conditional expectations. The definition below gives the continuous counterparts of the discrete formulas (10.7) and (10.9).

Definition 10.16. The conditional probability that $X \in A$, given $Y = y$, is

$$P(X \in A \mid Y = y) = \int_A f_{X|Y}(x \mid y)\, dx. \tag{10.15}$$

The conditional expectation of $g(X)$, given $Y = y$, is

$$E[g(X) \mid Y = y] = \int_{-\infty}^{\infty} g(x) f_{X|Y}(x \mid y)\, dx. \tag{10.16}$$

The quantities above are defined for y such that $f_Y(y) > 0$.

Here are the averaging identities for the continuous case.

Fact 10.17. Let X and Y be jointly continuous. Then

$$f_X(x) = \int_{-\infty}^{\infty} f_{X|Y}(x \mid y) f_Y(y)\, dy. \tag{10.17}$$

For any function g for which the expectations below make sense,

$$E[g(X)] = \int_{-\infty}^{\infty} E[g(X)|Y = y] f_Y(y)\, dy. \tag{10.18}$$

The integrals on the right of (10.17) and (10.18) appear to violate Definitions 10.15 and 10.16 by integrating over all $-\infty < y < \infty$, even though $f_{X|Y}(x \mid y)$ and $E[g(X)|Y = y]$ are defined only for those y that satisfy $f_Y(y) > 0$. However, no problem arises because if $f_Y(y) = 0$ we can regard the entire integrand as zero, and so these values y do not actually contribute to the integral.

Proof of Fact 10.17. Rearranging the terms in (10.14) gives the identity

$$f_{X,Y}(x, y) = f_{X|Y}(x \mid y) f_Y(y). \tag{10.19}$$

Integration over y and (6.13) give

$$f_X(x) = \int_{-\infty}^{\infty} f_{X,Y}(x, y)\, dy = \int_{-\infty}^{\infty} f_{X|Y}(x \mid y) f_Y(y)\, dy,$$

and (10.17) has been verified.

We deduce (10.18) with a string of equalities that utilize the definitions and identity (10.19).

$$\begin{aligned}
\int_{-\infty}^{\infty} E[g(X)|Y=y] f_Y(y)\,dy &= \int_{-\infty}^{\infty}\left(\int_{-\infty}^{\infty} g(x) f_{X|Y}(x\,|\,y)\,dx\right) f_Y(y)\,dy \\
&= \int_{-\infty}^{\infty}\int_{-\infty}^{\infty} g(x) f_{X|Y}(x\,|\,y) f_Y(y)\,dx\,dy \\
&= \int_{-\infty}^{\infty}\int_{\infty}^{\infty} g(x) f_{X,Y}(x,y)\,dx\,dy \;=\; E[g(X)].
\end{aligned}$$

▲

To illustrate the concepts introduced we revisit three examples from Section 6.2. We begin with uniform points on planar regions where pictures help us interpret the formulas.

Example 10.18. Let (X, Y) be a uniformly chosen random point on a disk D centered at $(0, 0)$ with radius r_0 (see Figure 10.1). From Example 6.19 we recall the joint density function

$$f_{X,Y}(x,y) = \begin{cases} \dfrac{1}{\pi r_0^2}, & \text{if } (x,y) \in D \\ 0, & \text{if } (x,y) \notin D, \end{cases}$$

and the marginal density functions

$$f_X(x) = \frac{2\sqrt{r_0^2 - x^2}}{\pi r_0^2} \quad \text{and} \quad f_Y(y) = \frac{2\sqrt{r_0^2 - y^2}}{\pi r_0^2} \qquad \text{for } -r_0 < x < r_0,\ -r_0 < y < r_0.$$

Formula (10.14) for the conditional density function of X, given $Y = y$, gives

$$f_{X|Y}(x\,|\,y) = \frac{f_{X,Y}(x,y)}{f_Y(y)} = \frac{1}{2\sqrt{r_0^2 - y^2}}.$$

To determine the domain, first we need $-r_0 < y < r_0$ to guarantee that $f_Y(y) > 0$. Second, to have $f_{X,Y}(x,y) > 0$ we need $(x,y) \in D$ which is equivalent to $-\sqrt{r_0^2 - y^2} < x < \sqrt{r_0^2 - y^2}$.

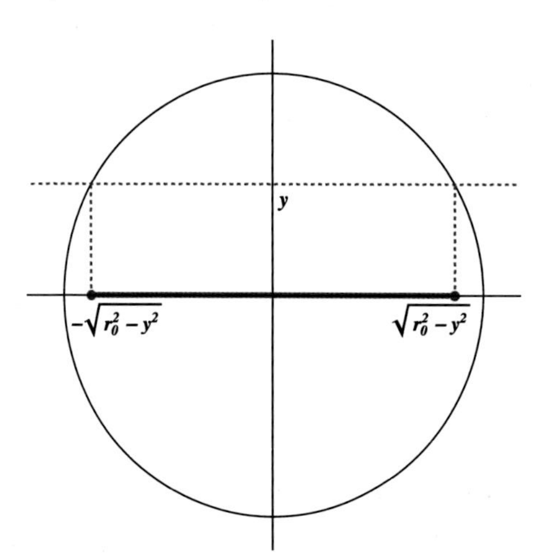

Figure 10.1. Conditional on $Y = y$, the point (X, Y) lies on the intersection of the blue dashed line with the disk, and the conditional distribution of X is uniform on the solid red line on the x-axis.

So the complete answer is that for $-r_0 < y < r_0$,

$$f_{X|Y}(x \mid y) = \frac{1}{2\sqrt{r_0^2 - y^2}} \qquad \text{for } -\sqrt{r_0^2 - y^2} < x < \sqrt{r_0^2 - y^2},$$

and zero otherwise. In other words, given $Y = y \in (-r_0, r_0)$, X is uniformly distributed on the interval $(-\sqrt{r_0^2 - y^2}, \sqrt{r_0^2 - y^2})$. ▲

Example 10.19. Let (X, Y) be uniformly distributed on the triangle D with vertices $(1, 0)$, $(2, 0)$ and $(0, 1)$, as in Example 6.20 (see Figure 10.2). Find the conditional density function $f_{X|Y}(x \mid y)$ and the conditional expectation $E[X|Y = y]$. Verify the averaging identity (10.18) for $E[X]$.

The area of the triangle is $\frac{1}{2}$, so the joint density function of (X, Y) is given by

$$f_{X,Y}(x, y) = \begin{cases} 2, & \text{if } (x, y) \in D \\ 0, & \text{if } (x, y) \notin D \end{cases} = \begin{cases} 2, & \text{if } 0 \le y \le 1 \text{ and } 1 - y \le x \le 2 - 2y \\ 0, & \text{otherwise.} \end{cases}$$

In the second version above we expressed $(x, y) \in D$ with inequalities in a way that is suitable for the sequel. In Example 6.20 we found the marginal density function of Y as

$$f_Y(y) = \begin{cases} 2 - 2y, & 0 \le y \le 1 \\ 0, & y < 0 \text{ or } y > 1. \end{cases}$$

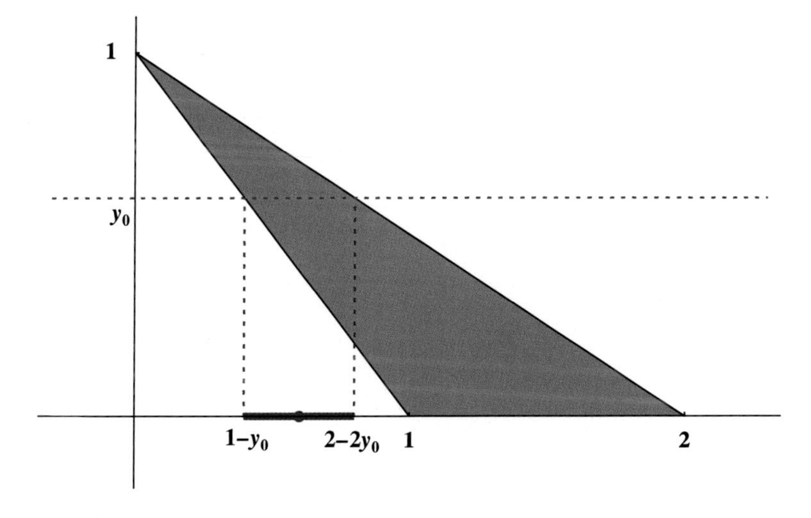

Figure 10.2. Conditional on $Y = y_0$, the point (X, Y) lies on the intersection of the blue dashed line with the shaded triangle. Since the conditional distribution of X is uniform on the interval $(1 - y_0, 2 - 2y_0)$ marked by the solid red line, its conditional expectation is the midpoint of this interval marked by the red bullet.

To have $f_Y(y) > 0$ we restrict to $0 \le y < 1$. Then

$$f_{X|Y}(x \mid y) = \frac{f_{X,Y}(x, y)}{f_Y(y)} = \frac{2}{2 - 2y} = \frac{1}{1 - y} \qquad \text{for } 1 - y \le x \le 2 - 2y. \tag{10.20}$$

This tells us that the conditional distribution of X, given that $Y = y$, is uniform on the interval $[1-y, 2-2y]$, provided that $0 \le y < 1$. This makes sense geometrically. If we choose a point (X, Y) from D, and we know that the y-coordinate is y_0, then this point (X, y_0) lies on the line segment $\{(x, y_0) : 1 - y_0 \le x \le 2 - 2y_0\}$. See Figure 10.2. Since (X, Y) is uniformly distributed on D, it is natural that we get a uniform distribution also on this smaller set.

The expectation of a uniform random variable on an interval is the midpoint of the interval, and so

$$E[X|Y=y] = \frac{1-y+2-2y}{2} = \frac{3-3y}{2} \quad \text{for} \quad 0 \le y < 1.$$

Averaging this with the marginal density function of Y gives

$$\int_{-\infty}^{\infty} E[X|Y=y] f_Y(y)\,dy = \int_0^1 \frac{3-3y}{2}(2-2y)\,dy = 1.$$

A glance at Figure 6.4 shows that $E[X] = 1$, and so we have verified the averaging identity (10.18) for $E[X]$. ▲

Example 10.20. Let the joint density function of the random variables X and Y be

$$f(x,y) = \begin{cases} 2xe^{x^2-y}, & \text{if } 0 < x < 1 \text{ and } y > x^2 \\ 0, & \text{else.} \end{cases} \tag{10.21}$$

Find the conditional density function $f_{Y|X}(y|x)$ and the conditional probability $P(Y \ge \frac{1}{4}|X = x)$. Verify the averaging identity

$$P(Y \ge \tfrac{1}{4}) = \int_{-\infty}^{\infty} P(Y \ge \tfrac{1}{4}|X = x) f_X(x)\,dx.$$

Example 6.17 derived the marginal density $f_X(x) = 2x$ for $0 < x < 1$ and zero elsewhere. Formula (10.14) then gives, for $0 < x < 1$,

$$f_{Y|X}(y|x) = e^{x^2-y} \qquad \text{for } y > x^2.$$

By formula (10.15), for $0 < x < 1$,

$$\begin{aligned} P(Y \ge \tfrac{1}{4}|X = x) &= \int_{1/4}^{\infty} f_{Y|X}(y|x)\,dy = \int_{\max(1/4,\,x^2)}^{\infty} e^{x^2-y}\,dy \\ &= \begin{cases} \int_{1/4}^{\infty} e^{x^2-y}\,dy = e^{x^2-\frac{1}{4}}, & \text{if } 0 < x \le \frac{1}{2} \\ \int_{x^2}^{\infty} e^{x^2-y}\,dy = 1, & \text{if } \frac{1}{2} < x < 1. \end{cases} \end{aligned} \tag{10.22}$$

The lower integration limit increased from $\frac{1}{4}$ to $\max(\frac{1}{4}, x^2)$, because $f_{Y|X}(y|x) = e^{x^2-y}$ only for $y > x^2$ and otherwise $f_{Y|X}(y|x) = 0$. Figure 10.3 illustrates this.

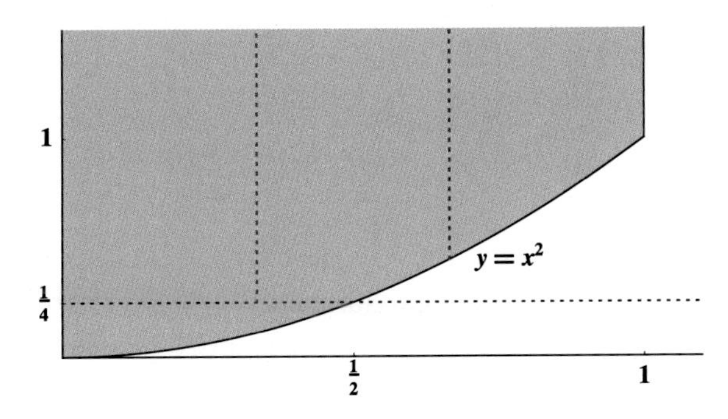

Figure 10.3. The blue line gives the portion of space over which $f_{Y|X}(y|x)$ is integrated in (10.22) when $0 < x \le \frac{1}{2}$. The red line gives the portion of space over which $f_{Y|X}(y|x)$ is integrated in (10.22) when $\frac{1}{2} < x < 1$.

In Example 6.17 we found the marginal density function

$$f_Y(y) = \begin{cases} 1 - e^{-y}, & 0 < y < 1 \\ e^{1-y} - e^{-y}, & y \ge 1 \\ 0, & \text{else.} \end{cases}$$

Use this to compute the probability

$$P(Y \ge \tfrac{1}{4}) = \int_{1/4}^{\infty} f_Y(y)\,dy = \int_{1/4}^{1} (1 - e^{-y})\,dy + \int_{1}^{\infty} (e^{1-y} - e^{-y})\,dy$$
$$= \tfrac{7}{4} - e^{-1/4} \approx 0.971.$$

On the other hand,

$$\int_{-\infty}^{\infty} P(Y \ge \tfrac{1}{4}|X = x) f_X(x)\,dx = \int_0^{1/2} e^{x^2 - \frac{1}{4}}\, 2x\,dx + \int_{1/2}^{1} 2x\,dx = \tfrac{7}{4} - e^{-1/4},$$

the same numerical answer. ▲

Constructing joint density functions

In complete analogy with the discrete case, we can use the identity $f_{X,Y}(x, y) = f_{X|Y}(x|y) f_Y(y)$ to define the joint density function $f_{X,Y}$ from a marginal and a conditional density function. The example below illustrates this and Exercise 10.38 continues the theme.

Example 10.21. Let Y be a standard normal random variable. Then let X be another normal random variable with variance 1 whose mean is the value Y just observed. Find the joint density function of (X, Y). Then, suppose we observe $X = x$. How is Y now distributed?

The problem statement gives these density functions for all real x, y:

$$f_Y(y) = \frac{1}{\sqrt{2\pi}} e^{-y^2/2} = \varphi(y) \quad \text{and} \quad f_{X|Y}(x|y) = \frac{1}{\sqrt{2\pi}} e^{-(x-y)^2/2} = \varphi(x - y)$$

where φ denotes the standard normal density function. The joint density function is given by

$$f_{X,Y}(x, y) = f_{X|Y}(x|y) f_Y(y) = \frac{1}{2\pi} e^{-\frac{1}{2}y^2 - \frac{1}{2}(x-y)^2} = \varphi(y)\varphi(x - y) \quad \text{for all real } x, y.$$

Once we observe $X = x$, the distribution of Y should be conditioned on $X = x$. First find the marginal density function of X:

$$f_X(x) = \int_{-\infty}^{\infty} f_{X,Y}(x,y)dy = \int_{-\infty}^{\infty} \varphi(y)\varphi(x-y)\,dy = \varphi * \varphi(x) = \frac{1}{\sqrt{4\pi}}e^{-x^2/4}.$$

We observed that the integral above is the convolution $\varphi * \varphi(x)$. By Fact 7.1 $\varphi * \varphi$ is the density function of the sum of two independent standard normals, which by Fact 7.9 is the $\mathcal{N}(0,2)$ density function. This way we can identify f_X above without explicit integration.

From the definition of the conditional density function we have, for all real x and y,

$$f_{Y|X}(y|x) = \frac{f_{X,Y}(x,y)}{f_X(x)} = \frac{1}{\sqrt{\pi}}e^{-\frac{1}{2}y^2-\frac{1}{2}(x-y)^2+\frac{1}{4}x^2} = \frac{1}{\sqrt{\pi}}e^{-(y-\frac{x}{2})^2}.$$

The conclusion is that, given $X = x$, we have $Y \sim \mathcal{N}(\frac{x}{2}, \frac{1}{2})$. ▲

As our last example we condition an exponential on a sum, and discover a familiar distribution.

Example 10.22. Let X and Y be independent exponential random variables with parameter μ and $Z = X + Y$. Find the conditional density function of X given $Z = z$ and the conditional expectation $E[X|Z = z]$.

A word problem formulation of this question could go as follows. Times between successive calls to the tech support line are independent exponential random variables with parameter μ. Given that the second call of the day came at time z, what is the distribution of the time of the first call?

To apply the formula $f_{X|Z}(x|z) = \frac{f_{X,Z}(x,z)}{f_Z(z)}$ we need the joint density function of (X, Z) and the marginal density function of Z. Example 7.12 deduced the gamma density $f_Z(z) = \mu^2 z e^{-\mu z}$ for $z > 0$. We give two approaches for finding $f_{X,Z}$.

(i) Find the joint cumulative distribution function $F_{X,Z}(x,z)$ and apply (6.25). Since $X, Y > 0$ with probability 1, it is enough to consider values $0 < x < z$ for (X, Z). Utilize the independence of X and Y below.

$$\begin{aligned} F_{X,Z}(x,z) &= P(X \le x, Z \le z) = P(X \le x, X + Y \le z) \\ &= \iint_{s \le x,\, s+t \le z} f_{X,Y}(s,t)\,ds\,dt = \iint_{s \le x,\, t \le z-s} f_X(s)f_Y(t)\,ds\,dt \\ &= \int_0^x \int_0^{z-s} \mu^2 e^{-\mu(t+s)}\,dt\,ds = 1 - e^{-\mu x} - \mu x e^{-\mu z}. \end{aligned}$$

By (6.25)

$$f_{X,Z}(x,z) = \frac{\partial^2}{\partial x \partial z}F_{X,Z}(x,z) = \mu^2 e^{-\mu z} \quad \text{for } 0 < x < z.$$

For other values we can take $f_{X,Z}(x,z) = 0$.

(ii) Here is a quick argument that requires very little computation that identifies the joint density function $f_{X,Z}$. We begin with the expectation formula

$$E[g(X, Z)] = \int_{-\infty}^{\infty} \int_{-\infty}^{\infty} g(x, z) f_{X,Z}(x, z)\, dz\, dx, \tag{10.23}$$

where g is an arbitrary function. Compute the left-hand side with the joint density of (X, Y):

$$\begin{aligned} E[g(X, Z)] = E[g(X, X+Y)] &= \int_0^{\infty} \int_0^{\infty} g(x, x+y) f_X(x) f_Y(y)\, dy\, dx \\ &= \int_0^{\infty} \left(\int_0^{\infty} g(x, x+y)\, \mu^2 e^{-\mu(x+y)}\, dy \right) dx \\ &= \int_0^{\infty} \left(\int_x^{\infty} g(x, z)\, \mu^2 e^{-\mu z}\, dz \right) dx \\ &= \iint_{0<x<z} g(x, z)\, \mu^2 e^{-\mu z}\, dz\, dx. \end{aligned}$$

The variable in the inner integral changed from y to $z = x + y$, for a fixed x. Comparison of the last representation of $E[g(X, Z)]$ with (10.23) identifies the joint density function as $f_{X,Z}(x, z) = \mu^2 e^{-\mu z}$ for $0 < x < z$ and zero elsewhere.

The conditional density function can now be written down:

$$f_{X|Z}(x|z) = \frac{\mu^2 e^{-\mu z}}{\mu^2 z e^{-\mu z}} = \frac{1}{z} \qquad \text{for } 0 < x < z.$$

This shows that, given $Z = z$, the conditional distribution of X is uniform on the interval $(0, z)$. Thus the conditional expectation is $E[X|Z = z] = \frac{z}{2}$. ▲

Justification for the formula of the conditional density function ♦

The infinitesimal reasoning based on Fact 3.7 for single variable density functions and Fact 6.39 for joint density functions gives a heuristic justification for Definition 10.15.

Let ε and δ be small positive numbers. By analogy with Fact 3.7, the conditional density function $f_{X|Y}(x \mid y)$ should satisfy an approximate identity

$$P(x < X < x + \varepsilon \mid Y = y) \approx f_{X|Y}(x \mid y)\varepsilon. \tag{10.24}$$

We circumvent the impossibility of conditioning on $Y = y$ by conditioning on $y < Y < y + \delta$. By definition,

$$P(x < X < x + \varepsilon \mid y < Y < y + \delta) = \frac{P(x < X < x + \varepsilon,\ y < Y < y + \delta)}{P(y < Y < y + \delta)}. \tag{10.25}$$

If the density functions are continuous then

$$\begin{aligned} P(x < X < x + \varepsilon,\ y < Y < y + \delta) &\approx f_{X,Y}(x, y)\varepsilon\delta \\ P(y < Y < y + \delta) &\approx f_Y(y)\delta. \end{aligned}$$

Substituting these into (10.25) gives

$$P(x < X < x + \varepsilon \,|\, y < Y < y + \delta) \approx \frac{f_{X,Y}(x, y)\varepsilon\delta}{f_Y(y)\delta} = \frac{f_{X,Y}(x, y)}{f_Y(y)}\varepsilon. \tag{10.26}$$

Comparing this with (10.24) shows why $\frac{f_{X,Y}(x,y)}{f_Y(y)}$ is the natural choice for the conditional density function.

10.3. Conditional expectation

This section studies a conditional expectation that achieves some degree of unification of the treatment of discrete and continuous random variables.

Conditional expectation as a random variable

The conditional expectation of X given $Y = y$, denoted by $E[X \,|\, Y = y]$, was defined in the previous two sections: for discrete random variables by (10.8) when $p_Y(y) > 0$, and for jointly continuous random variables by (10.16) when $f_Y(y) > 0$. For each legitimate y-value, $E[X \,|\, Y = y]$ is a real number, and we can think of it as a *function* of y. Denote this function by $v(y) = E[X \,|\, Y = y]$. By substituting the random variable Y into the function v we get a new random variable $v(Y)$. We denote this random variable $v(Y)$ by $E(X \,|\, Y)$ and call it the conditional expectation of X given Y. We can summarize the construction also by saying that the random variable $E(X \,|\, Y)$ takes the value $E[X \,|\, Y = y]$ when $Y = y$.

Definition 10.23. Let X and Y be discrete or jointly continuous random variables. The **conditional expectation of X given Y**, denoted by $E(X \,|\, Y)$, is by definition the random variable $v(Y)$ where the function v is defined by $v(y) = E[X \,|\, Y = y]$.♣

Note the distinction between $E[X \,|\, Y = y]$ and $E(X \,|\, Y)$. The first one is a number, the second one is a random variable. The possible values of $E(X \,|\, Y)$ are exactly the numbers $E[X \,|\, Y = y]$ as y varies. Note also the terminology. $E[X \,|\, Y = y]$ is the *conditional expectation of X given $Y = y$*, while $E(X \,|\, Y)$ is the *conditional expectation of X given Y*.

Example 10.24. In Example 10.8 the marginal probability mass function of Y was given by $p_Y(0) = \frac{4}{10}$ and $p_Y(1) = \frac{6}{10}$. We derived $E[X \,|\, Y = 0] = \frac{1}{4}$ and $E[X \,|\, Y = 1] = \frac{2}{3}$.

Thus in the notation of Definition 10.23, $v(0) = \frac{1}{4}$ and $v(1) = \frac{2}{3}$. The random variable $E(X \,|\, Y)$ is therefore defined by

$$E(X \,|\, Y) = \begin{cases} \frac{1}{4}, & \text{on the event } \{Y = 0\} \\ \frac{2}{3}, & \text{on the event } \{Y = 1\}. \end{cases} \tag{10.27}$$

As a discrete random variable, $E(X|Y)$ has the probability mass function

$$P(E[X|Y] = \tfrac{1}{4}) = \tfrac{4}{10}, \qquad P(E[X|Y] = \tfrac{2}{3}) = \tfrac{6}{10}. \qquad \blacktriangle$$

The random variable $E(X|Y)$ appears abstract. But it turns out to be quite useful. Examples below show that $E(X|Y)$ allows us to shorten certain computations. In Section 10.4 we see how $E(X|Y)$ provides the best predictor of X when Y is observed. Looking further ahead, this conditional expectation occupies an important role in the theory of stochastic processes.

Since $E(X|Y)$ is a random variable, we can ask for its expectation. It turns out that its expectation is always equal to the expectation of X:

$$E[E(X|Y)] = E[X]. \tag{10.28}$$

We verify this identity for the continuous case and leave the discrete case as Exercise 10.27. Utilize again the temporary notation $E(X|Y) = v(Y)$ from Definition 10.23 and then identity (10.18):

$$\begin{aligned} E\big[E(X|Y)\big] &= E[v(Y)] = \int_{-\infty}^{\infty} v(y) f_Y(y)\, dy \\ &= \int_{-\infty}^{\infty} E[X|Y=y] f_Y(y)\, dy = E(X). \end{aligned}$$

Example 10.25. We illustrate (10.28) in the context of Examples 10.8 and 10.24. On the one hand, from Example 10.8, $E(X) = 0 \cdot \frac{1}{2} + 1 \cdot \frac{1}{2} = \frac{1}{2}$. On the other hand, from Example 10.24

$$E[E(X|Y)] = E[X|Y=0]p_Y(0) + E[X|Y=1]p_Y(1) = \tfrac{1}{4} \cdot \tfrac{4}{10} + \tfrac{2}{3} \cdot \tfrac{6}{10} = \tfrac{1}{2}. \qquad \blacktriangle$$

Let us note that everything above applies if X is replaced by $g(X)$ for a function g. This is because $g(X)$ is just another random variable, like X itself. For example, $E(g(X)|Y)$, the *conditional expectation of $g(X)$ given Y*, is defined as $v(Y)$ for the function $v(y) = E[g(X)|Y=y]$, and it satisfies $E[E(g(X)|Y)] = E[g(X)]$.

We extend the definitions once more before turning to interesting examples.

Multivariate conditional distributions

The conditional probability mass functions, density functions, and expectations extend naturally to any number of random variables. For example, in the discrete case we can define the conditional joint probability mass function of $X_1, \dots, X_n$ given $Y = y$ by

$$p_{X_1,\dots,X_n|Y}(x_1,\dots,x_n|y) = P(X_1 = x_1, \dots, X_n = x_n \mid Y = y)$$

while in the jointly continuous case the joint density function of $X_1, \dots, X_n$ given $Y = y$ is given by

$$f_{X_1,\dots,X_n|Y}(x_1,\dots,x_n|y) = \frac{f_{X_1,\dots,X_n,Y}(x_1,\dots,x_n,y)}{f_Y(y)}.$$

We can now compute multivariate conditional expectations: in the discrete case

$$E[g(X_1,\dots,X_n) \mid Y = y] = \sum_{x_1,\dots,x_n} g(x_1,\dots,x_n)\, p_{X_1,\dots,X_n|Y}(x_1,\dots,x_n|y)$$

and in the jointly continuous case

$$E[g(X_1,\dots,X_n) \mid Y = y]$$
$$= \int_{-\infty}^{\infty} \cdots \int_{-\infty}^{\infty} g(x_1,\dots,x_n) f_{X_1,\dots,X_n|Y}(x_1,\dots,x_n|y)\, dx_1 \cdots dx_n.$$

The random variable $E[g(X_1,\dots,X_n) \mid Y]$ is again defined as $v(Y)$ with the function $v(y) = E[g(X_1,\dots,X_n) \mid Y = y]$. The expectation identity

$$E\left[E[g(X_1,\dots,X_n) \mid Y]\right] = E[g(X_1,\dots,X_n)] \tag{10.29}$$

is satisfied again.

For practical calculations an important point is that these conditional expectations satisfy the same properties as ordinary expectations. In particular, the following linearity properties always hold: for real numbers a and b

$$E(aX + b \mid Y) = aE(X \mid Y) + b$$

and for any random variables and functions,

$$E[g_1(X_1) + \cdots + g_n(X_n) \mid Y] = E[g_1(X_1) \mid Y] + \cdots + E[g_n(X_n) \mid Y]. \tag{10.30}$$

Exercise 10.48 asks you to verify a special case of the last identity.

Examples

The following two examples illustrate how conditional expectations can simplify computation of expectations.

Example 10.26. You hold a stick of unit length. Someone comes along and breaks off a uniformly distributed random piece. Now you hold a stick of length Y. Another person comes along and breaks off another uniformly distributed piece from the remaining part of the stick that you hold. You are left with a stick of length X. Find the probability density function f_X, mean $E(X)$ and variance $\mathrm{Var}(X)$ of X.

The problem tells us that $Y \sim \mathrm{Unif}(0,1)$, and conditional on $Y = y$, $X \sim \mathrm{Unif}(0,y)$. Thus,

$$f_Y(y) = \begin{cases} 1, & 0 < y < 1 \\ 0, & y \le 0 \text{ or } y \ge 1, \end{cases} \quad \text{and} \quad f_{X|Y}(x|y) = \begin{cases} \frac{1}{y}, & 0 < x < y < 1 \\ 0, & \text{otherwise.} \end{cases}$$

The density function of X is, for $0 < x < 1$,

$$f_X(x) = \int_{-\infty}^{\infty} f_{X|Y}(x|y) f_Y(y)\, dy = \int_x^1 \frac{1}{y}\, dy = -\ln x.$$

Note the integration limits that come from the case-by-case formula for $f_{X|Y}(x|y)$. For $x \notin (0,1)$ we have $f_X(x) = 0$.

We could compute the expectation and variance of X with the just computed density function f_X, but it is easier to use conditional expectations. Let $0 < y < 1$. Since the mean of a uniform random variable on an interval is the midpoint of the interval, $E(X \mid Y = y) = \frac{1}{2}y$. Next compute

$$E(X^2 \mid Y = y) = \int_{-\infty}^{\infty} x^2 f_{X|Y}(x|y)\,dx = \int_0^y \frac{x^2}{y}\,dx = \tfrac{1}{3}y^2.$$

From these we deduce first $E(X \mid Y) = \frac{1}{2}Y$ and $E(X^2 \mid Y) = \frac{1}{3}Y^2$. Then, by taking expectations of the conditional expectations,

$$E(X) = E[E(X \mid Y)] = \tfrac{1}{2}E(Y) = \tfrac{1}{4},$$

since $E[Y] = \frac{1}{2}$, and

$$E(X^2) = E[E(X^2 \mid Y)] = \tfrac{1}{3}E(Y^2) = \tfrac{1}{3}\int_0^1 y^2\,dy = \tfrac{1}{9}.$$

Finally, $\text{Var}(X) = E(X^2) - (E[X])^2 = \frac{1}{9} - \frac{1}{16} = \frac{7}{144} \approx 0.049$. Exercise 10.41 asks you to break off a third piece of the stick. ▲

Example 10.27. We roll a fair die repeatedly. How many fives do we see on average before seeing the first six? To be specific, let N denote the number of rolls needed to see the first six and Y the number of fives in the first $N - 1$ rolls. We want $E[Y]$. We calculate this expectation first by conditioning on N, and then again by a slick symmetry argument.

N is the familiar geometric random variable: $P(N = n) = (\frac{5}{6})^{n-1}\frac{1}{6}$ for $n \geq 1$ and $E(N) = 6$. The joint distribution of Y and N is readily deduced by thinking about the individual rolls:

$$\begin{aligned} P(Y = m, N = n) &= P(m \text{ fives and no sixes in the first } n-1 \text{ rolls, six at roll } n) \\ &= \binom{n-1}{m}\left(\tfrac{1}{6}\right)^m \left(\tfrac{4}{6}\right)^{n-1-m} \cdot \tfrac{1}{6}. \end{aligned}$$

The derivation above makes sense for $0 \leq m < n$. The conditional probability mass function of Y given $N = n$ is therefore

$$\begin{aligned} p_{Y|N}(m \mid n) &= \frac{P(Y = m, N = n)}{P(N = n)} = \frac{\binom{n-1}{m}\left(\frac{1}{6}\right)^m \left(\frac{4}{6}\right)^{n-1-m} \cdot \frac{1}{6}}{\left(\frac{5}{6}\right)^{n-1}\frac{1}{6}} \\ &= \binom{n-1}{m}\left(\tfrac{1}{5}\right)^m \left(\tfrac{4}{5}\right)^{n-1-m}, \qquad 0 \leq m \leq n-1. \end{aligned} \tag{10.31}$$

Thus given $N = n$, the conditional distribution of Y is $\text{Bin}(n-1, \frac{1}{5})$. If you did not foresee this before the computation, does it seem obvious afterwards?

From knowing the conditional distribution and the mean of a binomial,

$$E[Y \mid N = n] = \tfrac{1}{5}(n - 1).$$

Hence $E(Y \mid N) = \frac{1}{5}(N - 1)$ and using (10.28) we get

$$E[Y] = E[E[Y \mid N]] = E[\tfrac{1}{5}(N - 1)] = \tfrac{1}{5}(E[N] - 1) = \tfrac{1}{5}(6 - 1) = 1.$$

Example 10.32 below deduces the probability mass function of Y.

This problem can be solved quickly by utilizing symmetry arguments, without conditioning. For $1 \le i \le 5$ let Y_i denote the number of is in the first $N - 1$ rolls. Then

$$N = Y_1 + Y_2 + Y_3 + Y_4 + Y_5 + 1$$

from which, using symmetry in the labels,

$$6 = E[N] = E[Y_1] + E[Y_2] + E[Y_3] + E[Y_4] + E[Y_5] + 1 = 5E[Y_i] + 1$$

and we conclude that $E[Y_i] = 1$ for each $1 \le i \le 5$. ▲

The next example uses linearity of conditional expectations.

Example 10.28. (Bernoulli trials) Let $X_1, X_2, X_3, \dots$ be i.i.d. Bernoulli trials with success probability p and $S_k = X_1 + \cdots + X_k$. Find the conditional expectation $E[S_m | S_n]$ for $m < n$.

First use linearity to get

$$E[S_m | S_n = k] = E[X_1 + \cdots + X_m | S_n = k] = \sum_{i=1}^{m} E[X_i | S_n = k].$$

For $k = 0$ we have $E[X_i | S_n = 0] = 0$ because conditioning on $S_n = 0$ forces all $X_i = 0$. For $k > 0$ we compute the individual conditional expectations:

$$\begin{aligned} E[X_i | S_n = k] &= P(X_i = 1 | S_n = k) = \frac{P(X_i = 1, S_n = k)}{P(S_n = k)} \\ &= \frac{P(X_i = 1,\ k - 1 \text{ successes in the other } n - 1 \text{ trials})}{P(S_n = k)} \\ &= \frac{p \cdot \binom{n-1}{k-1} p^{k-1}(1 - p)^{n-1-(k-1)}}{\binom{n}{k} p^k (1 - p)^{n-k}} = \frac{k}{n}. \end{aligned}$$

Combining this with the above gives $E[S_m | S_n = k] = \frac{mk}{n}$ and then $E[S_m | S_n] = \frac{m}{n} S_n$. Example 10.30 below gives an alternative solution. Exercise 10.21 asks you to find this conditional expectation by first deducing the conditional probability mass function $p_{S_m | S_n}$. ▲

We turn to record some additional general features of conditional distributions.

Conditioning and independence

Recall from Section 6.3 that two discrete random variables X and Y are independent if and only if

$$p_{X,Y}(x, y) = p_X(x)\, p_Y(y)$$

for all values x, y of the random variables, and two jointly continuous random variables X and Y are independent if and only if

$$f_{X,Y}(x, y) = f_X(x) f_Y(y)$$

for all real x, y. On the other hand, we now have the identities

$$p_{X,Y}(x, y) = p_{X|Y}(x \mid y)\, p_Y(y)$$

and

$$f_{X,Y}(x, y) = f_{X|Y}(x \mid y) f_Y(y)$$

that are true without any assumptions.

Comparisons of the right-hand sides of these equations give the following criteria for independence in terms of conditional distributions. These statements amount to saying that X and Y are independent if and only if conditioning on Y does not alter the distribution of X.

Fact 10.29. Discrete random variables X and Y are independent if and only if $p_{X|Y}(x \mid y) = p_X(x)$ for all possible values x of X, whenever $p_Y(y) > 0$.

Jointly continuous random variables X and Y are independent if and only if $f_{X|Y}(x \mid y) = f_X(x)$ for all x, whenever $f_Y(y) > 0$.

Apply this to the computation of conditional expectations. The conclusion is that, for independent X and Y, and for any function g and all y for which the expectations below make sense,

$$E[g(X) \mid Y = y] = E[g(X)] \qquad \text{and} \qquad E[g(X)|Y] = E[g(X)]. \tag{10.32}$$

In particular, when X and Y are independent, $E[g(X)|Y]$ is no longer random, but is a constant equal to $E[g(X)]$.

Conditioning on the random variable itself

Conditioning X on Y that is independent of X is an extreme situation where the conditioning information makes no difference. The opposite extreme would be to condition X on X itself. The outcome is the following: for all random variables X,

$$E[g(X)|X] = g(X). \tag{10.33}$$

We can verify (10.33) in the discrete case. Apply definition (10.7) to the case $Y = X$ to get, when $p_X(y) > 0$,

$$p_{X|X}(x \mid y) = P(X = x|X = y) = \frac{P(X = x, X = y)}{P(X = y)}$$

$$= \begin{cases} 0, & x \neq y \\ \dfrac{P(X = y, X = y)}{P(X = y)} = \dfrac{P(X = y)}{P(X = y)} = 1, & x = y. \end{cases}$$

Then from (10.9)

$$E[g(X)|X = y] = \sum_x g(x)\, p_{X|X}(x \mid y) = g(y) \cdot 1 = g(y),$$

where the second equality holds because the only nonzero term in the sum over x is the one with $x = y$. Thus $E[g(X)|X = y] = g(y)$, and so by definition, $E[g(X)|X] = g(X)$. In particular, taking the identity function $g(x) = x$ gives the identity

$$E(X|X) = X. \tag{10.34}$$

Example 10.30. Suppose $X_1, \dots, X_n$ are independent and identically distributed and $S_n = X_1 + \cdots + X_n$. Find $E(X_i|S_n)$ for $1 \le i \le n$.

By symmetry, $E(X_i|S_n) = E(X_1|S_n)$ for each $1 \le i \le n$. Then by (10.34) and the linearity of conditional expectation,

$$\begin{aligned} S_n = E(S_n|S_n) &= E[X_1 + \cdots + X_n|S_n] \\ &= E(X_1|S_n) + \cdots + E(X_n|S_n) = nE(X_1|S_n). \end{aligned}$$

Hence $E(X_i|S_n) = E(X_1|S_n) = \frac{S_n}{n}$. In particular, we get an alternative solution to Example 10.28. ▲

Conditioning on a random variable fixes its value ♦

We take identity (10.33) a step further by conditioning a function of multiple random variables. Conditioning on $Y = y$ fixes the value of Y to be y. Then Y is no longer random and any occurrence of Y inside the conditional expectation can be replaced with the particular value y. In mathematical terms, for any function $h(x, y)$ we have

$$E[h(X, Y) \mid Y = y] = E[h(X, y) \mid Y = y]. \tag{10.35}$$

The formula above is valid in general. It is justified as follows in the discrete case. Assume $P(Y = y) > 0$.

$$\begin{aligned} E[h(X, Y) \mid Y = y] &= \sum_{x,w} h(x, w) P(X = x,\ Y = w \mid Y = y) \\ &= \sum_{x,w} h(x, w) \frac{P(X = x,\ Y = w,\ Y = y)}{P(Y = y)} \end{aligned}$$

$$= \sum_x h(x,y) \frac{P(X=x,\, Y=y)}{P(Y=y)}$$
$$= \sum_x h(x,y) P(X=x \mid Y=y) = E[h(X,y) \mid Y=y].$$

In the third equality above, we used that $P(X = x,\ Y = w,\ Y = y) = 0$ unless $w = y$.

Here is an interesting special case. Take two single variable functions a and b and apply (10.35) to $h(x,y) = a(x)b(y)$. Then

$$E[a(X)b(Y) \mid Y=y] = E[a(X)b(y) \mid Y=y] = b(y)\, E[a(X) \mid Y=y].$$

In the second step we took the constant $b(y)$ out of the expectation. If we switch to conditional expectations as random variables, the formula above becomes

$$E[\, a(X)\, b(Y) \mid Y] = b(Y)\, E[\, a(X) \mid Y]. \tag{10.36}$$

It seems like we broke the rules by moving the random variable $b(Y)$ outside the expectation. But this is not an ordinary expectation, it is a conditional expectation that is itself a random variable. *By conditioning on Y, we are allowed to treat $b(Y)$ as if it were a constant.*

Example 10.31. Recall the multinomial setting from Example 10.9. A trial has outcomes $1, 2$, and 3 that occur with probabilities p_1, p_2, and p_3, respectively. The random variable X_i is the number of times outcome i appears among n independent repetitions of the trial. We give a new computation of $\mathrm{Cov}(X_1, X_2)$, previously done in Example 8.34.

First recall that the marginal distribution of each X_i is $\mathrm{Bin}(n, p_i)$. Hence,

$$E[X_i] = np_i \quad \text{and} \quad E[X_i^2] = \mathrm{Var}(X_i) + (E[X_i])^2 = np_i(1-p_i) + n^2 p_i^2.$$

In Example 10.9 we deduced $E[X_2 \mid X_1 = m] = (n-m)\frac{p_2}{p_2+p_3}$, from which we get

$$E[X_2 \mid X_1] = (n - X_1)\frac{p_2}{p_2+p_3}.$$

We may now compute as follows,

$$E[X_1X_2] = E[\,E(X_1X_2|X_1)\,] = E[\,X_1\, E(X_2|X_1)\,] = E\left[X_1(n-X_1)\tfrac{p_2}{p_2+p_3}\right]$$
$$= \tfrac{p_2}{p_2+p_3}\left(nE[X_1] - E[X_1^2]\right) = -np_1p_2 + n^2p_1p_2.$$

The first equality above follows from (10.29), the second from an application of (10.36), and the third from the above expression for $E[X_2 \mid X_1]$. We also used that $p_1 + p_2 + p_3 = 1$, but omitted the algebra steps. From this we get the covariance: $\mathrm{Cov}(X_1, X_2) = E[X_1X_2] - E[X_1]E[X_2] = -np_1p_2 + n^2p_1p_2 - np_1np_2 = -np_1p_2$. ▲

10.4. Further conditioning topics ♦

This section develops further topics on conditioning and illustrates computational techniques. Individual parts of this section can be read independently of each other. We compute a moment generating function by conditioning, illustrate mixing discrete and continuous random variables, show how the conditional expectation provides the best estimate in a least-squares sense, and study random sums. After this, the last two parts each take a small step towards broad subjects that are natural follow-ups to an introductory probability course: statistics and stochastic processes.

Conditional moment generating function

The example below identifies an unknown probability mass function by finding its moment generating function. The moment generating function is computed by conditioning.

Example 10.32. We continue with the setting of Example 10.27. N is the number of rolls of a fair die needed to see the first six and Y is the number of fives in the first $N-1$ rolls. The task is to find the probability mass function $p_Y(m) = P(Y = m)$. We compute the moment generation function $M_Y(t)$ of Y by conditioning on N. Let $t \in \mathbb{R}$.

$$M_Y(t) = E[e^{tY}] = E\big[E[e^{tY} \mid N]\big] = \sum_{n=1}^{\infty} E[e^{tY} \mid N = n]\, P(N = n). \tag{10.37}$$

To compute the conditional expectation of e^{tY} needed above, we take the conditional probability mass function of Y from (10.31). The binomial theorem evaluates the sum below in a tidy formula.

$$\begin{aligned} E[e^{tY} \mid N = n] &= \sum_{m=0}^{n-1} e^{tm}\, p_{Y|N}(m \mid n) = \sum_{m=0}^{n-1} e^{tm} \binom{n-1}{m} \left(\tfrac{1}{5}\right)^m \left(\tfrac{4}{5}\right)^{n-1-m} \\ &= \left(\tfrac{1}{5}e^t + \tfrac{4}{5}\right)^{n-1}. \end{aligned}$$

Substitute this and $P(N = n) = \left(\tfrac{5}{6}\right)^{n-1} \tfrac{1}{6}$, for $n \geq 1$, into (10.37), rearrange, and apply the formula for a geometric series to find

$$\begin{aligned} E[e^{tY}] &= \sum_{n=1}^{\infty} \left(\tfrac{1}{5}e^t + \tfrac{4}{5}\right)^{n-1} \cdot \left(\tfrac{5}{6}\right)^{n-1} \tfrac{1}{6} = \tfrac{1}{6} \sum_{n=1}^{\infty} \left(\tfrac{4+e^t}{6}\right)^{n-1} \\ &= \tfrac{1}{6} \cdot \frac{1}{1 - \frac{4+e^t}{6}} = \frac{1}{2 - e^t}. \end{aligned} \tag{10.38}$$

In order for the series to converge we must have $\frac{4+e^t}{6} < 1$, which is equivalent to $t < \ln 2$.

We must now find the distribution whose moment generating function is $\frac{1}{2-e^t}$. Since the values of Y are nonnegative integers, its moment generating function is of the form $M_Y(t) = \sum_{k=0}^{\infty} P(Y = k)e^{tk}$. Expanding the function from (10.38) with the geometric series formula $(1-x)^{-1} = \sum_{k=0}^{\infty} x^k$ yields

$$\frac{1}{2-e^t} = \frac{1}{2} \cdot \frac{1}{1-e^t/2} = \frac{1}{2}\sum_{k=0}^{\infty}\left(\frac{e^t}{2}\right)^k = \sum_{k=0}^{\infty}\left(\tfrac{1}{2}\right)^{k+1} e^{tk}. \tag{10.39}$$

Since the moment generating function uniquely determines the distribution, we conclude that $P(Y = k) = (\frac{1}{2})^{k+1}$ for $k \in \{0, 1, 2, \dots\}$. Thus $Y = X - 1$ for $X \sim$ Geom(1/2). Y is called a *shifted geometric variable.*

What is the intuitive explanation for $P(Y = k) = \left(\frac{1}{2}\right)^{k+1}$? If we discard rolls of 1, 2, 3, and 4 and record only fives and sixes, then fives and sixes come with probability $\frac{1}{2}$ each. Among these recorded rolls the position of the first six is a Geom(1/2) random variable. All the earlier recorded rolls are fives. Thus the number of fives is a Geom(1/2) random variable minus one. ▲

Mixing discrete and continuous random variables

The conditioning machinery of this chapter is more flexible than is initially obvious. In particular, we can comfortably mix discrete and continuous random variables. This is illustrated in the following example.

Example 10.33. (Conditioning a continuous random variable on a discrete random variable) A factory produces lightbulbs with two different machines. Bulbs from machine 1 have an exponential lifetime with parameter $\lambda > 0$, while bulbs from machine 2 have an exponential lifetime with parameter $\mu > 0$. Suppose machine 1 produces 2/3 of the lightbulbs. Let T be the lifetime of a randomly chosen bulb from this factory. Find the probability density function f_T of T.

Our intuition tells us that the answer must be $f_T(t) = \frac{2}{3}\lambda e^{-\lambda t} + \frac{1}{3}\mu e^{-\mu t}$, for $t \geq 0$, and zero for $t < 0$. Let us see how this can be shown.

First note that for $t < 0$, we have $F_T(t) = P(T \leq t) = 0$. Hence $f_T(t) = \frac{d}{dt} F_T(t) = 0$.

Now we consider $t \geq 0$. Let $Y \in \{1, 2\}$ denote the machine that produced the bulb. The problem statement gives us the following information:

$$P(Y=1) = \tfrac{2}{3}, \quad P(Y=2) = \tfrac{1}{3}, \quad f_{T|Y}(t\,|\,1) = \lambda e^{-\lambda t} \quad \text{and} \quad f_{T|Y}(t\,|\,2) = \mu e^{-\mu t},$$

for $t \geq 0$. We can find the density function f_T by first finding the cumulative distribution function and then differentiating. We do this by conditioning on Y. For $t \geq 0$ we have

$$\begin{aligned}
F_T(t) &= P(T \leq t) = P(Y=1)P(T \leq t\,|\,Y=1) + P(Y=2)P(T \leq t\,|\,Y=2) \\
&= P(Y=1)\int_{-\infty}^{t} f_{T|Y}(s|1)\,ds + P(Y=2)\int_{-\infty}^{t} f_{T|Y}(s|2)\,ds \\
&= \tfrac{2}{3}\int_0^t \lambda e^{-\lambda s}\,ds + \tfrac{1}{3}\int_0^t \mu e^{-\mu s}\,ds \\
&= \int_0^t \left(\tfrac{2}{3}\lambda e^{-\lambda s} + \tfrac{1}{3}\mu e^{-\mu s}\right) ds.
\end{aligned}$$

We do not need to evaluate the integral since by the fundamental theorem of calculus,

$$f_T(t) = \frac{d}{dt}F_T(t) = \tfrac{2}{3}e^{-\lambda t} + \tfrac{1}{3}\mu e^{-\mu t} \qquad \text{for } t > 0. \tag{10.40}$$

What is the mean lifetime $E[T]$? Since the mean of an Exp(λ) random variable is λ^{-1}, we have $E[T|Y=1] = \lambda^{-1}$ and $E[T|Y=2] = \mu^{-1}$. Consequently

$$E[T] = E[E[T|Y]] = \tfrac{2}{3}E[T|Y=1] + \tfrac{1}{3}E[T|Y=2] = \tfrac{2}{3}\lambda^{-1} + \tfrac{1}{3}\mu^{-1}.$$

Of course, this could also have been computed directly from our probability density function (10.40).

Is there a conditional probability mass function for Y, given $T = t$ for $t \geq 0$? Yes, and we can find it with the infinitesimal reasoning illustrated on page 345.

$$\begin{aligned} p_{Y|T}(1\,|\,t) &= P(Y=1\,|\,T=t) \approx P(Y=1\,|\,t \leq T \leq t+dt) \\ &= \frac{P(Y=1,\, t \leq T \leq t+dt)}{P(t \leq T \leq t+dt)} = \frac{P(Y=1)P(t \leq T \leq t+dt\,|\,Y=1)}{P(t \leq T \leq t+dt)} \\ &\approx \frac{\tfrac{2}{3}\lambda e^{-\lambda t}\,dt}{\left(\tfrac{2}{3}\lambda e^{-\lambda t} + \tfrac{1}{3}\mu e^{-\mu t}\right)dt} = \frac{\tfrac{2}{3}\lambda e^{-\lambda t}}{\tfrac{2}{3}\lambda e^{-\lambda t} + \tfrac{1}{3}\mu e^{-\mu t}}. \end{aligned}$$

In the last step the dt factors were canceled. The final formula gives

$$p_{Y|T}(1\,|\,t) = \frac{p_Y(1)f_{T|Y}(t|1)}{p_Y(1)f_{T|Y}(t|1) + p_Y(2)f_{T|Y}(t|2)}.$$

Note that this is Bayes' formula in terms of densities. We started with conditional information on T, given Y, and ended up reversing the conditioning to obtain the conditional probability mass function of Y, given $T = t$.

To illustrate the qualitative behavior of the formula above, suppose $\lambda > \mu$. This means that bulbs from machine 1 die out faster on average. Thus, the longer the bulb lives, the less likely it is to have come from machine 1. This is consistent with the limit as $t \to \infty$:

$$\lim_{t\to\infty} p_{Y|T}(1\,|\,t) = \lim_{t\to\infty} \frac{\tfrac{2}{3}\lambda}{\tfrac{2}{3}\lambda + \tfrac{1}{3}\mu e^{(\lambda-\mu)t}} = 0.$$

▲

Conditional expectation as the best predictor

Suppose we wish to know a random quantity X, but we cannot observe X directly. What we know is another random variable Y, perhaps an inaccurate measurement corrupted by random noise. What is the best estimate of X in terms of Y? An estimate of X based on the value of Y must be $h(Y)$ for some function h. We measure the effectiveness of an estimate with the expected square error $E[(X - h(Y))^2]$. The theorem below states that in this sense the uniquely best estimate is the conditional expectation $h(Y) = E(X\,|\,Y)$.

Theorem 10.34. *Assuming that the expectations below are well defined, the inequality*

$$E\big[(X - h(Y))^2\big] \geq E\big[(X - E[X \mid Y])^2\big]$$

holds for any function h. Equality holds above only for the choice $h(Y) = E[X|Y]$.

Proof. The theorem is true completely generally. We prove it assuming that X, Y are jointly continuous. Exercise 10.57 asks for the proof for the discrete case.

For an arbitrary function h, add and subtract $E[X \mid Y]$ inside $E[(X - h(Y))^2]$ and expand the square:

$$\begin{aligned}
E\big[(X - h(Y))^2\big] &= E\big[(X - E[X \mid Y] + E[X \mid Y] - h(Y))^2\big] \\
&= E\big[(X - E[X \mid Y])^2\big] + E\big[(E[X \mid Y] - h(Y))^2\big] \\
&\quad + 2E\big[(X - E[X \mid Y])(E[X \mid Y] - h(Y))\big]. \qquad (10.41)
\end{aligned}$$

We check that the expectation on line (10.41) vanishes. It equals

$$\begin{aligned}
&\int_{-\infty}^{\infty}\int_{-\infty}^{\infty} (x - E[X \mid Y = y])\,\big(E[X \mid Y = y] - h(y)\big)\, f_{X,Y}(x, y)\, dx\, dy \\
&= \int_{-\infty}^{\infty}\int_{-\infty}^{\infty} (x - E[X \mid Y = y])\,\big(E[X \mid Y = y] - h(y)\big)\, f_{X|Y}(x \mid y) f_Y(y)\, dx\, dy \\
&= \int_{-\infty}^{\infty}\left[\int_{-\infty}^{\infty} (x - E[X \mid Y = y]) f_{X|Y}(x \mid y)\, dx\right]\big(E[X \mid Y = y] - h(y)\big)\, f_Y(y)\, dy.
\end{aligned}$$

The inner x-integral is calculated under a fixed y. It equals

$$\int_{-\infty}^{\infty} x f_{X|Y}(x \mid y)\, dx - E[X \mid Y = y] \;=\; E[X \mid Y = y] - E[X \mid Y = y] \;=\; 0.$$

Consequently the term on line (10.41) is zero, and we are left with the equality

$$E\big[(X - h(Y))^2\big] = E\big[(X - E[X \mid Y])^2\big] + E\big[(E[X \mid Y] - h(Y))^2\big]. \qquad (10.42)$$

Because the last term $E\big[(E[X \mid Y] - h(Y))^2\big]$ above is nonnegative, equation (10.42) tells us that the mean square error from any approximation $h(Y)$ is at least as large as the mean square error when X is approximated by $E[X \mid Y]$. That is,

$$E\big[(X - h(Y))^2\big] \geq E\big[(X - E[X \mid Y])^2\big],$$

which is what we had set out to show.

To see when the inequality is strict, observe that the last expectation on line (10.42) equals

$$\int_{-\infty}^{\infty} (E(X \mid Y = y) - h(y))^2 f_Y(y)\, dy.$$

Unless $h(y) = E(X \mid Y = y)$ for all y such that $f_Y(y) > 0$, this integral is strictly positive and leads to

$$E\big[(X - h(Y))^2\big] > E\big[(X - E(X \mid Y))^2\big]. \qquad \blacktriangle$$

Random sums

Suppose the average size of a claim on a certain car insurance policy is \$1000. Suppose further that on average the insurance company pays 20 claims per year. What is the average total amount paid out during a year? Linearity of expectation would seem to give the answer $20 \times \$1000 = \$20,000$. However, if *both* the sizes of individual claims *and* the number of claims are *random*, simple linearity does not apply. We need a new theorem called Wald's identity.

In the general setting $X_1, X_2, X_3, \dots$ are i.i.d. random variables with finite mean and $S_n = X_1 + \cdots + X_n$, with $S_0 = 0$. Let N be a nonnegative integer valued random variable with a finite expectation that is independent of the random variables X_i. Let $S_N = X_1 + \cdots + X_N$. This means that N tells us how many terms to add up from the sequence of X_is. The convention is that if $N = 0$ then $S_N = 0$. S_N is a *random sum* because both the individual terms and the number of terms are random. To connect with the question posed in the paragraph above, the random variables X_i are the values of the claims and N is the number of claims in a given year. Here is the formula for the mean of S_N.

Fact 10.35. (Wald's identity) Let $X_1, X_2, X_3 \dots$ be i.i.d. random variables with finite mean. Let N be a nonnegative integer-valued random variable independent of the X_is, also with finite mean. Let $S_N = X_1 + \cdots + X_N$. Then

$$E[S_N] = E[N] \cdot E[X_1].$$

The qualitative conclusion of Wald's identity is that linearity of expectation works in an averaged sense: the expected value of the random sum equals the expected value of the number of terms times the expected value of a single term. It is important to note that

$$E\left[\sum_{k=1}^{N} X_k\right] \neq \sum_{k=1}^{N} E[X_k] \qquad \text{when } N \text{ is random.}$$

The left-hand side is a number (an expectation) while the right-hand side is a random variable if N is random.

We give a proof of Wald's identity with the help of conditional expectations.

Proof of Wald's identity. Let $\mu = E[X_i]$. Conditioning on $\{N = n\}$ yields

$$E[S_N | N = n] = E[S_n | N = n] = E[S_n] = E[X_1] + \cdots + E[X_n] = n\mu.$$

First we replaced S_N with S_n since we conditioned on $N = n$ (an instance of (10.35)). Next we dropped the conditioning due to the independence of N and S_n (an instance of (10.32)). Finally, linearity of expectation applies because the number of terms is a constant n.

The equality above turns into $E[S_N|N] = N\mu$, and then we compute the expectation: $E[S_N] = E[E[S_N|N]] = E[N\mu] = \mu E[N]$. ▲

Example 10.36. Suppose the number of small accidents per year experienced by a fleet of cars is Poisson distributed with mean 20. 90% of the accidents require only a paint job that costs \$100, while 10% of the accidents need bodywork that costs \$5000. What is the average annual cost of these accidents? Assume that the costs of individual accidents and the number of accidents are all independent random variables.

The total cost is the random sum $S_N = X_1 + \cdots + X_N$ where $N \sim \text{Poisson}(20)$ and each X_i has probability mass function $P(X_i = 100) = 0.9$ and $P(X_i = 5000) = 0.1$. By Wald's identity,

$$E[S_N] = E[N] \cdot E[X_1] = 20 \cdot (0.9 \cdot 100 + 0.1 \cdot 5000) = 1180.$$ ▲

You might wonder whether Wald's identity is true even if N is not independent of the random variables X_i. The answer is "not necessarily," as the following example shows.

Example 10.37. (Breaking Wald's identity) Let $X_1, X_2, X_3, \ldots$ be independent Bernoulli random variables with parameter $0 < p < 1$. Let N be the number of failures before the first success. Then $N + 1$ is the position of the first success, and so $N + 1 \sim \text{Geom}(p)$ and $E[N] = \frac{1}{p} - 1 = \frac{1-p}{p}$. We have

$$S_N = X_1 + \cdots + X_N = 0,$$

since $X_i = 0$ for all $i \leq N$. This yields $E[S_N] = 0$, which is not equal to

$$E[N] \cdot E[X_1] = \frac{1-p}{p} \cdot p = 1 - p.$$ ▲

The story of Wald's identity does not end here. Independence of N from $\{X_i\}$ is in fact not necessary for Wald's identity, as long as N belongs to a class of random variables called *stopping times*. This notion is studied in the theory of stochastic processes.

Trials with unknown success probability

In Section 4.3 we used the observed frequency of successes $\widehat{p} = S_n/n$ to estimate the unknown success probability p of independent trials. Use of $\widehat{p}$ was justified by appeal to the law of large numbers and by the observation that $\widehat{p}$ is the maximum likelihood estimator of p (Remark 4.13).

In this section we take a different view of the same problem. We use probability theory to model our ignorance of the success probability and the accumulation of information from observed outcomes. Let $X_i \in \{0, 1\}$ be the outcome of the ith trial and $S_n = X_1 + \cdots + X_n$ the number of successes in n trials. Let ξ denote the unknown success probability. To represent absence of knowledge about ξ, assume that ξ is a uniform random variable on $(0, 1)$. Conditionally on $\xi = p$ the trials are independent with success probability p. Then, given $\xi = p$, we have $P(X_i = 1 \mid \xi = p) = p = 1 - P(X_i = 0 \mid \xi = p)$ and $S_n \sim \text{Bin}(n, p)$ with conditional probability mass function

$$P(S_n = k \mid \xi = p) = \binom{n}{k} p^k (1-p)^{n-k} \qquad \text{for } k = 0, 1, \dots, n. \tag{10.43}$$

The next three examples answer questions about this model.

Example 10.38. (Unconditional probability distribution of successes) First we observe that the unconditional success probability of each trial is $\frac{1}{2}$. Let f_ξ denote the density function of ξ, given by $f_\xi(p) = 1$ for $0 < p < 1$.

$$P(X_i = 1) = \int_0^1 P(X_i = 1 \mid \xi = p) f_\xi(p)\, dp = \int_0^1 p\, dp = \tfrac{1}{2}. \tag{10.44}$$

To determine the distribution of S_n we compute its moment generating function $M_{S_n}(t)$ by conditioning on ξ. Let $t \neq 0$ (we already know that $M_{S_n}(0) = 1$):

$$\begin{aligned}
M_{S_n}(t) &= E[e^{tS_n}] = \int_0^1 E[e^{tS_n} \mid \xi = p] f_\xi(p)\, dp = \int_0^1 \sum_{k=0}^{n} \binom{n}{k} e^{tk} p^k (1-p)^{n-k}\, dp \\
&= \int_0^1 (pe^t + 1 - p)^n\, dp = \frac{1}{e^t - 1} \int_1^{e^t} u^n\, du = \frac{1}{e^t - 1} \cdot \frac{u^{n+1}}{n+1} \Big|_{u=1}^{u=e^t} \\
&= \frac{1}{n+1} \cdot \frac{e^{(n+1)t} - 1}{e^t - 1}.
\end{aligned}$$

We used the binomial theorem and then the substitution $u = pe^t + 1 - p$ with $du = (e^t - 1)dp$.

Next we find the probability mass function whose moment generating function is the one above. Formula (D.2) for finite geometric sums gives us

$$\frac{1}{n+1} \cdot \frac{e^{(n+1)t} - 1}{e^t - 1} = \frac{1}{n+1} \sum_{k=0}^{n} e^{tk}.$$

Since a moment generating function (finite in an interval around 0) determines the probability distribution uniquely, we conclude that $P(S_n = k) = \frac{1}{n+1}$ for $k = 0, 1, \dots, n$. In other words, S_n is a uniform random variable on $\{0, 1, \dots, n\}$. (Exercise 10.50 gives an alternative route to this answer.)

The distribution of S_n is dramatically different from a binomial. As a consequence, the trials are *not* independent without the conditioning on ξ. If they were unconditionally independent, (10.44) would imply that $S_n \sim \text{Bin}(n, \frac{1}{2})$. However, the trials are exchangeable (Exercise 10.14). ▲

Example 10.39. (Estimating the unknown success probability) Repetitions of the trial give us information about the success probability. Given that we have seen k successes in n trials, what is the conditional density function of ξ?

While we do not have a definition of a conditional density function $f_{\xi|S_n}(p|k)$ that mixes discrete and continuous, we can proceed by analogy with our previous theory. Guided by equation (10.15), the conditional density function should give conditional probabilities, and so we expect the equality

$$P(a \le \xi \le b \,|\, S_n = k) = \int_a^b f_{\xi|S_n}(p|k)\, dp.$$

We can compute the left-hand side by moving the conditioning from S_n to ξ:

$$\begin{aligned} P(a \le \xi \le b \,|\, S_n = k) &= \frac{P(a \le \xi \le b,\, S_n = k)}{P(S_n = k)} \\ &= \frac{1}{P(S_n = k)} \int_a^b P(S_n = k \,|\, \xi = p) f_\xi(p)\, dp \\ &= (n+1) \int_a^b \binom{n}{k} p^k (1-p)^{n-k}\, dp. \end{aligned}$$

We conclude that the conditional density function of ξ given $S_n = k$ is

$$f_{\xi|S_n}(p|k) = \frac{(n+1)!}{k!(n-k)!} p^k (1-p)^{n-k} \quad \text{for} \quad 0 < p < 1. \tag{10.45}$$

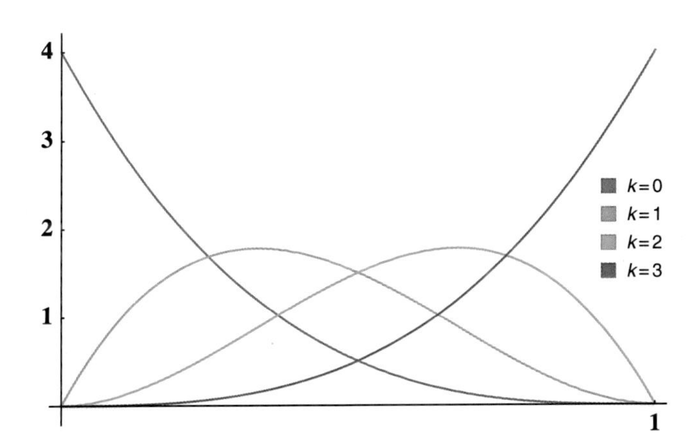

Figure 10.4. The conditional density function of ξ when $n = 3$ and k is 0, 1, 2 or 3. As the number of successes increases, the peak of the density function moves to the right.

The density function discovered above for ξ given $S_n = k$ is the beta distribution with parameters $(k+1, n-k+1)$. (Figure 10.4 illustrates.) The general Beta(a, b) density function with parameters $a, b > 0$ is defined as

$$f(x) = \frac{\Gamma(a+b)}{\Gamma(a)\Gamma(b)} x^{a-1}(1-x)^{b-1} \qquad \text{for } 0 < x < 1 \tag{10.46}$$

and zero otherwise. Recall from (4.26) that $\Gamma(t)$ is the gamma function defined for real $t > 0$ as

$$\Gamma(t) = \int_0^\infty x^{t-1} e^{-x}\, dx,$$

and $\Gamma(n) = (n-1)!$ for positive integers n. The fact that (10.46) is a probability density function follows from the identity

$$\int_0^1 x^{a-1}(1-x)^{b-1}\, dx = \frac{\Gamma(a)\Gamma(b)}{\Gamma(a+b)}$$

established in Example 7.29. ▲

Remark 10.40. (Bayesian versus classical statistics) Remark 4.13 and Example 10.39 illustrate a difference between *classical* and *Bayesian* statistics.

In classical statistics the unknown success probability p is regarded as a fixed quantity. Based on the information from the trials we formed the maximum likelihood estimator $\widehat{p} = k/n$ for p, where k is the observed number of successes.

In the Bayesian approach a prior probability distribution is placed on the unknown parameter, and then the posterior distribution is calculated with Bayes' rule. In the example above, the uniform distribution on ξ is the prior distribution, and the Beta$(k+1, n-k+1)$ distribution obtained in (10.45) is the posterior. The conditional expectation of ξ gives a Bayesian point estimate of p:

$$E(\xi \,|\, S_n = k) = \int_0^1 p\, f_{\xi|S_n}(p|k)\, dp = \frac{(n+1)!}{k!(n-k)!} \int_0^1 p^{k+1}(1-p)^{n-k}\, dp = \frac{k+1}{n+2}.$$

This calculation takes into account both the observed frequency k/n and the prior distribution, and hence does not agree with the maximum likelihood estimate $\widehat{p}$. ▲

Example 10.41. (Success probability of the next trial) Suppose we had k successes in the first n trials. What then is $P(X_{n+1} = 1 | S_n = k)$, the conditional probability that the next experiment is a success?

By the definition of conditional probability

$$P(X_{n+1} = 1 | S_n = k) = \frac{P(X_{n+1} = 1, S_n = k)}{P(S_n = k)}.$$

We already computed $P(S_n = k) = \frac{1}{n+1}$. To compute the numerator we condition on the success probability ξ:

$$\begin{aligned} P(X_{n+1} = 1, S_n = k) &= \int_0^1 P(X_{n+1} = 1, S_n = k | \xi = p) f_\xi(p)\, dp \\ &= \int_0^1 P(X_{n+1} = 1 | \xi = p) P(S_n = k | \xi = p) f_\xi(p)\, dp \\ &= \int_0^1 p \binom{n}{k} p^k (1-p)^{n-k} dp, \end{aligned} \tag{10.47}$$

where the second step follows from the conditional independence of the random variables $X_1, X_2, \ldots$. To evaluate the last integral we use the beta density function (10.46):

$$\int_0^1 p\binom{n}{k}p^k(1-p)^{n-k}dp = \frac{k+1}{(n+1)(n+2)}\int_0^1 \frac{(n+2)!}{(k+1)!(n-k)!}p^{k+1}(1-p)^{n-k}dp$$
$$= \frac{k+1}{(n+1)(n+2)},$$

where the last integral equals 1 since it is the integral of the Beta$(k+2, n-k+1)$ density function over the real line. From this

$$P(X_{n+1} = 1 | S_n = k) = \frac{\frac{k+1}{(n+1)(n+2)}}{\frac{1}{n+1}} = \frac{k+1}{n+2}. \qquad \blacktriangle$$

This last example is sometimes called the *sunrise problem* or *Laplace's law of succession.* In the eighteenth century Pierre-Simon Laplace computed the probability that the sun rises the next day, given that it had risen every day for the previous 5000 years (which was believed to be the age of the Earth at that time). Laplace assumed that sunrises happen according to independent Bernoulli random variables with a success probability that was "chosen" at the beginning of time according to a uniform distribution.

Conditioning on multiple random variables: a prelude to stochastic processes

A stochastic process in discrete time is a sequence of random variables $X_0, X_1, X_2, \ldots$. One can think of this sequence as the time evolution of a random quantity. The random variable X_n is called the *state* of the process at time n.

Assume that these random variables are discrete. Then for any $n > 0$ the joint probability mass function of $X_0, \ldots, X_n$ can be expressed by the multiplication rule (2.5) as a product of conditional probabilities:

$$\begin{aligned} &P(X_0 = x_0, X_1 = x_1, \ldots, X_n = x_n) \\ &\quad = P(X_0 = x_0)\, P(X_1 = x_1 \mid X_0 = x_0)\, P(X_2 = x_2 \mid X_0 = x_0, X_1 = x_1) \\ &\qquad \cdots P(X_n = x_n \mid X_0 = x_0, X_1 = x_1, \ldots, X_{n-1} = x_{n-1}). \end{aligned} \tag{10.48}$$

This identity is valid if the probability on the left-hand side is positive. The conditional probabilities $P(X_{k+1} = x_{k+1} \mid X_0 = x_0, X_1 = x_1, \ldots, X_k = x_k)$ in (10.48) give the distribution of the next state X_{k+1} of the process, given its past evolution up to time k.

Identity (10.48) is especially useful if the right-hand side can be simplified. In an extreme case the variables X_i are independent, and the right-hand side of (10.48) becomes the product of the marginal probability mass functions of $X_0, \ldots, X_n$.

A larger important class of stochastic processes have the property that, at any given time, the past influences the future only through the present state. Concretely speaking, all but the last state can be dropped from the conditioning side of each conditional probability in (10.48). This class of stochastic processes is amenable to deep mathematical analysis and is hugely successful in applications.

Definition 10.42. Let $X_0, X_1, X_2, \ldots$ be a stochastic process of discrete random variables. This process is a **Markov chain** if

$$P(X_{n+1} = x_{n+1} \mid X_0 = x_0,\ X_1 = x_1, \ldots, X_n = x_n) = P(X_{n+1} = x_{n+1} \mid X_n = x_n) \tag{10.49}$$

for all $n \geq 0$ and all $x_0, \ldots, x_n$ such that $P(X_0 = x_0, X_1 = x_1, \ldots, X_n = x_n) > 0$.

A sequence of independent random variables is an example of a Markov chain because both sides of (10.49) reduce to $P(X_{n+1} = x_{n+1})$.

The next example shows that the random walk $S_0, S_1, S_2, \ldots$ introduced in Section 4.3 is also a Markov chain. An intuitive explanation for this is the following: in order to compute the conditional distribution of the location of the walk at step $n+1$, it is sufficient to know the location at step n and the rest of the past is irrelevant.

Example 10.43. (Random walk) Fix a parameter $0 < p < 1$. Let $Y_1, Y_2, Y_3, \ldots$ be independent random variables with distribution $P(Y_j = 1) = p$ and $P(Y_j = -1) = 1 - p$. Define the simple random walk started at the origin by setting $S_0 = 0$ and then $S_n = Y_1 + \cdots + Y_n$ for $n \geq 1$.

We check that $S_0, S_1, \ldots$ satisfy the conditions of Definition 10.42. Let $x_0 = 0$ and let $x_1, \ldots, x_n$ be integer points such that $P(S_0 = x_0, S_1 = x_1, \ldots, S_n = x_n) > 0$. This means that $x_0, x_1, \ldots, x_n$ is a possible path of the walk, which happens if and only if $x_{k+1} - x_k = \pm 1$ for each k. We calculate the left and right sides of (10.49). This relies on the fact that the random variables $Y_k = S_k - S_{k-1}$ are independent:

$$\begin{aligned}
&P(S_0 = x_0, S_1 = x_1, \ldots, S_n = x_n, S_{n+1} = x_{n+1}) \\
&\qquad = P(Y_1 = x_1 - x_0, Y_2 = x_2 - x_1, \ldots, Y_{n+1} = x_{n+1} - x_n) \\
&\qquad = P(Y_1 = x_1 - x_0, \ldots, Y_n = x_n - x_{n-1})\, P(Y_{n+1} = x_{n+1} - x_n) \\
&\qquad = P(S_0 = x_0, S_1 = x_1, \ldots, S_n = x_n)\, P(Y_{n+1} = x_{n+1} - x_n)
\end{aligned}$$

which gives

$$\begin{aligned}
&P(S_{n+1} = x_{n+1} \mid S_0 = x_0,\ S_1 = x_1, \ldots, S_n = x_n) \\
&\qquad = \frac{P(S_0 = x_0, S_1 = x_1, \ldots, S_n = x_n, S_{n+1} = x_{n+1})}{P(S_0 = x_0, S_1 = x_1, \ldots, S_n = x_n)} \\
&\qquad = P(Y_{n+1} = x_{n+1} - x_n)
\end{aligned}$$

for the left side of (10.49). For the right side of (10.49) we have

$$P(S_{n+1} = x_{n+1} \mid S_n = x_n) = \frac{P(S_n = x_n,\, S_{n+1} = x_{n+1})}{P(S_n = x_n)}$$

$$= \frac{P(S_n = x_n,\, Y_{n+1} = x_{n+1} - x_n)}{P(S_n = x_n)} = \frac{P(S_n = x_n)\, P(Y_{n+1} = x_{n+1} - x_n)}{P(S_n = x_n)}$$

$$= P(Y_{n+1} = x_{n+1} - x_n).$$

This shows that the random walk S_n satisfies property (10.49), and hence it is a Markov chain. ▲

Markov chains can also be defined in continuous time. The most basic example is the Poisson process N_t introduced below Definition 4.34 in Section 4.6.

10.5. Finer points ♣

Conditional expectation

In Definition 10.23 in Section 10.3, the conditional expectation $E(X \mid Y)$ was defined as $v(Y)$ for the function $v(y) = E[X \mid Y = y]$. However, $v(y)$ is not necessarily defined for all y in the range of Y, and then the composition $v(Y)$ is not defined on all of Ω. The resolution to this problem is that in every case the set of y for which $v(y)$ is defined contains the values of Y with probability 1. Thus at most a zero probability subset of Ω has to be discarded, and this makes no difference to random variables.

When conditional expectation is developed in the measure-theoretic framework of probability theory, the primary object is the abstract conditional expectation $E(X \mid Y)$ that exists as a random variable on the probability space on which X and Y are defined. The conditional expectation $E(X \mid Y)$ is defined as the unique random variable that satisfies

$$E[X\, h(Y)] = E[\, E(X \mid Y)\, h(Y)] \tag{10.50}$$

for all bounded measurable functions h. This definition is valid for all random variables, as long as X has a finite expectation, and needs no special provisions such as $p_Y(y) > 0$ or $f_Y(y) > 0$. The more concrete conditional expectations $E(X \mid Y = y)$ are then derived from it.

We can deduce identity (10.50) from (10.36). Using the functions $a(x) = x$ and $b(y) = h(y)$ we get

$$E[X\, h(Y)] = E[\, E[Xh(Y) \mid Y]] = E[\, h(Y) E[X \mid Y]],$$

where the first step used (10.29) to introduce conditioning inside the expectation.

Exercises

We start with some warm-up exercises arranged by section.

Section 10.1

Exercise 10.1. The joint probability mass function of the random variables (X, Y) is given by the following table:

X \ Y	0	1	2
1	0	$\frac{1}{9}$	0
2	$\frac{1}{3}$	$\frac{2}{9}$	$\frac{1}{9}$
3	0	$\frac{1}{9}$	$\frac{1}{9}$

(a) Find the conditional probability mass function of X given $Y = y$.
(b) Find the conditional expectation $E[X|Y = y]$ for each of $y = 0, 1, 2$.

Exercise 10.2. Let X be a random variable with values $\{0, 1\}$ and Y a random variable with values $\{0, 1, 2\}$. Initially we have the following partial information about their joint probability mass function.

X \ Y	0	1	2
0			
1	$\frac{1}{8}$		$\frac{1}{8}$

Subsequently we learn the following.

(i) Given $X = 1$, Y is uniformly distributed.
(ii) $p_{X|Y}(0|0) = \frac{2}{3}$.
(iii) $E[Y|X = 0] = \frac{4}{5}$.

Use this information to fill in the missing values of the joint probability mass function table.

Exercise 10.3. Roll a fair die and denote the outcome by Y. Then flip Y fair coins and let X denote the number of tails observed. Find the probability mass function and expectation of X.

Exercise 10.4. Let N have a Bin$(100, \frac{1}{4})$ distribution. Given that $N = n$, flip a fair coin n times, and let X be the number of heads observed.

(a) Find the conditional distribution of X given that $N = n$. Be precise about the possible values and a proper conditional probability mass function.

(b) Find $E[X|N = n]$ for all the values of n for which it is defined.
(c) Use part (b) to compute $E[X]$.

Section 10.2

Exercise 10.5. Suppose that the joint density function of X and Y is

$$f(x, y) = \frac{12}{5}x(2 - x - y), \quad \text{for } 0 < x < 1 \text{ and } 0 < y < 1,$$

and zero otherwise.

(a) Find $f_{X|Y}(x|y)$.
(b) Compute $P(X > 1/2 \mid Y = 3/4)$ and $E[X \mid Y = 3/4]$.

Exercise 10.6. Let the joint density function of (X, Y) be

$$f(x, y) = \frac{x + y}{4}, \quad 0 < x < y < 2.$$

(a) Find $f_{X|Y}(x|y)$.
(b) Compute the conditional probabilities $P(X < 1/2 \mid Y = 1)$ and $P(X < 3/2 \mid Y = 1)$.
(c) Compute the conditional expectation $E[X^2|Y = y]$. Check the averaging identity

$$\int_{-\infty}^{\infty} E[X^2|Y = y] f_Y(y)\, dy = E[X^2].$$

Exercise 10.7. Let X, Y be two random variables with the following properties. Y has density function $f_Y(y) = 3y^2$ for $0 < y < 1$ and zero elsewhere. For $0 < y < 1$, given that $Y = y$, X has conditional density function $f_{X|Y}(x|y) = 2x/y^2$ for $0 < x < y$ and zero elsewhere.

(a) Find the joint density function $f_{X,Y}(x, y)$ of (X, Y). Be precise about the values (x, y) for which your formula is valid. Check that the joint density function you find integrates to 1.
(b) Find the conditional density function of Y, given $X = x$. Be precise about the values of x and y for which the answer is valid. Identify the conditional distribution of Y by name.

Section 10.3

Exercise 10.8. Let X be the number of rolls of a fair die until I see the first six. Next, I choose a sample, with replacement, of size X from an urn with 5 red and 4 green balls. Let Y be the number of green balls in my sample.

(a) Find $E[Y|X]$.
(b) Use your answer from part (a) to find $E[Y]$.

Exercise 10.9. Let the joint density function of (X, Y) be

$$f(x, y) = \frac{1}{y} e^{-x/y} e^{-y} \quad \text{for } 0 < x < \infty \text{ and } 0 < y < \infty.$$

(a) Find $f_Y(y)$ and $f_{X|Y}(x|y)$. Compute $E[Y]$.
(b) Find the conditional expectation $E[X \mid Y]$.
(c) Use parts (a) and (b) to compute $E[X]$.

Exercise 10.10. The number of customers that visit my store during a day is $N \sim \text{Poisson}(\lambda)$. Each customer buys something with probability p, independently of other customers. Let X be the number of customers that buy something during a day.

(a) From the description above, write down the conditional probability mass function $p_{X|N}(k \mid n) = P(X = k \mid N = n)$. Be careful about the values of n and k for which these are defined. Once you have the conditional probability mass function, use it to give the conditional expectations $E(X \mid N = n)$ and $E(X \mid N)$.
(b) Deduce $E[X]$ from the conditional expectation from part (a).
(c) Compute the covariance of N and X. Based on your intuition from the experiment, should the answer be positive or negative?
Hint. See Example 10.31.

Section 10.4

Exercise 10.11. Let $Y \sim \text{Exp}(\lambda)$. Given that $Y = y$, let $X \sim \text{Poisson}(y)$. Find the mean and variance of X.

Hint. Find $E(X|Y)$ and $E(X^2|Y)$ directly from knowledge of Poisson moments, and then $E(X)$ and $E(X^2)$ from knowledge of exponential moments.

Exercise 10.12. Let $X_1, X_2, X_3, \ldots$ be i.i.d. exponential random variables with parameter λ. Let N be a $\text{Geom}(p)$ random variable (with $0 < p < 1$) independent of the X_i random variables. Define the random sum $S_N = X_1 + \cdots + X_N$.

(a) Find the mean $E[S_N]$.
(b) Find the probability distribution of S_N.
Hint. By conditioning on N, compute either the cumulative distribution function $P(S_N \leq x)$ or the moment generating function of S_N. The answer is one of the named distributions.

Exercise 10.13. Consider this game of chance with a monetary payoff. First, a real number U is chosen uniformly at random from the interval $[0, 10]$. Next, an integer X is chosen according to the Poisson distribution with parameter U. The player receives a reward of $\$X$. What should be the fair price charged for playing this game? That is, how much should it cost to play so that expected net gain is zero?

Exercise 10.14. In the context of Example 10.38 of trials conditionally independent on the success probability $\xi = p$, show that the outcomes $X_1, \dots, X_n$ are exchangeable. Concretely, show that for any choice $t_1, \dots, t_n$ of zeros and ones,

$$P(X_1 = t_1, X_2 = t_2, \dots, X_n = t_n) = P(X_1 = t_{k_1}, X_2 = t_{k_2}, \dots, X_n = t_{k_n})$$

for any permutation $(k_1, k_2, \dots, k_n)$ of $(1, 2, \dots, n)$.

Further exercises

Exercise 10.15. Consider a large insurance company with two types of policies: policy A and policy B. Suppose that the number of claims the company sees in a given day has a Poisson distribution with a parameter of 12. Suppose further that each claim comes from policy A with probability 1/4, independently of other claims.

(a) Find the probability that the company will receive at least 5 claims from A policies tomorrow.
(b) Find the probability that the company will receive at least 5 claims from B policies tomorrow.
(c) Find the probability that the company will receive at least 10 total claims tomorrow.

Exercise 10.16. Continuing Problem 10.15, suppose that each claim from policy A is greater than \$100,000 with probability 4/5, each claim from policy B is greater than \$100,000 with probability 1/5, and these happen independently for each claim.

(a) Find the expected number of claims over \$100,000 tomorrow.
(b) Find the probability that there will be less than or equal to 2 claims over \$100,000 tomorrow.

Exercise 10.17. We flip a fair coin. If it is heads we roll 3 dice. If it is tails we roll 5 dice. Let X denote the number of sixes among the rolled dice.

(a) Find the probability mass function of X.
(b) Find the expected value of X.

Exercise 10.18. Consider a sequence of independent repetitions of a trial with $r \geq 3$ different outcomes. For two different outcomes s and t, how many outcomes t do we see on average before the first outcome s?

(a) Assume that the r outcomes are all equally likely.
(b) Assume that the outcomes $1, \dots, r$ in each trial have positive probabilities $p_1, \dots, p_r$ with $p_1 + \cdots + p_r = 1$.

Exercise 10.19. (Continuing Example 10.9) Repeat n times a trial with three outcomes $1, 2, 3$ that appear with probabilities p_1, p_2, p_3 with $p_1 + p_2 + p_3 = 1$. Let X_i be the number of times outcome i appears in the n trials.

(a) Find the conditional joint probability mass function of (X_2, X_3), given $X_1 = m$. That is, find the probabilities

$$P(X_2 = k, X_3 = \ell \mid X_1 = m) \qquad \text{for integers } k, \ell.$$

(b) Utilizing part (a), identify the conditional distribution of X_2 given $X_1 = m$.

Exercise 10.20. Let $X_1, X_2, X_3, \dots$ be i.i.d. Bernoulli trials with success probability p and $S_n = X_1 + \cdots + X_n$.

(a) Compute the conditional joint probability mass function of the Bernoulli(p) trials $(X_1, \dots, X_n)$ given that $S_n = k$. That is, find

$$P(X_1 = a_1, X_2 = a_2, \dots, X_n = a_n \mid S_n = k) \qquad (10.51)$$

for all vectors $(a_1, \dots, a_n)$ of zeros and ones.

(b) Show that, given $S_n = k$ with $0 < k < n$, the random variables $X_1, \dots, X_n$ are exchangeable but not independent. That is, show that the conditional joint probability mass function in (10.51) does not change under permutation of the a_is, but it is not equal to the product of the single variable conditional probabilities $P(X_i = a_i \mid S_n = k)$, $1 \le i \le n$.

(c) Without any computation, give a common sense reason for why $X_1, \dots, X_n$ cannot be independent under the condition $S_n = k$.

Exercise 10.21. Let $X_1, X_2, X_3, \dots$ be i.i.d. Bernoulli trials with success probability p and $S_k = X_1 + \cdots + X_k$. Let $m < n$. Find the conditional probability mass function $p_{S_m|S_n}(\ell|k)$ of S_m, given $S_n = k$.

(a) Identify the distribution by name. Can you give an intuitive explanation for the answer?

(b) Use the conditional probability mass function to find $E[S_m|S_n]$.

Exercise 10.22. We repeat independent identical copies of a trial with success probability $0 < p < 1$. Let X be the number of experiments needed until the first success and let Y denote the number of experiments needed until the first failure.

(a) Find the conditional probability mass function of X given Y. (It is a case by case defined function.)

(b) Find the expected value of $\max(X, Y)$.
Hint. Condition on the outcome of the first trial.

Exercise 10.23. We roll a die until we see the third six. Let X denote the number of rolls needed for the first six and Y the number of rolls needed for the third six.

(a) Find the conditional probability mass function of X given $Y = y$.
(b) Find $E[X\,|\,Y]$.

Exercise 10.24. Flip 10 coins and denote the number of heads by Y. Then roll Y dice. Let X denote the number of sixes in the dice. If $Y = 0$ then set $X = 0$.

(a) Find the conditional distribution of X conditioned on values of Y, the joint probability mass function of (X, Y), the unconditional probability mass function of X, $E[X\,|\,Y]$ and $E(X)$.
(b) Identify the joint probability mass function of $(X, Y - X, 10 - Y)$. Give an intuitive argument from the description of the experiment that derives this distribution without any computation.

Exercise 10.25. We roll a die until we get a six and denote the number of rolls by X. Then we take a fair coin and we repeatedly flip it until we get X heads. We denote the number of coin flips needed by Y.

(a) Find the conditional probability mass function of Y given $X = x$.
(b) Find the probability mass function of Y and show that it is geometric. Hint. The binomial theorem might be helpful.
(c) Find the conditional probability mass function of X given $Y = y$. What kind of distribution is that?

Exercise 10.26. Compute $E(N^2)$ for $N \sim \text{Geom}(p)$ by conditioning on the outcome of the first trial, as in Example 10.6. Then calculate $\text{Var}(N)$.

Exercise 10.27. Let X, Y be discrete random variables. Using Fact 10.10 and Definition 10.23 show that $E[E(X\,|\,Y)] = E[X]$.

Exercise 10.28. Generalize Example 10.13 as follows. Let the number of arrivals be $X \sim \text{Poisson}(\lambda)$. Mark each arrival independently as type i with probability p_i, $i \in \{1, \dots, r\}$, $p_1 + \cdots + p_r = 1$. Let X_i denote the number of arrivals marked type i. Find the joint distribution of $(X_1, \dots, X_r)$.

Exercise 10.29. Example 10.12 had random variables M, L and the probability mass functions

$$p_M(m) = \binom{n}{m} p^m (1-p)^{n-m} \quad \text{and} \quad p_{L|M}(\ell\,|\,m) = \binom{m}{\ell} r^\ell (1-r)^{m-\ell}$$

for integers $0 \le \ell \le m \le n$, with n fixed. Find the marginal probability mass function p_L of the random variable L. Before computing, try to deduce it with common sense intuition from the description of the example.

Exercise 10.30. Let $0 < \alpha, p < 1$. Suppose the number of arrivals is a random variable X with probability mass function $P(X = k) = (1-p)^k p$ for $k = 0, 1, 2, \dots$ Label each arrival independently as type 1 with probability α and type 2 with probability $1 - \alpha$. Let X_i be the number of customers labeled type i.

Show that each X_i is a shifted geometric random variable, but that X_1 and X_2 are not independent.
Hint. The following series expansion can be helpful:

$$(1+x)^\gamma = \sum_{n=0}^{\infty} \binom{\gamma}{n} x^n \tag{10.52}$$

valid for $|x| < 1$ and any real γ, where the meaning of the generalized binomial coefficient is

$$\binom{\gamma}{n} = \frac{\gamma(\gamma-1)\cdots(\gamma-n+1)}{n!}.$$

Exercise 10.31. Let B be an event with $P(B) > 0$ and I_B its indicator random variable, that is, $I_B = 1$ on the event B and $I_B = 0$ on the event B^c. Let X be another discrete random variable on the same sample space. What identity holds between the conditional probability mass functions $p_{X|B}$ and $p_{X|I_B}$? What identity holds between $p_{X|B^c}$ and $p_{X|I_B}$?

Exercise 10.32. Recall Exercise 6.34. Let (X, Y) be a uniformly distributed random point on the quadrilateral D with vertices $(0,0)$, $(2,0)$, $(1,1)$ and $(0,1)$. Find the conditional expectations $E(X|Y)$ and $E(Y|X)$. Find the expectation $E(X)$ and $E(Y)$ by averaging the conditional expectations. If you did not calculate the expectations in Exercise 8.15, confirm your answers by calculating the expectations by integrating with the joint density.

Exercise 10.33. Let (X, Y) be uniformly distributed on the triangle D with vertices $(1,0)$, $(2,0)$ and $(0,1)$, as in Example 10.19.

(a) Find the conditional probability $P(X \le \frac{1}{2} \mid Y = y)$. You might first deduce the answer from Figure 10.2 and then check your intuition with calculation.
(b) Verify the averaging identity for $P(X \le \frac{1}{2})$. That is, check that

$$P(X \le \tfrac{1}{2}) = \int_{-\infty}^{\infty} P(X \le \tfrac{1}{2} \mid Y = y) f_Y(y)\, dy.$$

Exercise 10.34. Deduce the identity $E[XY] = E[Y \cdot E(X|Y)]$ in both the discrete and the jointly continuous case.

Exercise 10.35. Let X and Y be independent standard normal random variables and $S = X + Y$.

(a) Find the conditional density function $f_{X|S}(x \mid s)$ of X, given $S = s$. Identify by name the distribution you found.
Hint. As in Example 10.22 begin by finding the joint density of (X, S).

(b) Find the conditional expectations $E(X \mid S)$ and $E(X^2 \mid S)$. (No calculation is necessary if you identified the conditional distribution of X correctly.)

Calculate $E[E(X \mid S)]$ and $E[E(X^2 \mid S)]$ and check that you got the answers that agree with the general identity $E[E(g(X) \mid Y)] = E[g(X)]$.

Exercise 10.36. Let X and Y be independent $\mathcal{N}(\mu, \sigma^2)$ random variables and $S = X + Y$. Find the conditional density $f_{X|S}(x \mid s)$ of X given $S = s$ and the conditional expectation $E[X|S]$.

Exercise 10.37. Let Y have standard normal distribution conditioned to be positive. Find the density function of Y.

Hint. The assumption means that $P(a \le Y \le b) = P(a \le Z \le b \mid Z > 0)$ for all $a < b$, where $Z \sim \mathcal{N}(0, 1)$.

Exercise 10.38. We generate a random variable $Y \sim \text{Exp}(1)$. Then we generate another exponential random variable X whose parameter is the value Y just observed.

(a) Find the joint density function of (X, Y).
(b) Suppose we observe $X = x$. How is Y now distributed? Identify the distribution by name.

Exercise 10.39. Suppose that X is a continuous random variable with density function $f_X(x) = 20x^3(1-x)$ on $[0, 1]$ and zero otherwise. We generate a random variable Y by choosing a random point uniformly from the (random) interval $[X, 1]$. Find the density function f_Y of Y.

Exercise 10.40. Let X and Y be two positive random variables with the following properties. The random variable X has density function given by

$$f_X(x) = \begin{cases} xe^{-x}, & \text{for } x > 0 \\ 0, & \text{otherwise.} \end{cases}$$

Given that $X = x$, the random variable Y is uniformly distributed on the interval $(0, 1/x)$.

(a) Find the probability $P(Y > 2|X = x)$ for all $x > 0$.
(b) Find the conditional expectation $E[Y|X]$ and the expectation of Y.

Exercise 10.41. The exercise continues the theme of Example 10.26. You hold a stick of unit length. Three pieces are broken off one after another, and each time the piece taken away is uniformly distributed on the length of the piece remaining in your hand. Let Z be the length of the piece that remains after the three pieces have been broken off. Find the density function f_Z, mean $E(Z)$ and variance $\text{Var}(Z)$ of Z.

Exercise 10.42. Break a stick of unit length at a uniformly chosen random point. Then take the shorter of the two pieces and break it into two again at a uniformly chosen random point. Let X denote the shortest of the final three pieces. Find the probability density function of X.

Exercise 10.43. Let (X, Y) be a uniformly chosen point from the triangle given by the points $(0, 0), (2, 0), (2, 2)$.

(a) Find the conditional density of X given $Y = y$. What is this distribution?
(b) Find the conditional expectation $E[X|Y]$.

Exercise 10.44. Let X and Y be independent continuous random variables with probability density functions $f_X(x)$ and $f_Y(y)$. Compute $P(X + Y \leq z)$ by first conditioning on $Y = y$ and then using the averaging principle of equation (10.29).

Exercise 10.45. Suppose that the joint distribution of X, Y is bivariate normal with parameters $\sigma_X, \sigma_Y, \rho, \mu_X, \mu_Y$ as described in Section 8.5.

(a) Compute the conditional probability density of X given $Y = y$.
(b) Find $E[X|Y]$.

Exercise 10.46. Let (X, Y) have joint density function f, let B be a subset of $\mathbb{R}$, and suppose $P(Y \in B) > 0$. Following the logic of this chapter, how would you define a conditional density function $f_X(x \mid Y \in B)$ of X, given that $Y \in B$? What integral identity should it satisfy to calculate probabilities $P(X \in A|Y \in B)$?

Exercise 10.47. Let X, Y be discrete random variables and g a function. Assuming that $P(Y = y) > 0$ and that the expectation below is well defined, show that $E[g(X) \mid Y = y] = \sum_x g(x) P(X = x \mid Y = y)$.

Hint. Use the calculation in (3.26) but with conditional probabilities.

Exercise 10.48. Let X, Y, Z be discrete random variables. Assuming that we have $P(Y = y) > 0$ and that the conditional expectations are finite, show that $E[X + Z \mid Y = y] = E[X \mid Y = y] + E[Z \mid Y = y]$.

Hint. Use the calculation in (8.5) but with conditional probabilities.

Exercise 10.49. The time T it takes me to complete a certain task is exponentially distributed with parameter λ. If I finish the job in no more than one time unit I get to flip a fair coin: heads I am paid one dollar, tails I am paid two dollars. If it takes me more than one time unit to complete the job I am simply paid one dollar. Let X be the amount I earn from the job.

(a) From the description above, write down the conditional probability mass function of X, given that $T = t$. Your answer will be a two-case formula that depends on whether $t \leq 1$ or $t > 1$.
(b) Find $E[X]$.

Exercise 10.50. In the context of the trials with uniformly distributed success probability in Section 10.4, by averaging the conditional probability mass function (10.43) of S_n we have the formula

$$P(S_n = k) = \int_0^1 P(S_n = k \mid \xi = p) f_\xi(p)\, dp = \binom{n}{k} \int_0^1 p^k (1-p)^{n-k}\, dp.$$

Show by integration by parts that $P(S_n = k) = P(S_n = k+1)$ for $0 \le k < n$, and conclude from this that $P(S_n = k) = \frac{1}{n+1}$ for $0 \le k \le n$.

Exercise 10.51. Let X and Z be independent random variables with distributions $Z \sim \mathcal{N}(0,1)$ and $X \sim \text{Bin}(n,p)$.

(a) Find $P(Z \in [-1,1], X = 3)$.
(b) Let $Y = X + Z$. Find $P(Y < 1 | X = 3)$.
(c) Find an expression for $P(Y < x)$ for any real number $x \in (-\infty, \infty)$.

Exercise 10.52. Let Y be a Bernoulli(1/2) distributed random variable. Suppose that the random variable X depends on Y in the following way: if $Y = 0$, X has Poisson distribution with mean 2, and if $Y = 1$, X has Poisson distribution with mean 3.

(a) Find $p_{Y|X}(y|k)$, that is, the conditional probability mass function of Y, given that $X = k$. Be precise about the range of admissible (y,k)-values.
(b) Find the limit $\lim_{k\to\infty} p_{Y|X}(1|k)$. Give an intuitive explanation for what you see in the limit.

Exercise 10.53. Let $Y_1, Y_2, Y_3, \dots$ be independent random variables with distribution $P(Y_j = 1) = p$ and $P(Y_j = 0) = 1 - p$, for example tosses of some biased coin. Let the random variable $X_n = (Y_{n-1}, Y_n)$ keep track of the last two tosses. Show that the random variables $X_2, X_3, X_4, \dots$ are not independent, but they are a Markov chain.

Note that the values of the random variables X_n are in the set $\{(0,0),(0,1),(1,0),(1,1)\}$ instead of a set of numbers, but this makes no difference. If desired, these values can be coded as $\{1,2,3,4\}$. Also the fact that the time index begins at 2 rather than 0 is an unimportant detail.

Challenging problems

Exercise 10.54. For a real number z let $\{z\}$ denote the *fractional part* of z, in other words, $\{z\} = z - \lfloor z \rfloor$, the difference between z and the largest integer less than or equal to z.

Now suppose that X and Y are independent random variables, $Y \sim \text{Unif}(0,1)$, and X has probability density function $f_X(x)$. Show that the fractional part $\{X + Y\}$ of the random variable $X + Y$ has Unif(0, 1) distribution.

Hint. Condition on X.

Exercise 10.55. Let $X, X_1, X_2, X_3, \dots$ be i.i.d. random variables and N an independent nonnegative integer-valued random variable. Let $S_N = X_1 + \cdots + X_N$

be the random sum. Assume that the moment generating functions $M_X(t)$ for the X_is and $M_N(t)$ for N are finite in some interval around the origin.

(a) Express the moment generating function $M_{S_N}(t)$ of S_N in terms of $M_X(t)$ and $M_N(t)$.
(b) Give an alternative proof of Wald's identity (Fact 10.35) by computing the expectation $E(S_N)$ as $M'_{S_N}(0)$.
(c) Express the second moment $E(S_N^2)$ in terms of the moments of X and N.
(d) Express the variance of S_N in terms of the means and variances of X and N.

Exercise 10.56. Let $X_1, X_2, \ldots$ be i.i.d. nonnegative integer valued random variables and N an independent nonnegative integer valued random variable. Let $S_N = X_1 + \cdots + X_N$. Compute the probability generating function of S_N in terms of the probability generating functions of X_1 and N. For the definition of the probability generating function see equation (5.18) and the surrounding discussion.

Exercise 10.57. Assume that X, Y are discrete random variables and h a function. Show that unless $h(y) = E(X \mid Y = y)$ for all y such that $p_Y(y) > 0$, we have the strict inequality

$$E\left[(X - h(Y))^2\right] > E\left[\left(X - E[X \mid Y]\right)^2\right].$$

Exercise 10.58. The *conditional variance* of a random variable X given $Y = y$ is defined as

$$\psi(y) = E\left[(X - E[X|Y = y])^2 \,\middle|\, Y = y\right]$$

and denoted by $\mathrm{Var}(X|Y = y)$. We can define the random variable $\mathrm{Var}(X|Y)$ as $\psi(Y)$.

(a) Show the following identity (sometimes called the conditional variance formula):

$$\mathrm{Var}(X) = E\left[\mathrm{Var}(X|Y)\right] + \mathrm{Var}(E[X|Y]). \tag{10.53}$$

Hint. You can try to prove this directly, for example by assuming that X and Y are discrete. Or you can start from identity (10.42) with the constant function $h(y) = E[X]$.

(b) Show that the variance of $E[X|Y]$ cannot be larger than the variance of X.

Exercise 10.59. Suppose that $X_1, X_2, X_3, \ldots$ are i.i.d. random variables with finite mean μ_X and variance σ_X^2. Let N be an independent nonnegative integer valued random variable with mean μ_N and variance σ_N^2. Use the conditional variance formula (10.53) to find the variance of the random sum $S_N = X_1 + \cdots + X_N$.

Exercise 10.60. (Gambler's ruin revisited) Recall the setup of Exercise 2.85. If a fair coin flip comes up heads, you win a dollar, otherwise you lose a dollar. You start with N dollars, and play repeatedly until you either reach M dollars or lose all your money, whichever comes first. M and N are fixed positive integers such that $0 < N < M$.

(a) Find the expected length of the game if the coin is fair.
(b) Find the expected length of the game if the coin is biased, in other words, if the probability of heads is $p \neq 1/2$.

Hint. The hint for Exercise 2.85 still applies.

Exercise 10.61. Suppose that the random variables (X, Y) have bivariate normal distribution with $\mu_X = \mu_Y = 0$. (See Section 8.5 for the definition.) Compute the conditional density function $f_{X|Y}(x|y)$ and identify the conditional distribution of X given $Y = y$. Your answer will depend on the parameters σ_X^2, σ_Y^2 and ρ.

Appendix A

Things to know from calculus

The list below covers the calculus topics used in the book. Review is suggested in case the reader lacks familiarity with some of them.

(i) The concept of a function, its domain and range.

(ii) The definition and basic properties of limits and continuity. In particular, the following limit shows up in Chapter 4:

$$\lim_{s\to\infty}\left(1+\frac{x}{s}\right)^{s} = e^{x} \qquad \text{for all real } x. \tag{A.1}$$

(iii) The definition, properties and basic techniques of derivatives.

(iv) The definition, properties and basic techniques of integrals, including integration by parts and improper integrals. The connection between differentiation and integration (fundamental theorem of calculus).

(v) Sequences, series and Taylor series expansions. The following series expansions are especially important:

$$\frac{1}{1-x} = 1 + x + x^2 + \cdots = \sum_{n=0}^{\infty} x^n \qquad \text{for all } x \in (-1, 1), \tag{A.2}$$

$$e^{x} = \sum_{n=0}^{\infty} \frac{x^n}{n!} \qquad \text{for all real } x. \tag{A.3}$$

(vi) Partial derivatives of multivariate functions.

(vii) Evaluation of multivariate integrals over regions defined by inequalities. Conversion of a multivariate integral into iterated single variable integrals. The reader should be able to carry out computations like this:

$$\iint\limits_{\substack{1\le x\le y\\ 1\le y\le 2}} x^2 y\,dx\,dy = \int_1^2 \int_1^y x^2 y\,dx\,dy = \int_1^2 (y^3/3 - 1/3)y\,dy = \frac{47}{30}.$$

Appendix B

Set notation and operations

This section reviews basic set theory to the extent needed for the book.

Let Ω be a set. In this text we typically think of Ω as the sample space of a particular experiment.

Subsets of Ω are collections of *elements*, or *points*, of Ω. For an element ω of Ω and a subset A of Ω, either $\omega \in A$ (ω is a member of A) or $\omega \notin A$ (ω is not a member of A). Between two subsets A and B of Ω, either $A \subseteq B$ (A is a subset of B, which means that every element of A is also an element of B) or $A \not\subseteq B$ (A is not a subset of B, which means there is at least one element of A that is not an element of B). If $A \subseteq B$ and $B \subseteq A$ then $A = B$, the sets A and B are equal (every element of A is an element of B, and vice versa). Alternative notation for $A \subseteq B$ is $A \subset B$.

The subset $A = \{\omega\}$ that consists of the single point ω must not be confused with the point ω itself. In particular, $\omega \in \{\omega\}$ but the point ω is not the same as the set $\{\omega\}$. A set with just one point is called a *singleton*. Two special subsets of Ω are the empty set, denoted by $\varnothing$, and Ω itself. The empty set contains no points: $\omega \notin \varnothing$ for every $\omega \in \Omega$. Every subset A of Ω satisfies $\varnothing \subseteq A \subseteq \Omega$.

Example B.1. Let $\Omega = \{1, 2, 3, 4, 5, 6\}$. Examples of subsets are

$$A = \{2, 4, 6\}, \quad B = \{1, 2, 3, 4\} \quad \text{and} \quad C = \{1\}.$$

There are alternative ways of specifying sets. We can equivalently write

$$A = \{\omega \in \Omega : \omega \text{ is even}\} \quad \text{and} \quad B = \{\omega \in \Omega : 1 \le \omega \le 4\}.$$

Check that you understand why each of these statements is true:

$$2 \in A, \quad 5 \notin A, \quad C \subseteq B, \quad C \not\subseteq A, \quad A \neq B.$$

▲

Elements of a set are never repeated. Writing $A = \{2, 4, 4, 6\}$ does not make sense and does not mean that there are somehow two "copies" of 4 in the set A.

Unions, intersections, complements and differences are operations that produce new sets from given sets. Here are the definitions for two subsets A and B of Ω:

union	$A \cup B = \{\omega \in \Omega : \omega \in A \text{ or } \omega \in B\}$
intersection	$A \cap B = \{\omega \in \Omega : \omega \in A \text{ and } \omega \in B\}$
complement	$A^c = \{\omega \in \Omega : \omega \notin A\}$
difference	$A \setminus B = \{\omega \in \Omega : \omega \in A \text{ and } \omega \notin B\}$.

Note that in mathematical usage "or" is an inclusive or: if S_1 and S_2 are two statements, then "S_1 or S_2" means "S_1 or S_2 or both." The definitions above imply these general facts: $A \setminus B = A \cap B^c$, $A^c = \Omega \setminus A$, and $(A^c)^c = A$. The complement of a set depends very much on the choice of the space in which the discussion is conducted.

Intersections can be also denoted by $AB = A \cap B$. This is analogous to the notation for multiplication that drops the operation symbol: $ab = a \cdot b$.

Two sets A and B are *disjoint* if they have no common elements, that is, if $A \cap B = \varnothing$.

Set operations on two or three sets can be visualized with Venn diagrams. Sets are represented by overlapping disks or ovals within a rectangle that represents Ω. The results of set operations are represented by shading or coloring. In the two Venn diagrams of Figure B.1 the shaded parts represent the sets $A \cap B^c$ and $(A \cap B^c) \cup (A \cap C)$.

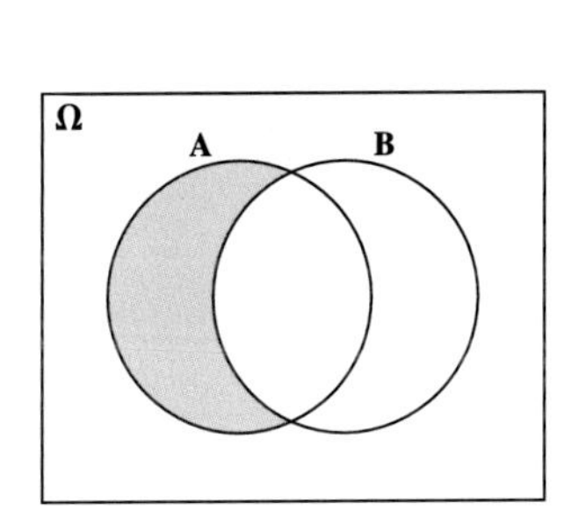

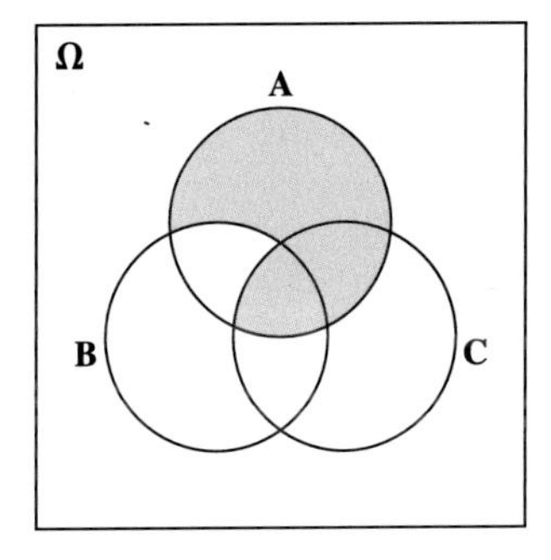

Figure B.1. Venn diagrams for $A \cap B^c$ and $(A \cap B^c) \cup (A \cap C)$.

Example B.2. Continuing from Example B.1, the following are true statements:

$$A \cup B = \{1,2,3,4,6\}, \quad A \cap B = \{2,4\}, \quad A \cap C = \varnothing, \quad A \setminus B = \{6\}, \quad B^c = \{5,6\}.$$

▲

When we form unions and intersections of many sets, it is convenient to distinguish different sets with a numerical index, instead of trying to use a different letter for each set. If A_1, A_2, A_3 and A_4 are subsets of Ω, their union can be expressed in these ways:

$$\begin{aligned} A_1 \cup A_2 \cup A_3 \cup A_4 &= \bigcup_{i=1}^{4} A_i \\ &= \{\omega \in \Omega : \omega \in A_i \text{ for at least one index value } i \in \{1,2,3,4\}\} \end{aligned}$$

The concise middle expression $\bigcup_{i=1}^4 A_i$ reads "the union of the sets A_i where i ranges from 1 to 4." This notation accommodates any number of sets: the union of sets $A_1, A_2, \dots, A_n$ is expressed as $\bigcup_{i=1}^n A_i$. It works just as well for infinitely many sets: if we have an infinite sequence $A_1, A_2, A_3, \dots$ of sets, their union is

$$\bigcup_{i=1}^{\infty} A_i = \{\omega \in \Omega : \omega \in A_i \text{ for at least one index } i = 1, 2, 3, \dots\}.$$

Example B.3. This example illustrates indexed sets. Let $\Omega = \{1, 2, 3, \dots, 10\}$ be the set of positive integers up to 10. Define these subsets of Ω:

$$A_1 = \{1, 2, 3\}, \ A_2 = \{2, 3, 4, 6\}, \ A_3 = \{2, 3, 6, 7\}, \ A_4 = \{6, 7, 8\}.$$

Then we have for example these two unions

$$\bigcup_{i=1}^{3} A_i = \{1, 2, 3, 4, 6, 7\} \quad \text{and} \quad \bigcup_{i=1}^{4} A_i = \{1, 2, 3, 4, 6, 7, 8\},$$

and these two intersections

$$\bigcap_{i=1}^{3} A_i = \{2, 3\} \quad \text{and} \quad \bigcap_{i=1}^{4} A_i = \varnothing.$$

▲

When the range of the index is understood from the context, or we want to leave it arbitrary, we can drop the range of the index from the notation:

$$\bigcup_{i} A_i = \{\omega \in \Omega : \omega \in A_i \text{ for at least one index } i\}$$

and

$$\bigcap_{i} A_i = \{\omega \in \Omega : \omega \in A_i \text{ for each index } i\}.$$

The last formula above reads "the intersection of the sets A_i."

Of fundamental importance are identities known as *de Morgan's laws*:

$$\left(\bigcup_{i} A_i\right)^c = \bigcap_{i} A_i^c \tag{B.1}$$

and

$$\left(\bigcap_{i} A_i\right)^c = \bigcup_{i} A_i^c. \tag{B.2}$$

Let us see why (B.1) is true. We prove it with a sequence of logical equivalences. For two statements S_1 and S_2, $S_1 \iff S_2$ reads "S_1 if and only if S_2" or "S_1 and S_2 are equivalent." This means that S_1 and S_2 are either simultaneously true or simultaneously false.

Here is the proof of identity (B.1). Let $\omega \in \Omega$. Then

$$\begin{aligned} \omega \in \Big(\bigcup_i A_i\Big)^c &\iff \omega \notin \bigcup_i A_i \\ &\iff \omega \notin A_i \text{ for each index } i \\ &\iff \omega \in A_i^c \text{ for each index } i \\ &\iff \omega \in \bigcap_i A_i^c. \end{aligned}$$

This establishes the equality of the two sides of (B.1) because any point ω is either a member of both sides or not a member of either side. A similar argument can be used to prove identity (B.2). Alternatively, (B.2) can be derived directly from (B.1).

Example B.4. Let us see de Morgan's law in action in Example B.1:

$$(A \cup B)^c = \{1, 2, 3, 4, 6\}^c = \{5\} \quad \text{and} \quad A^c \cap B^c = \{1, 3, 5\} \cap \{5, 6\} = \{5\}.$$

Thus $(A \cup B)^c = A^c \cap B^c$ as (B.1) asserts. ▲

Exercises

Exercise B.1. Let A, B and C be subsets of the set Ω.

(a) Let D be the set of elements that are in exactly two of the sets A, B and C. Using unions, intersections and complements express D in terms of A, B, and C.

(b) Let E be the set of elements that are in at least two of the sets A, B and C. Using unions, intersections and complements express E in terms of A, B, and C.

There may be more than one way to answer these questions.

Exercise B.2. Let A, B and C be subsets of the set Ω. Various other sets are described below in words. Use unions, intersections and complements to express these in terms of A, B and C. Drawing Venn diagrams may help.

(a) The set of elements that are in each of the three sets.
(b) The set of elements that are in A but neither in B nor in C.
(c) The set of elements that are in at least one of the sets A or B.
(d) The set of elements that are in both A and B but not in C.
(e) The set of elements that are in A, but not in B or C or both.

Exercise B.3. Let Ω be the set $\{1, 2, \dots, 100\}$, and let A, B and C be the following subsets of Ω:

$$A = \{\text{positive even numbers which are at most } 100\},$$

$$B = \{\text{two-digit numbers where the digit 5 appears}\},$$

$$C = \{\text{positive integer multiples of 3 which are at most 100}\},$$
$$D = \{\text{two-digit numbers such that the sum of the digits is 10}\}.$$

List the elements of each of the following sets.

(a) $B \setminus A$

(b) $A \cap B \cap C^c$

(c) $((A \setminus D) \cup B) \cap (C \cap D)$

Exercise B.4. Show that identity (B.2) is true.

Exercise B.5. If A and B are sets then let $A \triangle B$ denote their symmetric difference: the set of elements that are in exactly one of the two sets. (In words this is "A or B, but not both.")

(a) Express $A \triangle B$ from A and B using unions, intersections and complements.
(b) Show that for any three sets A, B and C we have

$$A \triangle (B \triangle C) = (A \triangle B) \triangle C.$$

Exercise B.6. Suppose that A and B are subsets of Ω. Let $E = A \cap B$ and $F = A \cap B^c$.

(a) Show that E and F are disjoint, i.e. that $E \cap F = \varnothing$.
(b) Show that $A = E \cup F$.

Exercise B.7. Consider the following sets:

$$A = \{1, 2, 3, 4, 5, 6, 7\}, \quad B = \{2, 4, 6, 8\}, \quad C = \{1, 3, 4, 7\}.$$

(a) Decide whether the set $D = \{1, 3, 7\}$ can be expressed from A, B and C using unions, intersections and complements.
(b) Decide whether the set $E = \{2, 4\}$ can be expressed from A, B and C using unions, intersections and complements.

Challenging problems

Exercise B.8. Let A, B and C be subsets of Ω. Give a list of *disjoint* sets $D_1, D_2, \dots$ so that *any nonempty set* constructed from A, B and C using unions, intersections and complements is the union of some of the sets D_i.

Hint. A Venn diagram could help.

Exercise B.9. Let $A_1, A_2, \dots, A_n$ be subsets of Ω. Give a list of *disjoint* sets $B_1, B_2, \dots$ so that *any nonempty set* constructed from $A_1, A_2, \dots, A_n$ using unions, intersections and complements is the union of some of the sets B_i.

Appendix C

Counting

The number of elements in a given set A is denoted by $\#A$. This is also called the *cardinality* of the set A. This section is a quick overview of simple counting techniques. We develop methods of finding $\#A$ from a description of the set A. We give some proofs, but on first reading we suggest focusing on the examples. For more details and refined counting ideas consult an introductory combinatorics textbook such as [Bru10].

The following obvious basic principle will be used over and over again. Its formalization in terms of a function may seem abstract but it is useful for building mathematics on this idea.

Fact C.1. Let A and B be finite sets and k a positive integer. Assume that there is a function f from A onto B so that each element of B is the image of exactly k elements of A. (Such a function is called *k-to-one.*) Then $\#A = k \cdot \#B$.

In plain English Fact C.1 seems almost too simple to state: if A can cover B exactly k times over, then the number of elements in A is k times the number of elements in B. In the case $k = 1$ the function f is one-to-one and onto (that is, a bijection), the elements of A and B can be paired up exactly, and the sets A and B are of the same size. See Figure C.1 for a graphical illustration.

Example C.2. Four fully loaded 10-seater vans transported people to the picnic. How many people were transported? Clearly the answer is $10 \cdot 4 = 40$. Here A is the set of people, B is the set of vans, and f maps a person to the van she rides in. $\#A = 40$, $\#B = 4$, and f is a 10-to-one function from A onto B. ▲

Most of our counting methods will be based on the idea of multiplying along a sequence of choices made one at a time. This is made precise in the next fact.

An *n-tuple* is an ordered sequence of n elements. $(7, 1, 4, 2)$ is an example of a 4-tuple of integers. The term *ordered* means here that *order matters*. In other words, $(7, 1, 4, 2)$ and $(7, 4, 1, 2)$ are distinct 4-tuples. To emphasize this we use also the term *ordered n-tuple*. $7, 1, 4, 2$ are the *entries* of the 4-tuple $(7, 1, 4, 2)$.

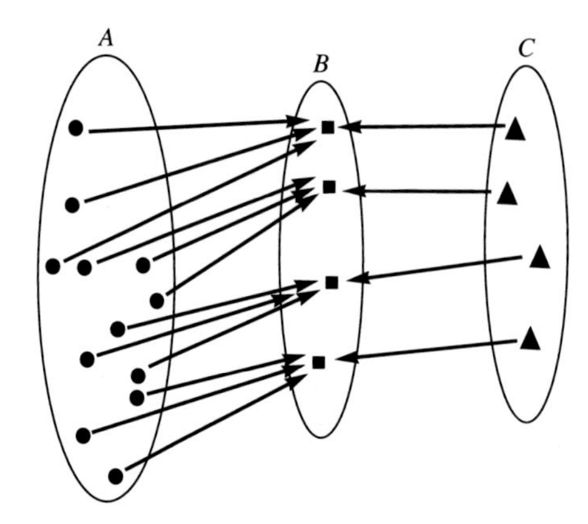

Figure C.1. A graphical illustration of Fact C.1. Set A has 12 elements represented by bullets, set B has 4 elements represented by squares, and set C has 4 elements represented by triangles. Each element of B is matched up with exactly three element of A and each element of C is matched up with one element of B. Thus $\#A = 3 \cdot \#B$. The elements of B and C are matched up one-to-one, so $\#B = \#C$.

Fact C.3. Suppose that a set of n-tuples $(a_1, \ldots, a_n)$ obeys these rules.

(i) There are r_1 choices for the first entry a_1.
(ii) Once the first k entries $a_1, \ldots, a_k$ have been chosen, the number of alternatives for the next entry a_{k+1} is r_{k+1}, regardless of the previous choices.

Then the total number of n-tuples is the product $r_1 \cdot r_2 \cdot r_3 \cdots r_n$.

Proof. The proof is based on induction on the length n of the n-tuple and on Fact C.1. A review of proof by induction is given at the end of this section.

The case $n = 1$ is clear because then we are not extending the n-tuple beyond the r_1 choices for a_1. So clearly r_1 is the number of 1-tuples (a_1) with r_1 choices of a_1.

Now assume that the statement is true for n-tuples. We prove it for $(n+1)$-tuples. Let A be the set of all $(n+1)$-tuples $(a_1, \ldots, a_n, a_{n+1})$ and let B be the set of all n-tuples $(a_1, \ldots, a_n)$, all obeying the scheme described in Fact C.3. Define the function f from A onto B by setting $f(a_1, \ldots, a_n, a_{n+1}) = (a_1, \ldots, a_n)$. The assumption that there are r_{n+1} choices for a_{n+1} regardless of $(a_1, \ldots, a_n)$ says exactly that f is an r_{n+1}-to-one function from A onto B. By Fact C.1, $\#A = r_{n+1} \cdot \#B$. By the induction assumption $\#B = r_1 \cdots r_n$. Together these imply that $\#A = r_1 \cdots r_n \cdot r_{n+1}$. In other words, the number of $(n+1)$-tuples is $r_1 \cdots r_{n+1}$.

We have extended the validity of Fact C.3 from n to $n+1$. This concludes the proof by induction. ▲

Fact C.3 is sometimes called the *general multiplication principle*. Note that while the number r_{k+1} of choices at step $k+1$ cannot be allowed to depend on the previous choices $(a_1, \ldots, a_k)$, the actual alternatives available to a_{k+1} can depend on $(a_1, \ldots, a_k)$.

Example C.4. To dress up for school in the morning, Joyce chooses from 3 dresses (red, yellow, or green), 3 blouses (also red, yellow, or green), and 2 pairs of shoes. She refuses to wear a dress and a blouse of matching colors. How many different outfits can she choose from?

We can imagine that Joyce goes through the choices one by one: first the dress, then the blouse, and finally the shoes. There are three choices for the dress. Once the dress is chosen, there are two choices for the blouse, and finally two choices for the shoes. By the multiplication principle, she has $3 \cdot 2 \cdot 2 = 12$ different outfits.

Note that one can get this result also by counting all the outfits without the color restriction and then subtracting the number of forbidden outfits. ▲

The construction of the n-tuples in Fact C.3 can be visualized by a *decision tree*, with arrows pointing to the available choices of the subsequent step. In Figure C.2 you can see the decision tree for Example C.4. The dresses are the triangles and the blouses the squares, labeled by the colors. The shoes are the circles. The possible outfits resulting from the three choices can be read off from the directed paths through the tree along the arrows.

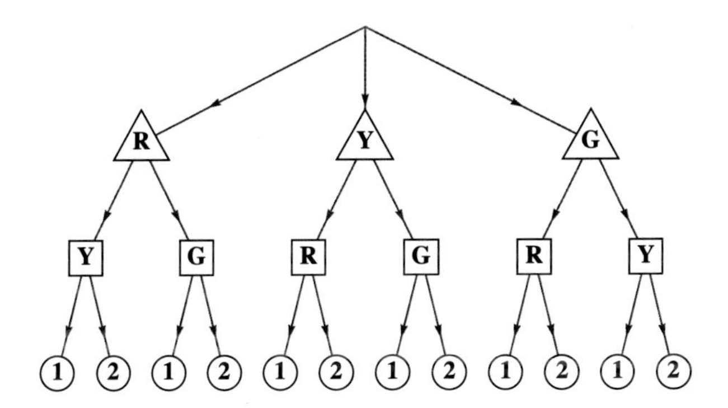

Figure C.2. Decision tree for Example C.4. The dresses are the triangles and the blouses the squares, labeled by the colors. The shoes are the circles. The different outfits are represented by the directed paths through the tree.

We turn to use Fact C.3 to derive counting formulas for various situations encountered in this book.

Let $A_1, A_2, \dots, A_n$ be finite sets. The *Cartesian product* of these sets is by definition the set of ordered n-tuples where the ith entry comes from A_i:

$$A_1 \times A_2 \times \cdots \times A_n = \{(x_1, \dots, x_n) : x_i \in A_i \text{ for } i = 1, \dots, n\}.$$

These n-tuples can be constructed by choosing one entry at a time, with $\#A_i$ choices for x_i. Hence as an immediate corollary of Fact C.3 we can count the number of n-tuples in a finite Cartesian product.

Fact C.5. Let $A_1, A_2, \ldots, A_n$ be finite sets.

$$\#(A_1 \times A_2 \times \cdots \times A_n) = (\#A_1) \cdot (\#A_2) \cdots (\#A_n) = \prod_{i=1}^{n} (\#A_i).$$

The capital pi notation $\prod$ is a shorthand for a product, reviewed in Appendix D. Here are examples of Cartesian products and their cardinalities.

Example C.6. In a certain country license plates have three letters followed by three digits. How many different license plates can we construct if the country's alphabet contains 26 letters?

Each license plate can be described as an element of the set $A \times A \times A \times B \times B \times B$ where A is the set of letters and B is the set of digits. Since $\#A = 26$ and $\#B = 10$, the answer is $26^3 \cdot 10^3 = 17{,}576{,}000$. ▲

Example C.7. We flip a coin three times and then roll a die twice. We record the resulting sequence of outcomes in order. How many different sequences are there?

The outcome of a coin flip is an element of the set $C = \{\mathrm{H}, \mathrm{T}\}$ and a die roll gives an element of the set $D = \{1, 2, 3, 4, 5, 6\}$. Each possible sequence is an element of the set $C \times C \times C \times D \times D$, which has $2^3 \cdot 6^2 = 288$ elements. ▲

Example C.8. How many distinct subsets does a set of size n have?

The answer is 2^n. Each subset can be encoded by an n-tuple with entries 0 or 1, where the ith entry is 1 if the ith element of the set is in the subset (and 0 if it is not). Thus the number of subsets is the same as the number of elements in the Cartesian product

$$\underbrace{\{0, 1\} \times \cdots \times \{0, 1\}}_{n \text{ times}} = \{0, 1\}^n,$$

which is 2^n. Note that the empty set $\varnothing$ and set itself are included in the count. These correspond to the n-tuples $(0, 0, \ldots, 0)$ and $(1, 1, \ldots, 1)$, respectively. ▲

Another classical application of the multiplication principle of Fact C.3 is counting permutations.

Fact C.9. Consider all k-tuples $(a_1, \ldots, a_k)$ that can be constructed from a set A of size n $(n \geq k)$ without repetition. So each $a_i \in A$ and $a_i \neq a_j$ if $i \neq j$. The total number of these k-tuples is

$$(n)_k = n \cdot (n-1) \cdots (n-k+1) = \frac{n!}{(n-k)!}.$$

In particular, with $k = n$, each n-tuple is an ordering or a *permutation* of the set A. So the total number of orderings of a set of n elements is $n! = n \cdot (n-1) \cdots 2 \cdot 1$.

Proof. Construct these k-tuples sequentially. Choose the first entry out of the full set A with n alternatives. If we have the first j entries chosen then the next one can be any one of the remaining $n - j$ elements of A. Thus the total number of k-tuples is the product $n \cdot (n-1) \cdots (n-k+1)$. ▲

The quantities that appeared in Fact C.9 are the *descending factorial* $(n)_k$ and the familiar *n factorial* $n!$. Note that this is an example of the multiplication principle where the choices available for a_{j+1} definitely depend on the previous choices $(a_1, \ldots, a_j)$, but the number of available alternatives is $n - j$ regardless of the previous choices.

Example C.10. I have a 5-day vacation from Monday to Friday in Santa Barbara. I have to specifically assign one day to exploring the town, one day to hiking the mountains, and one day for the beach. I can take up at most one activity per day. In how many ways can I schedule these activities?

In terms of Fact C.9, the underlying set has 5 elements and we choose a 3-tuple. For example, (`Mon`, `Fri`, `Tue`) means Monday in town, Friday for hiking, and Tuesday on the beach. The number of choices is $5 \cdot 4 \cdot 3 = 60$. ▲

Example C.11. Consider a round table with 8 seats.

(a) In how many ways can we seat 8 guests around the table?
(b) In how many ways can we do this if we do not differentiate between seating arrangements that are rotations of each other?

Number the seats from 1 to 8 and similarly number the guests. Part (a) asks for the number of ways we can assign customers $1, 2, \ldots, 8$ to seats $1, 2, \ldots, 8$. This is the same as the number of ways of ordering the set $\{1, 2, \ldots, 8\}$, which is $8!$.

For part (b) note that among the rotations of a given seating arrangement is a unique one in which guest 1 occupies seat 1. Thus we just have to count the number of arrangements where guest 1 sits in seat 1. This means seating the remaining 7 guests to 7 seats, which can be done in $7!$ different ways.

An alternative solution to (b) is to take the total number $8!$ of seatings from (a) and divide by the number 8 of distinct rotations of each arrangement. (Notice Fact C.1 here in action.) ▲

Next we turn to count unordered selections, that is, sets.

Fact C.12. Let n and k be nonnegative integers with $0 \leq k \leq n$. The number of distinct subsets of size k that a set of size n has is given by the **binomial coefficient**

$$\binom{n}{k} = \frac{n!}{k!(n-k)!}.$$

Note that the convention for zero factorial is $0! = 1$.

Proof of Fact C.12. Let A be a set of size n. By Fact C.9, $\frac{n!}{(n-k)!}$ ordered k-tuples without repetition can be constructed from elements of A. Each subset of A of size k has exactly $k!$ different orderings, and hence appears exactly $k!$ times among the ordered k-tuples. Thus by Fact C.1 the number of subsets of size k is $\frac{n!}{k!(n-k)!}$. ▲

Remark C.13. Convention $0! = 1$ gives $\binom{n}{0} = 1$ which makes sense because the empty set $\varnothing$ is the unique subset with zero elements. The binomial coefficient $\binom{n}{k}$ is defined as 0 for integers $k < 0$ and $k > n$. This is consistent with Fact C.12. ▲

Example C.14. In a class there are 12 boys and 14 girls. How many different teams of 7 pupils with 3 boys and 4 girls can be created?

The solution combines the multiplication principle Fact C.3 with Fact C.12. Imagine that first we choose the 3 boys and then the 4 girls. We can choose the 3 boys in $\binom{12}{3}$ different ways and the 4 girls in $\binom{14}{4}$ different ways. Thus we can form the team in $\binom{12}{3} \cdot \binom{14}{4}$ different ways. ▲

The example above illustrates that even though the multiplication principle Fact C.3 was expressed in terms of *ordered* n-tuples, it can be used in situations where the outcome is not ordered. In Example C.14 there is a "hidden" 2-tuple (B, G) where B is the set of boys chosen $\big(\binom{12}{3}$ choices$\big)$ and G is the set of girls chosen $\big(\binom{14}{4}$ choices$\big)$. Fact C.3 then says that the number of 2-tuples (B, G) is $\binom{12}{3} \cdot \binom{14}{4}$. This is the counting. It is then merely a matter of language that the final outcome is given as the set $B \cup G$ of boys and girls, rather than as the 2-tuple (B, G).

Practical advice

It can be helpful to practice counting techniques with examples small enough that you can verify your answer by listing all the possibilities. Counting problems can often be solved in several different ways, which can be used as checks against each other. The next example illustrates.

Example C.15. Let $A = \{1, 2, 3, 4, 5, 6\}$.

(a) How many different two-element subsets of A are there that have one element from $\{1, 2, 3, 4\}$ and one element from $\{5, 6\}$?

The exhaustive list shows that the answer is 8:

$$\{1,5\}, \{1,6\}, \{2,5\}, \{2,6\}, \{3,5\}, \{3,6\}, \{4,5\}, \{4,6\}.$$

We get the correct answer by first picking one element from $\{1,2,3,4\}$ (four choices) and then one element from $\{5,6\}$ (two choices) and multiplying: $4 \cdot 2 = 8$.

Another way to arrive at the same answer is to take

$$\binom{6}{2} - \binom{4}{2} - \binom{2}{2} = 15 - 6 - 1 = 8.$$

Here we start with the total number of 2-element subsets of A and subtract the number of those that lie entirely inside $\{1,2,3,4\}$ and those that lie entirely inside $\{5,6\}$.

(b) How many ordered pairs can be formed from elements of A that have one element from $\{1,2,3,4\}$ and one element from $\{5,6\}$? An ordered pair is the same as an ordered 2-tuple.

Here is the exhaustive list of 16:

$$(1,5), (1,6), (2,5), (2,6), (3,5), (3,6), (4,5), (4,6),$$
$$(5,1), (5,2), (5,3), (5,4), (6,1), (6,2), (6,3), (6,4).$$

We can calculate the answer in at least two ways.

(i) Take the answer from (a) and multiply by 2 since each 2-element set can be ordered in 2 ways.

(ii) Account for all the ways an ordered pair (a,b) can be constructed. If a comes from $\{1,2,3,4\}$ (4 choices) then b must come from $\{5,6\}$ (2 choices). That makes $4 \cdot 2 = 8$. The other possibility is that a comes from $\{5,6\}$ (2 choices) and b comes from $\{1,2,3,4\}$ (4 choices). That makes $2 \cdot 4 = 8$. Altogether $4 \cdot 2 + 2 \cdot 4 = 16$.

(c) Color the numbers 1, 2 red, the numbers 3, 4 green, and the numbers 5, 6 yellow. How many different two-element subsets of A are there that have two different colors?

The exhaustive list has 12 different possibilities:

$$\{1,3\}, \{1,4\}, \{1,5\}, \{1,6\}, \{2,3\}, \{2,4\}, \{2,5\}, \{2,6\}, \{3,5\}, \{3,6\}, \{4,5\}, \{4,6\}.$$

To count, choose first two colors: $\binom{3}{2} = 3$ choices. Then from both colors choose one element: 2 choices, and another 2 choices. Altogether $3 \cdot 2 \cdot 2 = 12$. ▲

One way to view $\binom{n}{k}$ is as the number of ways of painting n elements with two colors, red and yellow, with k red and $n-k$ yellow elements. Let us generalize to more than two colors.

> **Fact C.16.** Let n and r be positive integers and $k_1, \dots, k_r$ nonnegative integers such that $k_1 + \cdots + k_r = n$. The number of ways of assigning labels $1, 2, \dots, r$ to n items so that, for each $i = 1, 2, \dots, r$, exactly k_i items receive label i, is the **multinomial coefficient**
>
> $$\binom{n}{k_1, k_2, \dots, k_r} = \frac{n!}{k_1!\, k_2! \cdots k_r!}.$$

Proof. We assign the labels in stages and use the general multiplication principle.

The number of ways of choosing k_1 items to receive label 1 is

$$\binom{n}{k_1} = \frac{n!}{k_1!(n-k_1)!}.$$

After assigning labels 1 we have $n - k_1$ items left to label. The number of ways of labeling k_2 of these as 2 is

$$\binom{n-k_1}{k_2} = \frac{(n-k_1)!}{k_2!(n-k_1-k_2)!}.$$

After assigning labels 1 and 2 we have $n - k_1 - k_2$ items left to label. The number of ways of labeling k_3 of these as 3 is

$$\binom{n-k_1-k_2}{k_3} = \frac{(n-k_1-k_2)!}{k_3!(n-k_1-k_2-k_3)!}.$$

And so we continue. After assigning labels $1, 2, \dots, k_{r-2}$ we have $n - k_1 - k_2 - \cdots - k_{r-2}$ items left to label. The number of ways of labeling k_{r-1} of these as $r - 1$ is

$$\binom{n-k_1-k_2-\cdots-k_{r-2}}{k_{r-1}} = \frac{(n-k_1-k_2-\cdots-k_{r-2})!}{k_{r-1}!(n-k_1-k_2-\cdots-k_{r-2}-k_{r-1})!}$$
$$= \frac{(n-k_1-k_2-\cdots-k_{r-2})!}{k_{r-1}!k_r!}.$$

The remaining k_r unlabeled items are assigned label r. The labeling is complete. The number of ways of performing the entire labeling assignment is the product

$$\frac{n!}{k_1!(n-k_1)!} \cdot \frac{(n-k_1)!}{k_2!(n-k_1-k_2)!} \cdots \frac{(n-k_1-k_2-\cdots-k_{r-2})!}{k_{r-1}!k_r!}$$
$$= \frac{n!}{k_1!\, k_2! \cdots k_r!}$$

where the last formula comes from canceling the factorials of the form $(n - k_1 - \cdots - k_i)!$.

Alternative proof, utilizing permutation, the multiplication principle, and Fact C.1. Order the n items in some manner. The first k_1 items form group 1, the next k_2 items form group 2, and so on, with group i consisting of the items in the range $k_1 + \cdots + k_{i-1} + 1$ to $k_1 + \cdots + k_i$. Assign label i to the items in group i. A moment's thought should convince that considering all $n!$ orderings gives us every possible labeling. But how many orderings give the *same* labeling? As long as the groups do not change, the labeling remains the same. The number of orderings that keep the groups the same is $k_1!\, k_2! \cdots k_r!$, because we can reorder group 1 in $k_1!$ different ways, group 2 in $k_2!$ different ways, and so on. Thus every labeling arises from $k_1!\, k_2! \cdots k_r!$ different orderings. So by Fact C.1, the number of labelings is $n!/(k_1!\, k_2! \cdots k_r!)$ which is the multinomial coefficient. ▲

Example C.17. 120 students signed up for a class. The class is divided into 4 sections numbered 1, 2, 3 and 4, which will have 25, 30, 31 and 34 students, respectively. How many ways are there to divide up the students among the four sections?

This is an application of Fact C.16 with $n = 120$, $r = 4$, $k_1 = 25$, $k_2 = 30$, $k_3 = 31$ and $k_4 = 34$. The answer is

$$\binom{120}{25, 30, 31, 34} = \frac{120!}{25! \cdot 30! \cdot 31! \cdot 34!} \qquad ▲$$

Example C.18. In how many different ways can six people be divided into three pairs?

This sounds like a multinomial question, but there is a subtlety. Look again at the precise wording of Fact C.16: the multinomial coefficient counts the number of ways of assigning *specific labels* to the elements. So $\binom{6}{2,2,2}$ is the number of ways of dividing the six people into Pair 1, Pair 2 and Pair 3. Abbreviate the names of the six people as A, B, C, D, E and F. Among the assignments that are counted separately are for example these six:

Pair 1	AB	AB	CD	CD	EF	EF
Pair 2	CD	EF	AB	EF	AB	CD
Pair 3	EF	CD	EF	AB	CD	AB

However, notice that all six arrangements lead to the same pairs of people. We are simply permuting the three labels Pair 1, Pair 2 and Pair 3 on the three pairs AB, CD and EF. In other words, $\binom{6}{2,2,2}$ counts each division into pairs $3! = 6$ times. The number of distinct ways of dividing six people into pairs is then

$$\frac{1}{6}\binom{6}{2, 2, 2} = \frac{90}{6} = 15. \qquad ▲$$

Counting patterns in playing cards makes for good examples of the ideas of this section. In a standard deck of 52 cards each card has a *suit* (hearts ♡, diamonds ♢, spades ♠, or clubs ♣) and a *rank* (2, 3, ..., 9, 10, J, Q, K, A, in this order). So one mathematical view of a deck of cards is as the Cartesian product of the set of suits and the set of ranks. The *face cards* are the jack (J), the queen (Q), the king

(K) and the ace (A). The ace can be flexibly regarded as both the highest and the lowest rank card.

A *poker hand* is an unordered selection of five cards from the deck without replacement. Poker hands fall into various *categories* depending on the patterns in the hand. One category is to have *one pair*. This means that the hand has two cards of the same rank (the pair) and three cards that have ranks different from each other and from the pair. Exercise C.10 contains the descriptions of the other poker categories.

Example C.19. How many poker hands are in the category *one pair*?

While a poker hand is by definition unordered, we can count the number of hands with exactly one pair in two ways: we can build an ordered hand and then divide by 5! to remove the overcounting (Fact C.1 at work again), or we can build an unordered hand.

An ordered solution is $\binom{5}{2} \cdot 52 \cdot 3 \cdot 48 \cdot 44 \cdot 40$ divided by 5!. Think of filling 5 ordered positions or slots with the cards of the hand. First choose 2 slots out of the 5 for the pair, $\binom{5}{2}$ choices. Then assign cards to these two slots in order (any card out of the 52 for the first slot, and another one of the same rank for the second slot). Last fill in the 3 remaining slots in order but so that ranks are not repeated.

An unordered solution is $13 \cdot \binom{4}{2} \cdot \binom{12}{3} \cdot 4^3$. Namely, first choose the rank for the pair, then two suits for that rank, then the remaining three ranks, and finally a suit for each of the three ranks: first a suit for the lowest rank, then a suit for the middle rank, and finally a suit for the top rank of the three.

Both approaches give the numerical answer 1,098,240. ▲

Proof by induction

This is a method for verifying the truth of a statement for all values of a positive integer parameter. We give a formal statement of how proof by induction works and then illustrate it with an example.

Fact C.20. (Proof by induction) Let $S(n)$, $n = 1, 2, 3, \ldots$, be a sequence of statements indexed by positive integers n. Suppose we can verify these two claims.

(i) $S(1)$ is true.
(ii) For each n, the truth of $S(n)$ implies the truth of $S(n+1)$.

Then it follows that $S(n)$ is true for all positive integers $n = 1, 2, 3, \ldots$

The validity of Fact C.20 itself comes from the following train of reasoning. By (i) $S(1)$ is true. Since $S(1)$ is true, by (ii) also $S(2)$ is true. Since $S(2)$ is true, by (ii) also $S(3)$ is true. And so on, so that the truth of each $S(n)$ follows in turn.

In the induction proof scheme of Fact C.20, point (i) is called the *base case* and point (ii) the *induction step.*

Example C.21. Verify that for all positive integers n we have

$$1 + 2 + 3 + \cdots + n = \frac{n(n+1)}{2}. \tag{C.1}$$

Now $S(n)$ is the statement that inequality (C.1) holds for this particular n. Here is the inductive proof.

Base case. When $n = 1$, (C.1) reduces to $1 = \frac{1 \cdot 2}{2}$ which is true.

Induction step. Assume that (C.1) holds for n. This is the *induction assumption.* We show that then it also holds when n is replaced by $n + 1$.

$$\begin{aligned} 1 + 2 + 3 + \cdots + n + (n+1) &= (1 + 2 + 3 + \cdots + n) + (n+1) \\ &= \frac{n(n+1)}{2} + n + 1 = \frac{(n+1)(n+2)}{2}. \end{aligned}$$

We used the induction assumption in the second equality. The derivation proves the identity for $n+1$, which is exactly the statement $S(n+1)$. The proof is complete. We now know that (C.1) is true for all positive integers n. ▲

Exercises

Exercise C.1. In how many ways can we construct a license plate with three letters followed by three digits if we cannot have repetition among the letters and the digits?

Exercise C.2. In how many ways can we construct a license plate with three letters followed by three digits if we do not allow three zeros?

Exercise C.3. Suppose each license plate has three letters followed by three digits. Megan drives a car with the license plate UWU144. How many license plates differ from Megan's at exactly one position? (For example UWU194 or UWQ144.)

Exercise C.4. Find the number of (not necessarily meaningful) six letter words that can be constructed with the letters B, A, D, G, E, R.

Exercise C.5. Find the number of (not necessarily meaningful) five letter words that can be constructed with the letters A, P, P, L, E.

Exercise C.6. In a lottery 5 different numbers are chosen from the first 90 positive integers.

(a) How many possible outcomes are there? (An outcome is an unordered sample of five numbers.)

(b) How many outcomes are there with the number 1 appearing among the five chosen numbers?
(c) How many outcomes are there with two numbers below 50 and three numbers above 60?
(d) How many outcomes are there with the property that the last digits of all five numbers are different? (The last digit of 5 is 5 and the last digit of 34 is 4.)

Exercise C.7. Four tennis players would like to play a doubles match. (This means that there are two teams of two.) In how many ways can they set up the teams?

Exercise C.8. We organize a knock-out tournament among four teams, A, B, C and D. First two pairs of teams play each other, and then the two winners play in the final.

(a) In how many different ways can we set up the tournament? (We only care about which teams play with each other in the first round.)
(b) How many different outcomes does the tournament have? The outcome describes the pairs that play in each of the three games (including the final), and the winners for all these games.

Exercise C.9. In Texas Hold'em poker a player receives 2 cards first. These are called the player's pocket cards. Find the number of pocket card configurations (that is, the number of unordered samples of two cards from the deck of 52) that correspond to the following descriptions.

(a) Pair. (The two cards have the same rank.)
(b) Suited. (The two cards have the same suit.)
(c) Suited connectors. (The two cards have the same suit, and their ranks are next to each other. An ace has a special role: both $\{A, 2\}$ and $\{K, A\}$ count as pairs of neighbors.)

Exercise C.10. This exercise continues the poker hand theme of Example C.19. Find the number of poker hands that fall in the following categories.

(a) Two pairs. (Two cards of the same rank, two cards of another rank and a fifth card with a different rank.)
(b) Three of a kind. (Three cards of the same rank with two cards which have ranks different from each other and from the first three.)
(c) Straight. (Five cards with ranks in a sequence and which are not all in the same suit. A, 2, 3, 4, 5 and 10, J, Q, K, A both count.)
(d) Flush. (Five cards of the same suit which are not in sequential rank.)
(e) Full house. (Three cards with the same rank and two other cards with another common rank.)

(f) Four of a kind. (Four cards of the same rank and another card.)
(g) Straight flush. (Five cards in the same suit and whose ranks are in a sequence.)

Exercise C.11. Prove the following identity:

$$\binom{n}{k} = \binom{n-1}{k} + \binom{n-1}{k-1}.$$

Hint. You can prove this directly from the definition or by coming up with a counting problem that can be solved two different ways.

Exercise C.12. In the card game bridge the game begins by dividing up a deck of 52 cards among four players with each receiving 13. (We can assume that the players are numbered from 1 to 4.) Find the number of ways the following outcomes can happen.

(a) Player 1 receives all four aces.
(b) Each player receives 13 cards of the same suit.
(c) Player 1 and Player 2 together receive all the $\heartsuit$ cards.

Exercise C.13. In how many ways can we color the sides of a square with the colors red, blue, yellow, and green so that each side has a different color? We treat colorings that can be rotated into each other the same.

Exercise C.14. In how many ways can we color the sides of a square with the colors red, blue, and yellow so that we use each color at least once? We treat colorings that can be rotated into each other the same.

Exercise C.15. In how many ways can we color the faces of a cube with the colors red, blue, yellow, green, black and purple so that each face has a different color? We treat colorings that can be rotated into each other the same.

Exercise C.16. We construct a necklace using 7 green and 9 red beads. How many different necklaces can we construct? Necklaces that can be rotated to each other are considered to be the same.

Exercise C.17. Prove the following identity by coming up with a counting problem that can be solved two different ways:

$$\binom{2n}{n} = \binom{n}{0}^2 + \binom{n}{1}^2 + \cdots + \binom{n}{n}^2.$$

Exercises to practice induction

Exercise C.18. Verify that

$$(1+x)^n \geq 1 + nx \qquad \text{(C.2)}$$

for all real $x \geq -1$ and all positive integers n.

Exercise C.19. Show that $11^n - 6$ is divisible by 5 for all positive integers n.

Exercise C.20. We stated Fact C.20 with $n = 1$ as the base case for convenience. Sometimes the base case may have to be something other than $n = 1$. Here is an example.

For which positive integers n is the inequality $2^n \geq 4n$ true? Identify the correct range of n-values by demonstrating when the inequality fails and by doing an inductive proof from the correct base case.

Appendix D

Sums, products and series

Sum and product notation

A sum with finitely many terms is represented as

$$\sum_{k=1}^{n} a_k = a_1 + a_2 + \cdots + a_n.$$

There is a counterpart of the summation symbol for products:

$$\prod_{k=1}^{n} a_k = a_1 \cdot a_2 \cdot a_3 \cdots a_n.$$

To illustrate the use of this symbol, the factorial is $n! = 1 \cdot 2 \cdot 3 \cdots n = \prod_{k=1}^{n} k$, and exponentials and logarithms convert between sums and products:

$$\prod_{k=1}^{n} a_k = \exp\left(\ln \prod_{k=1}^{n} a_k\right) = e^{\sum_{k=1}^{n} \ln a_k}, \qquad \text{for } a_k > 0.$$

Conventions that arise frequently are that an empty sum has value zero while an empty product has value 1. This is because 0 is the additive identity and 1 is the multiplicative identity. (Adding an empty sum to something must not change that something, and multiplying something by an empty product must not change that something.) An instance of this convention is $0! = 1$.

Infinite series

Discrete probability models with infinitely many outcomes require us to be comfortable with infinite series, that is, sums of infinitely many terms. A convergent series of infinitely many terms is typically written as

$$\sum_{k=1}^{\infty} a_k = a_1 + a_2 + a_3 + \cdots$$

Note that in contrast with the notation for a finite sum, there is no a_∞ term at the end of the sum, even though the notation might suggest that. The proper definition of the series is the limit:

$$\sum_{k=1}^{\infty} a_k = \lim_{n\to\infty} \sum_{k=1}^{n} a_k.$$

An alternative notation for $\sum_{k=1}^{\infty} a_k$ is $\sum_{1\le k<\infty} a_k$.

Two particular series appear many times during the course, namely the geometric series and the series for the exponential function.

For geometric series the formula to remember is

$$\sum_{n=0}^{\infty} ax^n = \frac{a}{1-x} \quad \text{for any } x \text{ such that } |x| < 1. \tag{D.1}$$

If you forget the formula (D.1), derive it like this. Let

$$s = a + ax + ax^2 + \cdots + ax^{n-1} + ax^n.$$

Then

$$sx = ax + ax^2 + ax^3 + \cdots + ax^n + ax^{n+1}.$$

Lots of cancellation on the right-hand side gives

$$s - sx = a - ax^{n+1}.$$

If $x \neq 1$ divide by $1 - x$ to get

$$\sum_{k=0}^{n} ax^k = s = \frac{a(1-x^{n+1})}{1-x} \quad \text{for any } x \neq 1. \tag{D.2}$$

Formula (D.2) itself is very valuable. If $|x| < 1$, then $x^n \to 0$ as $n \to \infty$ and we can take the limit to obtain (D.1).

The series for the exponential function is

$$e^x = \sum_{k=0}^{\infty} \frac{x^k}{k!}, \quad \text{for any } x \in \mathbb{R}. \tag{D.3}$$

This does not come from a simple calculation like the geometric formula above. It is often taken as the definition of the exponential function. Or, if the exponential function is defined in some other way, then (D.3) is obtained as the Taylor series of the exponential function.

Changing order of summation

Manipulating summation symbols is useful. A trick that can help evaluate multiple sums is interchanging the order of summation. Let $\{a_{i,j}\}$ be a finite collection of real numbers indexed by two indices i and j. Then

$$\sum_i \sum_j a_{i,j} = \sum_j \sum_i a_{i,j}. \tag{D.4}$$

This equation is a consequence of the associativity and commutativity of addition. To make explicit what is going on here, consider this small example.

$$\sum_{i=1}^{2}\sum_{j=1}^{3} a_{i,j} = \sum_{i=1}^{2}(a_{i,1} + a_{i,2} + a_{i,3})$$
$$= (a_{1,1} + a_{1,2} + a_{1,3}) + (a_{2,1} + a_{2,2} + a_{2,3})$$
$$= (a_{1,1} + a_{2,1}) + (a_{1,2} + a_{2,2}) + (a_{1,3} + a_{2,3})$$
$$= \sum_{j=1}^{3}(a_{1,j} + a_{2,j}) = \sum_{j=1}^{3}\sum_{i=1}^{2} a_{i,j}.$$

Equation (D.4) can fail for some infinite series. See Exercise D.2 for an example. However equation (D.4) is true for infinite series under either one of these assumptions:

$$\text{(i) if all } a_{i,j} \geq 0 \qquad \text{or} \qquad \text{(ii) if } \sum_i \sum_j |a_{i,j}| < \infty.$$

Proof of this last claim requires techniques beyond calculus.

Switching the order of summation in (D.4) was straightforward because the ranges of the indices i and j did not depend on each other. A bit trickier is the case where the range of the inner sum depends on the index of the outer sum. Consider first this small example where we can list and rearrange the terms explicitly.

$$\sum_{i=1}^{4}\sum_{j=1}^{i} a_{i,j} = \sum_{j=1}^{1} a_{1,j} + \sum_{j=1}^{2} a_{2,j} + \sum_{j=1}^{3} a_{3,j} + \sum_{j=1}^{4} a_{4,j}$$
$$= a_{1,1} + (a_{2,1} + a_{2,2}) + (a_{3,1} + a_{3,2} + a_{3,3}) + (a_{4,1} + a_{4,2} + a_{4,3} + a_{4,4})$$
$$= (a_{1,1} + a_{2,1} + a_{3,1} + a_{4,1}) + (a_{2,2} + a_{3,2} + a_{4,2}) + (a_{3,3} + a_{4,3}) + a_{4,4}$$
$$= \sum_{j=1}^{4}\sum_{i=j}^{4} a_{i,j}.$$

Note that the sum $\sum_{i=1}^{4}\sum_{j=1}^{i} a_{i,j}$ cannot be written as $\sum_{j=1}^{i}\sum_{i=1}^{4} a_{i,j}$ because then the outer sum is not well defined.

More generally, it is true that

$$\sum_{i=1}^{N}\sum_{j=1}^{i} a_{i,j} = \sum_{j=1}^{N}\sum_{i=j}^{N} a_{i,j}. \tag{D.5}$$

One way to justify (D.5) is to combine the two ranges into an inequality and then separate the ranges again:

$$\sum_{i=1}^{N}\sum_{j=1}^{i} a_{i,j} = \sum_{(i,j):1\leq j\leq i\leq N} a_{i,j} = \sum_{j=1}^{N}\sum_{i=j}^{N} a_{i,j}.$$

The middle sum ranges over all index pairs (i,j) that satisfy $1 \leq j \leq i \leq N$. The logic above is that the initial ranges $1 \leq i \leq N$ and $1 \leq j \leq i$ combine into

$1 \le j \le i \le N$. This double inequality in turn can be separated into $1 \le j \le N$ and $j \le i \le N$.

Handling multiple sums is analogous to handling multiple integrals. For example (D.5) corresponds to the identity

$$\int_0^1 \int_0^x f(x, y)\, dy\, dx = \int_0^1 \int_y^1 f(x, y)\, dx\, dy.$$

Summation identities

The following identities are useful in some calculations. Analogous identities exist for sums of higher powers.

> **Fact D.1.** Let n be a positive integer. Then
>
> $$1 + 2 + \cdots + n = \frac{n(n+1)}{2}, \tag{D.6}$$
>
> $$1^2 + 2^2 + \cdots + n^2 = \frac{n(n+1)(2n+1)}{6}, \tag{D.7}$$
>
> $$1^3 + 2^3 + \cdots + n^3 = \frac{n^2(n+1)^2}{4}. \tag{D.8}$$

Proofs of these identities are exercises in induction. Identity (D.6) appeared already in Example C.21. Identities (D.7) and (D.8) appear as Exercises D.6 and D.7.

The first identity of Fact D.1 can be expressed by saying that the sum equals the number of terms times the average of the first and last terms. This continues to hold for arbitrary upper and lower summation limits:

$$\sum_{k=a}^{b} k = a + (a+1) + (a+2) + \cdots + b = (b - a + 1)\frac{a+b}{2}. \tag{D.9}$$

Another pair of identities which comes up many times are the binomial theorem and its generalization, the multinomial theorem. For example, they are applied in the text to show that the probabilities of the binomial and multinomial distributions sum to 1. Recall the definitions of the binomial and multinomial coefficients from Facts C.12 and C.16.

> **Fact D.2.** (Binomial theorem) Let n be a positive integer. Then for any x, y,
>
> $$(x + y)^n = \sum_{k=0}^{n} \binom{n}{k} x^k y^{n-k}. \tag{D.10}$$

Fact D.3. (Multinomial theorem) Let n and r be positive integers. Then for any $x_1, \dots, x_r$,

$$(x_1 + x_2 + \cdots + x_r)^n = \sum_{\substack{k_1 \geq 0, k_2 \geq 0, \dots, k_r \geq 0 \\ k_1 + k_2 + \cdots + k_r = n}} \binom{n}{k_1, k_2, \dots, k_r} x_1^{k_1} x_2^{k_2} \cdots x_r^{k_r}. \quad \text{(D.11)}$$

The sum runs over r-tuples $(k_1, k_2, \dots, k_r)$ of nonnegative integers that add up to n.

Proof of the binomial theorem. To prove (D.10) we expand $(x+y)^n$. The case $n = 2$ shows the way:

$$(x+y)^2 = (x+y)(x+y) = x^2 + xy + yx + y^2 = x^2 + 2xy + y^2.$$

If it helps, write out also $n = 3$. In the general n case, when we multiply out $(x+y)^n = (x+y) \cdots (x+y)$, each term of the resulting expansion is of the form $x^k y^{n-k}$ for some $0 \leq k \leq n$. This is because we pick up either x or y from each of the n factors $(x+y)$. The remaining question is, how many terms $x^k y^{n-k}$ does the expansion have for a particular k? The answer is $\binom{n}{k}$ because there are this many ways of specifying which k factors supply an x and which $n-k$ factors supply a y. This explains the right-hand side of (D.10). ▲

A similar counting argument, using Fact C.16, proves the multinomial theorem Fact D.3.

Integral test for convergence of infinite series

There is no general recipe for checking whether an infinite series $\sum_{n=1}^{\infty} a_n$ converges or not. For series with special structure useful tools are available. The next test works if a_n is given by $a_n = f(n)$ for a decreasing nonnegative function f. A *decreasing* function (also called *nonincreasing*) satisfies $f(x) \geq f(y)$ for all $x < y$.

Fact D.4. (Integral test for infinite series) Let f be a decreasing nonnegative function on $[1, \infty)$. Then the series $\sum_{k=1}^{\infty} f(k)$ and the integral $\int_1^{\infty} f(x)dx$ converge together or diverge together.

Outline of the proof. Both the infinite series and the improper integral are by definition limits:

$$\sum_{k=1}^{\infty} f(k) = \lim_{n\to\infty} \sum_{k=1}^{n} f(k) \quad \text{and} \quad \int_1^{\infty} f(x)\,dx = \lim_{n\to\infty} \int_1^n f(x)\,dx.$$

Since f is nonnegative these limits do exist by monotonicity: as n grows, the partial sums $\sum_{k=1}^{n} f(k)$ and the integrals $\int_1^n f(x)dx$ can never decrease. But one or both of the limits could be equal to infinity. We must show that either both are finite or both are infinite. We do this by showing that $\sum_{k=1}^{n} f(k)$ stays close to $\int_1^n f(x)dx$ even as n tends to infinity.

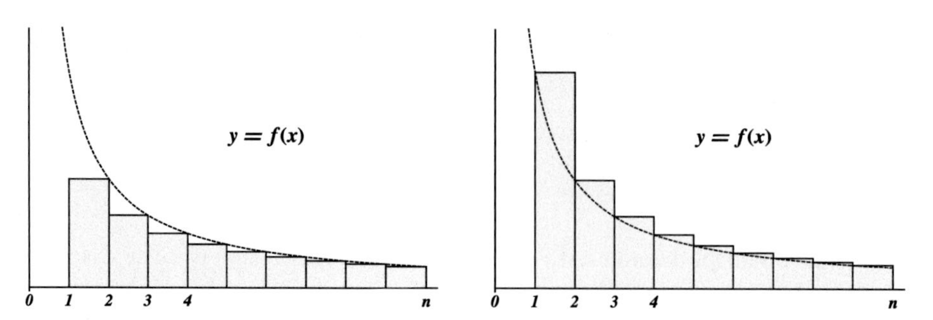

Figure D.1. Riemann-sum approximation of $\int_1^n f(x)dx$. On the left the total area of the blue rectangles is below the integral and on the right it is above. This gives the bounds $\sum_{k=2}^{n} f(k) \le \int_1^n f(x)dx \le \sum_{k=1}^{n-1} f(k)$.

Figure D.1 shows how the upper and lower Riemann approximation of the integral $\int_1^n f(x)\,dx$, together with the fact that f is decreasing, gives the inequalities

$$\sum_{k=2}^{n} f(k) \le \int_1^n f(x)\,dx \le \sum_{k=1}^{n-1} f(k).$$

Rearranging the terms gives

$$0 \le f(n) \le \sum_{k=1}^{n} f(k) - \int_1^n f(x)\,dx \le f(1).$$

This tells us that the absolute difference $\left|\sum_{k=1}^{n} f(k) - \int_1^n f(x)dx\right|$ is bounded by the fixed constant $f(1)$ for *all* n. Thus $\lim_{n\to\infty} \sum_{k=1}^{n} f(k) = \infty$ if and only if $\lim_{n\to\infty} \int_1^n f(x)dx = \infty$. The proof is complete. ▲

Example D.5. A simple application of the integral test checks whether the infinite series $\sum_{n=1}^{\infty} \frac{1}{n^\gamma}$ converges or not.

If $\gamma \le 0$ then $\frac{1}{n^\gamma} = n^{-\gamma} \ge 1$ for all n, and hence $\sum_{n=1}^{\infty} \frac{1}{n^\gamma}$ is infinite. If $\gamma > 0$ we can use the integral test with the decreasing nonnegative function $f(x) = \frac{1}{x^\gamma}$. First evaluate the integral:

$$\int_1^n \frac{1}{x^\gamma}\,dx = \begin{cases} \frac{1}{1-\gamma}\left(n^{1-\gamma} - 1\right), & \text{if } \gamma \ne 1, \\ \ln n, & \text{if } \gamma = 1. \end{cases}$$

From the right-hand side we see that the limit $\lim_{n\to\infty} \int_1^n \frac{1}{x^\gamma} dx$ is finite for $\gamma > 1$ and it is infinite for $0 < \gamma \le 1$.

By Fact D.4, $\sum_{n=1}^{\infty} \frac{1}{n^\gamma}$ is finite if $\gamma > 1$, and it is infinite for $\gamma \le 1$. Commonly used special cases include $\sum_{k=1}^{\infty} \frac{1}{k} = \infty$ and $\sum_{k=1}^{\infty} \frac{1}{k^2} < \infty$. Note that the integral test does not give the value of $\sum_{k=1}^{\infty} \frac{1}{k^2}$, only that it is finite. It can be shown that $\sum_{k=1}^{\infty} \frac{1}{k^2} = \frac{\pi^2}{6}$. Exercise 7.41 gives a derivation with basic probability tools. ▲

Exercises

Exercise D.1. Let n be a positive integer. Evaluate the sum $\sum_{k=1}^{n}(n + 2k)$.

Exercise D.2. Verify that equation (D.4) does not hold for the following collection of real numbers, where i and j range over the positive integers:

$$a_{i,j} = \begin{cases} 1, & \text{if } j = i \\ -1, & \text{if } j = i + 1 \\ 0, & \text{if } j \notin \{i, i + 1\}. \end{cases}$$

Exercise D.3. Let n be a positive integer. Evaluate the following sums.

(a) $\sum_{k=1}^{n} \sum_{\ell=1}^{k} \ell$

(b) $\sum_{k=1}^{n} \sum_{\ell=1}^{k} k$

(c) $\sum_{k=1}^{n} \sum_{\ell=1}^{k} (7 + 2k + \ell)$

Exercise D.4. Let n be a positive integer. Evaluate the sum $\sum_{i=1}^{n} \sum_{j=i}^{n} j$.

Exercise D.5. Let $-1 < x < 1$. Evaluate the series below.

(a) $\sum_{i=1}^{\infty} \sum_{j=i}^{\infty} x^j$

(b) $\sum_{k=1}^{\infty} kx^k$

Hint. Write $k = \sum_{i=1}^{k} 1$ and switch the order of summation.

Exercise D.6. Prove identity (D.7).

Exercise D.7. Prove identity (D.8).

Exercise D.8. Use the binomial theorem to show that

$$\binom{n}{0} + \binom{n}{2} + \binom{n}{4} + \cdots = \binom{n}{1} + \binom{n}{3} + \binom{n}{5} + \cdots$$

Recall that $\binom{n}{k} = 0$ for integers $k > n$ by Remark C.13.

Use this to find the number of even subsets of a set of size n. (An even subset is a subset with even number of elements.)

Exercise D.9. Prove the binomial theorem by induction on n. Exercise C.11 can be helpful.

Exercise D.10. Prove the multinomial theorem by induction on r.

Exercise D.11. Prove the multinomial theorem by induction on n.

Exercise D.12. Prove the multinomial theorem by generalizing the combinatorial proof given for the binomial theorem.

Challenging problems

Exercise D.13. Prove the following version of Stirling's formula: the limit $\lim_{n\to\infty} \dfrac{n!}{n^{n+1/2}e^{-n}}$ exists and is a finite positive number. (The complete Stirling's formula would also identify the limit as $\sqrt{2\pi}$.)

Hint. Let $d_n = \ln\left(\frac{n!}{n^{n+1/2}e^{-n}}\right) = \ln(n!) - (n+1/2)\ln(n) + n$. Show that

$$0 < d_n - d_{n+1} < \frac{1}{12n} - \frac{1}{12(n+1)}.$$

For this you might use the series expansion of the function $\ln \frac{1+t}{1-t}$ near $t = 0$. Then show that the limit of d_n exists and is finite.

Appendix E

Table of values for $\Phi(x)$

$\Phi(x) = P(Z \leq x)$ is the cumulative distribution function of the standard normal random variable Z.

	0.00	0.01	0.02	0.03	0.04	0.05	0.06	0.07	0.08	0.09
0.0	0.5000	0.5040	0.5080	0.5120	0.5160	0.5199	0.5239	0.5279	0.5319	0.5359
0.1	0.5398	0.5438	0.5478	0.5517	0.5557	0.5596	0.5636	0.5675	0.5714	0.5753
0.2	0.5793	0.5832	0.5871	0.5910	0.5948	0.5987	0.6026	0.6064	0.6103	0.6141
0.3	0.6179	0.6217	0.6255	0.6293	0.6331	0.6368	0.6406	0.6443	0.6480	0.6517
0.4	0.6554	0.6591	0.6628	0.6664	0.6700	0.6736	0.6772	0.6808	0.6844	0.6879
0.5	0.6915	0.6950	0.6985	0.7019	0.7054	0.7088	0.7123	0.7157	0.7190	0.7224
0.6	0.7257	0.7291	0.7324	0.7357	0.7389	0.7422	0.7454	0.7486	0.7517	0.7549
0.7	0.7580	0.7611	0.7642	0.7673	0.7704	0.7734	0.7764	0.7794	0.7823	0.7852
0.8	0.7881	0.7910	0.7939	0.7967	0.7995	0.8023	0.8051	0.8078	0.8106	0.8133
0.9	0.8159	0.8186	0.8212	0.8238	0.8264	0.8289	0.8315	0.8340	0.8365	0.8389
1.0	0.8413	0.8438	0.8461	0.8485	0.8508	0.8531	0.8554	0.8577	0.8599	0.8621
1.1	0.8643	0.8665	0.8686	0.8708	0.8729	0.8749	0.8770	0.8790	0.8810	0.8830
1.2	0.8849	0.8869	0.8888	0.8907	0.8925	0.8944	0.8962	0.8980	0.8997	0.9015
1.3	0.9032	0.9049	0.9066	0.9082	0.9099	0.9115	0.9131	0.9147	0.9162	0.9177
1.4	0.9192	0.9207	0.9222	0.9236	0.9251	0.9265	0.9279	0.9292	0.9306	0.9319
1.5	0.9332	0.9345	0.9357	0.9370	0.9382	0.9394	0.9406	0.9418	0.9429	0.9441
1.6	0.9452	0.9463	0.9474	0.9484	0.9495	0.9505	0.9515	0.9525	0.9535	0.9545
1.7	0.9554	0.9564	0.9573	0.9582	0.9591	0.9599	0.9608	0.9616	0.9625	0.9633
1.8	0.9641	0.9649	0.9656	0.9664	0.9671	0.9678	0.9686	0.9693	0.9699	0.9706
1.9	0.9713	0.9719	0.9726	0.9732	0.9738	0.9744	0.9750	0.9756	0.9761	0.9767
2.0	0.9772	0.9778	0.9783	0.9788	0.9793	0.9798	0.9803	0.9808	0.9812	0.9817
2.1	0.9821	0.9826	0.9830	0.9834	0.9838	0.9842	0.9846	0.9850	0.9854	0.9857
2.2	0.9861	0.9864	0.9868	0.9871	0.9875	0.9878	0.9881	0.9884	0.9887	0.9890
2.3	0.9893	0.9896	0.9898	0.9901	0.9904	0.9906	0.9909	0.9911	0.9913	0.9916
2.4	0.9918	0.9920	0.9922	0.9925	0.9927	0.9929	0.9931	0.9932	0.9934	0.9936
2.5	0.9938	0.9940	0.9941	0.9943	0.9945	0.9946	0.9948	0.9949	0.9951	0.9952
2.6	0.9953	0.9955	0.9956	0.9957	0.9959	0.9960	0.9961	0.9962	0.9963	0.9964
2.7	0.9965	0.9966	0.9967	0.9968	0.9969	0.9970	0.9971	0.9972	0.9973	0.9974
2.8	0.9974	0.9975	0.9976	0.9977	0.9977	0.9978	0.9979	0.9979	0.9980	0.9981
2.9	0.9981	0.9982	0.9982	0.9983	0.9984	0.9984	0.9985	0.9985	0.9986	0.9986
3.0	0.9987	0.9987	0.9987	0.9988	0.9988	0.9989	0.9989	0.9989	0.9990	0.9990
3.1	0.9990	0.9991	0.9991	0.9991	0.9992	0.9992	0.9992	0.9992	0.9993	0.9993
3.2	0.9993	0.9993	0.9994	0.9994	0.9994	0.9994	0.9994	0.9995	0.9995	0.9995
3.3	0.9995	0.9995	0.9995	0.9996	0.9996	0.9996	0.9996	0.9996	0.9996	0.9997
3.4	0.9997	0.9997	0.9997	0.9997	0.9997	0.9997	0.9997	0.9997	0.9997	0.9998

Appendix F

Table of common probability distributions

Discrete distributions				
Name	Parameters	Probability mass function	Expectation	Variance
Bernoulli	$0 \leq p \leq 1$	$p_X(0) = 1 - p, \quad p_X(1) = p$	p	$p(1-p)$
Binomial	$0 \leq p \leq 1$ $n \geq 1$	$p_X(k) = \binom{n}{k} p^k (1-p)^{n-k}$, $0 \leq k \leq n$	np	$np(1-p)$
Geometric	$0 < p \leq 1$	$p_X(k) = p(1-p)^{k-1}$, $k = 1, 2, \ldots$	$\frac{1}{p}$	$\frac{1-p}{p^2}$
Poisson	$\lambda > 0$	$p_X(k) = \frac{\lambda^k}{k!} e^{-\lambda}$, $k = 0, 1, 2, \ldots$	λ	λ
Hypergeometric	$N \geq 1$ $N_A \geq 0$ $n \geq 1$	$p_X(k) = \frac{\binom{N_A}{k}\binom{N-N_A}{n-k}}{\binom{N}{n}}$, $0 \leq k \leq n$	$\frac{nN_A}{N}$	$\frac{N-n}{N-1} n \frac{N_A(N-N_A)}{N^2}$
Negative binomial	$k \geq 1$ $0 < p \leq 1$	$p_X(n) = \binom{n-1}{k-1} p^k (1-p)^{n-k}$, $n \leq k$	$\frac{k}{p}$	$k\frac{1-p}{p^2}$

Continuous distributions				
Name	Parameters	Probability density function	Expectation	Variance
Uniform	$a < b$	$f_X(t) = \frac{1}{b-a}$ for $t \in [a, b]$	$\frac{a+b}{2}$	$\frac{1}{12}(b-a)^2$
Normal	μ real, $\sigma^2 > 0$	$f_X(t) = \frac{1}{\sqrt{2\pi\sigma^2}} e^{-\frac{(t-\mu)^2}{2\sigma^2}}$	μ	σ^2
Exponential	$\lambda > 0$	$f_X(t) = \lambda e^{-\lambda t}$ for $t \geq 0$	$\frac{1}{\lambda}$	$\frac{1}{\lambda^2}$
Gamma	$r \geq 1, \quad \lambda > 0$	$f_X(t) = \frac{\lambda^r x^{r-1}}{\Gamma(r)} e^{-\lambda t}$ for $t \geq 0$	$\frac{r}{\lambda}$	$\frac{r}{\lambda^2}$
Beta	$r, s > 0$	$f_X(t) = \frac{\Gamma(r+s)}{\Gamma(r)\Gamma(s)} t^{r-1}(1-t)^{s-1}$ for $0 \leq t \leq 1$	$\frac{r}{r+s}$	$\frac{rs}{(r+s)^2(r+s+1)}$

Answers to selected exercises

Chapter 1

1.1 $\frac{5}{12}$.

1.3 (b) The number of elements in the sample space is 12^{10}. (c) The number of outcomes in which nobody rolls a 5 is 10^{10}. The number of outcomes where at least 1 person rolls a five is $12^{10} - 10^{10}$.

1.5 (b) $\frac{475}{1443}$.

1.7 (a) $\frac{4}{35}$, (b) $\frac{12}{35}$.

1.9 $\frac{2}{5}$.

1.11 $\frac{\pi}{100}$.

1.13 (a) $\frac{2}{5}$, (b) $\frac{3}{20}$.

1.15 $\frac{13}{16}$.

1.17 (a) Possible values of Z are $\{0, 1, 2\}$. $p_Z(0) = \frac{2}{7}$, $p_Z(1) = \frac{4}{7}$, $p_Z(2) = \frac{1}{7}$.
(b) Possible values of W are $\{0, 1, 2\}$. $p_W(0) = \frac{16}{49}$, $p_W(1) = \frac{24}{49}$, $p_W(2) = \frac{9}{49}$.

1.19 The possible values are 5 and 1. The probability mass function is $p_X(1) = \frac{8}{9}$ and $p_X(5) = \frac{1}{9}$.

1.21 $\frac{12}{35}$.

1.23 (a) $\frac{1}{3}$, (b) $\frac{2}{3}$.

1.25 (a) $\frac{5}{6}$, (b) $\frac{4}{7}$.

1.27 (a) $\frac{1}{7}$, (b) $\frac{3}{7}$.

1.29 (b) $\frac{59}{143}$.

1.31 (a) $\frac{1}{442}$, (b) $\frac{33}{221}$.

1.33 $\frac{25}{648}$.

1.35 (a) $\frac{5}{9}$, (b) $1 - \frac{\pi}{18}$.

1.37 (a) $\frac{1}{7}$, (b) $\frac{6}{7}$, (c) Peter starts $\frac{4}{7}$, Mary starts $\frac{1}{7}$.

1.39 (a) $\frac{83}{100}$, (b) $\frac{1547}{4950}$.

1.41 $\frac{5}{9}$.

Chapter 2

2.1 $\frac{3}{5}$.

2.3 $\frac{6}{19}$.

2.5 $\frac{4}{15}$.

2.7 (b) 0.2.

2.9 $\frac{3}{4}$.

2.11 0.5.

2.13 A and B are independent.

2.15 $\frac{729}{10000}$.

2.17 $\frac{3}{8}$.

2.19 (a) With replacement: $P(X_1 = 4) = \frac{1}{7}$, $P(X_2 = 5) = \frac{1}{7}$, and $P(X_1 = 4, X_2 = 5) = \frac{1}{49}$.
(b) Without replacement: $P(X_1 = 4) = \frac{1}{7}$, $P(X_2 = 5) = \frac{1}{7}$, and
$P(X_1 = 4, X_2 = 5) = \frac{1}{42}$.

2.21 (a) $\frac{512}{625}$, (b) $\frac{112}{125}$.

2.23 (a) 0.6983, (b) 0.2586, (c) 0.2158.

2.25 $\frac{2}{7}$.

2.27 (a) $\frac{1}{2}$, (b) $\frac{1}{4}$, (c) $\frac{5}{18}$, (d) $\frac{2}{9}$.

2.29 0.185.

2.31 (b) $\frac{2}{3}$, (c) $\frac{1}{2}$.

2.33 (a) $\frac{3}{10}$, (b) $\frac{k}{15}$.

2.35 $\frac{1}{52}$.

2.37 (a) $\frac{11}{120}$, (b) $\frac{4}{15}$.

2.39 (a) $\frac{47}{480}$.

2.41 0.3721.

2.43 (a) $\frac{2}{5}$, (b) $\frac{7}{10}$, (c) $\frac{4}{7}$.

2.45 (a) 0.03925, (b) 0.9554.

2.47 $\frac{\binom{78}{53}}{\binom{80}{55}}$.

2.49 (b) $(\frac{2}{3})^{k-1}$.

2.51 $\frac{5}{8}$.

2.53 (a) 0.9, (b) 0.72.

2.55 (a) $\frac{(1-p)r}{p+r-pr}$. (c) Given that Mary wins, $X \sim \text{Geom}(p + r - pr)$.

2.57 (a) 0.942, (b) 0.905.

2.59 $\frac{931}{1000}$.

2.63 The maximal probability is $\frac{3}{8}$.

2.65 $\frac{1}{2}$.

2.69 $P(A_1A_2) = \frac{2655}{10000}$. The events are not independent.

2.71 (a) After one flip, $\frac{2}{17}$. After two flips, $\frac{4}{29}$. (b) 25.

2.73 (a) 0.00003122, (b) 0.4458.

2.75 0.7273.

2.79 0.0316.

2.81 56.

Chapter 3

3.1 (a) $\frac{3}{7}$, (b) $\frac{3}{14}$, (c) $\frac{2}{3}$.

3.3 (b) $1 - e^{-3}$, (c) $1 - e^{-15}$, (d) $\frac{e^{-6}-e^{-12}}{1-e^{-15}}$.

3.5 The possible values are $\{1, \frac{4}{3}, \frac{3}{2}, \frac{9}{5}\}$. The probability mass function is

$$p_X(1) = \tfrac{1}{3}, \quad p_X(4/3) = \tfrac{1}{6}, \quad p_X(3/2) = \tfrac{1}{4}, \quad p_X(9/5) = \tfrac{1}{4}.$$

3.7 (c) $\frac{1}{4}$.

3.9 (a) $\frac{1}{3}$, (b) 3.

3.11 $\frac{7}{18}$.

3.13 (a) 4 and $\ln(2)/3$, (b) $\ln(10)/3$.

3.15 (a) 11, (b) 13, (c) 97, (d) 64.

3.17 (a) 0.0188, (b) 0.032, (c) 0.9345, (d) 0.0013, (e) 0.0116.

3.19 $F_Z(2) \approx 0.2991$ and $F_Z(8) \approx 0.9996$.

3.21 (a) $p_X(0) = \frac{1}{4}$, $p_X(1) = \frac{1}{2}$, and $p_X(2) = \frac{1}{4}$. (b) $\frac{3}{4}$ and $\frac{1}{4}$. (c) $E[X] = 1$ and $\mathrm{Var}(X) = \frac{1}{2}$.

3.23 (b) $\frac{1}{10000}$. (c) $E[X] = -0.094$, $\mathrm{Var}(X) \approx 4918.21$.

3.25 (a) There is no choice of b that works. (b) Only $b = \frac{\pi}{6}$ works.

3.27 (a) $\frac{3}{5}$, (b) $\frac{2}{3}$, (c) $\frac{34}{25}$.

3.29 $E[X] = \frac{209}{9}$, $\mathrm{Var}(X) = \frac{1274}{81}$.

3.31 (a) $c = 3$, (b) 0, (c) $\frac{7}{8}$, (d) $\frac{7}{64}$, (e) $1 - x^{-3}$ for $x \geq 1$, and 0 for $x < 1$, (f) $E(X) = \frac{3}{2}$ and $\mathrm{Var}(X) = \frac{3}{4}$, (g) $\frac{39}{2}$, (h) $E(X^n) = \begin{cases} \infty, & n \geq 3 \\ \frac{3}{3-n}, & n \leq 2. \end{cases}$

3.33 (a) $\frac{3}{8}$, (b) $\frac{1}{2}$, (c) $\frac{27}{8}$.

3.35 $E(X^4) = 29$.

3.37 (a) $(1+x)^{-2}$ for $x \geq 0$, and 0 otherwise, (b) $\frac{1}{12}$, (c) $\frac{1}{2}$.

3.43 (a) $F(x) = 0$ if $x < 0$, $F(x) = 1$ if $x > \frac{1}{2}$, and $F(x) = 4x - 4x^2$ for $0 \leq x \leq \frac{1}{2}$.

3.45 (a) The cumulative distribution function is

$$F(t) = \begin{cases} 0, & t < 0 \\ \frac{1}{2}t, & 0 < t < 1 \\ 1 - \frac{1}{2t}, & t \geq 1. \end{cases}$$

(b) The probability density is

$$f(t) = \begin{cases} 0, & t < 0 \\ \frac{1}{2}, & 0 < t < 1 \\ \frac{1}{2t^2}, & t \geq 1. \end{cases}$$

3.47 (a) The cumulative distribution function is

$$F(x) = \begin{cases} 0, & x < 0 \\ \frac{x^2}{900}, & 0 \leq x < 30 \\ 1, & x \geq 30. \end{cases}$$

(b) The density function is

$$f(x) = \begin{cases} \frac{x}{450}, & 0 \leq x < 30 \\ 0, & \text{otherwise.} \end{cases}$$

(c) 20.

3.49

$$F(x) = \begin{cases} 0, & x < 0 \\ \frac{1}{2}x - \frac{1}{16}x^2, & 0 \leq x < 4 \\ 1, & x \geq 4. \end{cases}$$

and

$$f(x) = \begin{cases} \frac{1}{2} - \frac{1}{8}x, & 0 < x < 4 \\ 0, & \text{otherwise.} \end{cases}$$

3.53 (a) $\frac{3}{8}$, (b) $\frac{39}{64}$.

3.55 $\frac{2-p}{rp-r-p}$.

3.57 $\frac{p\ln(1/p)}{1-p}$.

3.61 (a)

$$F_X(s) = \begin{cases} 0, & s < 0 \\ \frac{2s}{M} - \frac{s^2}{M^2}, & 0 \leq s \leq M \\ 1, & s > M. \end{cases}$$

(b) $Y = g(X)$ where

$$g(x) = \begin{cases} x, & 0 \leq x \leq \frac{M}{2} \\ \frac{M}{2}, & \frac{M}{2} < x \leq M. \end{cases}$$

(d) $P(Y < M/2) = \frac{3}{4}$.

3.65 (a) 12, (b) 31.

3.67 (a) 0, (b) $3\sigma^2\mu + \mu^3$.

3.71 0.2033.

Chapter 4

4.1 (a) 0.0009, (b) 0.0018.

4.3 0.0478.

4.5 (a) 1, (b) 0.

4.7 $(0.426, 0.488)$.

4.9 (a) 0.8699, (b) 0.844.

4.11 0.8488.

4.13 (a) e^{-1}, (b) $e^{-1/3} - e^{-8/3}$, (c) e^{-1}.

4.15 (a) Approximately 0.7619. (b) Approximately 0.01388.

4.17 Without continuity correction: 0.3578. With continuity correction: 0.3764.

4.19 Let X be number of people who prefer cereal A. Without the continuity correction: $P(X \geq 25) \approx 0.1056$.

4.21 200 games: 0.5636 with continuity correction. 300 games: 0.1762 with continuity correction.

4.23 0.1292.

4.25 63.2%.

4.27 68.

4.31 $\frac{1-e^{-\mu}}{\mu}$.

4.33 Approximately 5.298.

4.35 (b) 0.1603.

4.37 20.45%.

4.39 (a) $1 - \left(\frac{349}{350}\right)^{400} - 400\left(\frac{1}{350}\right)\left(\frac{349}{350}\right)^{399} \approx 0.3167$. (b) $1 - e^{-8/7} - \frac{8}{7}e^{-8/7} \approx 0.3166$.

4.41 Exact 0.00095, Poisson 0.0018, Normal 0.0023.

4.43 $P(X \geq 48) \approx 0.1056$. $P(Y \geq 2) \approx 0.2642$.

4.45 0.0906.

4.47 (a) 0.00579, (b) 0.2451 (without continuity correction).

4.49 All (C, r) for which $0 = Ce^{-r/10} + (C - 800)(1 - e^{-r/10})$.

4.51

$$P(T_1 \leq s \mid N_t = 1) = \begin{cases} 0, & s < 0 \\ s, & 0 \leq s \leq t \\ 1, & s > t. \end{cases}$$

4.53 $E[X] = \frac{r}{\lambda}$ and $\text{Var}(X) = \frac{r}{\lambda^2}$.

Chapter 5

5.1 $M(t) = \frac{4}{9}e^{-6t} + \frac{1}{9}e^{-2t} + \frac{2}{9} + \frac{2}{9}e^{3t}$.

5.3

$$M(t) = \begin{cases} 1, & t = 0 \\ \frac{e^t - 1}{t}, & t \neq 0. \end{cases}$$

5.5 $e^{-3} \cdot \frac{3^4}{4!}$.

5.7 $f_Y(t) = \lambda e^{t - \lambda e^t}$ for all t.

5.9 (a) $(q+pe^t)^n$ with $q = 1-p$.

5.11 $M(t) = \frac{1}{(1-t)^2}$ for $t < 1$.

5.13 27.53.

5.15 (a) $M_X(t) = \frac{1}{10}e^{-2t} + \frac{1}{5}e^{-t} + \frac{3}{10} + \frac{2}{5}e^t$.

5.17 (a)

$$M_X(t) = \begin{cases} 1, & t = 0 \\ \frac{1+e^{2t}(2t-1)}{2t^2}, & t \neq 0. \end{cases}$$

(b) $E(X^k) = \frac{2^{k+1}}{k+2}$.

5.19 (a)

$$M_X(t) = \begin{cases} \frac{3e^t-8}{15e^t-20}, & t < \ln(4/3) \\ \infty, & \text{else.} \end{cases}$$

5.21 $M_Y(t) = e^{bt}M_X(at)$.

5.23 $2\exp(-2)$.

5.25

$$f_Y(t) = \begin{cases} \frac{2}{5}, & 0 < t < 2 \\ \frac{1}{5}, & 2 \leq t < 3 \\ 0, & \text{otherwise.} \end{cases}$$

5.27

$$f_Y(t) = \begin{cases} \lambda t^{-(\lambda+1)}, & t > 1 \\ 0, & \text{else.} \end{cases}$$

5.29

$$f_{|Z|}(y) = \begin{cases} \frac{2}{\sqrt{2\pi}}e^{-\frac{y^2}{2}}, & y \geq 0, \\ 0, & \text{else.} \end{cases}$$

5.31

$$f_Y(t) = \begin{cases} \frac{1}{t(1+\ln(t))^2}, & t > 1 \\ 0, & \text{else.} \end{cases}$$

5.35 $P(\lfloor X \rfloor = k) = e^{-\lambda k}(1-e^{-\lambda})$ for $k = 0, 1, \ldots$

5.37 $\lambda e^{-\lambda t}\frac{1}{1-e^{-\lambda}}$ for $0 < t < 1$ and zero otherwise.

Chapter 6

6.1 (a) $p_X(1) = 0.3$, $p_X(2) = 0.5$, $p_X(3) = 0.2$.

(b) The values of $Z = XY$ are given by the table.

$X \backslash Y$	0	1	2	3
1	0	1	2	3
2	0	2	4	6
3	0	3	6	9

The probability mass function of Z is given in the next table:

k	0	1	2	3	4	5	6	7	8	9
$p_Z(k)$	0.35	0.15	0.05	0.05	0.05	0	0.3	0	0	0.05

(c) 16.3365.

6.3 (a) $\frac{63}{625}$, (b) $\frac{5}{512}$.

6.5 (b) The marginals are

$$f_X(x) = \begin{cases} \frac{6}{7}x + \frac{4}{7}, & 0 \le x \le 1 \\ 0, & \text{else} \end{cases} \quad \text{and} \quad f_Y(y) = \begin{cases} \frac{12}{7}y^2 + \frac{6}{7}y, & 0 \le y \le 1 \\ 0, & \text{else.} \end{cases}$$

(c) $P(X < Y) = \frac{9}{14}$. (d) $E[X^2 Y] = \frac{2}{7}$.

6.7 (a) The marginals are

$$f_X(x) = \begin{cases} 2(1-x), & 0 \le x \le 1 \\ 0, & \text{else} \end{cases} \quad \text{and} \quad f_Y(y) = \begin{cases} 2(1-y), & 0 \le y \le 1 \\ 0, & \text{else.} \end{cases}$$

(b) $E[X] = E[Y] = \frac{1}{3}$. (c) $E[XY] = \frac{1}{12}$.

6.9 $p_{X,Y}(a, b) = \binom{3}{a}\frac{1}{48}$ for $a \in \{0, 1, 2, 3\}$ and $b \in \{1, 2, 3, 4, 5, 6\}$.

6.11 The joint density function is

$$f_{x,y}(x, y) = \begin{cases} 2x, & 0 < x < 1,\ 1 < y < 2 \\ 0, & \text{else.} \end{cases}$$

$P(Y - X \ge \frac{3}{2}) = \frac{1}{24}$.

6.13 The random variables X and Y are not independent.

6.17 The joint density is

$$f_{X,Y}(x, y) = \begin{cases} \lambda^2 e^{-\lambda(x+y)}, & x, y > 0 \\ 0, & \text{else.} \end{cases}$$

6.19 (a) $p_X(0) = \frac{1}{3}$, $p_X(1) = \frac{2}{3}$. $p_Y(0) = \frac{1}{6}$, $p_Y(1) = \frac{1}{3}$, $p_Y(2) = \frac{1}{2}$.

(b)

$Z \backslash W$	0	1	2
0	$\frac{1}{18}$	$\frac{1}{9}$	$\frac{1}{6}$
1	$\frac{1}{9}$	$\frac{2}{9}$	$\frac{1}{3}$

6.21 They are not independent.

6.23 $\frac{4}{9}$.

6.25 (a)

		Y		
		0	1	2
X	0	$\frac{1}{8}$	$\frac{1}{8}$	0
	1	$\frac{1}{8}$	$\frac{1}{4}$	$\frac{1}{8}$
	2	0	$\frac{1}{8}$	$\frac{1}{8}$

(b) $p_{XY}(0) = \frac{3}{8}$, $p_{XY}(1) = \frac{1}{4}$, $p_{XY}(2) = \frac{1}{4}$, $p_{XY}(4) = \frac{1}{8}$.

6.29 $\frac{p(1-r)}{p+r-pr}$.

6.31 $p_{X_1,X_2,X_3}(a,b,c) = \frac{\binom{10}{a}\binom{15}{b}\binom{20}{c}}{\binom{45}{8}}$ for nonnegative integers a, b, and c satisfying $a+b+c=8$.

6.33 (a)

$$f_X(x) = \begin{cases} \frac{3}{2}x^2(2-x), & 0 < x \leq 1 \\ \frac{3}{2}(2-x)^3, & 1 < x \leq 2 \\ 0, & \text{otherwise} \end{cases}$$

$$f_Y(y) = \begin{cases} 6y(1-y), & 0 < y < 1 \\ 0, & \text{else.} \end{cases}$$

(b) $\frac{3}{16}$.

6.35 (b) $\frac{7}{12}$.

(c)

$$f_Y(y) = \begin{cases} \frac{3y^2}{8}, & 0 \leq y \leq 2 \\ 0, & \text{else.} \end{cases}$$

6.37 $f_X(x) = \frac{h(x)}{\int_a^b h(s)ds}$ for $a < x < b$, and zero otherwise.

6.39 $P(a < X < b, c < Y < d) = F(b,d) - F(b,c) - F(a,d) + F(a,b)$.

6.41 $\frac{2r}{3h}$.

6.43

$$f_{T,V}(u,v) = \begin{cases} 2u^2v + \sqrt{v} + 2v^2u + \sqrt{u}, & \text{if } 0 < u < v < 1 \\ 0, & \text{else.} \end{cases}$$

6.45 (b) The density functions are

$$f_V(v) = f_X(v)F_Y(v) + F_X(v)f_Y(v)$$
$$f_T(z) = f_X(z)(1 - F_Y(z)) + f_Y(z)(1 - F_X(z)).$$

Chapter 7

7.1 $\frac{50}{27}e^{-2}$.

7.3 $\frac{1}{8}$.

7.5 (a) $\mathcal{N}(-7,25)$, (b) 0.1587.

7.7 $\frac{1}{4}$.

7.9 (a) $1-e^{-1}$, (b) $1-\frac{5}{2}e^{-3/2}$.

7.11 (a) $\frac{pr}{p+r-pr}$.
(b) $P(Z=n)=pr\frac{(1-r)^{n-1}-(1-p)^{n-1}}{p-r}$ for $n\geq 2$.

7.13

$$P(X+Y=n)=\begin{cases}\binom{n-1}{k+m-1}p^{k+m}(1-p)^{n-k-m}, & n\geq k+m\\ 0, & \text{else.}\end{cases}$$

7.15

$$p_{X+Y}(a)=\begin{cases}\frac{a-1}{mn}, & 2\leq a\leq n\\ \frac{1}{m}, & n+1\leq a\leq m+1\\ \frac{m+n+1-a}{mn}, & m+2\leq a\leq m+n.\end{cases}$$

7.19 (a) $\frac{1}{2}e^{-4}$, (b) $f_{X-Y}(z)=\frac{1}{2}e^{-|z|}$.

7.21 (a) $\mathcal{N}(10,59)$, (b) 0.2578.

7.23 0.0793.

7.25

$$f_{X+Y}(z)=\begin{cases}\frac{z-8}{2}, & 8\leq z<9\\ \frac{1}{2}, & 9\leq z<10\\ \frac{11-z}{2}, & 10\leq z\leq 11\\ 0, & \text{otherwise.}\end{cases}$$

7.27 $f_{X+Y}(t)=\int_{t-1}^{t}f(s)ds$.

7.29 $\frac{1}{100}$.

7.31 $\frac{100}{473}$.

7.33 $\frac{2}{11}$.

7.35 $\frac{1}{6}$.

Chapter 8

8.1 (a) $\frac{1}{p}+nr$. (b) Not enough information. (c) $\frac{2-p}{p^2}+nr(1-r)+n^2r^2$. (d) Not enough information.

8.3 $E[X]=1.4$.

8.5 The expectation of the perimeter is 16. The expectation of the area is 15.

8.7 $E[X]=1.4$, $\mathrm{Var}(X)=0.66$.

8.9 (a) 6, (b) 30, (c) 83.

8.11 $M_X(t) = 0.126 + 0.432e^t + 0.358e^{2t} + 0.084e^{3t}$.

8.13 Let $X_1, \ldots, X_{36}$ be independent random variables, each with probability mass function

$$p(-1) = \tfrac{1}{2}, \quad p(0) = \tfrac{2}{5}, \quad p(\tfrac{1}{2}) = \tfrac{1}{10}.$$

Then Z has the same distribution as $X_1 + \cdots + X_{36}$.

8.15 $\mathrm{Cov}(X, Y) = -\frac{13}{324}$.

8.17 0.41.

8.21 0.381.

8.23 (a) $\frac{24}{49}$, (b) 24.

8.25 (a) $\frac{1}{2}$, (b) $\frac{9}{2}$.

8.27 $\frac{n-1}{2}$.

8.29 $\frac{210}{13}$.

8.31 18.

8.33 (a) $\frac{45}{64}$, (b) 30, (c) $\frac{10!}{(10-k)!}(\frac{1}{2})^{k-1}$.

8.35 $E[X] = \frac{781}{256}$, $\mathrm{Var}(X) \approx 0.4232$.

8.37 (a) $\frac{738}{125}$, (b) $\mathrm{Var}(X) \approx 0.8092$.

8.39 $E[X] = \frac{51}{8}$, $\mathrm{Var}(X) = \frac{1955}{832}$.

8.41 $\frac{b}{n^2}$.

8.43 (a) $E[Y] = n$ and $\mathrm{Var}(Y) = 2n$.
(b)

$$M_Y(t) = \begin{cases} (1-2t)^{-n/2}, & t < \frac{1}{2} \\ \infty, & \text{else.} \end{cases}$$

8.45 $\mathrm{Cov}(X, Y) = \mathrm{Corr}(X, Y) = 0$.

8.47 $E[X] = 1.4$, $\mathrm{Var}(X) = 0.9$.

8.49 0.

8.51 14.

8.53 (a) -18, (b) $\frac{-1}{\sqrt{2}}$.

8.57 It is not possible.

Chapter 9

9.1 (a) $\frac{3}{8} = 0.375$, (b) $\frac{3}{10}$, (c) $\left(\frac{5}{6}\right)^{15} \approx 0.0649$.

9.3 $\lim_{n\to\infty} P(C_n > C_0) = 0$.

9.5 (a) $\frac{2}{3}$, (b) $\frac{3}{25}$, (c) 0.1587.

9.7 Chebyshev \$1569.21, CLT \$1055.80.

9.9 (a) $\frac{5}{7}$, (b) $\frac{9}{8000}$, (c) 36.

9.11 (a) $\frac{3}{8}$, (b) $\frac{6}{25}$.

9.13 $\frac{7}{48}$.

9.17 (a) $\frac{5}{6}$, (b) $\frac{1}{4}$, (c) 0.0228.

9.19 With Chebyshev's inequality $\frac{4}{9}$, with the CLT 0.0668.

9.21 0.1587.

Chapter 10

10.1 (a) $p_{X|Y}(2|0) = 1$,

$$p_{X|Y}(x|1) = \begin{cases} \frac{1}{4}, & x = 1 \\ \frac{1}{2}, & x = 2 \\ \frac{1}{4}, & x = 3 \end{cases} \quad \text{and} \quad p_{X|Y}(x|2) = \begin{cases} \frac{1}{2}, & x = 2 \\ \frac{1}{2}, & x = 3. \end{cases}$$

(b) $E[X|Y = 0] = E[X|Y = 1] = 2$ and $E[X|Y = 2] = \frac{5}{2}$.

10.3 $p_X(x) = \sum_{y=1}^{6} \binom{y}{x} \frac{1}{2^y} \cdot \frac{1}{6}$ for $0 \le x \le 6$, with the usual convention $\binom{y}{x} = 0$ if $y < x$. $E[X] = \frac{7}{4}$.

10.5 (a) $f_{X|Y}(x|y) = \frac{6x(2-x-y)}{4-3y}$, $0 < x < 1$ and $0 < y < 1$.
(b) $P(X > \frac{1}{2}|Y = \frac{3}{4}) = \frac{17}{28}$, $E[X|Y = \frac{3}{4}] = \frac{4}{7}$.

10.7 (a)

$$f_{X,Y}(x, y) = \begin{cases} 6x, & 0 < x < y < 1 \\ 0, & \text{else.} \end{cases}$$

(b) For $0 < x < 1$, the conditional density is

$$f_{Y|X}(y|x) = \begin{cases} \frac{1}{1-x}, & x < y < 1 \\ 0, & \text{else.} \end{cases}$$

Hence, Y is uniformly distributed on the interval $(x, 1)$.

10.9 (a) $f_Y(y) = e^{-y}$ for $y > 0$. $f_{X|Y}(x|y) = \frac{1}{y}e^{-x/y}$ for $x > 0$ and $y > 0$. $E[Y] = 1$.
(b) $E[X|Y] = Y$. (c) $E[X] = E[Y] = 1$.

10.11 $E[X] = \frac{1}{\lambda}$, $\text{Var}(X) = \frac{1}{\lambda} + \frac{1}{\lambda^2}$.

10.13 \$5.

10.15 (a) ≈ 0.1847, (b) ≈ 0.9450, (c) ≈ 0.7576.

10.17 (a) $P(X = k) = \binom{3}{k}(\frac{1}{6})^k(\frac{5}{6})^{3-k} \cdot \frac{1}{2} + \binom{5}{k}(\frac{1}{6})^k(\frac{5}{6})^{5-k} \cdot \frac{1}{2}$ for $k \in \{0, 1, \ldots, 5\}$, where $\binom{a}{b} = 0$ if $b > a$.
(b) $E[X] = \frac{2}{3}$.

10.19 (a)

$$P(X_2 = k, X_3 = \ell|X_1 = m) = \begin{cases} \frac{(n-m)!}{(n-m-k)!k!} \left(\frac{p_2}{p_2+p_3}\right)^k \left(\frac{p_3}{p_2+p_3}\right)^\ell, & k + \ell + m = n \\ 0, & \text{else.} \end{cases}$$

(b) Conditioned upon $X_1 = m$, the distribution of X_2 is $\text{Bin}(n - m, \frac{p_2}{p_2+p_3})$.

10.21 (a) Hypergeom(n, m, k). (b) $E[S_m|S_n] = S_n \frac{m}{n}$.

10.23 (a)

$$p_{X|Y}(x|y) = \begin{cases} \frac{2(y-x-1)}{(y-1)(y-2)}, & 1 \le x \le y-2 \\ 0, & \text{else.} \end{cases}$$

(b) $E[X|Y] = \frac{Y}{3}$.

10.25 (a) $p_{Y|X}(y|x) = \binom{y-1}{x-1}\left(\frac{1}{2}\right)^y$, for $1 \le x \le y$.
(b) $P(Y = y) = (\frac{11}{12})^{y-1}\frac{1}{12}$.
(c) $p_{X|Y}(x|y) = \binom{y-1}{x-1}\left(\frac{6}{11}\right)^{y-x}\left(\frac{5}{11}\right)^{x-1}$, for $1 \le x \le y$.

10.29 For $0 \le \ell \le n$,

$$p_L(\ell) = \binom{n}{\ell}(pr)^\ell (1-pr)^{n-\ell}.$$

10.33 (a)

$$P(X \le \tfrac{1}{2} \mid Y = y) = \begin{cases} 0, & y < \frac{1}{2} \\ \dfrac{y - \frac{1}{2}}{1-y}, & \frac{1}{2} \le y < \frac{3}{4} \\ 1, & y \ge \frac{3}{4}. \end{cases}$$

10.35 (a) $f_{X|S}(x|s) = \frac{1}{\sqrt{\pi}} \exp\left\{-(x - \frac{s}{2})^2\right\}$.
(b) $E[X \mid S] = \frac{S}{2}$ and $E[X^2 \mid S] = \frac{1}{2} + \frac{S^2}{4}$.

10.37

$$f_Y(y) = \begin{cases} \sqrt{2/\pi}\, e^{-y^2/2}, & y > 0 \\ 0, & y \le 0. \end{cases}$$

10.39 $f_Y(y) = 5y^4$ if $0 \le y \le 1$, and zero otherwise.

10.41 $f_Z(z) = \frac{1}{2}(\ln z)^2$ for $0 < z < 1$, $E(Z) = \frac{1}{8}$, and $\mathrm{Var}(Z) = \frac{37}{1728} \approx 0.0214$.

10.43 (a) For $0 \le y < 2$, the marginal density is

$$f_{X|Y}(x|y) = \begin{cases} \frac{1}{2-y}, & y \le x \le 2 \\ 0, & \text{else.} \end{cases}$$

(b) $E[X|Y] = \frac{2+Y}{2}$.

10.45 (a) Given $Y = y$, X is normally distributed with mean $\frac{\sigma_X}{\sigma_X}\rho(y - \mu_Y) + \mu_X$ and variance $\sigma_X^2(1-\rho^2)$.
(b) $\frac{\sigma_X}{\sigma_X}\rho(Y - \mu_Y) + \mu_X$.

10.49 (a) For $t < 1$

$$p_{X|T}(1|t) = \frac{1}{2} \quad \text{and} \quad p_{X|T}(2|t) = \frac{1}{2}.$$

For $t \ge 1$, $p_{X|T}(1|t) = 1$.
(b) $\frac{3}{2} - \frac{1}{2}e^{-\lambda}$.

10.51 (a) $(2\Phi(1) - 1) \cdot \binom{n}{3} p^3 (1-p)^{n-3}$.

(b) $\Phi(-2)$.

(c) $P(Y < x) = \sum_{k=0}^{n} \Phi(x-k)\binom{n}{k}p^k(1-p)^{n-k}$.

Appendix B

B.1 (a) $D = ABC^c \cup ACB^c \cup BCA^c$.

(b) $E = ABC^c \cup BCA^c \cup ACB^c \cup ABC$. Another way to write this is $AB \cup AC \cup BC$.

B.3 (a) $B \setminus A = \{15, 25, 35, 45, 51, 53, 55, 57, 59, 65, 75, 85, 95\}$.

(b) $A \cap B \cap C^c = \{50, 52, 56, 58\}$.

(c) $((A \setminus D) \cup B) \cap (C \cap D) = \varnothing$.

B.5 (a) $A \triangle B = AB^c \cup A^cB$.

B.7 (a) Yes. (b) No.

Appendix C

C.1 11,232,000.

C.3 102.

C.5 60.

C.7 3.

C.9 (a) 78, (b) 312, (c) 52.

C.13 6.

C.15 30.

Appendix D

D.1 $2n^2 + n$.

D.3 (a) $\frac{1}{6}n^3 + \frac{1}{2}n^2 + \frac{1}{3}n$. (b) $\frac{1}{3}n^3 + \frac{1}{2}n^2 + \frac{1}{6}n$. (c) $\frac{5}{6}n^3 + 5n^2 + \frac{25}{6}n$.

D.5 (a) $\frac{x}{(1-x)^2}$. (b) $\frac{x}{(1-x)^2}$.

Bibliography

[Bil95] Patrick Billingsley, *Probability and Measure*, third edition, Wiley Series in Probability and Mathematical Statistics, John Wiley & Sons, New York, 1995.

[Bru10] Richard A. Brualdi, *Introductory Combinatorics*, fifth edition, Pearson Prentice Hall, Upper Saddle River, NJ, 2010.

[Dur10] Rick Durrett, *Probability: Theory and Examples*, fourth edition, Cambridge Series in Statistical and Probabilistic Mathematics, Cambridge University Press, Cambridge, 2010.

[Fol99] Gerald B. Folland, *Real Analysis: Modern Techniques and Their Applications*, second edition, Pure and Applied Mathematics, John Wiley & Sons, New York, 1999.

[Kah11] Daniel Kahneman, *Thinking, Fast and Slow*, Farrar, Straus and Giroux, New York, 2011.

[Pac11] Luigi Pace, *Probabilistically proving that* $\zeta(2) = \pi^2/6$. *The American Mathematical Monthly*, 118(7) (2011), 641–643.

[Rud87] Walter Rudin, *Real and Complex Analysis*, third ediion, McGraw-Hill, New York, 1987.

[SC13] Leila Schneps and Coraliel Colmez, *Math on Trial: How Numbers Get used and Abused in the Courtroom*, Basic Books, New York, 2013.

Index

Printed in the United States
by Baker & Taylor Publisher Services